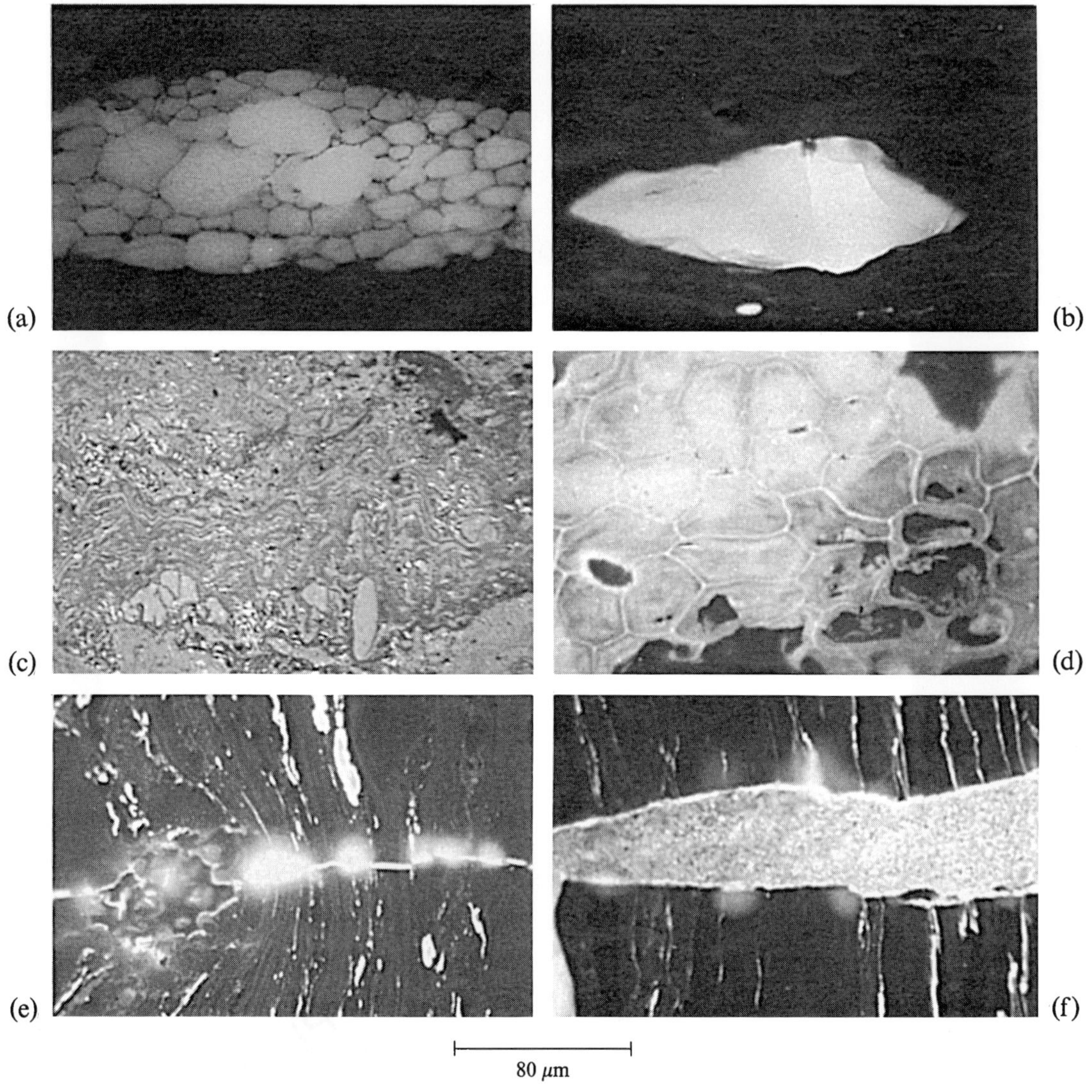

PLATE 1. Normal liptinite macerals of the Walloon coals and expulsion of liquid hydrocarbons form perhydrous vitrinites: (a) lipid resinite, (b) terpene resinite, (c) suberinite, (d) cutinite, (e) expulsion of liquid hydrocarbons under UV irradiation, and (f) exudatinite; (a, b, d, e, f): UV fluroescence mode; (c) white reflected light. (From paper by G. K. Khorasani, page 305)

GEOLOGICAL SOCIETY SPECIAL PUBLICATION CLASSICS

Coal and Coal-bearing Strata: Recent Advances

Keynote addresses and
invited papers to a conference held at
Royal Holloway and Bedford New College
University of London
8–10 April 1986

EDITED BY
ANDREW C. SCOTT
Geology Department
Royal Holloway and Bedford New College
University of London

Published by
The Geological Society
LONDON

THE GEOLOGICAL SOCIETY

The Society was founded in 1807 as The Geological Society of London and is the oldest geological society in the world. It received its Royal Charter in 1825 for the purpose of 'investigating the mineral structure of the Earth'. The Society is Britain's national learned society for geology with a membership of 7374 (1992). It has countrywide coverage and approximately 1000 members reside overseas. The society is responsible for all aspects of the geological sciences including professional matters. The Society has its own publishing house which produces the Society's international journals, books and maps, and which acts as the European distributor for publications of the American Association of Petroleum Geologists and the Geological Society of America.

Fellowship is open to those holding a recognized honours degree in geology or a cognate subject and who have at least two years relevant postgraduate experience, or who have not less than six years relevant experience in geology or a cognate subject. A Fellow who has not less than five years relevant postgraduate experience in the practice of geology may apply for validation and, subject to approval, may be able to use the designatory letters C. Geol (Chartered Geologist).

Further information about the Society is available from the Membership Manager, The Geological Society, Burlington House, Piccadilly, London W1V 0JU, UK.

Published by The Geological Society from:
The Geological Society Publishing House
Unit 7
Brassmill Enterprise Centre
Brassmill Lane
Bath BA1 3JN
UK

(*Orders*: Tel. 0225 445046
Fax 0225 442836)

First published 1987
as Geological Society Special Publication No 32
Paperback edition 1993

British Library Cataloguing in Publication Data
A catalogue record for this book is available from the British Library

ISBN 1-897799-01-2

Printed in Great Britain by
The Alden Press, Oxford

Distributors
USA
AAPG Bookstore
PO Box 979
Tulsa
Oklahoma 74101-0979
USA
(*Orders*: Tel: (918) 584-2555
Fax (918) 584-0469)

Australia
Australian Mineral Foundation
63 Conyngham Street
Glenside
South Australia 5065
Australia
(*Orders*: Tel: (08) 379-0444
Fax (08) 379-4634)

India
Affiliated East-West Press PVT Ltd
G-1/16 Ansari Road
New Delhi 110 002
India
(*Orders*: Tel: (11) 327-9113
Fax (11) 326-0538)

Japan
Kanda Book Trading Co.
Tanikawa Building
3-2 Kand Surugadai
Chiyoda-Ku
Tokyo 101
Japan
(*Orders*: Tel: (03) 3255-3497
Fax (03) 3255-3495)

Contents

Preface

This volume comprises the invited keynote papers given at the International Symposium on Coal and Coal-bearing Strata held at Royal Holloway and Bedford New College (University of London) from April 8th to 10th 1986. The authors were asked to consider progress in their field over the past 15 years in particular. Some additional contributions were solicited to fill gaps in the coverage. In this volume I have tried to put together a balanced view of the field and where possible asked the authors to use a similar terminology and to make reference to other contributions in the volume. Other contributions at the conference have been published in the *Journal of the Geological Society of London*. It is hoped that this volume will not only be of interest to researchers but also to lecturers and students alike.

I thank the Geological Society of London, the Geology Department of Royal Holloway and Bedford New College and British Coal for financial help for organizing the conference. I thank Professor A. J. Smith and Miss S. Clay for their support. Finally I thank all the authors for their presentations at the conference and all the referees who undertook the task of reviewing the papers.

January 1987

ANDREW C. SCOTT
Geology Department
Royal Holloway and
Bedford New College

Coal and coal-bearing strata: recent advances and future prospects

Andrew C. Scott

ABSTRACT: Recent advances in coal geology are highlighted. The increase in our knowledge of peat formation is emphasized and the application of hydrological models of mire systems to coal-bearing strata is advocated. The importance of coalification studies to the geological community as a whole should not be underestimated. Consideration of the original peat-forming vegetation by coal geologists is advocated. Both conceptual and technical advances in coal geology are reviewed. Integration of different disciplines is likely to yield further insights into the study of coal and coal-bearing strata.

Keywords: peat, coal, Tertiary, Permian, Carboniferous, petrology, geochemistry, palaeobotany, palynology, sedimentology

Over the past 15 or so years there has been a considerable increase in interest in all aspects of coal and coal-bearing strata. Most conferences and volumes have, however, either concentrated on coal as a rock or on coal mining. In 1968, however, following a student inter-university geological congress, a volume, edited by D. G. Murchison and T. S. Westoll, was published which dealt with a diversity of studies on coal and coal-bearing strata. Another influential text has been *Stach's Textbook of Coal Petrology* which is now in its third edition (Stach *et al.* 1982). The increase of interest over recent years in coal is well illustrated by the publications of Ward (1984) and Rahmani & Flores (1984). Despite the continuing interest in coal geology and exploration there have been few attempts to integrate different disciplines so that some might regard coal as an economic mineral whilst others regard it as a fossil peat. There is often, however, little communication between the two groups.

In an attempt to promote interdisciplinary discussion a conference covering a broad spectrum of coal geology and biology was organized by A. C. Scott under the auspices of the Geological Society of London and held at Royal Holloway and Bedford New College of the University of London between 8 and 10 April 1986. Ten distinguished keynote speakers were invited to cover these significant research areas and were asked to give special place to recent advances in their fields. In addition, there were numerous contributed paper sessions over the three days of the meeting. This volume contains nine of the keynote addresses plus seven other invited contributions. A selected number of other papers have been published in the *Journal of the Geological Society of London*, volume 144 (1987).

Coal is a widespread and readily available fossil fuel. Often, however, its biological origin is forgotten and potentially useful data neglected when considering the distribution and uses of coal. The contributions in this volume by Moore and by Clymo emphasize the controls on peat formation. It has become clear over recent years that hydrological models of peat formation are of major significance for the coal geologist. In particular two integrating peat-forming (mire) environments are important; rainwater fed (ombrotrophic) and groundwater fed (rheotrophic). Ombrotrophic peats would typically yield low ash coals. Both Moore and Clymo point out that typical Upper Carboniferous coal seams from Euramerica are best compared with tropical ombrotrophic mire complexes. Clymo compares Australian Permian coals with temperate ombrotrophic peats and emphasizes selective decay in the peat-forming process. Despite the rapid

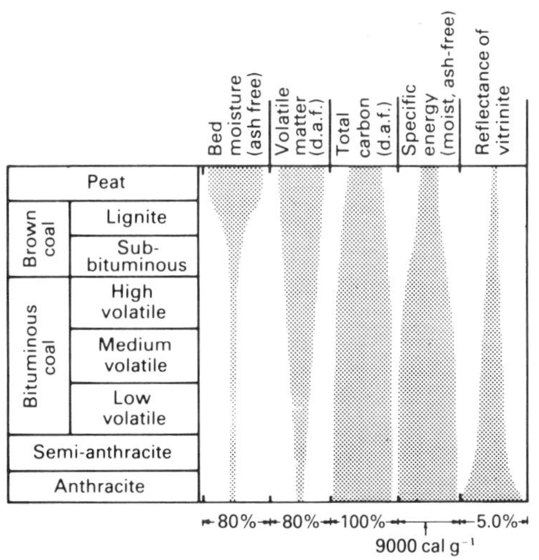

FIG. 1. Changes in coal composition with rank advance (from Ward 1984).

TABLE 1. *Macerals and group macerals recognized in hard coals. Based on Stach* et al. *(1982) (from McCabe 1984)*

Group maceral	Maceral	Morphology	Origin
Vitrinite (Huminite)	Telinite	Cellular structure	Cell walls of trunks, branches, roots, leaves, etc.
	Collinite	Structureless	Reprecipitation of dissolved organic matter in a gel form
	Vitrodetrinite	Fragments of vitrinite	Very early degradation of plant and humic peat particles
Exinite (Liptinite)	Sporinite	Fossil form	Mega- and microspores
	Cutinite	Bands which may have appendages	Cuticles—the outer layer of leaves, shoots and thin stems
	Resinite	Cell filling, layers or dispersed	Plant resins, waxes and other secretions
	Alginite	Fossil form	Algae
	Liptodetrinite	Fragments of exinite	Degradation residues
Intertinite	Fusinite	Empty or mineral filled cellular structure. Cell structure usually well preserved	Oxidized plant material—mostly charcoal due to burning of vegetation
	Semifusinite	Cellular structure	Partly oxidized plant material
	Macrinite	Amorphous 'cement'	Oxidized gel material
	Inertodetrinite	Small particles of fusinite, semi-fusinite and/or macrinite	Redeposited inertinites
	Micrinite	Granular: rounded grains ~ 1 μm in diameter	Degradation of macerals, especially exinites, during coalification
	Sclerotinite	Fossil form	Mainly fungal remains

increase in our knowledge of peat formation the occurrence of very thick coals in the fossil record still causes problems for the modern peat ecologist. The importance of rainfall in peat formation is also emphasized by Ziegler, Raymond, Gierlowski, Horrell, Rowley and Lottes. These authors discuss climatic models of both the present and the Cretaceous and show the importance of rainfall belts in controlling the distribution of peats and coals. They note that at present there is an equatorial and two temperate rainfall belts. In the Cretaceous, however, they show that there are only temperate rain belts with abundant coals and that in the tropics precipitation was markedly seasonal. This paper emphasizes the use of predictive climatological modelling for coal exploration and is an area of further development.

For many years the cyclothem concept of coal measure sediments was popular. Recently more sophisticated sedimentological studies have shown major problems in its use. Facies analysis and the development of sedimentological models has come of age. McCabe (1984) made a significant contribution in reviewing the sedimentology of coal-bearing sequences. In this volume McCabe extends his earlier paper and considers the use of facies modelling in subsurface exploration. The author advocates considering not only clastic but also coal facies in developing models and points to the need of predictive modelling. The distinction of different mire types has necessitated a revision of our view of coal-bearing sequences. McCabe, following Cohen (see Cohen *et al.*, *this volume*), emphasizes the potential hiatus between clastic sediments and coals in contact. Together with Cohen *et al.*, McCabe also points to the deficiencies of the popular deltaic models for coal depositional systems. The integration of classical sedimentological with palaeontological and coal studies will be an important element in coal exploration.

The importance of considering the peat-forming vegetation is a theme developed by numerous authors. Collinson and Scott discuss the biology of peat-forming plants and compare Carboniferous and Tertiary peat-forming vegetation and consider the implications for coal geology. In particular they stress implications for the recognition of seat earths, coal petrological interpretations and in the calculation of peat to coal compression ratios. These authors also emphasize that the fossil plant compressions found in clastic sediments between the coals do not necessarily represent the peat-forming flora. The use of permineralized peats, palaeobotanical and palynological analyses of coal is also advocated.

TABLE 2. *The lithotypes of humic and sapropelic coals (from McCabe 1984)*

Lithotype	Description	Composition	
Vitrain	Black, very bright lustre, thin layers break cubically, thick layers have conchoidal fracture	Vitrinite macerals with <20% exinite macerals	Humic coals
Clarain	Finely stratified layers of vitrain, durain and, in some cases fusain; medium lustre	Variable	
Durain	Black or grey, dull, rough fracture surfaces	Mainly inertinite and exinic macerals	
Fusain	Black, silky lustre, friable and soft	Mainly fusinite	
Cannel coal	Black, dull, lustre may be 'greasy', breaks with conchoidal fracture	Fine maceral particles, usually dominated by sporinite	Sapropelic coals
Boghead coal	Black or brown, dull, homogeneous, breaks with conchoidal fracture, lustre may be 'greasy'	Dominated by alginite	

Unfortunately we have, for example, little data on many southern hemisphere Permo–Triassic coals. In the more high rank coals (Fig. 1), extraction of acid-resistant plant microfossils may prove difficult. The development of an argon ion etching technique by Kizilshtein & Shpitzgluz (1982) offers potential in allowing the identification of the plant constituents of such coals.

The relationships between hydrology and vegetation are demonstrated by numerous authors. Likewise the occurrence of mineral matter such as sulphur is also shown to be in part related to the original depositional setting. Coal sulphur content is a major consideration in coal marketing, being a major cause of acid rain. Casagrande discusses the development of sulphur geochemical models. Sulphur takes three main forms in coal: as organic sulphur, as sulphate and as iron pyrite. Some sulphur originates during the peat-forming process, other is diagenetic. Microorganisms reduce sulphate to hydrogen sulphide which then reacts with available ferrous ions or organic matter to produce pyrite and organic sulphur respectively.

As also shown by Cohen *et al.*, marine and brackish water peats have higher sulphur levels. The vegetation of these peats is also characteristic. The studies by Cohen and his co-authors on recent peats mainly from the southeastern USA have generated many new ideas for coal geologists. These include: the identification of progenitors of coal macerals; the dissolution of mineral matter from peat; the importance of marine waters in introducing sulphur into peat; the doming of peats during their development; the development of splits following fires; mechanisms for the stratification of coal seams and the importance of back barrier coal-forming environments world-wide. Cohen *et al.* downplay deltaic models of coal bearing strata and emphasize the importance of the botanical constituents of peats and coals. The recognition of fire splay spits in recent peat-forming environments has significance for coal sedimentological models and typically the sands have abundant charcoal at the base. Also Scott (1979), Scott & Collinson (1978) and Cope & Chaloner (1985) have emphasized the importance of recording fusain-rich (Tables 1 and 2) horizons in coal-bearing strata.

The burial of peat to form coal has received considerable attention in recent years. The work of Teichmüller in coalification studies is particularly notable. In her paper Teichmüller reviews the significant progress in coalification studies and considers their wide application to many geological problems. In particular studies on coal particles, more especially the maceral vitrinite (Table 1) but also in recent years on liptinite, have implications for tectonics, heat flow assessments, hydrocarbon as well as for coal exploration. Teichmüller also highlights new techniques including the use of fluorescence particularly helpful for the study of liptinite. It has also been shown that bituminization is significant between subbutiminous and bituminous coal rank (Fig. 1) and that it is responsible for the softening of coal during carbonization. Teichmüller integrates new work on spore and conodont colouration indices with illite crystallinity and vitrinite reflectance to allow heat flow analyses to be measured. This work also has major significance in the assessment of crustal thicknesses as well as for the understanding of plate tectonic processes.

Considerable changes affect peats during the coalification process. The growth of diagenetic minerals other than sulphur is discussed by Spears who uses data from the British Upper Carboniferous Pennine coalfield in particular. Spears recognizes a depth burial sequence of minerals in cleat, those are microfractures developed in coal during burial. Spears also concludes that residual diagenetic pore fluids resulting from burial diagenesis are the source of economically important high chlorine concentrations in the coal.

There has been recent increased activity in investigating the vegetation of Upper Carboniferous Euramerican coals (see Scott 1977, DiMichele *et al.* 1985). The classic work of Smith (1968) on the palynology of British coals has only recently been developed. Phillips & Peppers (1984) has clearly demonstrated that most Euramerican Westphalian coals were dominated by lycopods and that these were heterosporous producing both microspores and megaspores. Following an initial investigation by Scott & King (1981) Bartram describes here the quantitative distribution of megaspores in the Barnsley Seam of Yorkshire, one of Britain's most important coals. Bartram demonstrates the common association of spores in phases and also clearly documents their repeated development and considers that they represent the building of a raised bog. Bartram concludes, perhaps unexpectedly, that there is no simple relationship of phase to coal petrology. Importantly she demonstrates that megaspore studies pick out significantly herbaceous assemblages not noted using other techniques. Fulton, in his study of the Warwickshire Thick Coal, only considers the miospore component. He shows the existence of four assemblages each with a single dominant. Each leaf of coal shows a similar succession of assemblages which he interprets as being due to the variation of the water table in a tropical ombrotrophic bog. Fulton considers that miospore cycles can be used to interpret the proximity of seam splits, a potentially useful exploration method. It is clear from these investigations that detailed palynological studies of coals from other ages and areas could be very significant.

In most coal textbooks there is much emphasis on Upper Carboniferous Euramerican coals. Gondwana coals receive much less attention and yet, both in S Africa and Australia in particular, there is considerable research and exploration activity. Much has been made of the fact that these Permian coals were formed under temperate conditions and yet we know little of the original peat-forming plants. Most authors have assumed that plant compressions found in associated clastic sediments represent the peat forming flora (but see Collinson & Scott). Interestingly, Clymo has speculated on the occurrence of *Sphagnum*-like plants forming some coals but palaeobotanical studies are urgently needed. Hobday highlights these Gondwana coals and shows that those from Australia are more variable than those from South Africa. These coals were formed under temperate conditions following widespread glaciations. The range of depositional systems in which economic coals are found include proximal conglomeratic alluvial fans and occasionally blanket peat mires in sediment-starved basins. Hobday emphasizes the importance of tectonic subsidence, changes in groundwater regimes and of marine transgressions in controlling peat deposition.

Many of the studies outlined in this volume relate to or have direct relevance to coal exploration. Coal exploration procedures and case histories are rarely documented. Land and Jones illustrate the development and execution of a coal exploration programme undertaken in Indonesia. Tertiary coal basins are extensive in Indonesia and the Tertiary Kutai Basin in E Kalimantan (Borneo) was chosen for second stage exploration. The drilling programme identified 1000 million tons of coal reserves in 43 seams between 1.5 and 13 m thick in a 3000 m coal-bearing sequence. This work is of particular interest as in the lateral marine rocks there are significant oil deposits. The possibility that the vegetation was potentially oil generating has recently been discussed by Thompson *et al.* (1985). The possibility that some oil fields may have been sourced from coal-bearing strata has been of increasing interest (Thomas 1982).

The interest in oil derived from coals can be seen in the extensive recent literature (Saxby & Shibaoka 1986, Teichmüller 1986). In his contribution, Murchison outlines advances in organic petrology and organic geochemistry with particular reference to oil from coal. In this he considers the integration of coal petrological methods together with organic geochemical methods and emphasizes recent work on terrestrial sourced oils, particularly from Australia, Canada, Indonesia, China and from the Niger Delta. Murchison points out the importance of new work on quantitative fluorometric methods, on biomarkers from coals and on dispersed organic matter (Bertrand 1984, Bertrand *et al.* 1986a, b, MacKenzie 1984, Radke *et al.* 1984). Murchison rightly emphasizes the advances which have been made possible by interdisciplinary research.

The work on Australian coals (Saxby & Shiboaka 1986) is given prominence. In her contribution, Khorisani describes the Jurassic Walloon coal measures of the Surat Basin, Australia and investigates their oil-prone coals. In this case the occurrence of oil is considered to be due in part to a process of liptinite enrichment and controlled by both the original vegetation and by biochemical changes.

Whilst there is no doubt that coal and coal-bearing strata are a major source of oil, contrary to some widely held beliefs, we still have little palaeobotanical evidence of the plants responsible. For the future, integrated palaeobotanical, petrological and organic geochemical studies should help in our understanding of the origin and history of oil-prone coals.

The advances using fluorescence spectrometry have been made possible by more sophisticated apparatus. Likewise important new data concerning the structure of high rank coals is being obtained from the correlation of optical and transmission electron microscopy with Raman microspectroscopy (Beny-Bassez & Rouzaud 1985). The laser Raman microspectroscope examines vibrations in the molecular bonding which are caused by the organization of, and defects in, the lattice structure. This technique is still under development and has much to contribute towards our understanding of the molecular structure of coals and changes which occur during coalification and carbonization.

Despite many advances we still have large gaps in our knowledge of coal and coal-bearing strata. New technology is playing its part but a basic problem is the lack of integrated interdisciplinary studies. In addition there is often a communication problem between specialists. In this volume some integration has been attempted. It is important for geologists, technologists, chemists and biologists to have an input in problem solving. The conference stimulated many important contacts and discussions and it is hoped that this volume might also promote a wide range of fresh insights and new research into the study of coal and coal-bearing strata.

References

BERTRAND, P. 1984. Geochemical and petrographic characterization of humic coals considered as possible oil source rocks. *Organic Geochemistry*, **6**, 481–8.

BERTRAND, P., BEHAR, F. & DURAND, B. 1986a. Composition of potential oil from humic coals in relation to their petrographic nature. *Organic Geochemistry*, **10**, 601–8.

BERTRAND, P., PITTION, J.-L. & BERNAUD, C. 1986b. Fluorescence of sedimentary organic matter in relation to its chemical composition. *Organic Geochemistry*, **10**, 641–7.

BENY-BASSEZ, C. & ROUZAUD, J. M. 1985. Characterization of carbonaceous materials by correlated electron and optical microscopy and Raman microspectroscopy. *Scanning Electron Microscopy 1985/1*, 119–32.

COPE, M. J. & CHALONER, W. G. 1985. Wildfire: an interaction of biological and physical processes. *In:* TIFFNEY, B. (ed.), *Geological Factors and the Evolution of Plants*, pp. 257–277. Yale University Press.

DIMICHELE, W. A., PHILLIPS, T. L. & PEPPERS, R. A. 1985. The influence of climate and depositional environment on the distribution and evolution of Pennsylvanian coal-swamp plants. *In:* TIFFNEY, B. (ed.), pp. 223–256.

KIZILSHTEIN, L. YA & SHPITZGLUZ, A. L. 1982. New method of petrographic analysis of anthracite. *Doklady Akademic Nauk SSSR*, **263**, 175–179.

MACKENZIE, A. S. 1984. Applications of biological markers in petroleum geochemistry. *In:* BROOKS, J. & WELTE, D. (eds) *Advances in Petroleum Geochemistry*, volume 1, pp. 115–214. Academic Press, London.

MCCABE, P. J. 1984. Depositional environments of coal and coal-bearing strata. *In:* RHAMANI, R. A. & FLORES, R. M. (eds), pp. 13–42.

MURCHISON, D. G. & WESTOLL, T. S. (eds) 1968. *Coal and Coal-bearing Strata*. Oliver & Boyd, Edinburgh.

PHILLIPS, T. L. & PEPPERS, R. A. 1984. Changing patterns of Pennsylvanian coal-swamp vegetation and implications of climatic control on coal occurrence. *International Journal of Coal Geology*, **3**, 205–255.

RADKE, M., LEYTHAEUSER, D. & TEICHÜLLER, M. 1984. Relationship between rank and composition of aromatic hydrocarbons for coals of different origins. *Organic Geochemistry*, **6**, 423–30.

RAHMANI, R. A. & FLORES, R. M. (eds) 1984. *Sedimentology of Coal and Coal-bearing Sequences*. Special Publication of the International Association of Sedimentologists 7. Blackwell Scientific Publications, Oxford.

SAXBY, J. D. & SHIBAOKA, M. 1986. Coal and coal macerals as source rocks for oil and gas. *Applied Geochemistry*, **1**, 25–36.

SCOTT, A. C. 1977. A review of the ecology of Upper Carboniferous plant assemblages with new data from Strathclyde. *Palaeontology*, **20**, 447–473.

—— 1979. The ecology of Coal Measure floras from northern Britain. *Proceedings of the Geologists Association*, **90**, 97–116.

—— & COLLINSON, M. E. 1978. Organic sedimentary particles: results from scanning electron microscope studies of fragmentary plant material. *In:* WHALLEY, W. B. (ed.), *Scanning Electron Microscopy in the Study of Sediments*, pp. 137–167. Geoabstracts, Norwich.

—— & KING, G. R. 1981. Megaspores and coal facies: an example from the Westphalian A of Leicestershire. *Review of Palaeobotany and Palynology*, **34**, 107–113.

SMITH, A. H. V. 1968. Seam profiles and seam

characters. *In:* MURCHISON, D. G. & WESTOLL, T. S. (eds), pp. 31–40.

STACH, E., MACKOWSKY, M.-TH., TEICHMÜLLER, M., TAYLOR, G. H., CHANDRA, D. & TEICHMÜLLER, R. 1982. *Stach's Textbook of Coal Petrology*, 3rd edn. Gebrüder Borntraeger, Berlin.

TEICHMÜLLER, M. 1986. Organic petrology of source rocks, history and state of the art. *Organic Geochemistry*, **10**, 581–99.

THOMAS, B. M. 1982. Land-plant source rocks for oil and their significance in Australian basins. *Australian Petroleum Exploration Association Journal*, **22**, 164–178.

THOMPSON, S., COOPER, B. S., MORLEY, R. J. & BARNARD, P. C. 1985. Oil-generating coals. *In:* THOMAS, B. *et al.* (eds), *Petroleum Geochemistry in Exploration of the Norwegian Shelf*, pp. 59–73. Graham & Trotman, London.

TIFFNEY, B. (ed.) 1985. *Geological Factors and the Evolution of Plants*. Yale University Press.

WARD, C. R. (ed.) 1984. *Coal Geology and Coal Technology*. Blackwell Scientific Publications, Oxford.

ANDREW C. SCOTT, Geology Department Royal Holloway and Bedford New College, University of London, Egham Hill, Egham, Surrey TW20 0EX, England.

Ecological and hydrological aspects of peat formation

P. D. Moore

ABSTRACT: Peat formation is dependent upon an imbalance in the energetic relations of an ecosystem such that the total energy fixation by photosynthesis exceeds the total respiration on the part of the plants, animal consumers, detritivores and microbial components of the ecosystem. The imbalance comes about through the impedence of some detritivore and microbial activity as a result of the waterlogging of the environment. A knowledge of hydrological relationships in different types of mire (any freshwater wetland ecosystem in which soil organic matter accumulates) is necessary for the understanding of the rate of formation and the nature of the peat formed.

A hydrological scheme for mire classification is discussed in relation to the type of peat developed from the mire types described, particularly the proportion of inorganic material incorporated into various peatland types. Two major hydrological divisions can be distinguished—those depending entirely upon rainfall for their water intake and those receiving water both by rainfall and overland drainage. The differentiation is of profound importance in terms of the inorganic (loss on ignition; ash) content of the peat, since the rain-based (ombrotrophic) system generally has a far more limited supply of inorganic material than the flow-fed (rheotrophic) one.

One must seek models for coal-forming systems among ombrogenous tropical mires. Elevated peat-forming surfaces equivalent to the temperate raised mires are recorded from tropical areas and some of these bear vegetation whose physiognomy (general form and structure) resembles that of reconstructed coal forests. In these bog forests, peats of low ash content accumulate and such mires may prove to be the closest modern analogy to coal-forming systems.

Keywords: peat, mire, bog, swamp, coal, ash

Peat-forming ecosystems

Peat is a material consisting largely of organic detritus but details of how much inorganic material is permissible before the term peat becomes unacceptable varies from one author to another (see Clymo 1983); figures vary from about 1% to over 50% in materials which are still commercially regarded as peats. An upper limit of 25% ash by dry weight is perhaps the most generally accepted figure for peats. The term 'muck' is sometimes applied to organic soils with an inorganic content higher than this. Although some of the organic components of peats are of animal and microbial origin (such as fungal hyphae), the bulk of peat normally consists of the accumulated detritus from former plant growth.

The precise botanical constituents of peat vary with the nature of the peat-forming vegetation and with the degree of decomposition to which the peat has been subjected. Thus the rapidly accumulating peats of temperate bogs are often rich in mosses, particularly the bog mosses, *Sphagnum*, but the leaves and stems of these plants are lost more rapidly under oxidizing conditions than are the harder woody materials of dwarf shrubs, which prove less palatable to microbes and which therefore became proportionately more abundant in bog peats as decomposition proceeds. So the composition of peats does not always reflect accurately the nature of the original peat-forming vegetation.

Those habitats in which peat accumulates can be given the general term 'mires'. All such peat-forming ecosystems have certain characteristics in common which are best expressed in terms of their energy flow patterns (see Fig. 1). There are, however, many different types of mires which differ largely in their patterns of nutrient movements, which in turn are largely determined by the nature of their hydrology, especially their source of water input. These differences provide one of the most ecologically meaningful ways of classifying mires into different types.

Energetics of mires

(1) *General features*

In almost all terrestrial ecosystems the initial input of energy is derived from the photosynthetic activity of green plants. The plants use a proportion of this energy in their own metabolism and the residue becomes available as a food resource for grazing animals, or falls to the ground

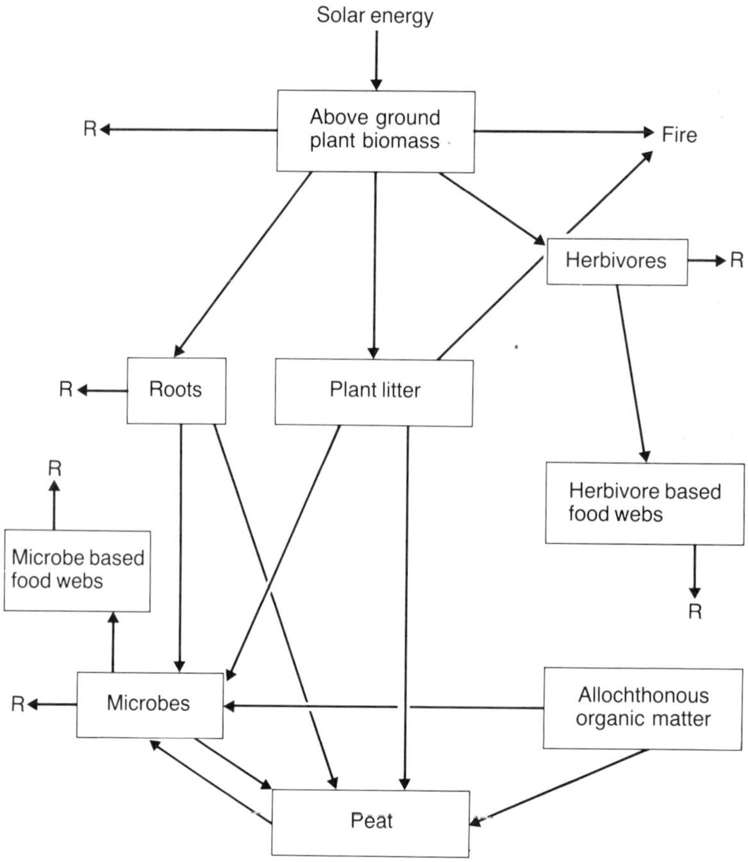

FIG. 1. Pattern of energy flow in a mire ecosystem. R = respiration.

as litter where it is attacked by detritivorous animals or is decomposed by the microbial components of the ecosystem. The main feature which all mires have in common is the fact that some of the organic productivity of the ecosystem is retained within the system and accumulates as peat. This means that the entire ecosystem is increasing in its energy content.

This process is best illustrated by means of some simple equations.

$$GPP = NPP + Ra$$

where GPP = gross primary productivity, NPP = net primary productivity, Ra = autotroph (green plant) respiration. The net primary productivity may then be consumed by herbivores and respired by them and the food webs which they support, it may be retained by the plant and used in growth, or it may enter the so-called decomposer food web and be used as an energy resource by detritivore and decomposer organisms.

$$NPP = PG + Rgr + Rdet + Rdec$$

where PG = plant growth, Rgr = respiration of grazer food web, Rdet = respiration of detritivores, Rdec = respiration of decomposers.

These formulae as they stand can be applied to any type of ecosystem (Odum 1969). In a steady-state ecosystem, where there are no net changes taking place in the vegetation or other biota, the total amount of energy fixed by photosynthesis is consumed by the respiratory processes occurring within the ecosystem, such that:

$$GPP = Ra + Rgr + Rdet + Rdec$$

or

$$GPP = \text{ecosystem respiration.}$$

The alternative is that energy can be stored in the system, either in increased biomass (usually mainly in the form of plant material), or as litter and soil organic matter. In such a case the total productivity of the ecosystem is in excess of the total respiration of the ecosystem.

GPP > eosystem respiration.

The continued storage of energy in the system means that it is in the course of succession and has not achieved a state of stability. In peatlands the respiration of the detritivore and decomposer food webs is curtailed by waterlogging and consequent limited oxygen availability, leading to the accumulation of organic detritus.

$$(NPP - Rgr) > (Rdet + Rdec)$$

Figure 1 summarizes diagrammatically the energy relations of a mire.

Allochthonous organic matter (derived from outside the ecosystem) is only likely to be of importance in swamps receiving redistributed peat from other, eroding mires in their catchments.

Clymo (1965) has conducted experiments on the breakdown of plant litter at different depths in a bog profile and has shown that the rate of weight loss in upper layers (the acrotelm) of a profile (see Ingram 1983), where waterlogging is intermittent, is considerably greater than in permanently waterlogged, anaerobic layers (the catotelm) (see also Clymo, *this volume*). Nevertheless, anaerobic microbial decay occurs very slowly even in the saturated deeper layers. Plant organs, such as leaves and fruits, which fall upon the surface of the peat may spend a considerable time in the acrotelm until the upward growth of the peat surface has resulted in their burial and their becoming part of the catotelm zone. At that stage the rate of decomposition declines and the survival of the remaining tissue becomes more likely. Some materials, like roots, however, penetrate the catotelm layers even before death and are hence not subjected to decomposition in the acrotelm. Such tissues have a far greater chance of long-term intact survival and are often major constituents of peat.

The formation of peat in an ecosystem is thus clearly related to the hydrology of a site and, in particular, to its state of waterlogging. Hydrology, in turn, is affected by such climatic factors as precipitation and evaporation (and hence temperature), as well as local factors of topography and catchment land use or vegetation cover (see e.g. Ziegler *et al.*, *this volume*).

Figure 2 shows some data which illustrate the relative importance of different pathways in energy flow for one type of mire. In this case the figures are collated and simplified from a number of investigations on British blanket mires (data from IBP work at Moor House, northern Pennines). Less than 1% of the net primary production is used by grazers (including sheep), about 5% is taken up by detritivores, 80% by microbes and the remaining 15% remains as peat.

Peat is not an inert material which, once formed, is free from any chance of further involvement in energy flow. On the contrary, there is evidence to suggest that peat may constitute the energetic basis of elaborate food webs in some ecosystems, especially if it is physically disturbed by erosion. Thus the fisheries of the Mekong delta in SE Asia are believed to be based to some considerable extent on the energy resources of redistributed peat (Maltby 1986). One remarkable and direct piece of evidence for such energetic recycling is provided by the radiocarbon date of a long-tailed (old squaw) duck in the Arctic (Schell 1983). Its date of 1300 years old, although only recently killed, is a clear indication that it had been feeding on ancient materials probably derived from old peat deposits.

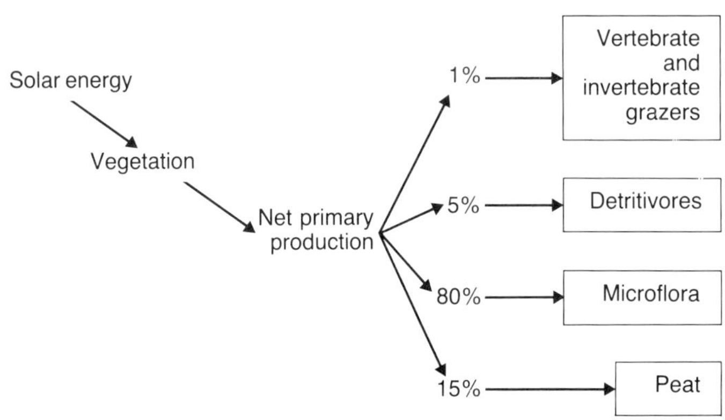

FIG. 2. Relative importance of various energy flow pathways in a mire (derived from various data relating to British blanket mires).

(2) *Productivity and peat accumulation in mires*

Many measurements have been made of net primary productivity of various types of mire in different parts of the world and a great deal of variation has been found (see tabulations in Moore & Bellamy 1973, Leith 1975a, Bradbury & Grace 1983). Swamps of cat-tail (*Typha*) are among the world's most productive ecosystems with plant dry matter being produced at rates of 5–6 (even up to 15) kg m^{-2} a^{-1}. Swamp forests in Florida achieve 2–3 kg but the bogs of temperate and boreal regions rarely exceed 1 kg m^{-2} a^{-1}.

Yet when we examine peat accumulation rates, the swamps are not as far ahead of the high latitude bogs as one might expect from the differential in production rates. Cat-tail swamps form peats at about 0.5–1.0 mm a^{-1}, compared with Sphagnum bogs at 0.2–0.5 mm a^{-1} (but up to 10 mm a^{-1} in some blanket peats over the last 100 years, Bowler & Bradshaw 1985). High rates of peat formation are observed in swamp forests, with 1–2 mm in the Florida swamps (Gower *et al.* 1985) and 2–3 mm in the raised forested bogs of Sarawak (Anderson 1983). It must be noted, however, that rates of peat accumulation are very difficult to interpret since they are derived from data which are not always comparable in any simple way. Some figures are obtained from shallow deposits of recently formed peat, whereas others are derived from deeper sections in which such forces as compression and continued anaerobic decomposition have been taking place over several millenia (see also comments of Collinson & Scott, *this volume*). Care must therefore be used in the interpretation of such data.

The rates of peat accumulation are thus seen to be dependent on factors other than productivity, as one might expect from the equations given above. Clymo (1983) has made calculations concerning the effects of factors such as temperature on the decomposition rates of peats. But the application of such data to climatic models would be premature. Mohr & van Bahren (1959) have developed a very simple, speculative model relating peat formation to global mean temperatures. This is given in a modified and extended form in Fig. 3. The relationship between productivity and mean annual temperature generated by Leith (1975b) is given and superimposed are two speculative scenarios of ecosystem respiration which differ in their degree of wetness. Under dry climatic and topographic conditions (i.e. sloping ground and plateaux) peat is able to form only where low temperature reduces the rate of microbial activity substantially. Where climate is wet, or where topography permits water to accumulate in sites such as basins and valleys, however, peat can form even in hot climates.

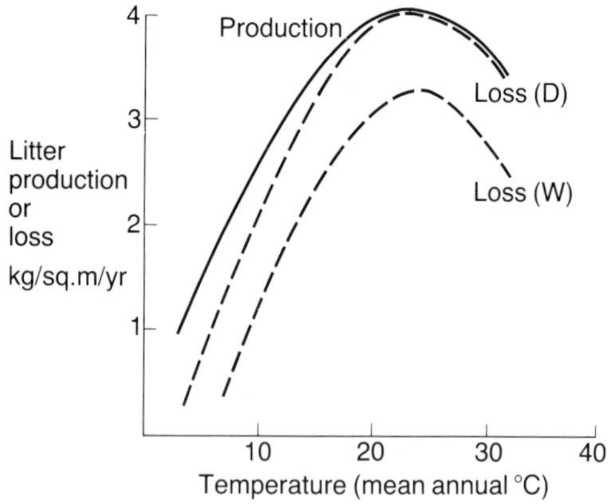

FIG. 3. A tentative model relating primary productivity and decomposition to mean annual temperature for two topographic situations. D = water shedding, W = water receiving.

An additional model is proposed in Fig. 4, where ecosystem production and respiration rates are related to annual precipitation. Again two scenarios are envisaged, one for cool temperate or boreal environments where ecosystem respiration lags behind productivity even at relatively low precipitation rates. The effect of this is the possibility of peat formation on sloping ground even with an annual precipitation of only 1.5 m. In tropical ecosystems a very high rate of precipitation is necessary for peat to form in situations other than valleys and basins. In tropical Africa, for example, Thompson & Hamilton (1983) regard 2 m annual rainfall as sufficient only to maintain swamps in situations which are topographically suitable.

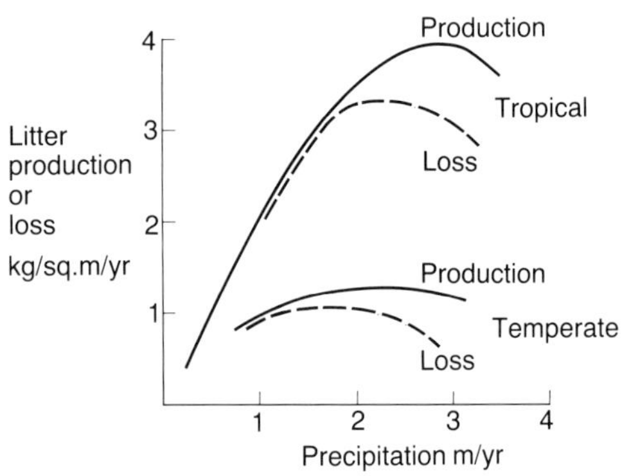

FIG. 4. A simplified model relating primary productivity to annual precipitation for high temperature (tropical) and low temperature (temperate) regimes.

Nutrient relations of mires

(1) *General features*

Nutrients (i.e. those elements required for the healthy growth of plants and animals) enter and leave mires mainly in solution or in suspension in the mire waters or in the precipitation and dust landing on the mire surface. The presence of an insulating layer of peat usually renders the surface vegetation relatively independent of the underlying soil or bedrock and results in the weathering of these inorganic materials being of little importance in the nutrient relationships of most mires. This may not be the case if the surface vegetation has a sufficiently deep rooting system to tap the resources of the underlying mineral substrate. A schematic representation of the nutrient cycling of mires is given in Fig. 5.

The most important feature to emerge from this diagram is the two fold nature of the ionic inputs to the mire surface. In general the ionic content of rainfall is low, but clearly this varies very considerably with geographical location, particularly the oceanicity of the site in question (see Moore & Bellamy 1973 for a detailed consideration of this point). Only in a few very distinctive mires is dust of major importance, as in those which are periodically affected by volcanic eruptions and receive an input of airborne ash. In contrast to rainfall, the ionic content of ground water is likely to be relatively high, though this varies with geology. In either case the actual supply of elements to the surface vegetation of a mire will be a function of ionic concentration and the rate of water flow through the system. It is entirely possible for a mire system to be fed by nutrient poor, oligotrophic waters and yet to support nutrient demanding vegetation because of a relatively rapid rate of water movement. It is this process which results in richer vegetation developing around the central stream courses in valley mires. Nutrient supply and cycling in mires can thus be seen to be closely related to hydrology.

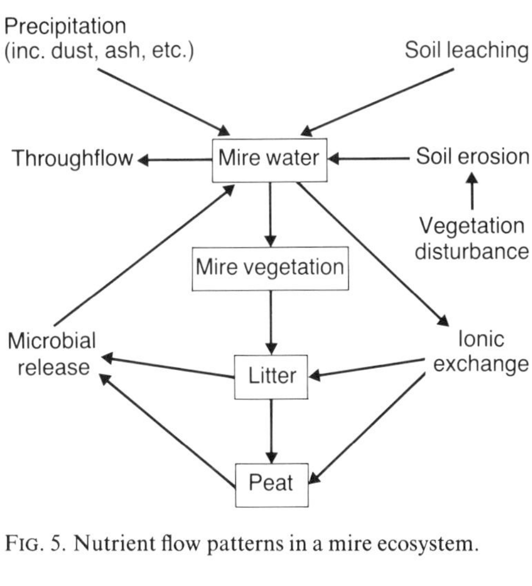

FIG. 5. Nutrient flow patterns in a mire ecosystem.

It is both convenient and ecologically meaningful, therefore, to classify mires on the basis of their hydrology or, more specifically, on their main source of ionic input. Mires fed by precipitation alone are termed ombrotrophic, and the expression 'bog' should be confined to this type of mire. This term and others used in this account are generally accepted by peatland ecologists and have been adopted as standard in the definitive international work on mire ecosystems (Gore 1983) and in the recent survey of European mires (Moore 1984). Mires fed by both precipitation and groundwater movement are termed rheotrophic and include fens, swamps and wooded swamps or 'carr'. Misuse, or irregular use of such terms, has led to a great deal of confusion in the literature. In particular, some of the tropical forested mires which have been compared with ancient coal-forming ecosystems (e.g. Moore 1983 and McCabe 1984) are actually bogs rather than swamps. Such mires should strictly be termed 'bog forest'. Many low latitude forested mires, however, such as those of Maryland described by Richardson (1985), are rheotrophic ecosystems and the term 'swamp forest' is entirely accurate. This hydrological distinction is not mere pedantry: it is of critical importance when considering the ash content of the peats formed by these mires.

Ombrotrophic mires have their surfaces elevated above the influence of ground water flow, usually as a result of the development of masses of insulating peat, though some, the so-called blanket mires, develop in topographic locations which shed rather than receive incoming water, so are in no position to be influenced by ground water movements. But it is not always a simple matter to separate ombrotrophic and rheotrophic mires, for they are often found in close association and may indeed be derived from different portions of a single hydrological unit. In such cases it is advisable to use the term 'mire complex'.

A further variable is imposed by scale considerations. When one examines the surface of a mire in detail, minor patterns of water (and therefore nutrient) movement becomes apparent. When considering scales of metres rather than kilometres, features such as hummocks of bog-moss become important. They form small ombrotrophic units and shed water into drainage runnels, thus enriching them by increasing the flow of water through them (Bellamy & Rieley 1967). Such runnels may be regarded as minor units of relatively rheotrophic conditions. Some German workers have found it convenient to consider mires in terms of 'Grossform' and 'Kleinform' features.

Etherington (1983) has constructed a model relating the precipitation/evaporation ratio of mires to the influence of groundwater, and a modified version of this is shown in Fig. 6. Ombrotrophic mires are confined to the upper left of the diagram, where there is always an excess of incoming precipitation over evaporation and where ground water has no influence. Rheotrophic mires occupy a central position in which water supply from both sources is important. In the lower right part of the diagram are found those wetland systems in which ground water is the most influential hydrological factor and which include certain types of swamp, some marshes with saturated soils but in which peat does not accumulate, and some estuarine and maritime wetlands, such as saltmarshes, and the tropical mangroves. This sequence, from top left to bottom right, also reflects the importance of inorganic components in the accumulating sediments, which is of special interest when seeking modern analogues for the coal-forming ecosystems of former times. These were in the main freshwater rather than saline and were low in their inorganic components. The modern successors to the 'coal swamps' are clearly not swamps at all, but are bog forests of an ombrotrophic nature.

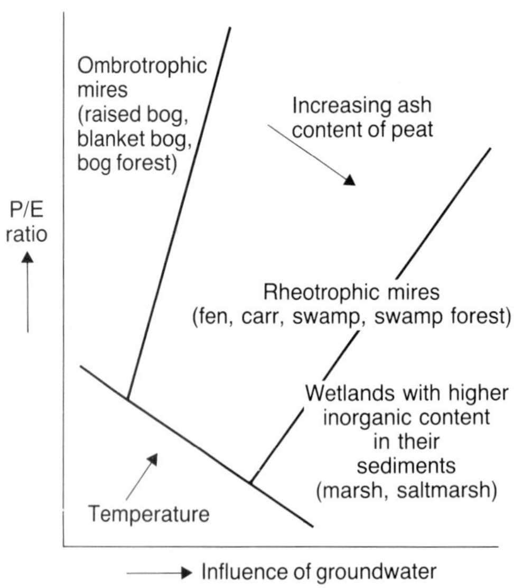

FIG. 6. Proposed relationship between mires in terms of the relative influence of rainwater and groundwater in their hydrological input (modified from Etherington 1983).

(2) *Hydrological types of mire ecosystem*

The precise terms used in peatland ecology for different types of mire is of more than mere semantic interest. The imprecise use of terms such as 'swamp' in the past has led to very considerable confusion among ecologists and palaeoecologists. It may be useful, therefore, to define some of the more widely used terms here in the context of the two major hydrological categories of mire.

Mire has been extensively adopted as a general term for peat-forming ecosystems of all types (Gore 1983). Certain wetland ecosystems, such as some marshes, are not strictly mires according to this definition in that true peat may not develop within them. It is preferable to define mire, therefore, as a freshwater ecosystem in which there is a net accumulation of organic matter in the soil. This term then includes peat-forming ecosystems and those producing muck soils.

Swamps are rheotrophic wetland ecosystems in which the dry season water table is almost always above the surface of the 'soil' or sediment. They are thus essentially aquatic ecosystems and are typically dominated by emergent plants (Spence 1982, Osborne & Polunin 1986).

Floating swamps often develop around the fringes of lakes in both temperate and tropical areas. A platform of roots, rhizomes or stolons of the emergent vegetation is formed and extends out over open water. In the case of tropical papyrus swamps, these platforms may be both thick and extensive (Gaudet 1979).

Swamp forest is a specific type of swamp in which trees are an important constituent of the vegetation, but which is essentially rheotrophic in nature. The term is usually applied to tropical and sub-tropical swamps with a strong wooded component in their vegetation. 'Mangrove swamp' is a specific type of swamp forest. Temperate wooded mires of a rheotrophic type are usually given the Scandinavian term 'carr' (see Daubenmire 1978, p. 96).

Fen is a term used by Tansley (1939) to denote a rheotrophic ecosystem in which the dry season water table may be below the surface of the peat. It is regarded as a successional stage following swamp. Moore & Bellamy (1973) and Gore (1983), however, have used the term in a more general sense to include all rheotrophic mires and equate it to the German term *Niedermoor*. Here the term is used *sensu stricto*.

Marsh is a term which has been used in many

different senses and has almost lost any specific connotation. Many wetland scientists in the United States use the term to denote wetlands which are dominated by floating or emergent, herbaceous aquatics, such as cat-tail, reeds or sedges rather than trees (for which the term swamp is sometimes reserved). An example of this use of the term is the work of Weller (1978). Others, such as Reader (1978) have introduced further confusion by the use of the expression 'bog marsh', which is a hydrological contradiction in terms if 'bog' is used solely for ombrotrophic mires. Gleason & Cronquist (1964) use the term marsh of habitats in which there is permanently standing water, though they imply that a herbacious vegetation and a rheotrophic hydrology is also required. In the European literature, however, the term marsh is used of ecosystems dominated by herbaceous vegetation and situated on seasonally waterlogged mineral soils in which the water table is close to the soil surface but aeration is (at least periodically) adequate to ensure very little if any development of peat. It can apply both to fresh and saline (salt marsh) situations. This use of the term is considered preferable in that it is more closely defined than the American usage and has a longer historical claim to precedence (Tansley 1911).

Bog is a term generally confined to ombrotrophic peat-forming ecosystems.

Bog forest consists of ombrotrophic forested vegetation. Many raised mires in eastern Europe and in temperate North America are of this type and their vegetation usually consists of an upper storey of coniferous trees, a dwarf shrub layer usually rich in ericaceous species and a ground layer of *Sphagnum* moss. The bog forests of the tropics, however, are dominated by angiosperm trees, such as *Shorea* species.

Succession in mires

The fact that mires develop masses of peat within the ecosystem means that they are constantly modifying their own hydrology and that they are recording within those peats their own history. Both of these facts are of interest to the peatland ecologist. The autogenic changes in hydrology mean that there is a tendency within a developing mire to move from a rheotrophic to an ombrotrophic type of ecosystem. The second feature, the recording of such change in the peat, means that the progress of such changes can be reconstructed from stratigraphic studies.

Figure 7 shows a stratigraphic section (idealized) of a western European raised mire in which the development stages are recorded in the

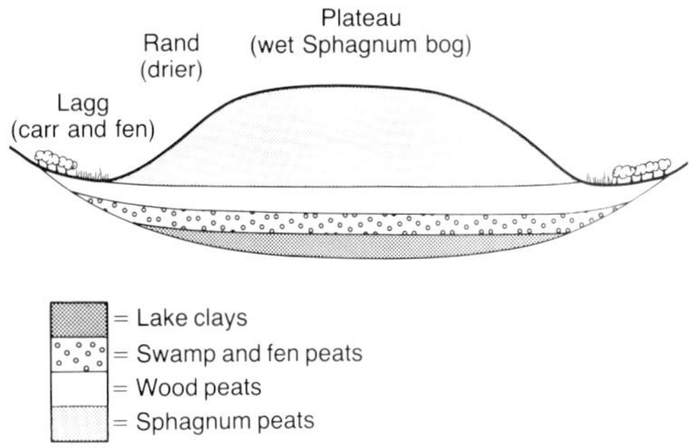

FIG. 7. Simplified stratigraphic section of a raised bog. The term 'fen' is here used in its strict sense (see text).

stratification of the sediment types. This type of sequence is very frequent in the raised mires of Europe (Succow & Lange 1984). The process of infilling of a former lake basin is recorded in the sediments as a progression from open water sediments, through swamp and fen peats into wooded swamp, or carr. This is then superseded by bog peats, dominated by *Sphagnum* as the peat mass is raised above the influence of groundwater and enters its ombrotrophic stage of development.

As a result of collating the stratigraphic profiles of many such sites, Walker (1970) produced a diagram of general trends in successional sequences in northern Europe, and this is shown in a simplified form in Fig. 8. Tallis (1983) has repeated the exercise for N American sequences and has produced a scheme which is also simplified in Fig. 8. The main difference is in the greater tendency for open water sites to become invaded by floating swamps producing 'sedimen-

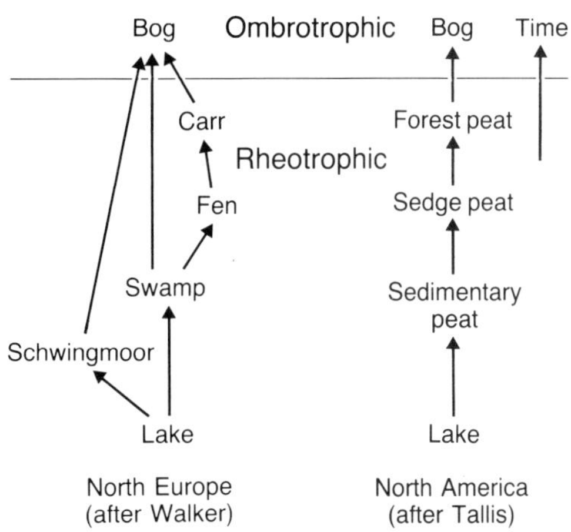

FIG. 8. Major successional pathways in mire development in northern Europe and eastern N America (after Walker 1970 and Tallis 1983).

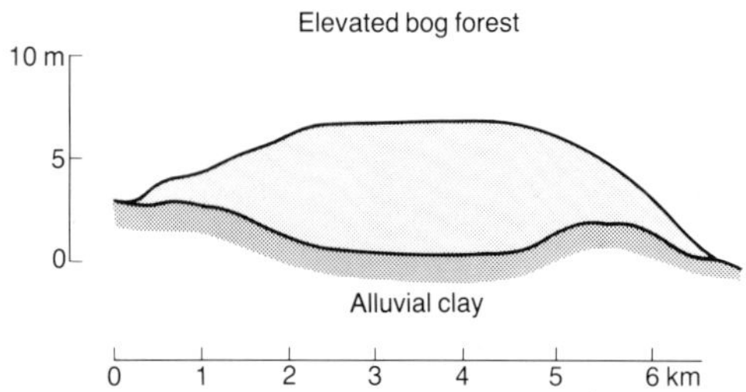

FIG. 9. Simplified stratigraphic section of a Sarawak bog forest (after Anderson 1983).

tary peat' in North America. Such a sequence leads to the formation of a quaking bog or '*Schwingmoor*' on which ombrotrophic vegetation develops.

Very little work has been published on the stratigraphic profiles of the ombrotrophic bog forests of Sarawak or on their successional development. Anderson (1983) has published some diagrams, one of which is simplified in Fig. 9, but these do not provide sufficient information for the reconstruction of successional pathways. From this diagram, however, it is possible to discern the domed structure of the mire which leads to its ombrotrophic hydrological regime.

Modern analogues of coal-forming ecosystems

Clearly, there can be no precise modern analogues of the coal-forming ecosystems of the past since the continuation of the evolutionary process has led to a very different flora in the modern world, with different ecophysiological properties and forming very different vegetation types (see, e.g. Collinson & Scott, *this volume*). Atmospheric and climatic changes have also taken place, some of which, such as the atmospheric changes in carbon dioxide, are in part a product of the formation of the organic geological reservoirs themselves. These changes also render the precise reconstruction of past conditions impossible. But there are some features about which one can speculate or generalize.

In the case of the coals of the Carboniferous, for example, one is often dealing with organic materials with a low ash content, formed under tropical conditions and consisting largely of the remains of woody vegetation (see Bartram, *this volume*; Fulton, *this volume*). Such material can only have developed, in an ecosystem of a tropical 'forest' type in which the abundant supply of water was entirely by precipitation, i.e. ombrogenous. In the modern world we only know of the ombrogenous bog forests of such areas as Sarawak which approach these conditions.

The accumulation of any thickness of coal is also dependent upon the basin within which the peat develops undergoing subsidence, otherwise there are limits to the depth of peat which can develop (Ingram 1982, 1983). An estuarine situation rather than a hard rock basin is therefore indicated and corresponds again to the Sarawak forests. It is still difficult to envisage, however, the hydrological conditions which permitted the very extensive development of bog forests such as those indicated by the vast spatial areas of some coal seams.

There also remains the puzzle of some more recent coal types apparently developed under temperate rather than tropical climates. There do not appear to be modern temperate bog forests which could give rise to the thicknesses of peat found in some of these deposits.

Although it may never be possible to make complete reconstructions of the peat-forming habitats of the past, the major processes at work within those ecosystems can be assessed with a degree of confidence from modern counterparts. A closer study of the process of peat accumulation within modern mires, particularly the tropical bog forests, would therefore be a valuable contribution to the understanding of coal formation.

References

ANDERSON, J. A. R. 1983. The tropical peat swamps of western Malesia. *In:* GORE, A. J. P. (ed.) *Ecosystems of the World. Vol. 4B, Mires: Swamp, Bog, Fen and Moor, Regional Studies*, pp. 181–199. Elsevier, Amsterdam.

BELLAMY, D. J. & RIELEY, J. 1967. Some ecological statistics of a 'miniature bog'. *Oikos*, **18**, 33–40.

BOWLER, M. & BRADSHAW, R. 1985. Recent accumulation and erosion of blanket peat in the Wicklow Mountains, Ireland. *New Phytologist*, **101**, 543–550.

BRADBURY, I. K. & GRACE, J. 1983. Primary production in wetlands. *In:* GORE, A. J. P. (ed.) *Ecosystems of the World. Vol. 4A, Mires: Swamp, Bog, Fen and Moor, General Studies*, pp. 285–310. Elsevier, Amsterdam.

CLYMO, R. S. 1965. Experiments on breakdown of *Sphagnum* in two bogs. *Journal of Ecology*, **53**, 747–758.

—— 1983. Peat. *In:* GORE, A. J. P. (ed.) *Ecosystems of the World. Vol. 4A, Mires: Swamp, Bog, Fen and Moor, General Studies*, pp. 159–224. Elsevier, Amsterdam.

DAUBENMIRE, R. 1978. *Plant Geography with Special Reference to North America*. Academic Press, New York.

ETHERINGTON, J. R. 1983. *Wetland Ecology*. Studies in Biology No. 154, Arnold, London.

GAUDET, J. J. 1979. Seasonal changes in nutrients in a tropical swamp: North Swamp, Lake Naivasha, Kenya. *Journal of Ecology*, **67**, 953–981.

GLEASON, H. A. & CRONQUIST, A. 1964. *The Natural Geography of Plants*. Columbia University Press, New York.

GORE, A. J. P. 1983. Introduction. *In:* GORE, A. J. P. (ed.) *Ecosystems of the World. Vol. 4A, Mires: Swamp, Bog, Fen and Moor, General Studies*, pp. 1–34. Elsevier, Amsterdam.

GOWER, S. T., LEA, R., FREDRICK, D. J., CLARK, A. & PHILIPS, D. R. 1985. Above ground energy production and distribution of southeastern hardwood swamp forests. *Biomass*, **7**, 185–197.

INGRAM, H. A. P. 1982. Size and shape in raised mire ecosystems: a geophysical model. *Nature, London*, **297**, 300–303.

—— 1983. Hydrology. *In:* GORE, A. J. P. (ed.) *Ecosystems of the World. Vol. 4A, Mires: Swamp, Bog, Fen and Moor, General Studies*, pp. 67–158. Elsevier, Amsterdam.

LEITH, H. 1975a. Primary production of the major vegetation units of the world. *In:* LEITH, H. & WHITTAKER, R. H. (eds) *Primary Productivity of the Biosphere*, pp. 203–215. Springer-Verlag, New York.

—— 1975b. Modeling the primary productivity of the world. *In:* LEITH, H. & WHITTAKER, R. H. (eds) *Primary Productivity of the Biosphere*, pp. 237–263. Springer-Verlag, New York.

MALTBY, E. 1986. *Waterlogged Wealth*. Earthscan, London.

MCCABE, P. J. 1984. Depositional environments of coal and coal-bearing strata. *In:* RAHMANI, R. A. & FLORES, R. M. (eds) *Sedimentology of Coal and Coal-bearing Sequences*, pp. 13–42. Special Publication of the International Association of Sedimentologists 7. Blackwell Scientific Publications, Oxford.

MOHR, E. C. J. & VAN BAHREN, F. A. 1959. *Tropical Soils*. Van Hoeve, The Hague.

MOORE, P. D. 1983. Plants and the palaeoatmosphere. *Journal of the Geological Society of London*, **140**, 13–25.

—— 1984. (ed.) *European Mires*. Academic Press, London.

—— & BELLAMY, D. J. 1973. *Peatlands*. Elek, London.

ODUM, E. P. 1969. The strategy of ecosystem development. *Science*, **164**, 262–270.

OSBORNE, P. L. & POLUNIN, N. V. C. 1986. From swamp to lake: recent changes in a lowland tropical swamp. *Journal of Ecology*, **74**, 197–210.

READER, R. J. 1978. Primary production in northern bog marshes. *In:* GOOD, R. E., WHIGHAM, D. F. & SIMPSON, R. L. (eds) *Freshwater Wetlands*, pp. 53–62. Academic Press, New York.

RICHARDSON, C. J. 1985. Mechanisms controlling phosphorus retention capacity in freshwater wetland. *Science*, **228**, 1242–1427.

SCHELL, D. M. 1983. Carbon-13 and carbon-14 abundances in Alachan aquatic organisms; delayed production from peat in Arctic food webs. *Science*, **219**, 1068–1071.

SPENCE, D. H. N. 1982. The zonation of plants in freshwater lakes. *Advances in Ecological Research*, **12**, 37–125.

SUCCOW, M. & LANGE, E. 1984. The mire tyes of the German Democratic Republic. *In:* MOORE, P. D. (ed.) *European Mires*, pp. 149–175. Academic Press, London.

TALLIS, J. H. 1983. Changes in wetland communities. *In:* GORE, A. J. P. (ed.) *Ecosystems of the world. Vol. 4A, Mires: Swamp, Bog, Fen and Moor, General Studies*, pp. 311–347. Elsevier, Amsterdam.

TANSLEY, A. G. 1911. *Types of British Vegetation*. Cambridge University Press.

—— 1939. *The British Islands and their Vegetation*. Cambridge University Press.

THOMPSON, K. & HAMILTON, A. C. 1983. Peatlands and swamps of the African continent. *In:* GORE, A. J. P. (ed.) *Ecosystems of the World. Vol. 4B, Mires: Swamp, Bog, Fen and Moor, Regional Studies*, pp. 331–373. Elsevier, Amsterdam.

WALKER, D. 1970. Direction and rate in some British post-glacial hydroseres. *In:* WALKER, D. & WEST, R. G. (eds) *Studies in the Vegetational History of the British Isles*, pp. 117–139. Cambridge University Press.

WELLER, M. W. 1978. Management of freshwater marshes for wildlife. *In:* GOOD, R. E., WHIGHAM, D. F. & SIMPSON, R. L. (eds) *Freshwater Wetlands*, pp. 267–284. Academic Press, New York.

P. D. MOORE, Department of Biology, King's College (KQC), University of London, Campden Hill Road, Kensington, London W8 7AH, England.

Rainwater-fed peat as a precursor of coal

R. S. Clymo

ABSTRACT: About 3% of the Earth's land surface is peat-covered at present. Most peat forms where there is waterlogging, because oxygen diffuses down through water more slowly than it is consumed by micro-organisms. The anaerobic micro-organisms in peat continue the decay process, but only slowly.

A great variety of rheotrophic (usually minerotrophic and eutrophic) peats form if the waterlogging is caused by groundwater. Of greater extent, but less varied, are the ombrotrophic peats which are dependent on precipitation. Most of these peatlands are in Russia, Canada, Fennoscandia and the British Isles. The main peat-forming plants in such places are bogmosses (*Sphagnum*), sedges and heaths, with pines (*Pinus*) and larch (*Larix*) in some places. These peats may be analogous to those which formed the Permian coals of Australia. In western Malesia are large areas of coastal ombrotrophic forest-bogs. These tropical forest peats may be analogous to those which formed the Carboniferous coals. The organic chemistry of newly dead plants differs considerably amongst all these plant groups, and they in turn may be very different from the majority of plants which formed the starting materials for the formation of coals in the past.

In spite of these differences the study of present-day moss and sedge bogs may be helpful and particularly in the study of Permian coals. There is usually a two-layered structure. In the 5–50 cm thick surface layer aerobic decay may remove as much as 90% of the original matter and do so selectively. Below is the much thicker waterlogged anaerobic layer in which microbial decay is of a different sort and is much slower. But there is evidence that it continues for thousands of years at least. The extent to which peat is a product of microbial activity operating on a slowly changing substrate is unknown. Equally unknown is the balance between microbially mediated and chemical changes. One consequence of bacterial decay is that there seems to be an inherent limit to the depth of such peat-forming systems. At least one (unusual) case is known which has resulted in the production of a lignitic peat, which breaks with a conchoidal fracture, and which is barely 12 000 years old. The limiting depth, assessed from growth during the last 10 000 years, seems to be too little to account for the observed thickness of coals.

Keywords: Peat, coal, ombrotrophic mire, lignite, Permian coals, *Sphagnum*

Coals are often assumed to have passed through a stage when they would have been called peat. But the botanical remains in the wide variety of present-day peats are very different from those in coals. This observation prompts the question: to what extent does the organic chemistry of coals depend on that of the plants from which they formed, and to what extent is it determined by the *processes* of peat formation, and independent of the original botanical composition?

The types and distribution of present-day peats

Most, but not all, peats form as a consequence of waterlogging (see Moore, *this volume*). Dead plant matter is attacked by microorganisms and, where oxygen is available, usually decays quickly. The rate of diffusion of oxygen in water is only 10^{-4} that in air, so where the dead plant matter is waterlogged and microorganisms are active then oxygen is used up faster than it can be replaced by diffusion. The incipient peat thus becomes anoxic. This does not mean that there is no further decomposition by microorganisms, but it does mean that different species—those able to grow in anoxic conditions—survive. The rate at which they operate seems to be much slower than that of aerobic microorganisms, and they are probably less abundant, so peat accumulates even though decay and consequent changes do continue. In general, both fungi and bacteria contribute to aerobic decay, but anaerobic decay is predominantly due to bacteria (Table 1). General reviews of peat formation and properties are given in Moore & Bellamy (1974), Fuchsman (1980), Clymo (1983), Gore (1983), Hobbs (1986) and Moore (*this volume*).

There are numerous ways of classifying peats and peat-forming systems: see, for example, Kivinen *et al.* (1979). For the present purpose the

TABLE 1. *Abundance of microorganisms in the surface layers of blanket peat in the Moor House National Nature Reserve (54° 65′ N, 2° 45′ W). The fungi were estimated by direct observation; the bacteria (of many different groups) by plating on nutrient agars of various sorts (from Collins* et al. *1978).*

Depth (cm)	Zone	Fungi: stained mycelium (m g^{-1})	Bacteria (10^5 g^{-1}) [range]	
			Aerobic	Anaerobic
0–5	Litter	2450	9–260	9–250
5–12	Dark brown	1030	6–150	32–200
12–20	Green brown	750	11–76	16–500
20–32	Red brown	200	0.7–42	28–260

most important distinction is between peats formed from plants influenced by groundwater and peats entirely dependent on precipitation.

Peats dependent for inorganic solutes on ground water are rheotrophic: their nutrition depends on flowing water even if the rate of flow is very slow. The concentration of inorganic solutes in such waters is usually, but not invariably, markedly greater than that in precipitation. Peats formed in these conditions are therefore sometimes called minerotrophic, and as the supply of solutes is relatively high the peat may also be called eutrophic: that is it formed under 'good' nutritional conditions. Most rheotrophic peats are also minerotrophic and eutrophic. Peats which depend entirely on precipitation for their supply of inorganic solutes are called ombrotrophic (Greek ombros = rainstorm). The term 'swamp' is usually reserved for perennially flooded areas where rheotrophic peat may be accumulating. 'Marsh' is best reserved for situations where there is a large proportion of solid inorganic matter in the substrate. The cavalier use of these terms not only destroys useful distinctions but obscures understanding of important features of peat (see Moore, *this volume*). The vegetation of minerotrophic peats is very varied, ranging from that of reedswamp and calcareous fen dominated by grasses, sedges, and other monocotyledonous plant species to freshwater swamps. These very varied minerotrophic peats may be found in small amounts in all those parts of the world, including the tropics, which provide an assured supply of groundwater.

Ombrotrophic or weakly minerotrophic, peats cover far larger areas today than do strongly minerotrophic peats: a single peatland in western Siberia is 1800 × 800 km (Walter 1977). Because they usually depend on precipitation exceeding evaporation for most of the year they are geographically more restricted than minerotrophic peats: most ombrotrophic peats are in, or close to the Boreal Zone. The dominant plants are bog-mosses (*Sphagnum*), cotton sedges (*Eriophorum*) and low shrubs of the heather family (*Ericaceae*). Such peats are usually acid with pH 4.0 or less—*Sphagnum* makes them so (Clymo & Hayward 1983)—and poor in inorganic nutrients. *Sphagnum* and the heathers, at least, are intolerant of solute-rich groundwater.

Minerotrophic peats frequently contain the remains of trees. Some ombrotrophic or weakly minerotrophic peatlands may support trees—pine (*Pinus sylvestris*) in Europe, tamarack and black spruce (*Larix laricina* and *Picea mariana*) in North America. Of special interest are the ombrotrophic tropical forest peats formed of the remains of trees, particularly *Shorea albida*, alongside estuaries in Sarawak, Brunei and Malaya (Anderson 1964).

Estimates of the area covered by peat have tended to increase. Taylor (1964) gave a world total of 121 Mha, with 10 Mha in Canada. Twenty years later (Taylor 1983) the total was 240 Mha, with 130 Mha in Canada. Kivinen & Parkarinen (1981) reached 480 Mha—about 3% of the Earth's land surface. These differences reflect genuine difficulties of definition and mapping. Even the 1:5 000 000 UNESCO/FAO Soil Map of the World is unsuitable for it shows no histosols (predominantly organic soils) in Newfoundland. All estimates agree on the joint premier position of the USSR and Canada, with Fennoscandia next, followed by the USA, British Isles, Germany, Poland and surprisingly large areas of coastal Malaysia and Indonesia. Remote sensing has been used to make an accurate map of peatlands in Ontario province (Pala 1984) and holds promise for the future.

In summary, there is a lot of peatland at present, but its botanical composition is very varied. Most ombrotrophic peats are Boreal and contain moss or herb remains, with a few shrubs and occasional trees. They may perhaps be analogous to the precursors of the Permian coals of Australia, which were geographically Boreal at the time they were deposited, and which show underlying 'stone-roll swarms' reminiscent of the structures formed today beneath string mires and palsas in the northern Boreal zone (Conaghan 1984). It may be significant that fossils of *Protosphagnum*, having two types of cell in the leaf, are known from the Permian in Russia (Lacey 1969), whilst *Sphagnum* is one of the commonest sub-fossils in modern peats.

Analogues of tropical Carboniferous coals may perhaps be sought in the present-day coastal ombrotrophic forest-bogs of west Malesia. The peat in such bogs is mostly formed from wood (see also Collinson & Scott, *this volume*).

What then of the chemical composition of peat? As an illustration consider the bog-moss (*Sphagnum*) which is one of the most abundant plants in Boreal peatlands and which, because it decays relatively slowly, comes to be over-represented in peat (Clymo 1984).

The bog-mosses are an ancient and structurally distinct group of plants. About 20–30% of the cell wall is uronic acid residues in long polymers, possibly mixed (Clymo 1963, Theander 1954). Some of the uronic acids are unusual. For example, D-*lyxo*-5-hexosulopyranuronic acid has been found by Painter (1983). The 'lignins' of *Sphagnum* seem to be odd too (Bland *et al.* 1968), and the walls also contain two remarkable phenolic compounds: *Sphagnum* acid (p-hydroxy-β-carboxymethyl-cinnamic acid) and a number of sphagnorubins (based on a unique phenyl-phenanthropyran skeleton) which give several of the species their characteristic red colour (e.g. Rudolph & Johnk 1982). All of these substances form 0.5–5% of the plant dry mass, and are normal constituents. We thus know that one of the most important components of modern Boreal ombrotrophic peats is chemically very different from other modern components. The same distinctive chemical variety *may* have characterized the vegetation which formed coals, but it would only by chance have been the same compounds in similar amounts to those in modern peats.

The processes of formation of ombrotrophic peat

It is convenient to consider the surface of 5–50 cm thick aerobic layer—the acrotelm of Ingram (1978)—from the much thicker anoxic catotelm below. Suppose plant matter (measured as dry mass on a unit area basis) is added to this layer at a constant rate, p, and that the rate of decay, α, is proportional to the remaining mass. Then the mass, x, of plant matter accumulated on a unit area is given by $dx/dt = p - \alpha x$. The solution of this is $x = \frac{p}{\alpha}(1 - e^{-\alpha t})$. The realism of these assumptions is discussed by Clymo (1984). For large values of t then $x \rightarrow p/\alpha$, and if α is large then a negligible amount of peat accumulates. But if α is small—say $0.10\,a^{-1}$, as it is for *Sphagnum*—and particularly if the decay process is cut off (t small) because the plant matter has been submerged by a rising water table, then the steady-state thickness of the acrotelm may not be negligible. Even more important is that a proportion of the plant matter survives its passage through the acrotelm and reaches the catotelm. A limited number of measurements on ombrotrophic peatlands indicate that about 10% of the original material passes into the catotelm (Clymo 1984). This steady 'rain' of organic matter from the acrotelm into the catotelm is exactly analogous to the constant addition of new plant matter at the surface. If one assumes that the rate of decay, which is known to be much smaller in the catotelm (Clymo 1965), is still at a rate proportional to what is left, then one gets exactly the same defining equation and solution. The variable p now represents the rate at which matter enters the catotelm (about 10% of the rate at which it enters the acrotelm), α is the decay rate constant for the catotelm, and t is time in the catotelm. The effects are shown in Fig. 1, which uses realistic parameter values derived in the way explained later. This hypothesis, for so it is, has two interesting consequences.

The first is that in such circumstances there is a limit to the thickness of peat which can accumulate. Formally it is p/α. At this asymptotic value the integrated loss at all depths exactly balances the injection of matter at the surface. Because the model concerns mass the effects of compaction and compression are irrelevant, though these processes certainly occur and are important (Clymo 1978, Hobbs 1986). Provided only that decay does continue—however slowly—then there must be a limit to the depth of peat.

The second consequence of this hypothesis, which can be used to test it, is to be seen at the right side of Fig. 1. Although the rate at which matter has been injected has remained constant (p) the age versus depth (as mass) relationship is clearly a hollow curve. The five cases of suitable data were examined (Clymo 1984) and found to show such a hollow curve. Since then three more sets of data which satisfy the criteria of being from ombrotrophic peat, with at least ten ^{14}C dates and a complete set of bulk density measurements, have appeared (Lewis Smith & Clymo 1984, Middeldorp 1984, Dupont 1985). One of them is shown in Fig. 2. A minimization technique (Clymo 1984) allows p and α to be

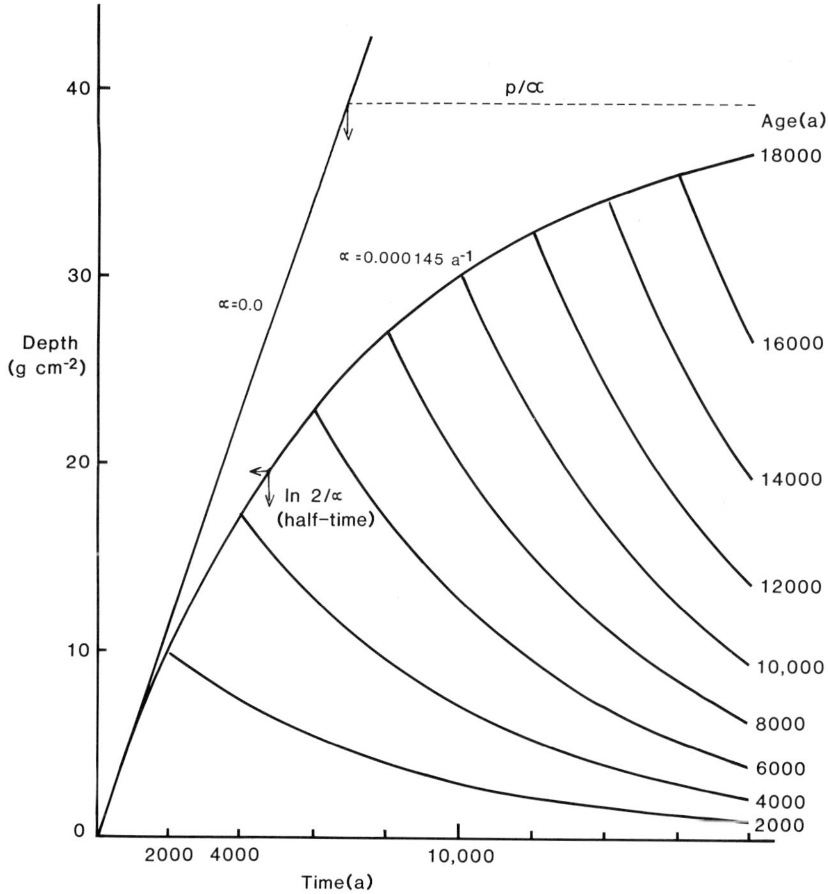

FIG. 1. Growth in depth of a peat mass, modelled by $dx/dt = p - \alpha x$, where x, is the organic dry mass, p is the rate of addition of dry mass, α is the proportional rate of decay and t is time. Both x and p are on a unit area basis. The parameter values, estimated from the data in Fig. 2, are: $p = 0.0057$ g cm^{-2}a^{-1} $\alpha = 0.000145$ a^{-1}. The depth scale is given as mass/area. If the bulk density were 0.05 g cm^{-3} then multiplying the values given by 20 would give the depth in cm linear measure (e.g. the limiting depth of $p/\alpha = 39.3$ g cm^{-2}/786 cm depth). The straight diagonal line shows what would happen if there were no decay. It intersects p/α at $1/\alpha$. The convex curve is for the surface. The concave curves show what would happen to peat which was at the surface at 2000 year intervals. The right margin shows that although both p and α are assumed constant the age versus depth profile is *not* linear (as is usually supposed).

estimated: in essence the general slope governs p and the curvature governs α. For the eight cases the value of α is in the range $0.1\text{–}1.0 \times 10^{-3}$ a^{-1}. This implies half-lives (ln $2/\alpha$) of 700–7000 years. Very few present-day peat deposits have had the luxury of 20 000 years relatively undisturbed development, but some might hope for this. By that time there would be left between 2×10^{-9} and 0.13 of the mass which entered the catotelm. Even the larger of these values (corresponding to $\alpha = 0.0001$), when combined with a 10% survival into the catotelm, implies that barely 1% of the original plant matter survives.

Survival is certainly selective. The leaves of species such as *Rubus chamaemorus* and *Andromeda polifolia* are often abundant on the surface of northern ombrotrophic peatlands but rarely survive in recognizable form into peat, whereas *Sphagnum* and *Eriophorum* shoot bases survive unusually well. Indeed it was only Ockham's advice 'Nunquam ponenda est pluralitas sine necessitate' which persuaded me in the face of such selective survival to try the simple single component model which appears to be sufficient to account for results such as those in Fig. 2. Some of the consequences of applying the same

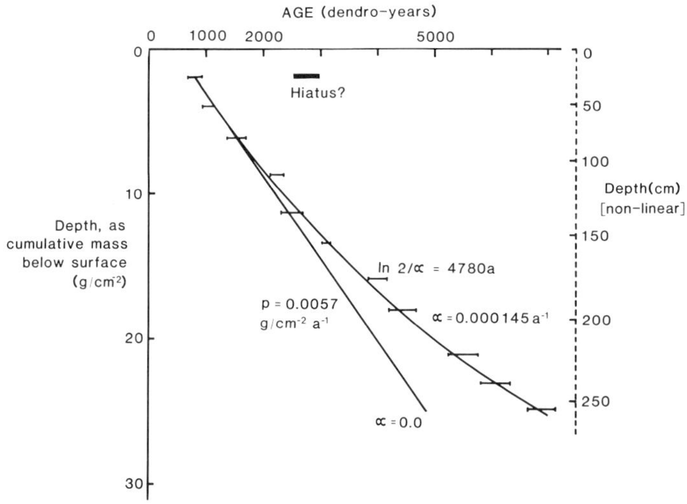

FIG. 2. Age versus depth (as cumulative mass below the surface) for K2 Hollow in the Meerstalblok of Bargerveen on the Dutch–German border (52°41′N, 7°02′E). The data are those of Dupont (1985) supplemented by privately communicated bulk density values. I acknowledge, with thanks, Dr Dupont's help. The right vertical axis shows actual linear depths at 50 cm intervals. The ages are derived from ^{14}C dates after calibration to 'true' (dendro) age. Horizontal bars are 5% confidence limits. The curve was fitted to the model specified in the text and Fig. 1 by methods described in Clymo (1984): its parameters are $p = 0.0057$ g cm^{-2} a^{-1}, $\alpha = 0.000145$ a^{-1}. The straight diagonal line shows what would have occurred if $\alpha = 0$. There may be a hiatus from 2500 to 3000 years ago. This would slightly increase the value of p and slightly decrease the value of α.

ideas to mixtures are considered in Clymo (1984).

These ideas apply to chemical structures as well as to macroscopic or microscopic visible structures. An illuminating example is that of the very unusual peat formed almost entirely of the remains of the tussock grass *Poa flabellata* (Lewis Smith & Clymo 1984). On Beauchêne Island, in the South Atlantic, the peat is at least 11 m deep. Near the surface plant structures are visible, but further down the peat becomes more humified and the basal 1.5 m, which are about 8000–12 000 years old, have a hard cheesy texture. When exposed to the air this peat becomes black and hard, and breaks with a conchoidal fracture. It can only be described as lignitic, though it is barely 10 000 years old and has not been subjected to geothermal heat. The age versus depth profile shows the same sort of concave curve as that in Fig. 2.

What are the sources of old, often structureless, peats? There seem to be three main possibilities (Fig. 3): (a) material, produced by the plants which grew on the surface, which has survived chemically unchanged; (b) products of the continued slow activity of microorganisms—probably mostly bacteria; (c) products of slow chemical reactions, not requiring the presence of microorganisms.

The idea that microorganisms may still be living and causing changes at such depths is not yet generally accepted. It is known that the concentration of methane in the depths of the peat is near or above saturation (Svensson & Rosswall 1984, Clymo 1984, Claricoates *in lit*) and methane efflux, which is known to be large (Clymo & Reddaway 1971, Svensson 1980, Harriss *et al.* 1985), could easily account for the mass loss during decay. The existence of a methane concentration gradient from the base to the surface of a peat mass (Claricoates *in lit*) is explicable only if there is *continued* production. Is this a microbiological process—it seems likely—or could it be a consequence of slow chemical reactions? It should be simple enough to distinguish these possibilities, but what *proportion* of the peat at the base of a deposit is produced in these three ways? Where 99% of the original material has been disposed of the possibility—though not the probability—of a degree of Procrustean tailoring cannot be ignored.

Other important questions may be asked. What is the maximum depth to which ombrotrophic peat may accumulate? The analysis presented here gives the result as p/α, which is a *mass* per unit area. The linear depth will depend on the dry bulk density of the peat. Values between 7 and 15 m have been derived so far. Compaction during coal formation might reduce this to

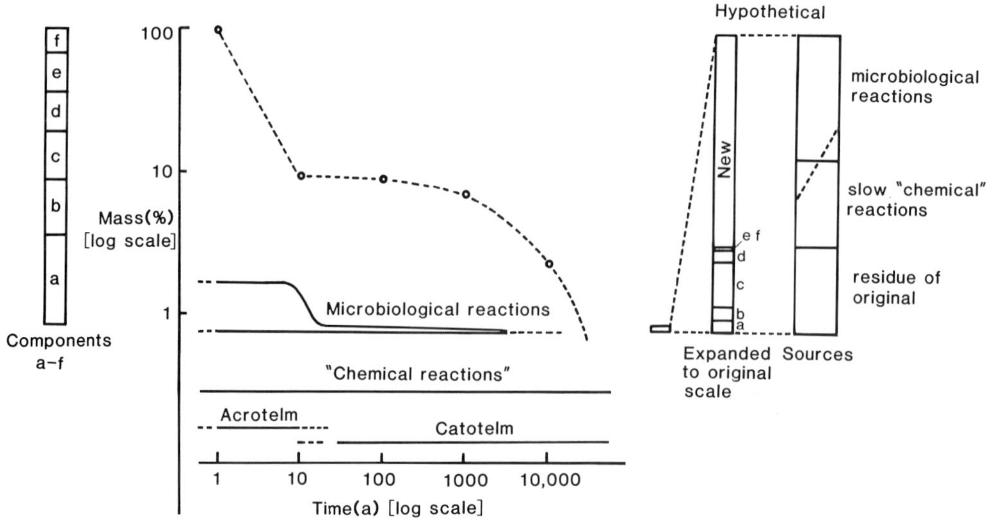

FIG. 3. Summary of known and hypothetical features of peat formation. The central graph shows the probable change in mass with time (both axes have been log transformed). The left column shows hypothetical components a–f. At the right of the graph is what is left after 10 000 years or so. This is expanded to show the selective change of components and the possibility of new materials produced by microbiological or chemical change, or by their interaction (diagnonal dashed line).

perhaps 1/5 of these values. If the concentration of acid-insoluble ash at the top of the catotelm of an ombrotrophic peat is about 10 mg g^{-1}, then towards the base after a very long time it might approach 10% of the dry mass. Further compression (without loss of mass) would not affect the concentration. These calculations imply that the concentration of ash might increase downwards in a coal seam to as much as 10%, and that the maximum thickness of such a seam might be about 3 m. The Permian coals of the Sydney basin do not seem to have these characteristics: it may be necessary to postulate more complicated origins.

We have no idea whether these values, or even the same model, are applicable to the tropical ombrotrophic forest peat. Finding out is simple in principle, but may be difficult in practice. Even more difficult is to assess the probability that any of these peats may survive to become coal.

Whatever the answers to these questions it seems likely that much of the material which is eventually subject to coalification proper has already undergone extensive changes, both quantitative and qualitative.

Conclusions

(1) Ombrotrophic peats cover about 3% of the Earth's land surface today.

(2) There is a limit to the depth which such deposits may grow.

(3) As much as 99% of the original plant materials may have disappeared during peat formation, and losses are selective. It is this highly modifed residue which may be the material in which coal formation proper begins.

References

ANDERSON, J. A. R. 1964. The structure and development of the peat swamps of Sarawak and Brunei. *Journal of Tropical Geography*, **18**, 7–16.

BLAND, D. E., LOGUN, A. & MENSHUN, M. 1968. The lignin of *Sphagnum*. *Phytochemistry*, **7**, 1373–1377.

CLYMO, R. S. 1963. Ion exchange in *Sphagnum* and its relation to bog ecology. *Annals of Botany, London*, New Series, **27**, 309–327.

—— 1965. Experiments on breakdown of *Sphagnum* in two bogs. *Journal of Ecology*, **53**, 747–758.

—— 1978. A model of peat bog growth. *In:* HEAL, O. W. & PERKINS, D. F. with BROWN, W. M. (eds) *Production Ecology of British Moors and Montane Grasslands*, pp. 187–223. Springer-Verlag, Berlin.

—— 1983. Peat. *In:* GORE, A. J. P. (ed.) *Ecosystems of the World. Vols 4A, B, Mires: Swamp, Bog, Fen and*

Moor, General and Regional Studies, pp. 159–224. Elsevier, Amsterdam.

—— 1984. The limits to peat bog growth. *Philosophical Transactions of the Royal Society of London*, **B303**, 605–654.

—— & HAYWARD, P. M. 1983. The ecology of *Sphagnum*. *In:* SMITH, A. J. E. (ed.) *Bryophyte Ecology*, pp. 229–289. Chapman & Hall, London.

—— & REDDAWAY, E. J. F. 1971. Productivity of *Sphagnum* (bog-moss) and peat accumulation. *Hydrobiologia*, **12**, 181 192. (Reproduced without arbitrary cuts as: *A tentative dry matter balance sheet for the wet blanket bog on Burnt Hill, Moor House NNR*. Moor House Occasional Paper No. 3, Institute of Terrestrial Ecology, 1972.)

COLLINS, V. G., D'SYLVA, B. T. & LATTER, P. M. 1978. Microbial populations in peat. *In:* HEAL, O. W. & PERKINS, D. F. with BROWN, W. M. (eds) *Production Ecology of British Moors and Montane Grasslands*, pp. 94–112. Springer-Verlag, Berlin.

CONAGHAN, P. J. 1984. Aapamire (string-bog) origin for stone-roll swarms and associated 'fluvio-deltaic' coals in the late Permian Illawarra coal measures of the southern Sydney basin: climatic geomorphic, and tectonic implications. *Geological Society of Australia 7th Congress Abstracts* No. 12, 1–4.

DUPONT, L. M. 1985. Temperature and rainfall variation in a raised bog ecosystem. *Thesis*, pp. 1–62. University of Amsterdam.

FUCHSMAN, C. H. 1980. *Peat: Industrial Chemistry and Technology*. Academic Press, New York.

GORE, A. J. P. (ed.) 1983. *Ecosystems of the World. Vols 4A, B, Mires: Swamp, Bog, Fen and Moor, General and Regional Studies*. Elsevier, Amsterdam.

HARRISS, R. C., GORHAM, E., SEBACHER, D. I., BARTLETT, K. B. & FLEBBE, P. 1985. Methane flux from northern peatland. *Nature, London*, **315**, 652–654.

HOBBS, N. B. 1986. Mire morphology and the properties and behaviour of some British and foreign peats. *Quarterly Journal of Engineering Geology*, **19**, 7–80.

INGRAM, H. A. P. 1978. Soil layers in mires: function and terminology. *Journal of Soil Science*, **29**, 224–227.

KIVINEN, E., HEIKURAINEN, L. & PAKARINEN, P. (eds) 1979. *Classification of Peat and Peatlands*. International Peat Society, Helsinki.

—— & PAKARINEN, P. 1981. Geographical distribution of peat resources and major peatland complex types in the world *Annales Academiae Scientarum Fennicae. Series A III Geologia—Geographica*, **132**, 1–28.

LACEY, W. S. 1969. Fossil bryophytes. *Biological Reviews*, **44**, 189–205.

LEWIS SMITH, R. I. & CLYMO, R. S. 1984. An extraordinary peat-forming community on the Falkland Islands. *Nature, London*, **309**, 617–620.

MIDDELDORP, A. A. 1984. Functional ecology of raised bogs. *Thesis*, pp. 1–124. University of Amsterdam.

MOORE, P. D. & BELLAMY, D. J. 1974. *Peatlands*. Elek, London.

PAINTER, T. J. 1983. Residues of D-*lyxo*-5-hexosulopyranuronic acid in *Sphagnum* holocellulose, and their role in cross-linking. *Carbohydrate Research*, **124**, C18–C21.

PALA, S. 1984. A *LANDSAT*-based method for the survey and inventory of peat resources over extensive regions. *In:* ROBERTSON, R. S. (ed.) *Remote Sensing in Peat and Terrain Resource Surveys*, pp. 156–168. International Peat Society, Helsinki.

RUDOLPH, H. & JOHNK, A. 1982. Physiological aspects of phenolic compounds in the cell walls of *Sphagna*. *Journal of the Hattori Botanical Laboratory*, **53**, 195–203.

SVENSSON, B. 1980. Carbon dioxide and methane fluxes from the ombrotrophic parts of a subarctic mire. *In:* SONESSON, M. (ed.) *Ecology of a Subarctic Mire*, pp. 235–250. Ecological Bulletins 30, Swedish Natural Science Research Council, Stockholm.

—— & ROSSWELL, T. H. 1984. *In situ* methane production from acid peat in plant communities with different moisture regimes in a subarctic mire. *Oikos*, **43**, 341–350.

TAYLOR, J. A. 1964. Distribution and development of the World's peat deposits. *Nature, London*, **201**, 454–456.

—— 1983. The peatlands of Great Britain and Ireland. *In:* GORE, A. J. P. (ed.) *Ecosystems of the World. Vol. 4B, Mires: Swamp, Bog, Fen and Moor, General Studies*, pp. 1–46. Elsevier, Amsterdam.

THEANDER, O. 1954. Studies on *Sphagnum* peat. III. A quantitative study on the carbohydrate constituents of *Sphagnum* mosses and *Sphagnum* peat. *Acta Chemica Scandinavica*, **8**, 989–1000.

WALTER, H. 1977. The oligotrophic peatlands of Western Siberia—the largest peino-helobiome in the world. *Vegetatio*, **34**, 167–178.

R. S. CLYMO, School of Biological Sciences, Queen Mary College, University of London, Mile End Road, London E1 4NS, England.

Coal, climate and terrestrial productivity: the present and early Cretaceous compared

A. M. Ziegler, A. L. Raymond, T. C. Gierlowski, M. A. Horrell, D. B. Rowley & A. L. Lottes

ABSTRACT: At the present time, rainfall sufficient and consistent enough for swamp formation and peat preservation occurs in the equatorial and temperate belts. Evaporite deposits and aeolian sandstones occur in the consistently dry areas, while the intermediate areas of seasonal rainfall tend not to have any of these climatically significant sediments. In the Cretaceous, the temperate belts are well represented by coals as today, but the tropical belt is not, and the same can be said of other Mesozoic and early Cenozoic periods. Instead, the tropics are represented by low diversity floras and sediments, like the Nubian Sandstone, which can be interpreted as representative of climates in which precipitation was markedly seasonal. The inference is that the Intertropical Convergence Zone was less latitudinally confined during the warmer 'greenhouse' periods, and that this might have been due to weaker polar fronts.

Keywords: peat, coal, climate, palaeogeography, Cretaceous

Preliminary maps showing the world-wide distribution of coal have been published for periods ranging from the Carboniferous to the Neogene (Ziegler *et al.* 1979, Parrish *et al.* 1982). The resulting patterns have been compared by these authors to supposed rainfall zones reconstructed on uniformitarian grounds with satisfactory results. However, the connection between present rainfall patterns and peat occurrence has not been examined in recent years, so we begin this paper with maps comparing peat, evaporites and desert sands with forest formations, and in turn, with various aspects of precipitation. Secondly, we have chosen the early Cretaceous to show the distribution of the same sedimentary and floral indicators. Most would agree that the Cretaceous climates contrasted sharply with the Present, and our objectives are to see how these purported contrasts are reflected in the climatic indicators, and to go on to use the early Cretaceous as a model for times in the past that lacked strong latitudinal temperature gradients. Finally, the distribution of peat and coal, as well as floras and fossil floras, provide the basis for speculations on past variations in land surface productivity, and indeed its possibly important interaction with climates through time.

Present peat distribution

The world-wide distribution of peat is not very well known, particularly for the tropical zone where swamps, until recently, have been regarded as more of a nuisance than a resource (see Clymo, *this volume*). Yet, the majority of the vast Carboniferous coal swamps were clearly based in tropical rainforest situations, so an important question is: Do modern counterparts exist in degree? For the dry climate indicators, sand dunes and evaporites, the opposite problem exists, for their distribution can be determined by remote sensing techniques (McKee 1979), while they have relatively low preservation potentials. In the following paragraphs the discussion will be confined to the distribution and depositional settings of these most useful climatic tracers and, in subsequent sections, the floral and climatic correlations with peat will be examined. The map (Fig. 1) was compiled from a great number of sources and only the more general ones are included in the following text. Care has been taken to exclude Pleistocene bogs as well as those deposits with admixtures of clastic sediments (cf. Gore 1983, p. 3).

Temperate zone peat is abundant in the northern hemisphere and in the few southern hemisphere land areas that extend into this zone (between the Antarctic Circle and the Tropic of Capricorn). Drainage disruptions related to Pleistocene glaciation are obviously responsible for many peat swamps in Canada (Tarnocai 1984), northeastern United States (Soper & Osborn 1922), Europe (Zurek 1984) and European Russia (Vinogradov 1967); and the same may be true for New Zealand and southern S American localities. Some have argued that peats would not occur in temperate zone settings without the prerequisite

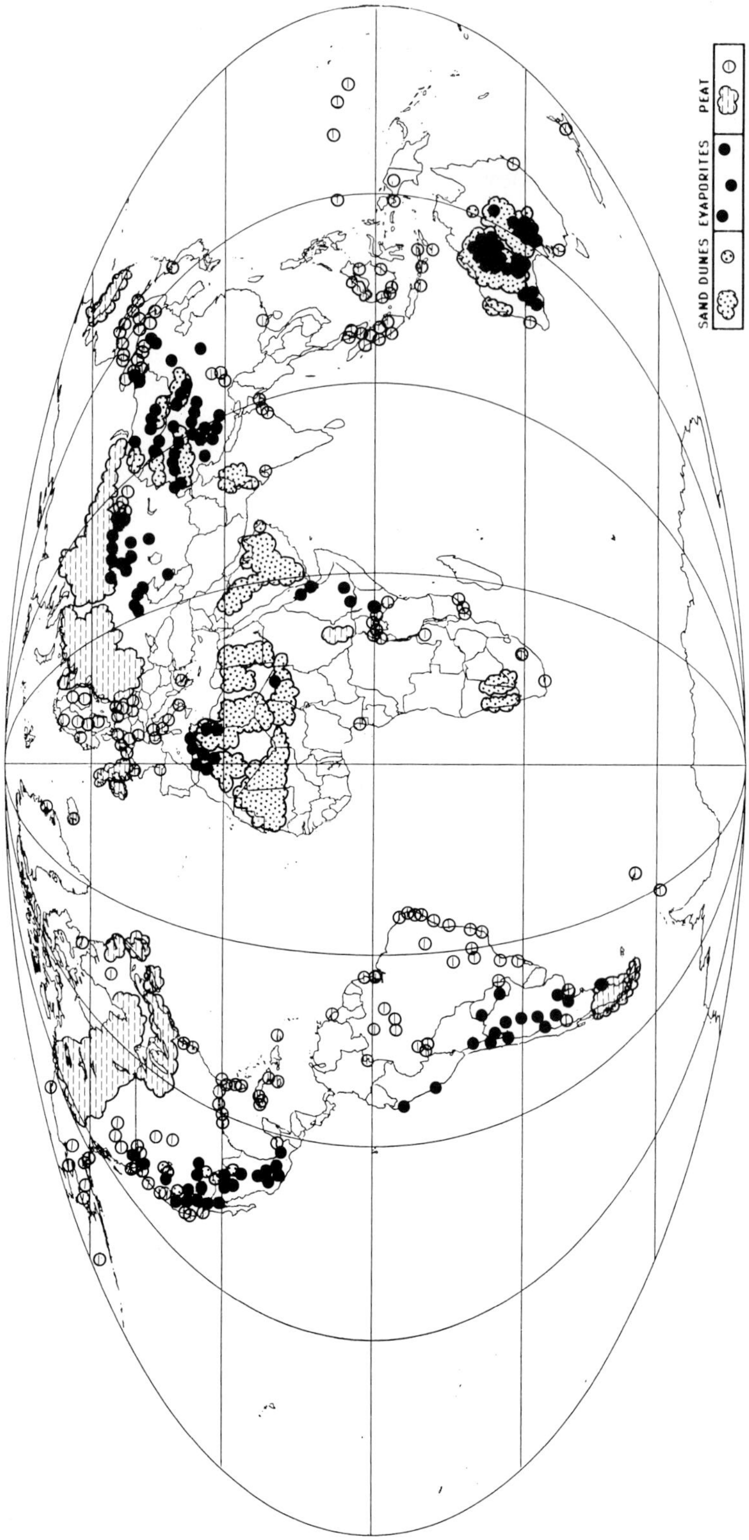

FIG. 1. Present-day climatically sensitive sediments. Some of the literature sources are quoted in the text.

of glacial scouring but, the W Siberian Lowlands, a subsiding basin since early Mesozoic times, contains many of the Soviet Union's peat swamps and it has never been more than peripherally glaciated (Vinogradov 1967). Perhaps the largest continuous area of peat accumulation (approximately 300 × 1000 km) lies along the SW coast of Hudson's Bay (Monenco Ontario Ltd 1981) in an area that owes its poor drainage to near level conditions resulting from it being a marine sedimentary platform that has isostatically rebounded after deglaciation. So, temperate peat swamps can indeed occur in non-glaciated settings, and this is true of most of the Mesozoic coal deposits as will be discussed later.

Tropical peat bogs are being discovered and described in increasing numbers in Brazil (Suszczynski 1984), central Africa (Thompson & Hamilton 1983) and SE Asia (Anderson 1983). As recently as 1979, Frakes wrote, 'The modern distribution of peat bogs suggests generally high latitude sites of accumulation, and this is thought to result from destruction of woody materials in low latitudes by oxidation and bacterial activity'. But Suszczynski (1984) speaks of the 'recent discovery of the tropical facies' of peat and claims that it represents a 'major natural resource' for Brazil. In equatorial Africa, 'Peat deposits are found in high-altitude bogs, medium-altitude swamps and in brackish coastal swamps' (Thompson & Hamilton 1983), and they show the 'main peatland' areas to be on the high ground associated with the E African Rift System. In Sumatra, peat swamps cover about one-fifth of the island and attain thicknesses of 16 m, and peat appears to be nearly as abundant in New Guinea, Borneo and Malaya (Anderson 1983). In these SE Asian localities, most of the peat is of coastal and lowland origin. Thus, tropical peats are now known to be abundant, and to form in a variety of depositional settings. The climatic correlations of tropical and temperate peats will be discussed in a later section of this paper.

The sand dunes and evaporite deposits of subtropical climes have been included on Fig. 1 because they occur in depositional settings that compare, in many ways, to peat, but they, of course, represent diametrically different climatic conditions. As in the case of peat, sand dunes of Pleistocene age, the 'fixed dunes' of authors, have been excluded from the map, although there seems to be some difference of opinion as to where to draw the line between Pleistocene and Recent dunes. Sarnthein (1978) seems to map the modern dunes more parcimoniously than Talbot (1984) in Africa or Bowler (1976) in Australia, and it seems that dune ridges of Pleistocene age may retain active crests, and therefore be difficult to classify. These authors agree on the important point that sand dunes were much more extensive during the Pleistocene glacial maxima; Sarnthein (1978) states that they covered 'almost 50% of the land between 30°N and 30°S versus 10% today'. This expansion occurred mainly on the equatorial side of the deserts of all low latitude continents, indicating an areal confinement of savannah environments and in some areas, the tropical rainforest in the Amazon (Tricart 1975) and Congo Basins (Butzer 1971).

The evaporite deposits were compiled mainly from the *Handbook of World Salt Resources* (Lefond 1969). They occur in a great variety of depositional settings, with continental interior basins seeming to dominate. At times in the geological past, the marginal marine evaporite basins were much more impressive; some covering up to a million square kilometres (see Ziegler 1982, Permian and Triassic maps), while the largest today of any sort is the Kara Bogaz Gulf on the Caspian Sea at about 15 000 km^2 (Valyashko 1972).

Present productivity patterns

A discussion of land-surface productivity is included in this paper for two reasons; first, forest formations correlate well with peat deposits, and so the occurrence of fossil floras may be taken to indicate the proximity of coal-forming environments and, second, land-surface productivity directly influences climate and hence the amount and distribution of coal deposits. A map of the 'natural vegetation regions of the world' has been presented by Strahler & Strahler (1983, plate M.2; cf. Schmithüsen 1976) in which the biomes are arranged in a spectrum from rainforest to desert and tundra. We have replotted this information (Fig. 2) with the productive end of the spectrum, the forest biomes, indicated by dashes, the desert and semidesert biomes indicated by dots, and the savannah, grassland and tundra biomes as an intermediate category. The sclerophyll forests of the Mediterranean (winter rain) climate are included with the intermediate category because, unlike the more productive forest areas, they do not generally correlate with areas of peat accumulation. Depositional sites within the forest biomes are indicated by symbols with shapes that correspond to the various forest types. Thus, the symbols are intended to approximate what a palaeobiogeographer working in the future might find preserved of the present forest biomes.

The land surface portion of Lieth's (1964) classic primary productivity map (Fig. 3) is

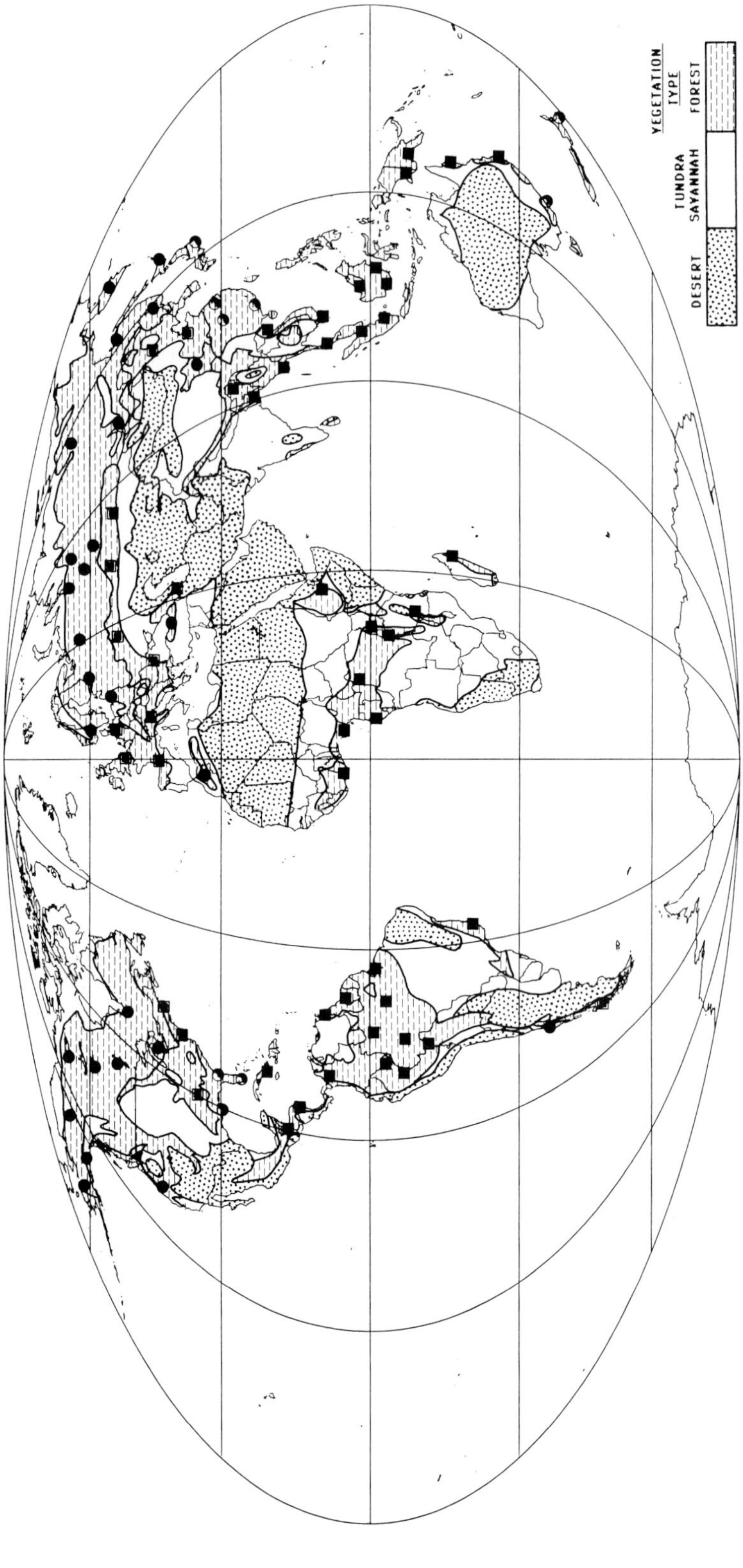

FIG. 2. Present-day vegetation types (simplified from Strahler & Strahler 1983, plate M2). Symbols are placed at depositional sites such as swamps and deltas and are as follows: squares = low-latitude rainforest, yin-yangs = subtropical evergreen forests, squares within boxes = mid-latitude deciduous forests, triangle = 'coastal forest' (one symbol only) and circles = cold needleleaf forests.

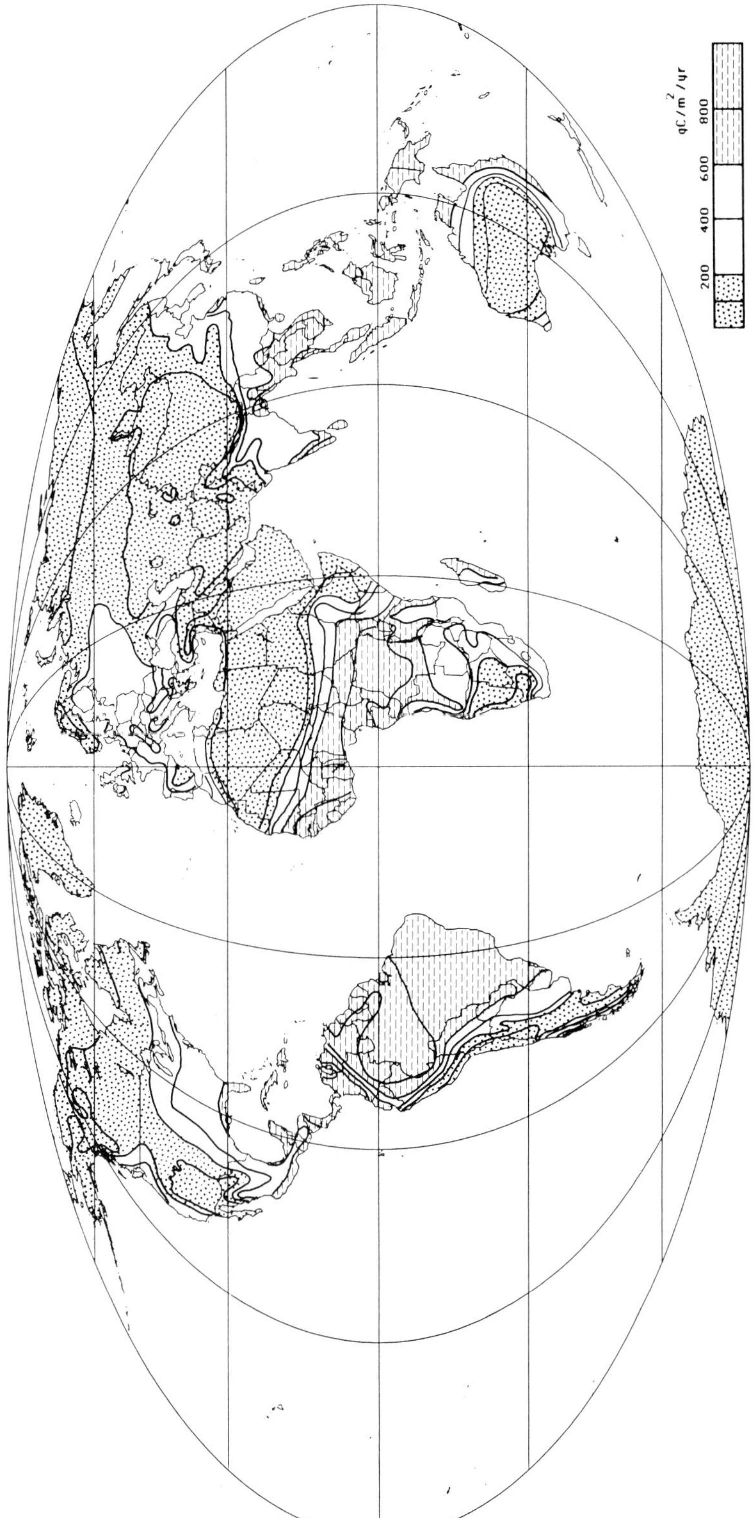

FIG. 3. Present-day land surface productivity (reproduced from Lieth 1964).

redrawn on our standard Mollweide (equal area) base map. Lieth later wrote (1975, p. 209), '. . . the primary productivity data available at the time were supplemented with agricultural and forest yield data using estimated corrections to community productivity to produce a world productivity map' (cf. Rodin *et al.* 1975). The map differs from the vegetative biomes map (Fig. 2) because temperate forests are limited in productivity to the warmer months. This means that the highly productive areas are confined almost entirely to the tropical zone, while the cooler conditions of the temperate zone are more favourable for peat preservation. The highly productive areas of the tropics include all of the rainforest and much of the savannah (summer-green) vegetation.

In conclusion, coal deposits and fossil floras of the past can be expected to correlate with each other and with the distribution of forest biomes, but there is a heavy bias toward lowland (depositional) sites. However, caution must be used in interpreting past land-surface productivity patterns, because many of the fossil floras will represent low productivity situations. Finally, conclusions on the influence of productivity on climate must be based on the tropics because the majority of the world's primary production occurs in this rather limited area (Whittaker & Likens 1975, p. 306).

Present climate

Obviously, rainfall is important in controlling the distribution of peat-forming swamps, but annual precipitation (Fig. 4) *per se* is a poor predictor for the distribution of peat at present. Temperature, as it affects evaporation, is certainly very important and Sellers (1965, p. 84) shows precipitation to exceed evaporation above 38°N and 38°S (within the temperate zones) and between 13°N and 7°S along the equator. So, although annual precipitation is highest at the equator, the temperate zones have greater 'total runoff'. The coals of the Palaeozoic (Ziegler *et al.* 1981a) and Mesozoic–Cenozoic (Parrish *et al.* 1982) in fact conform to these latitudinal ranges.

The monthly continuity of rainfall is probably the single most important factor in peat distribution, particularly in the tropics where temperatures are high. 'The frequency and seasonal distribution of the rainfall is far more important than the total volume of precipitation received in one year' (Thompson & Hamilton 1983, writing on Africa). 'The total precipitation, which may frequently exceed 3000 mm per year, is of less importance than its distribution throughout the year which greatly influences the potential evapotranspiration' (Anderson 1983, writing on 'Malesia'). Accordingly, a map showing precipitation continuity has been prepared (Fig. 5) using the climatic data tables of Müller (1982). Over 1000 stations are included and the number of months during which precipitation exceeded 20 mm was recorded and contoured. The selection of the 20 mm cut off point was arbitrary, but in any case this number generally lies along the steep portion of the yearly precipitation curve for most areas, so the selection of another number would not alter the overall map pattern. The resulting map (Fig. 5) correlates well with both forest biomes and peat swamps. Essentially, 12 months of precipitation are needed to support most forests and all swamps, with the exception of high latitude areas, such as northern Canada and Siberia, where the cold and dry seasons coincide and biological activity in the form of growth and decay, is suspended.

Also important is the season during which the most rainfall occurs. Most stations around the world, particularly in the continental interiors, experience summer maxima and this is a favourable situation for both plant growth and preservation. The opposite is true of the Mediterranean climates where the sclerophylls, the 'hard-leaved evergreen trees and shrubs capable of enduring a long, dry summer' (Strahler & Strahler 1983) have evolved. Winter rains are also common to many temperate zone maritime regions, such as the N Atlantic bordering continents. However, the presence of the thermal anomaly of the Gulf Stream results in storm activity throughout the year and this affects eastern N America and northern Europe. A parallel situation occurs in the N Pacific but the area affected is limited to the coasts by the generally high mountains of the region.

Precipitation continuity is favoured in a narrow belt along the equator because the Intertropical Convergence Zone is of sufficient width that, during the solstices, the equator still receives rainfall. The situation is clearly seen at the present time in Africa where this zone of overlap is less than 10° wide (Fig. 6). Another situation in the tropics where continuity exists is along mountains and plateau margins on east-facing coasts (Fig. 7) where warm moist air rains out due to orographic effects. Continuity can also be aided by the swamps and trees themselves through evaporation and transpiration, respectively. In the Amazon Basin, for instance, over 50% of the water vapour available for precipitation is derived internally by these processes (Salati & Marques 1984).

Generally speaking, seasonality in precipita-

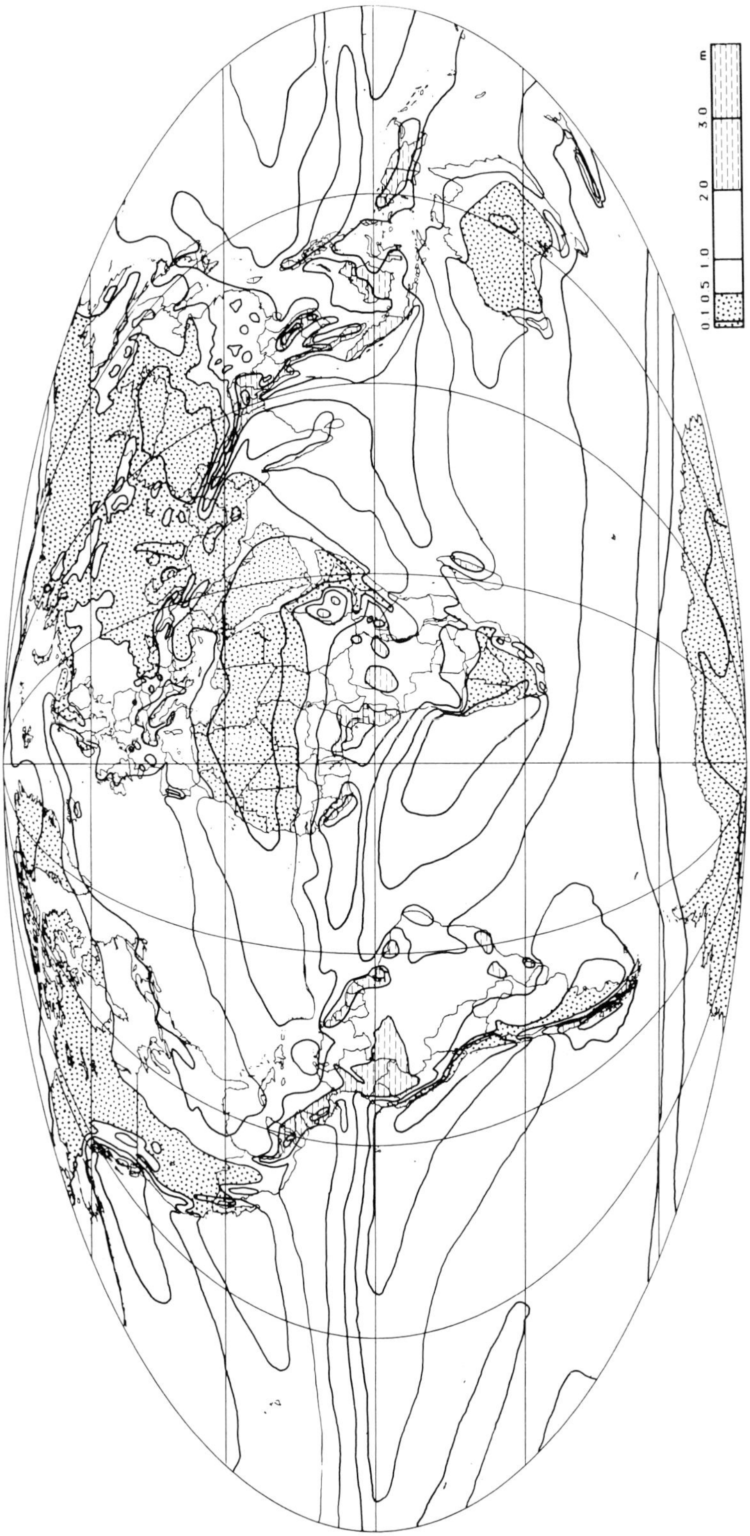

FIG. 4. Present-day annual precipitation (reprojected from an unpublished map of T T. Fujita, University of Chicago). This map incorporates satellite (photographic) data and is therefore more accurate than published maps over the oceans and in areas with few recording stations.

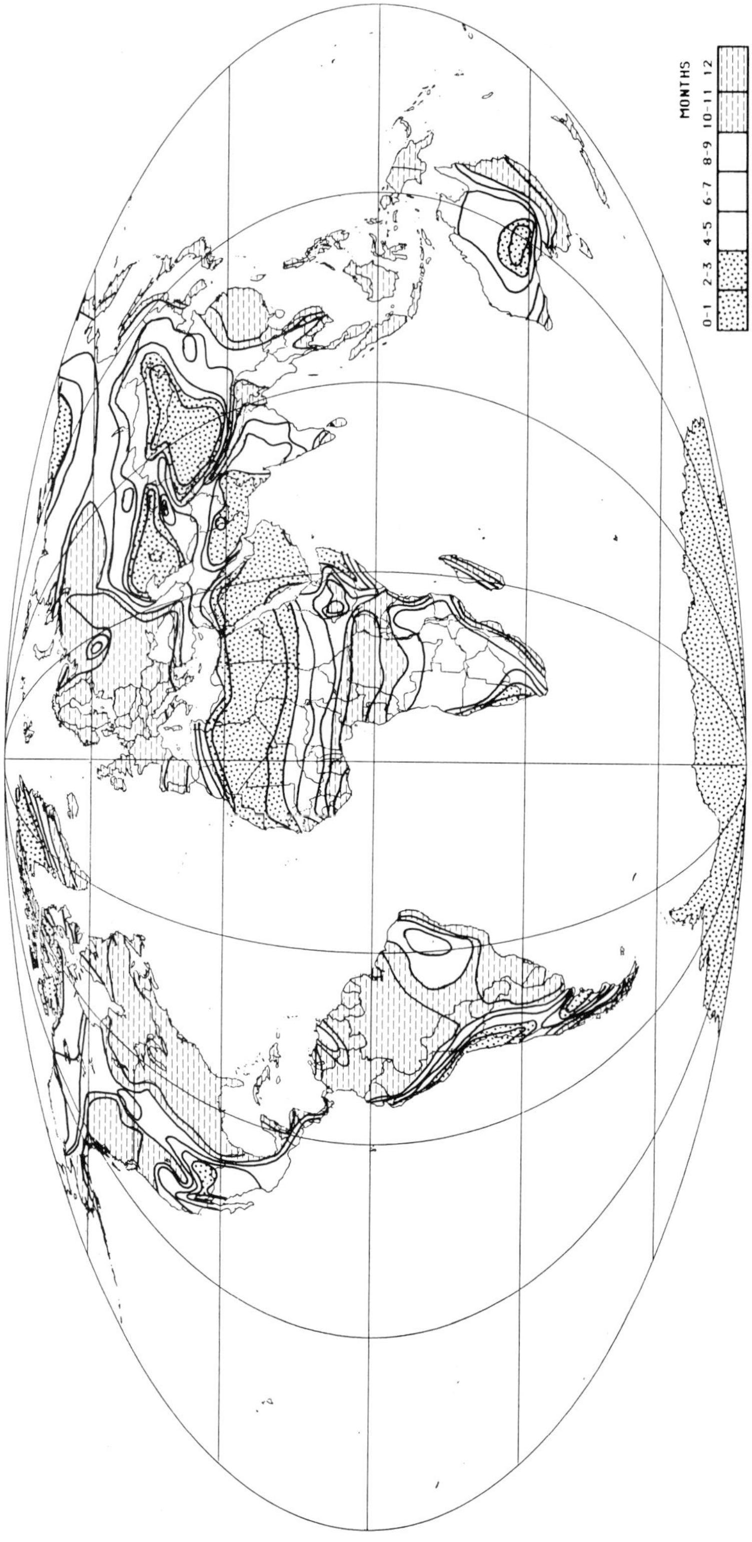

FIG. 5. Present-day 'precipitation continuity'. This map was contoured from data in Müller (1982) and shows the number of months during which precipitation exceeds 20 mm.

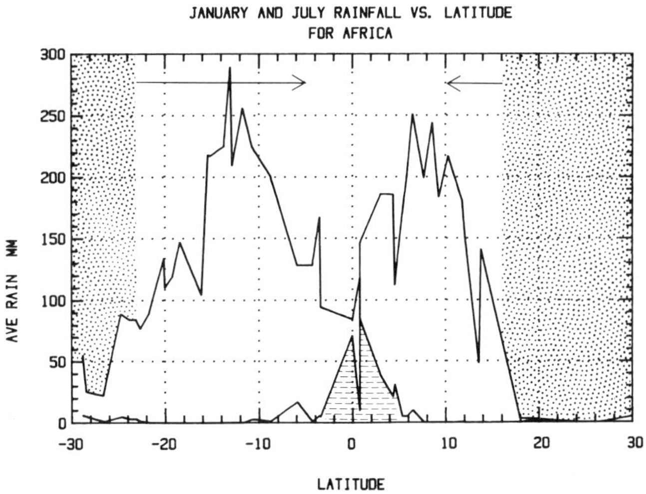

FIG. 6. Latitudinal precipitation profiles for the summer months in each hemisphere, showing the poleward excursions of the equatorial rainy belt. Data are from Müller (1982) and are confined to the 10°–30° E longitude zone in Africa. Three coastal stations were not used in this plot (115, 116 and 144) because they are adjacent to cold upwelling currents and are therefore exceptionally dry, and one upland station (104) because it receives an unusual amount of orographic rain. The zone of overlap is shaded with dashes and represents that latitudinal span which receives precipitation during all 12 months. The zone shaded with dots represents the latitudinal extent of sand dunes while the arrows show the equatorward extent of dunes during the Pleistocene (18 000 yr BP).

tion adversely affects plant growth and this seasonality can be translated into water-level fluctuations. In the Amazon Basin, the larger catchment area on the south results in an annual fluctuation of the main river by up to 20 m (Sioli 1984, p. 133), so that the vegetation of the flooded areas is subjected to oxidation every year. As we have seen, peat does occur in the Amazon Basin, but the water-level fluctuations probably prevent extensive deposits (Junk 1984, pp. 230–232). It may be observed that the Carboniferous coal swamps, being marginal to broad epeiric seas, were not subject to water table fluctuations and so were preserved in this tropical setting. Finally, rainy seasons in source and depositional settings can be offset in a complementary way and sustain a permanent swamp such as in the Okavango Basin on the margin of the Kalahari Desert (Thompson & Hamilton 1983, p. 332).

From the preceding discussion, it is clear that year-round continuity of precipitation is the most important factor in forest growth and swamp preservation. Such continuity is expectable in areas near the latitudinal centre of both the Intertropical Convergence Zone and the two tropical/polar air mass contacts. Elsewhere, orographic effects in low latitudes and areas of 'frontogenesis' (areas of storm generation) in high latitudes will be areas where the normal summer (convective) rains are supplemented during other seasons.

As the discussion in this paper shifts to earlier periods it is interesting to comment on the relative youth of the areas where precipitation is now continuous. The lowlands of eastern N America and northern Europe derive moisture from the N Atlantic, which did not exist prior to the Tertiary and the central Atlantic, which began opening in the Jurassic but was probably not effective as a moisture source before the Cretaceous. The S Atlantic is the moisture source for the equatorial rainforests of Africa and S America but it did not exist until the Cretaceous. Dry contours that extend westward from the Sahara and Kalahari Deserts still cross the Atlantic and impinge on S America, suggesting that even with the present width of the S Atlantic, S America is still in a sense in the rain shadow of the African deserts (Fig. 4). The uplift of the Andes is a late Tertiary effect and these mountains stimulate orographic precipitation in the headwater regions of the Amazon. Finally, northern Australia and New Guinea are late Tertiary arrivals in the tropical rainy zone, and were preceded in the late Mesozoic by SE Asia. In conclusion, there has been a nearly complete change in the areas of continents affected by precipitation since the late Mesozoic.

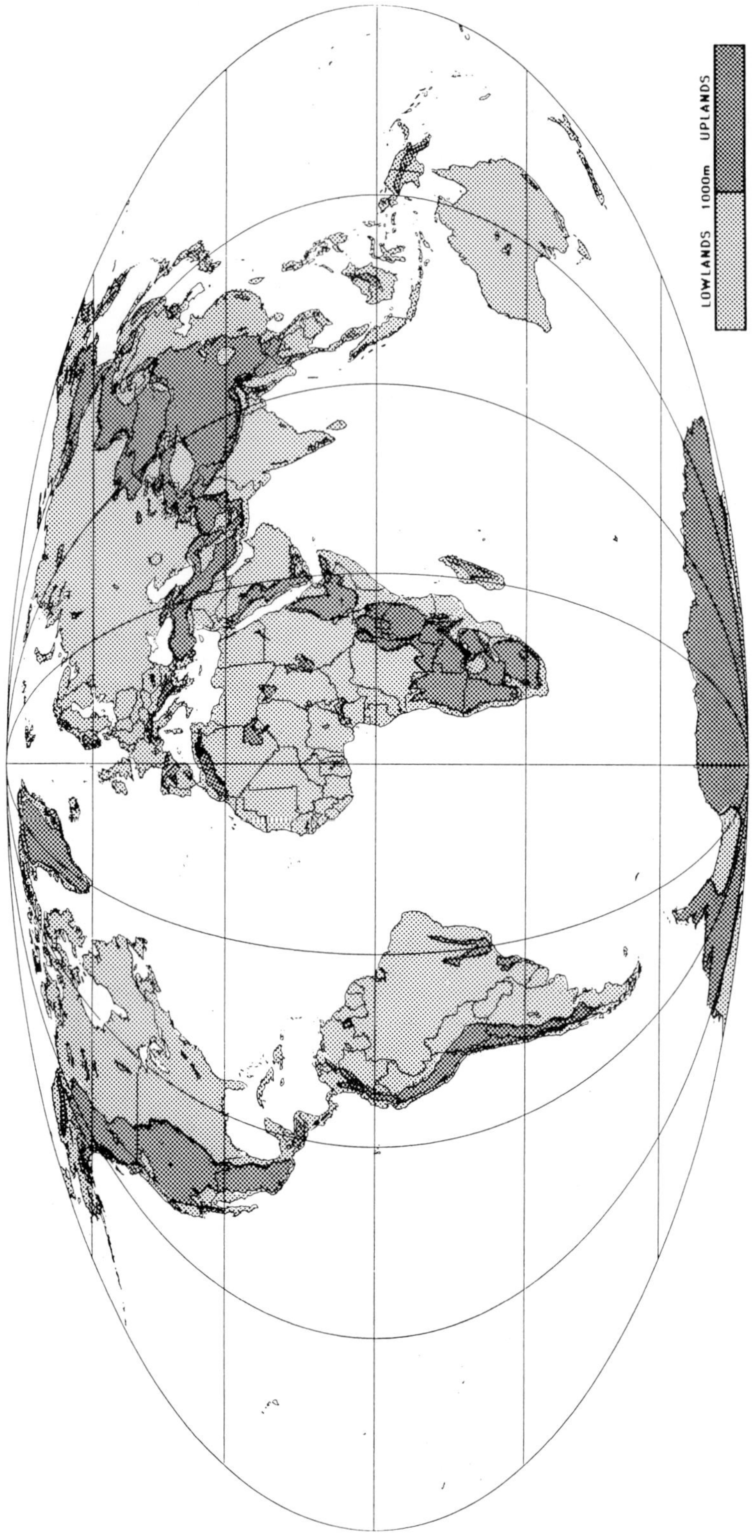

FIG. 7. Present-day orography.

Early Cretaceous palaeogeography

The early Cretaceous (Valanginian) palaeogeographic map (Fig. 8) shows a reconstruction of the continents intermediate between the latest Jurassic (Volgian) and middle Cretaceous (Cenomanian) reconstructions published by Ziegler *et al.* (1983). The reconstruction is based on our own analysis of the available linear magnetic anomaly, fracture zone and onland geological data using interactive computer graphic technology. The reconstruction of palaeolatitudes is based on combined palaeomagnetic (primary using McElhinney's compendium of published results) and constraints imposed by sediments (such as coals, evaporites, sand dunes, and reefs) that are palaeolatitudinally controlled.

The Valanginian reconstruction shows an anomaly M10 fit appropriate for ~130 Ma based on the Kent & Gradstein (1985) time-scale. We have used identifications of anomaly M10 in the central Atlantic (Klitgord & Schouten 1984), off Dronning Maud Land (Bergh 1977), in the Mozambique (Ségoufin 1978, Simpson *et al.* 1979), and Somalian (Ségoufin & Patriat 1980) basins, and the west coast of Australia (Larson *et al.* 1979). Fracture zone data from these sources and Haxby (1985) were used. All available and relevant data are satisfied by this reconstruction. The salient aspects of this reconstruction are discussed in the following paragraphs.

The N Atlantic reconstruction is that of Rowley *et al.* (1985), which modifies all existing fits by accommodating non-seafloor spreading extension along the various margins, and uses *Seasat*-derived fracture zone traces to better constrain the pre-rift position of N America and Europe. In particular, the reconstruction accommodates approximately 150+ km of post-Valanginian extension within the Rockall–Mohrs–Voring basin. Iberia is shown approximately 150 km closer to the Grand Banks margin of N America than on most reconstructions (e.g. Keen & Hyndman 1979), based on restoration of extension observed along the Galicia Bank and Grand Banks margins. Overall this results in a much tighter fit.

In the central Atlantic, anomaly M10 is well defined, and our fit is very similar to that of Klitgord & Schouten (1984). The only other regions where M10 has been proposed is in the Cape basin and adjacent to the Falkland Plateau in the S Atlantic (e.g. Rabinowitz & La Breque 1979), and in the Agulhas Basin south of S Africa (Martin *et al.* 1982). M10 in the Cape Basin sequence has been shown subsequently to overlie extended continental crust (Austin & Uchupi 1982) implying that seafloor spreading in this region post-dates M10, so our reconstruction shows no ocean within this region. The reconstruction of S America and Africa is based on Rowley *et al.* (1986). Africa is shown as three separate elements that have moved relative to one another since the Valanginian. Pindell & Dewey (1982) and Rowley *et al.* (1986) point out that it is geometrically impossible simultaneously to fit the equatorial and S Atlantic margins of Africa and S America if only these two plates are used. Fits utilizing only these two plates result in a triangular 200+ km wide gap between either the Guinea Plateau–Demerara (Guiana) Rise region where both geometric (Pindell & Dewey 1982) and reflection seismic data (Mascle *et al.* 1986) demonstrate that such a gap did not exist, or in the Cape–Argentine basin area where a wide array of geological and geophysical data negate such a gap (e.g. Dingle *et al.* 1983). This geometric dilemma is resolved in Fig. 8 by recognizing that Africa did not behave rigidly during the Early Cretaceous (see, e.g. Burke & Dewey 1974, Pindell & Dewey 1982, Fairhead 1986). Africa comprises at least three separate blocks bounded by major early Cretaceous strike-slip faults and associated fault-bounded basins, including the E–W striking central African shear zone (Browne *et al.* 1985) and N–S striking Elbiod fault (Bertrand & Caby 1978) systems (Rowley *et al.* 1986). Restoration of the integrated motions on these major structures allows a perfect simultaneous fit of the equatorial and S Atlantic. Figure 8 displays a three 'plate' Africa solution in which all of the motion is accommodated along the central African and Elbiod fault systems; the reconstruction therefore shows the maximum offsets, although somewhat smaller offsets on these structures might also satisfy the geometrical constraints.

The relative positions of E Gondwana (Antarctica, Australia, India and Madagascar) to Africa are constrained by fracture zone and anomaly fits in the Mozambique and Somali basins. Our reconstruction of Antarctica to Africa and Madagascar requires approximately 300+ km of dextral strike-slip between India and Madagascar, similar to that of Smith & Hallam (1970). India, Australia and Antarctica are shown in essentially their pre-drift configuration. The time of initiation of seafloor spreading between India and Antarctica is poorly constrained, as no M series magnetic anomalies have yet been identified between their conjugate margins, but M10 and younger anomalies are recognized along the west coast of Australia. It is therefore possible that seafloor spreading had just begun between India and Antarctica as suggested by Veevers *et al.* (1980).

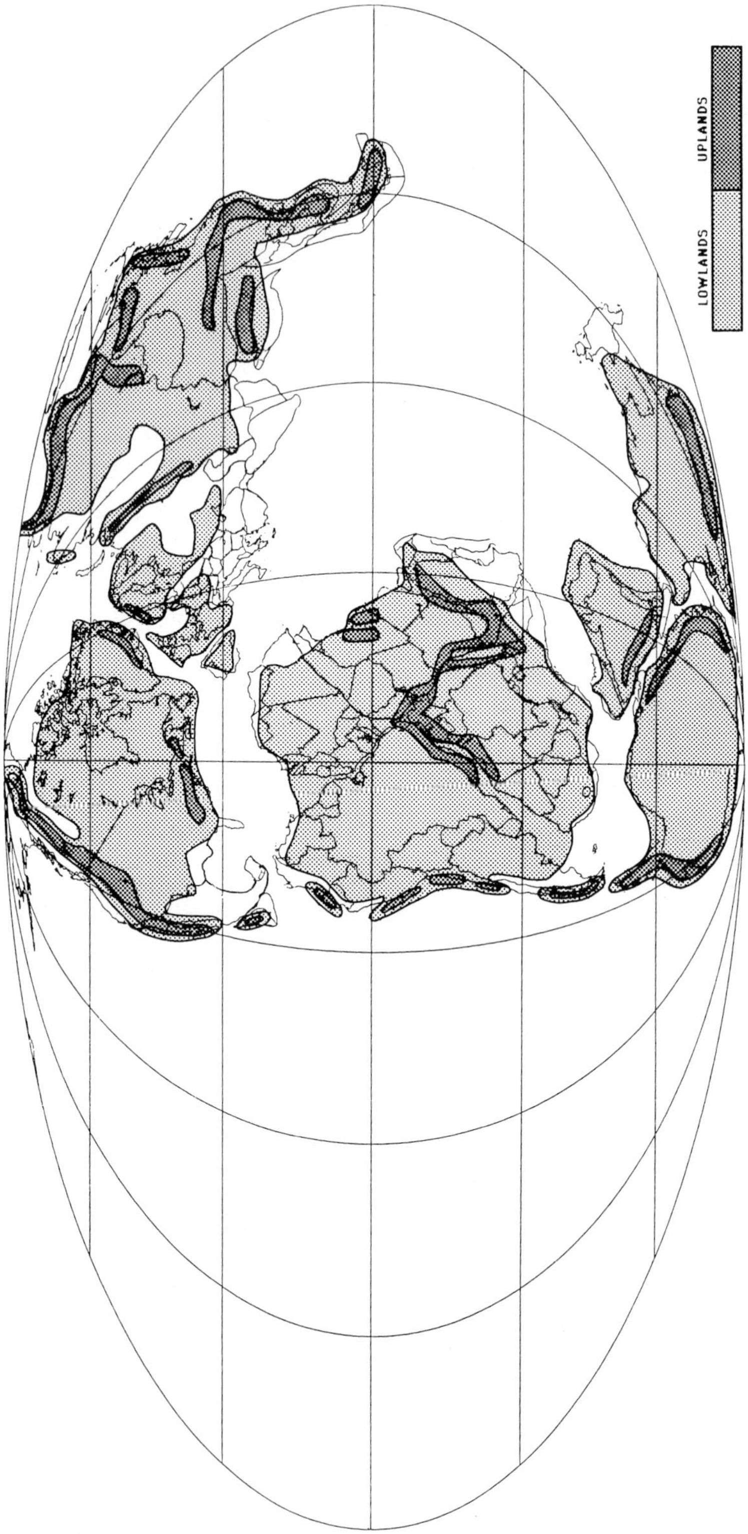

FIG. 8. Early Cretaceous (Valanginian) palaeogeography and orography.

Although Asia comprises a large number of micro-continental fragments (see, e.g. Sengör 1984) almost all can be demonstrated to have collided with Eurasia by or during the early Cretaceous. Asia has subsequently been substantially moulded by later deformation, particularly during the Tertiary. This subsequent deformation is not palinspastically restored in Fig. 8, and therefore China and Indochina are shown in their present configuration relative to Asia.

The reconstruction of palaeolatitudes was accomplished using a combination of two independent data sets: palaeolatitudinally sensitive sediments and palaeomagnetic pola data. Coal, evaporites and reefs are, to a first-order approximation, controlled by latitude (Ziegler *et al.* 1984, Parrish *et al.* 1982). The distribution of these sediments has been compiled by the Palaeogeographic Atlas Project into a computerized database, that was plotted on our reconstructed basemap (a portion of these data is plotted on Fig. 9) and used to determine the best fit palaeolatitudinal orientation of the continents to these data. Palaeomagnetic pole data from McElhinney's compendia of published results from the latest Jurassic to early Cretaceous for cratonal portions of N America, Europe and Africa were compared with the orientation derived from the palaeolatitudinally sensitive sediments and found to be completely compatible, yielding essentially identical results, and giving confidence in the overall reconstruction shown in Fig. 8.

Early Cretaceous climate indicators

The climate indicators plotted on the early Cretaceous map (Fig. 9) are restricted to the Valanginian Stage of that interval and were compiled from sources given in the Appendix of this paper. Coal, evaporite and aeolian sandstone deposits are included as before (Fig. 1) and, in addition, sandstone-dominated terrestrial deposits of Africa are shown. These sandstones include the extensive Nubian Sandstone of NE Africa and are significant because they generally lack coal, evaporite or aeolian sandstone and, as will be argued later, probably indicate a highly seasonal rainfall pattern.

The temperate zones in the early Cretaceous are well represented by coal, as today, and these deposits include both coastal and inland sites. Comparisons with the Present can also be made with the subtropical distribution of the evaporite and aeolian sandstone deposits. However, few coals attributable to the equatorial rainy zone are known from this, or any other Mesozoic interval (Parrish *et al.* 1982). Those Valanginian coals that are known from low latitudes are interbedded with marine deposits (Peru, Algeria, Tibet) or are geographically near the coast (Malaya) and so all probably represent coastal swamps. Some have assumed that most of N and central Africa was covered by 'lowland rainforest' in the Cretaceous (Axelrod & Raven 1978). Coal deposits are generally lacking, but the early Cretaceous is abundantly represented by the Nubian Sandstone, the Continental Intercalaire and their equivalents and, since these rocks are good petroleum reservoirs, they have received much attention recently.

The modern study of the Nubian Sandstone began with McKee (1962) who went to Africa to determine if its reported aeolian origin was correct. He concluded that aeolian deposits are not present, but that '. . . fluviatile deposits probably are represented in many parts of the Nubian Sandstone, in association with the widespread lagoonal or lacustrine deposits'. Silicified wood and leaf impressions are locally abundant and the dominant lithology is clean sand. 'The ubiquitous tabular-planar cross strata probably were formed as foreset beds deposited in shallow standing bodies of water', and directions of sand transport were found to be 'reasonably consistent' and generally northerly in NE Africa. Recent studies have confirmed these observations in the Messak Sandstone of SW Libya (Lorenz 1980), the Cabao Sandstone of NW Libya (El-Zouki 1980), the Desert Rose Unit of SW Egypt (Klitzsch *et al.* 1979), the Selima Formation of NW Sudan (Wycisk 1984), the Kornub Sandstone of N Jordan (Abed 1982) and the Biyadh Sandstone of central Saudi Arabia (Moshrif & Kelling 1984). All of these units represent the Lower Cretaceous part of the succession, but this magnafacies is also known from earlier and later intervals. Everywhere, large to medium scale planar crossbedding is the dominant sedimentary structure and the 'low variance unidirectional current flow' is generally taken to indicate 'braided or low sinuosity streams'. Flood activity is assumed by Lorenz (1980), Klitzsch *et al.* (1979) and Moshrif & Kelling (1984) and the latter authors point out that flooding is '. . . indicated by the repetition of channel-base scoured surfaces and their lag deposits'. Klitzsch *et al.* (1979) envisage well vegetated alluvial plains 'based on the prevalence of soil horizons and the common preservation of flora', but Lorenz (1980) states 'The common fossil tree *Dadoxylon* was probably restricted to areas near channels, swamps and lakes, and did not forest the land'. Floral diversity, at least as preserved, is not high (Klitzsch & Lejal-Nicol 1984) and the ubiquitous braided

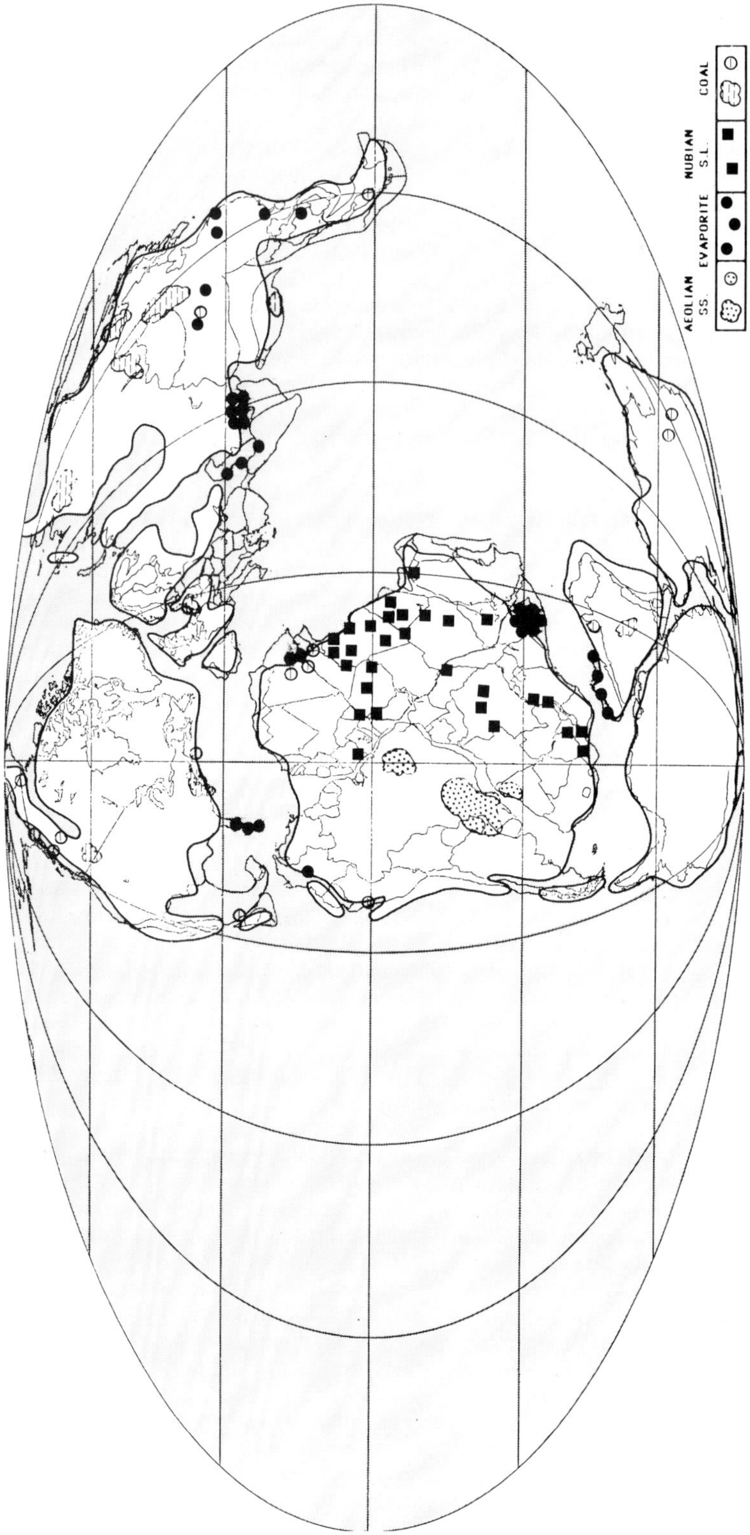

FIG. 9. Early Cretaceous (Valanginian) climatic indicators. References are given in the Appendix.

stream deposits would seem to rule out the possibility of forest vegetation and especially the lowland rainforest envisaged by Axelrod & Raven (1978).

We propose a savannah environment with its typical seasonal rainfall pattern for the Nubian and its equivalents in NE Africa and Arabia. We infer similar climates elsewhere in Africa, where sand-dominated terrestrial sequences of early Cretaceous age are widespread, although, unlike the Nubian, significant amounts of shale are present. Unfortunately, contemporary descriptions are lacking, but it can be said that dinosaur, crocodile and turtle remains are quite common in W Africa (de Lapparent 1960) and E Africa (Cooper & Dixey 1956). Support for a 'hot climate with strong contrasts in seasonal humidity' comes from the Atlantic basin where smectite, thought to have originally formed in badly drained coastal areas under these conditions, is abundant in Cretaceous black shales (Chamley & Robert 1982). Hallam (1984) also concluded for the Mesozoic that '. . . the main seasonal change through the year in many regions involved precipitation more than temperature'.

Early Cretaceous phytogeography

'Ever since the pioneering days in the study of fossil plants, palaeobotanists have held the opinion that the floras of the Jurassic period and during the part of the early Cretaceous were more globally uniform than at almost any other time in the history of land plants' (Batten 1984). Vachrameev (1978) stated, 'Two major paleofloristic regions may be distinguished on the basis of paleobotanical data: a more northerly Siberian–Canadian region and a more southerly Indo–European region'. The northerly flora has been termed 'humid temperate-warm' and was dominated by gingkos and members of the gingko-like order Czekanowskiales, all of which had wood of pycnoxylic type (able to withstand subzero temperatures). This flora had one half the diversity of the 'humid subtropical' Indo–European flora which had ferns, including *Weichselia*, the tree-fern *Tempskya* and Cycadeoids, the latter two of which possessed trunks of manoxylic-type (incapable of withstanding subzero temperatures). The palaeolatitudinal positions of the division between these two provinces was about 50°N (Fig. 10), that is, about 20° poleward of the present frost-line. We accept Vachrameev's location of the early Cretaceous frost-line, but feel that the word 'subtropical' should not be applied to a flora that extended as far north of the Tropic of Cancer, and prefer the terms warm and cool temperate for these two great floral provinces.

The 'arid subtropical' zone was delimited not by compression floras, but by the abundance of *Classopolis* pollen and dry climate lithofacies indicators (Vachrameev 1978). However, the boundary of the humid and arid regions was characterized by 'the xeromorphic fern *Weichselia*'. Doubt has been cast on this dry climate adaptation of *Weichselia* by Daber (1968) who proposed a tropophytic adaptation (to an alteration of dryness and moisture). However that may be, *Weichselia* is extremely common in subtropical and tropical latitudes, including areas of coal formation such as Tibet (Chen 1984, p. 184), and we basically follow Vachrameev's system of delineating the arid zone on the basis of high percentages of *Classopolis* pollen in concert with evaporites and aeolian sand deposits. This means that some of the diverse *Weichselia*-dominated floras would be classified as warm temperate, while some would be humid equatorial. *Frenelopsis*, the generator of *Classopolis* pollen (Vachrameev 1978, p. 144), 'was well adapted to conditions of high osmotic stress with its extremely thick cuticle, deeply sunken guard cells, highly reduced leaves and long persistent epidermis' (Daghlian & Person 1977), so this cypress like conifer was clearly xeromorphic (see also Francis 1983).

The term 'Northern Gondwana flora' has been applied to the equatorial zone (between 15°N and S) and this is also best characterized by its palynofloras (Doyle *et al.* 1982). These authors point out that '. . . the nature of tropical plant communities before and during the rise of angiosperms . . . is hard to visualize because of the paucity of low-latitude megafossil floras and the almost total dominance of analogous modern tropical environments by angiosperms'. Like Axelrod & Raven (1978), they argue on strictly uniformitarian grounds that there must have been a permanently wet zone near the equator, but admit that these floras are of low taxonomic diversity, have few ferns, and are dominated by *Classopolis* and ephedroid pollen (living *Ephedra* is a widely distributed xerophitic gymnosperm). Doyle *et al.* (1982) are correct to point out the pitfalls in modern analogies, but we feel that the plants, like the sediments, indicate arid conditions with seasonal rainfall, and that tropical rainforests, and indeed forests of any description, were lacking from the equatorial zone in the Mesozoic.

In the southern hemisphere temperate zone, 'warm humid' types have been identified from S America, India and Australia (Vachrameev 1972). A 'warm to cool-temperate' biota occurs

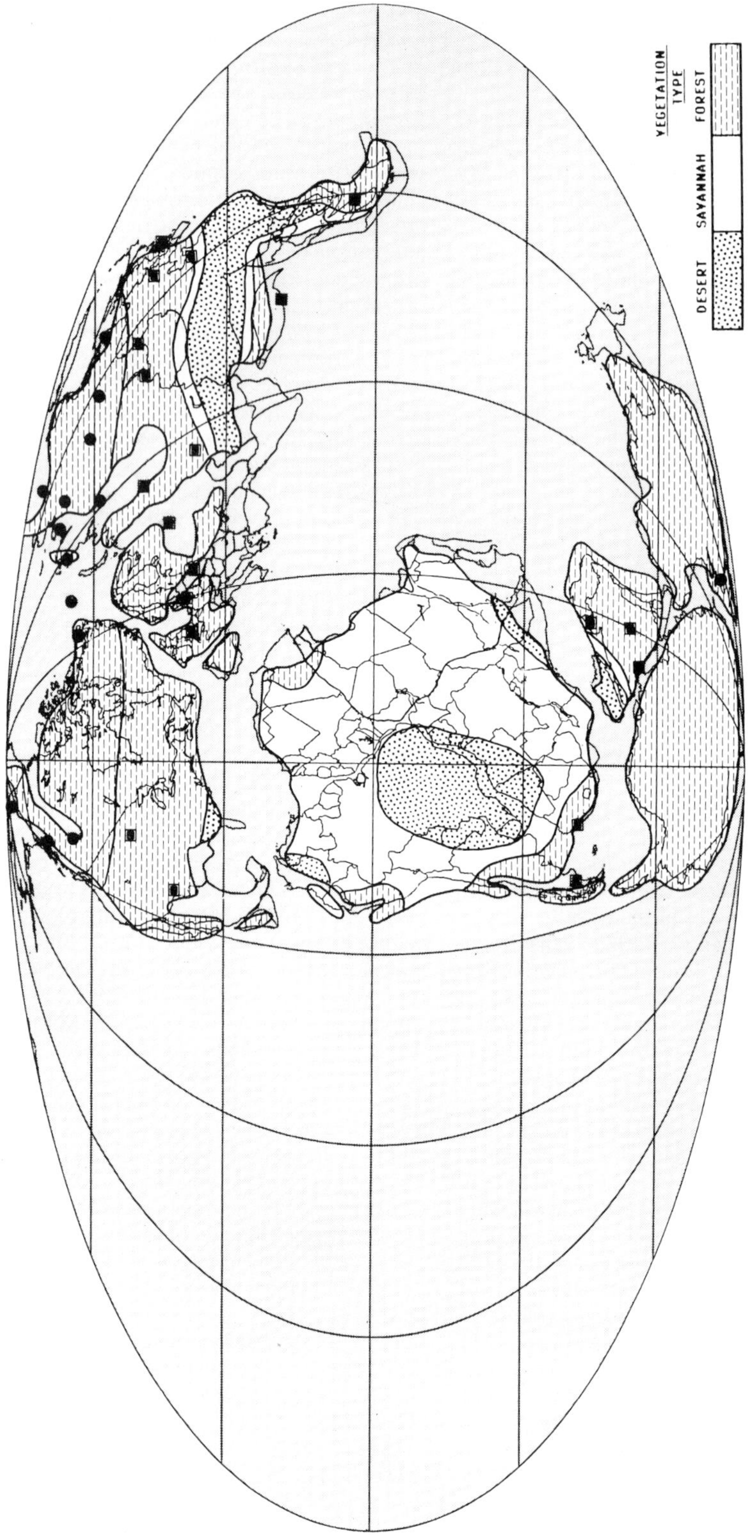

FIG. 10. Early Cretaceous vegetation type (see text). Symbols are as follows: squares = low latitude forests, squares within boxes = warm temperate forests and circles = cool temperate forests.

about 80°S in Victoria, Australia (Douglas & Williams 1982), and this may be the counterpart to the Siberian–Canadian floras. Dinosaurs and lungfish occur here and the latter would seemingly rule out the possibility of frost, while the former 'no doubt wandered widely in search of food' (Axelrod 1984).

We conclude that the early Cretaceous floras were not strongly differentiated but that high latitudes are represented by low diversity deciduous trees with prominent growth lines (Creber & Chaloner 1985). Diversity was highest in the warm temperate zone and perhaps in certain coastal swamps along the equator, and lowest in the subtropical dry belts.

Early Cretaceous and Present climates compared

Obviously the Cretaceous was warmer than the Present, but a quick examination of Pleistocene precipitation effects is helpful in understanding climates during extreme conditions. During the last glacial maximum at 18 000 years ago, sand dunes and associated deserts extended closer to the equator than today (Sarnthein 1978). This has been widely interpreted to indicate ice age aridity, although recent computer model results indicate a decrease in soil moisture only in a zone south of the northern hemisphere ice sheets (Manabe & Broccoli 1985). Another interpretation of the sand dune data is that the Intertropical Convergence Zone was more confined during the Pleistocene, so that the deserts would expand but so would the zone of seasonal overlap of the equatorial rainy belt (see Fig. 6). Thus, these contrasting zones would expand at the expense of the savannah or seasonal rainfall belt, and the result would be increased biological productivity in low latitudes. Flohn (1984) has concluded that polar temperatures influence the position of the Hadley Cells from the theory that, today, the colder south pole displaces the Earth's thermal equator to a yearly average of 6°N of the equator. 'In the case of the Pleistocene glaciations, the extension of permanent drift ice toward the equator, . . . causes an intensification of extratropical westerlies, a shift of the subtropical belts towards the equator and thus an intensification of both Hadley Cells' (Flohn 1984, p. 16).

Were rainforests more extensive in the Pleistocene? The carbon dioxide content of the atmosphere was lower, although Flohn relates this to increased upwelling due to the intensified circulation. Some evolutionary theories require '. . . that the lowland rainforest was indeed divided into separate blocks by the extension of savannahs in the Pleistocene', but direct evidence of this is slim (Flenley 1979, pp. 71–72). Finally, Pleistocene peat has been dredged from a number of offshore areas of SE Asia (Biswas 1973, Gerard & Oesterle 1973, Woollands & Haw 1976) where lowered sea-levels at that time would have doubled the potential area for tropical rain forests. We feel that it is quite possible that tropical rainforests were indeed more widespread during the glacial maxima and that this would have been true especially of Africa and SE Asia.

If the polar fronts constrained the Hadley Cells during glacial extremes, then what about periods that were warmer than today? During the 'climatic optimum', 9000–6000 years BP, there was widespread wetness in the Sahara and other deserts (Street-Perrott & Harrison 1984, Sarnthein 1978). Schneider (1984) has cautioned '. . . that the extra warmth of the Altithermal was primarily a summer phenomenon, not over the whole year' and Kutzbach & Street-Perrot (1985) relate this to an amplified seasonal cycle of solar radiation in the northern hemisphere due to the position of the Earth in the Milankovitch orbital cycle. Precipitation patterns for periods before, during and after this interval have been determined using the Community Climate Model (Kutzbach & Guetter 1986) and indicate that '. . . summer monsoon precipitation of North Africa–South Asia was increased by 10–20% between 12 000 and 6000 yr BP'. Importantly both the lake-level and computer studies indicate that the dryest deserts can be subject to at least seasonal rain on relatively short time-scales and during interglacial conditions. The implications for erosion and deposition, and for floral and faunal migrations are obvious. Whether the equatorial belt 'lost' some of the precipitation being spread more widely at this time is unclear to us. It seems possible that the whole of the tropical zone experienced seasonal rainfall patterns, and that it was at this time that the isolation and evolution of the Amazonian species could have occurred.

Flohn (1978) has proposed for the Cretaceous '. . . a reasonable climatic interpretation would be a (seasonally shifting) tropical Hadley circulation with convective summer rainfall controlling about 70–80% of the global surface'. He has 'conservatively' estimated the position of the subtropical anticyclones to average near 50° latitude which, in view of Mesozoic evaporite data (Parrish *et al.* 1982), is too extreme. In any case, the Intertropical Convergence Zone today migrates about 20° annually as compared to the 47° potential as represented by the declination of the sun (Fig. 11) so, without cold polar air masses,

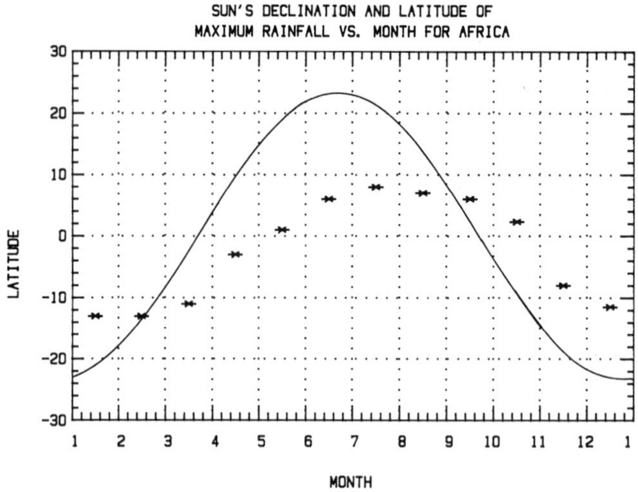

FIG. 11. Modal position of the Intertropical Convergence Zone each month. Data from Müller (1982) for the same stations listed in Fig. 6. The mode was chosen by subjecting the latitudinal profiles to a 15 point weighted moving average.

it would seem reasonable for it to have fluctuated more widely in the Cretaceous. Computer simulations of Cretaceous rainfall patterns have been made and indicate a 30° annual shift in the position of the Intertropical Convergence Zone, with no overlap of the seasonal extremes (Barron & Washington 1982, p. 114). We construe this, and the widespread nature of the Nubian Sandstone, as evidence in favour of a highly seasonal rainfall pattern in Africa.

Implications for climate control

The geological norm is represented by warm climates and low climatic gradients and this is generally assumed to be due to higher atmospheric carbon dioxide contents (Chen & Drake 1986). The control on carbon dioxide is further assumed by many to be due to high rates of volcanism (Budyko 1981), volcanism and raised sea-levels (Fischer 1981) or the balance of metamorphism and weathering as it ultimately relates to rates of seafloor spreading (the 'BLAG' model as developed by Berner *et al.* 1983 and Lasaga *et al.* 1985). There is little question that these effects are all correlated with temperature change, but the problem lies in distinguishing between cause and effect. For instance, the BLAG authors justified their geochemical model by relating their predicted CO_2 curve to temperature curves based on planktonic carbonate and palaeobotanical data. However, their CO_2 curve is based on the seafloor spreading curve which in turn affects sea-level, and a world with high sea-levels is bound to be warmer and more equable anyway. This is simply because of the heat capacity of water, and the role of the ocean in absorbing and reradiating heat, and in damping the fluctuations in air temperature. In fact, during the Cretaceous, epeiric seas brought ameliorating effects into the heart of N America and Eurasia (Fig. 8; and to a more extreme degree in the later Cretaceous). Despite these problems, the climatic modellers claim that elevated CO_2 levels must have existed in the Cretaceous to result in warm enough polar temperatures (Barron & Washington 1985, Schneider *et al.* 1985), and the geochemists assure us '. . . that CO_2 in the atmosphere could easily have changed in the past as a result of geological, as opposed to biological processes' (Berner *et al.* 1983).

The purpose of this section is to introduce still another correlate of seafloor spreading, sea-level and temperature, and that is the incidence of tropical rain forests through time. These developed during the late Palaeozoic, became extinct in the Mesozoic and redeveloped in the middle late Tertiary, and each of these intervals is characterized by cold poles and steep climatic gradients as shown by the glacial tillites (Ziegler *et al.* 1979) and the increased number of latitude related floral and faunal provinces (Ziegler *et al.* 1981b). Also, floras show bursts of evolutionary activity during these more climatically differentiated intervals, to judge by the diversity curve of Niklas (1986).

As was argued earlier in this paper, the tropical

floras account for over half the present land surface productivity, so the implication is that there have been profound changes in terrestrial productivity through time. It seems likely that there is a feedback loop involving temperatures, surface productivity and CO_2 levels, such that as polar temperatures drop, seasonal excursions of the Intertropical Convergence Zone are damped, resulting in a concomitant restriction of the Intertropical Convergence Zone-controlled rainfall. This restriction yields rainforests, which in turn reduce CO_2 levels and hence affect temperatures. Again, cause and effect are uncertain. In the case of the Pleistocene, most would agree that the temperature changes were driven by orbital variations and so any biologically induced CO_2 change must have followed the initial temperature drop. In the longer term, it seems conceivable that with the panoply of palaeogeographic change, the distribution of land areas and moisture sources would be occasionally optimized for equatorial productivity, and this might result in CO_2 reduction and cooler poles. In any case, modellers should not assume that 'the long-term organic cycle is perfectly inbalance' (e.g. the BLAG model).

Conclusions

Rainfall consistency through the year is essential for forest growth and peat formation, particularly in the tropics where evaporation rates are high. At the present time, the Intertropical Convergence Zone migrates north and south, being driven by, but lagging somewhat behind, the solar declination. The width of the Intertropical Convergence Zone is such that there is a narrow zone, along the equator, of consistent rainfall through the year, and this belt, approximately 10° wide, is where the tropical swamps occur. This zone is represented by abundant peat and coal deposits in the Carboniferous–Permian and in the Oligocene to Present, but not during the intervening Mesozoic and early Cenozoic periods. Nor do diverse floras characterize the tropics during this long intervening span, and the sediments that do occur are suggestive of highly seasonal precipitation patterns. The conclusion is that the Intertropical Convergence Zone during the early Cretaceous exhibited a wider latitudinal excursion during the year, and that there was no area in the tropics with year-round rainfall. This was possibly due to the warmer poles and a reduction in the effect of the polar front that at present seems to confine the Hadley Cells and thereby to channel the tropical rainy belt.

It is paradoxical that the warm 'greenhouse' climate of Fischer (1981) is correlated with times of lower floral diversity, provinciality, productivity and peat preservation. It is also noteworthy that the cooler 'icehouse' periods exhibit simultaneously the driest and wettest climate indicators meaning, again, that precipitation is better channelled during glacial times. Finally, the generalization may be made that seasonality as expressed in temperature alternates in time with seasonality as expressed in precipitation, and we suggest that the concept of seasonality be addressed systematically in future palaeoclimatological studies.

ACKNOWLEDGEMENTS: This paper has greatly benefited from the comments of W. S. McKerrow, D. Jablonski, V. Ramanathan and two anonymous reviewers. The Palaeogeographic Atlas Project at the University of Chicago is supported by Amoco, Arco, British Petroleum, Chevron, Cities Service, Conoco, Elf Aquitaine, Exxon, Marathon, Mobil, Pennzoil, Phillips, Shell, Sohio, Sun and Texaco.

References

ABED, A. M. 1982. Depositional environments of the Early Cretaceous Kurnub (Hatira) Sandstones, north Jordan. *Sedimentary Geology*, **31**, 267–279.

ANDERSON, J. A. R. 1983. The tropical peat swamps of western Malesia. *In:* GORE, A. J. P. (ed.) *Ecosystems of the World. Vol. 4B, Mires: Swamp, Bog, Fen and Moor, Regional Studies*, pp. 181–199. Elsevier, Amsterdam.

AUSTIN, J. A. & UCHUPI, E. 1982. Continental-oceanic crustal transition off southwest Africa. *American Association of Petroleum Geologists Bulletin*, **66**, 1328–1347.

AXELROD, D. I. 1984. An interpretation of Cretaceous and Tertiary biota in polar regions. *Palaeogeography, Palaeoclimatology, Palaeoecology*, **45**, 105–147.

—— & RAVEN, P. H. 1978. Late Cretaceous and Tertiary vegetation history of Africa. *In:* WERGER, M. J. A. (ed.) *Biogeography and Ecology of Southern Africa*, pp. 77–130. Monographiae Biologicae 31.

BARRON, E. J. & WASHINGTON, W. M. 1982. Cretaceous climate: a comparison of atmospheric simulations with the geologic record. *Palaeogeography, Palaeoclimatology, Palaeoecology*, **40**, 103–133.

—— & —— 1985. Warm Cretaceous climates: high atmospheric CO_2 as a plausible mechanism. *In:* SUNDQUIST, E. T. & BROECKER, W. S. (eds) *The*

Carbon Cycle and Atmospheric CO_2: Natural Variation Archean to Present, pp. 546–553. Geophysical Monograph 32.

BATTEN, D. J. 1984. Palynology, climate and the development of Late Cretaceous floral provinces in the Northern Hemisphere; a review. *In:* BRENCHLEY, P. J. (ed.) *Fossils and Climate*, pp. 127–164. Wiley, Chichester.

BERGH, H. W. 1977. Mesozoic sea-floor off Dronning Maud Land, Antarctica. *Nature, London*, **269**, 686–687.

BERNER, R. A., LASAGA, A. C. & GARRELS, R. M. 1983. The carbonate-silicate geochemical cycle and its effect on atmospheric carbon dioxide over the past 100 million years. *American Journal of Science*, **283**, 641–683.

BERTRAND, J. M. L. & CABY, R. 1978. Geodynamic evolution of the Pan-African orogenic belt: a new interpretation of the Hoggar Shield (Algerian Sahara). *Geologische Rundschau*, **67**, 357–383.

BISWAS, B. 1973. Quarternary changes in sea level in the South China Sea. *Geological Society of Malaysia Bulletin*, **6**, 229–256.

BOWLER, J. M. 1976. Aridity in Australia: age, origins and expression in aeolian landforms and sediments. *Earth-Science Reviews*, **12**, 279–310.

BROWNE, S. E., FAIRHEAD, J. D. & MOHAMED, I. J. 1985. Gravity study of the White Nile Rift, Sudan, and its regional tectonic setting. *Tectonophysics*, **113**, 123–137.

BUDYKO, M. I. 1981. Environmental changes and the formation of successive faunas. *Paleontological Journal*, **15**, 1–8.

BURKE, K. & DEWEY, J. F. 1974. Two plates in Africa during the Cretaceous. *Nature, London*, **249**, 313–316.

BUTZER, K. W. 1971. *Environment and Archeology*. Aldine, Atherton, Inc., Chicago.

CHAMLEY, H. & ROBERT, C. 1982. Paleoenvironmental significance of clay deposits in Atlantic black shales. *In:* SCHLANGER, S. O. & CITA, M. B. (eds) *Nature and Origin of Cretaceous Carbon-rich Facies*, pp. 101–112. Academic Press, London.

CHEN, C. T. A. & DRAKE, E. T. 1986. Carbon dioxide increase in the atmosphere and oceans and possible effects on climate. *Annual Review of Earth and Planetary Sciences*, **14**, 201–235.

CHEN, C. Z. 1984. Biru region. *In:* ACADEMIA SINICA QINGHAI-XIZANG PLATEAU EXPEDITION GROUP (ed.) *Stratigraphy of Xizang (Tibetan) Plateau*, pp. 180–185. Series of the Comprehensive Scientific Expedition to the Qinghai-Xizang Plateau.

COOPER, W. G. G. & DIXEY, F. 1956. *Afrique-Nyassaland-Nyasaland*. Lexique Stratigraphique International 4.

CREBER, G. T. & CHALONER, W. G. 1985. Tree growth in the Mesozoic and Early Tertiary and the reconstruction of palaeoclimates. *Palaeogeography, Palaeoclimatology, Palaeoecology*, **52**, 35–60.

DABER, R. 1968. A *Weichselia-Stiehleria*-Matoniaceae community within the Quedlinburg Estuary of Lower Cretaceous age. *Journal of the Linnaean Society (Botany)*, **61**, 75–85.

DAGHLIAN, C. P. & PERSON, C. P. 1977. The cuticular anatomy of *Frenelopsis varians* from the Lower Cretaceous of central Texas. *American Journal of Botany*, **64**, 564–569.

DE LAPPARENT, A. F. 1960. Les dinosauriens du "continental intercalaire" du Sahara central. *Mémoires de la Societe Géologique de France*, **88A**, 9–22.

DINGLE R. V., SIESSER, W. G. & NEWTON, A. R. 1983. *Mesozoic and Tertiary Geology of Southern Africa*. Balkema, Rotterdam.

DOUGLAS, J. G. & WILLIAMS, G. E. 1982. Southern polar forests: the early Cretaceous floras of Victoria and their palaeoclimatic significance. *Palaeogeography, Palaeoclimatology, Palaeoecology*, **39**, 171–185.

DOYLE, J. A., JARDINE, S. & DOERENKAMP, A. 1982. *Afropollis*, a new genus of early angiosperm pollen, with notes on the Cretaceous palynostratigraphy and paleoenvironments of northern Gondwana. *Bulletin des Centres de Rechèrches Exploration-Production Elf-Aquitaine*, **6**, 39–117.

EL-ZOUKI, A. Y. 1980. Stratigraphy and lithofacies of the continental clastics (Upper Jurassic and Lower Cretaceous) of Jabal Nafusah, NW Libya. *In:* SALEM, M. J. & BUSREWIL, M. T. (eds) *The Geology of Libya, Volume 2*, pp. 393–418. Academic Press, London.

FAIRHEAD, J. D. 1986. Geophysical controls on sedimentation within the African rift systems. *In:* FROSTICK, L. E. *et al.* (eds) *Sedimentation in the African Rifts*, pp. 19–27. Geological Society Special Publication 25.

FISCHER, A. G. 1981. Climatic oscillations in the biosphere. *In:* NITECKI, M. H. (ed.) *Biotic Crises in Ecological and Evolutionary Time*, pp. 103–131. Academic Press, New York.

FLENLEY, J. R. 1979. *The Equatorial Rain Forest: a Geological History*. Butterworths, London.

FLOHN, H. 1978. Comparison of Antarctic and Arctic climate and its relevance to climatic evolution. *In:* VAN ZINDEREN BAKKER, E. M. (ed.) *Antarctic Glacial History and World palaeoenvironments*, pp. 3–13. Balkema, Rotterdam.

—— 1984. Climate evolution in the Southern Hemisphere and the equatorial region during the late Cenozoic. *In:* VOGEL, J. C. (ed.) *Late Cainozoic Palaeoclimates of the Southern Hemisphere*, pp. 5–20. Balkema, Rotterdam.

FRAKES, L. A. 1979. *Climates Throughout Geologic Time*. Elsevier, Amsterdam.

FRANCIS, J. E. 1983. The dominant conifer of the Jurassic Purbeck Formation, England. *Palaeontology*, **26**, 277–294.

GERARD, J. & OESTERLE, H. 1973. Facies study of the offshore Mahakam area. *Proceedings of the Indonesian Petroleum Association, Second Annual Convention, Jakarta*, pp. 187–194.

GORE, A. J. P. 1983. Introduction. *In:* GORE, A. J. P. (ed.) *Ecosystems of the World. Vol. 4B, Mires: Swamp, Bog, Fen and Moor*, pp. 1–34. Elsevier, Amsterdam.

HALLAM, A. 1984. Continental humid and arid zones during the Jurassic and Cretaceous. *Palaeogeography, Palaeoclimatology, Palaeoecology*, **47**, 195–223.

HAXBY, W. F. 1985. *Gravity Field of the World's Oceans*. Lamont-Doherty Geological Observatory.

JUNK, W. J. 1984. Ecology of the *várzea*, floodplain of Amazonian whitewater rivers. *In:* SIOLI, H. (ed.) *The Amazon*, pp. 215–243. Monographiae Biologicae 56.

KEEN, C. E. & HYNDMAN, R. D. 1979. Geophysical review of the continental margins of eastern and western Canada. *Canadian Journal of Earth Sciences*, **16**, 712–747.

KENT, D. V. & GRADSTEIN, F. M. 1985. A Cretaceous and Jurassic geochronology. *Geological Society of America Bulletin*, **96**, 1419–1427.

KLITGORD, K. D. & SCHOUTEN, H. 1984. Plate kinematics of the Central Atlantic. *In:* TUCHOLKE, B. E. & VOGT, P. R. (eds) *The Geology of North America: The Western Atlantic Region*. Geological Society of America DNAG Series 1.

KLITZSCH, E., HARMS, J. C., LEJAL-NICOL, A. & LIST, F. K. 1979. Major subdivisions and depositional environments of Nubia strata, southwestern Egypt. *American Association of Petroleum Geologists Bulletin*, **63**, 967–974.

—— & LEJAL-NICOL, A. 1984. Flora and fauna from strata in southern Egypt and northern Sudan. *Berliner Gewissenschaftliche Abhandlungen Reihe A*, **50**, 47–79.

KUTZBACH, J. E. & GUETTER, P. J. 1986. The influence of changing orbital parameters and surface boundary conditions on climate simulations for the past 18 000 years. *Journal of Atmospheric Sciences*, **43**, 1726–1759.

—— & STREET-PERROTT, F. A. 1985. Milankovitch forcing of fluctuations in the level of tropical lakes from 18 to 0 kyr BP. *Nature, London*, **317**, 130–134.

LARSON, R. L., MUTTER, J. C., DIEBOLD, J. B. & CARPENTER, G. B. 1979. Cuvier Basin: a product of ocean crust formation by Early Cretaceous rifting off Western Australia. *Earth and Planetary Science Letters*, **45**, 105–114.

LASAGA, A. C., BERNER, R. A. & GARRELS, R. M. 1985. An improved model of atmospheric CO_2 fluctuations over the past 100 million years. *In:* SUNDQUIST, E. T. & BROECKER, W. S. (eds) *The Carbon Cycle and Atmospheric CO_2: Natural Variations Archean to Present*, pp. 397–411. Geophysical Monograph 32.

LEFOND, S. J. 1969. *Handbook of World Salt Resources*. Plenum Press, New York.

LIETH, H. 1964. Versuch einer kartographischen Darstellung der Produktivität der Pflanzendecke auf der Erde. *Geographisches Taschenbuch 1964/65*, pp. 72–80.

—— 1975. Primary production of the major vegetation units of the world. *In:* LIETH, H. & WHITTAKER, R. H. (eds) *Primary Productivity of the Biosphere*, pp. 203–215. Springer-Verlag, New York.

LORENZ, J. 1980. Late Jurassic–Early Cretaceous sedimentation and tectonics of the Murzuq Basin, southwestern Libya. *In:* SALEM, M. J. & BUSREWIL, M. T. (eds) *The Geology of Libya, Volume 2*, pp. 383–392. Academic Press, London.

MANABE, S. & BROCCOLI, A. J. 1985. The influence of continental ice sheets on the climate of an ice age. *Journal of Geophysical Research*, **90**, 2167–2190.

MARTIN, A. K., GOODLAD, S. W., HARTNADY, C. J. H. & DU PLESSIS, A. 1982. Cretaceous palaeopositions of the Falkland Plateau relative to southern Africa using Mesozoic seafloor spreading anomalies. *Geophysical Journal of the Royal Astronomical Society*, **71**, 567–579.

MASCLE, J., MARINHO, M. & WANNESSON, J. 1986. The structure of the Guinean continental margin: implications for the connection between the central and the South Atlantic Oceans. *Geologische Rundschau*, **75**, 57–70.

MCKEE, E. D. 1962. Origin of the Nubian and similar sandstones. *Geologische Rundschau*, **52**, 551–587.

—— (ed.) 1979. A study of global sand seas. *United States Geological Survey Professional Paper* 1052.

MONENCO ONTARIO LTD 1981. Evaluation of the potential of peat in Ontario. *Ontario Ministry of Natural Resources, Occasional Paper* 7.

MOSHRIF, M. A. & KELLING, G. 1984. Stratigraphy and sedimentary history of Upper-Lower and Middle Cretaceous rocks, central Saudi Arabia. *Mineral Resources Bulletin* 28.

MÜLLER, M. J. 1982. *Selected Climatic Data for a Global Set of Standard Stations for Vegetation Science*. Tasks for Vegetation Science 5. Junk, The Hague.

NIKLAS, K. J. (1986). Large scale changes in animal and plant terrestrial communities. *In:* JABLONSKI, D. & RAUP, D. M. (eds) *Pattern and Process in the History of Life*, pp. 383–405. Springer-Verlag, Berlin.

PARRISH, J. T., ZIEGLER, A. M. & SCOTESE, C. R. 1982. Rainfall patterns and the distribution of coals and evaporites in the Mesozoic and Cenozoic. *Palaeogeography, Palaeoclimatology, Palaeoecology*, **40**, 67–101.

PINDELL, J. L. & DEWEY, J. F. 1982. Permo–Triassic reconstruction of western Pangea and the evolution of the Gulf of Mexico/Caribbean region. *Tectonics*, **1**, 179–211.

RABINOWITZ, P. D. & LABRECQUE, J. 1979. The Mesozoic South Atlantic Ocean and evolution of its continental margins. *Journal of Geophysical Research*, **84**, 5973–6002.

RODIN, L. E., BAZILEVICH, N. I. & RONOV, N. N. 1975. Productivity of the world's main ecosystems. *In: Productivity of World Ecosystems*, pp. 13–26. Proceedings of a Symposium, National Academy of Science, Washington, DC.

ROWLEY, D. B., LOTTES, A. L. & ZIEGLER, A. M. 1985. North America–Greenland–Eurasian relative motions: implications for Circum-Arctic tectonic reconstructions. *American Association of Petroleum Geologists Bulletin*, **69**, 303.

——, PINDELL, J., LOTTES, A. L. & ZIEGLER, A. M. 1986. Phanerozoic reconstructions of northern South America, West Africa, North America and Caribbean region. *Geological Society of America Abstracts with Programs* 735.

SALATI, E. & MARQUES, J. 1984. Climatology of the Amazon region. *In:* SIOLI, H. (ed.) *The Amazon*, pp. 85–126. Monographiae Biologicae 56.

SARNTHEIN, M. 1978. Sand deserts during glacial

maximum and climatic optimum. *Nature, London*, **272**, 43–46.

SCHMITHÜSEN, J. 1976. *Atlas zur Biogeographie*. Bibliographisches Institut Mannheim, Wien.

SCHNEIDER, S. H. 1984. On the empirical verification of model-predicted CO_2-induced climatic effects. *In:* HANSEN, J. E. & TAKAHASHI, T. (eds) *Climate Processes and Climate Sensitivity*, pp. 187–201. Geophysical Monograph 29, Maurice Ewing Volume 5.

——, THOMPSON, S. L. & BARRON, E. J. 1985. Mid-Cretaceous continental surface temperatures: are high carbon dioxide concentrations needed to simulate above-freezing winter conditions? *In:* SUNDQUIST, E. T. & BROECKER, W. S. (eds) *The Carbon Cycle and Atmospheric CO_2: Natural Variations Archean to Present*, pp. 554–559. Geophysical Monograph 32.

SÉGOUFIN, J. 1978. Anomalies magnétiques mésozoïques dans le bassin de Mozambique. *Comptes Rendus des Séances de l'Académie des Sciences Serie D*, **287**, 109–112.

—— & PATRIAT, P. 1980. Existence d'anomalies mésozoïques dans le bassin de Somalie. Implications pour les relations Afrique–Antarctique–Madagascar. *Comptes Rendus des Séances de l'Académie des Sciences Serie B*, **291**, 85–88.

SELLERS, W. D. 1965. *Physical Climatology*. The University of Chicago Press.

SENGÖR, A. M. C. 1984. The Cimmeride orogenic system and the tectonics of Eurasia. *Geological Society of America Special Paper* 195.

SIMPSON, E. S. W., SCLATER, J. G., PARSONS, B., NORTON, I. & MEINKE, L. 1979. Mesozoic magnetic lineations in the Mozambique Basin. *Earth and Planetary Science Letters*, **43**, 260–264.

SIOLI, H. 1984. The Amazon and its main affluents: hydrography, morphology of the river courses and river types. *In:* SIOLI, H. (ed.) *The Amazon*, pp. 127–165. Monographiae Biologicae 56.

SMITH, A. G. & HALLAM, A. 1970. The fit of the southern continents. *Nature, London*, **225**, 139–144.

SOPER, E. K. & OSBORN, C. C. 1922. The occurrence and uses of peat in the United States. *United States Geological Survey Bulletin* 728.

STRAHLER, A. N. & STRAHLER, A. H. 1983. *Modern Physical Geography*, 2nd edn. Wiley, New York.

STREET-PERROTT, F. A. & HARRISON, S. P. 1984. Temporal variations in lake levels since 30,000 yr BP—an index of the global hydrological cycle. *In:* HANSEN, J. E. & TAKAHASHI, T. (eds) *Climate Processes and Climate Sensitivity*, pp. 118–129. Geophysical Monograph 29, Maurice Ewing Volume 5.

SUSZCZYNSKI, E. F. 1984. The peat resources of Brazil. *Proceedings of the 7th International Peat Congress, Dublin*, **1**, 468–492.

TALBOT, M. R. 1984. Late Pleistocene rainfall and dune building in the Sahel. *In:* COETZEE, J. A. & VAN ZINDEREN BAKKER, E. M. (eds) *Palaeoecology of Africa*, pp. 203–214. Balkema, Rotterdam.

TARNOCAI, C. 1984. *Peat Resources of Canada*. Land Resource Research Institute, Research Branch, Agriculture Canada.

THOMPSON, K. & HAMILTON, A. C. 1983. Peatlands and swamps of the African continent. *In:* GORE, A. J. P. (ed.) *Ecosystems of the World. Vol. 4B, Mires: Swamp, Bog, Fen and Moor, Regional Systems*, pp. 331–373. Elsevier, Amsterdam.

TRICART, J. 1975. Existence de périodes sèches au Quaternaire en Amazonie et dans les régions voisines. *Revue de Geomorphologie Dynamique*, **24**, 145–158.

VACHRAMEEV, V. A. 1972. Mesozoic floras of the southern hemisphere and their relationships to the floras of the northern continents. *Paleontological Journal*, **6**, 409–421.

—— 1978. The climates of the northern hemisphere in the Cretaceous in light of paleobotanical data. *Paleontological Journal*, **12**, 143–154.

VALYASHKO, M. G. 1972. Scientific works in the field of geochemistry and the genesis of salt deposits in the U.S.S.R. *In:* UNESCO (ed.) *Geology of Saline Deposits*, pp. 289–311. Proceedings of the Hanover Symposium.

VEEVERS, J. J., POWELL, C. M. & JOHNSON, B. D. 1980. Seafloor constraints on the reconstruction of Gondwanaland. *Earth and Planetary Science Letters*, **51**, 435–444.

VINOGRADOV, A. P. (ed.) 1967. *Atlas of the Lithological–Paleogeographical Maps of the U.S.S.R., Volume 4, Paleogene, Neogene and Quaternary*. Akademia Nauk SSSR, Moscow.

WHITTAKER, R. H. & LIKENS, G. E. 1975. The biosphere and man. *In:* LIETH, H. & WHITTAKER, R. H. (eds) *Primary Productivity of the Biosphere*, pp. 305–328. Springer-Verlag, New York.

WOOLLANDS, M. A. & HAW, D. 1976. Tertiary stratigraphy and sedimentation in the Gulf of Thailand. *SEAPEX Program Offshore South East Asia Conference Paper* 7.

WYCISK, P. 1984. Depositional environments of Mesozoic strata from northwestern Sudan. *Berliner Gewissenschaftliche Abhandlungen Reihe A*, **50**, 81–97.

ZIEGLER, A. M., BAMBACH, R. K., PARRISH, J. T., BARRETT, S. F., GIERLOWSKI, E. H., PARKER, W. C., RAYMOND, A. & SEPKOSKI, J. J. 1981b. Paleozoic biogeography and climatology. *In:* NIKLAS, K. J. (ed.) *Paleobotany, Paleoecology and Evolution, Volume 2*, pp. 231–266. Praeger, New York.

——, BARRETT, S. F. & SCOTESE, C. R. 1981a. Palaeoclimate, sedimentation and continental accretion. *Philosophical Transactions of the Royal Society of London Series A*, **301**, 253–264.

——, HULVER, M. L., LOTTES, A. L. & SCHMACHTENBERG, W. F. 1984. Uniformitarianism and palaeoclimates: influences from the distribution of carbonate rocks. *In:* BRENCHLEY, P. J. (ed.) *Fossils and Climate*, pp. 3–25. Wiley, Chichester.

——, SCOTESE, C. R. & BARRETT, S. F. 1983. Mesozoic and Cenozoic paleogeographic maps. *In:* BROSCHE, P. & SUNDERMANN, J. (eds) *Tidal Friction and the Earth's Rotation II*, pp. 240–252. Springer-Verlag, Berlin.

——, SCOTESE, C. R., MCKERROW, W. S., JOHNSON, M. E. & BAMBACH, R. K. 1979. Paleozoic paleogeography. *Annual Review of Earth and Planetary Sciences*, **7**, 473–502.

ZIEGLER, P. A. 1982. *Geological Atlas of Western and Central Europe*. Shell International Petroleum Maatschappij B.V.

ZUREK, S. 1984. Organic matter accumulation in European peatlands (on the basis of ^{14}C data). *In: Proceedings of the 7th International Peat Congress, Dublin*, **2**, 68–87.

A. M. ZIEGLER, University of Chicago, Department of Geophysical Sciences, Chicago, IL 60637, USA.

A. L. RAYMOND, Texas A & M University, Department of Geology, College Station, TX 77843-3115, USA.

T. C. GIERLOWSKI, M. A. HORRELL, D. B. ROWLEY & A. L. LOTTES, University of Chicago, Department of Geophysical Sciences, Chicago, IL 60637, USA.

Appendix: Bibliography of papers from which Valanginian coal, evaporite, Nubian sandstone and sand dune data points were derived

Coal

BENAVIDES-CACERES, V. E. 1956. Cretaceous system in northern Peru. *American Museum of Natural History Bulletin*, **108**, 355–493.

BROWN, P. M., MILLER, J. A. & SWAIN, F. M. 1972. Structural and stratigraphic framework and spatial distribution of permeability of the Atlantic Coastal Plain, North Carolina to New York. *United States Geological Survey Professional Paper* 796.

GIBSON, D. W. 1977. Sedimentary facies in the Jura-Cretaceous Kootenay Formation, Crowsnest Pass area, southwestern Alberta and southeastern British Columbia. *Canadian Petroleum Geology Bulletin*, **25**, 767–791.

—— 1979. The Morrissey and Mist Mountain Formations—newly defined lithostratigraphic units of the Jura-Cretaceous Kootenay Group, Alberta and British Columbia. *Canadian Petroleum Geology Bulletin*, **27**, 183–208.

GOBBETT, D. J. & HUTCHISON, C. S. 1973. *Geology of the Malay Peninsula*. Wiley (Interscience), New York.

GUPTA, V. J. 1975. *Indian Mesozoic Stratigraphy*. Hindustan Publishing Corporation, Delhi.

IMLAY, R. W. 1944. Correlation of the Cretaceous Formations of the Greater Antilles, Central America and Mexico. *Geological Society of America Bulletin*, **55**, 1005–1046.

JELETZKY, J. A. 1958. Uppermost Jurassic and Cretaceous rocks of Aklavik Range, northeastern Richardson Mountains, Northwest Territories. *Geological Survey of Canada Paper* 58-2.

NEI MONGGOL STRATIGRAPHIC GROUP, 1978. *Regional Stratigraphic Data of Northern China (Nei Monggol)*. Geology Press, Beijing.

POAG, W. 1982. Stratigraphic reference section for Georges Bank Basin-depositional model for New England passive margin. *American Association of Petroleum Geologists Bulletin*, **66**, 1021–1041.

RUDKIN, R. A. 1966. Lower Cretaceous. *In:* MCCROSSAN, R. G. & GLAISTER, R. P. (eds) *Geological History of Western Canada*, pp. 156–168. Alberta Society of Petroleum Geologists, Calgary.

SCHOTT, W. (ed.) 1969. *Paläogeographischer Atlas der Unterkreide von Nordwestdeutschland mit einer Übersichtdarstellung des nördlichen Mitteleuropa.* Bundesanstalt für Bodenforschung, Hannover.

SMART, J. & SENIOR, B. R. 1980. Jurassic-Cretaceous basins of northeastern Australia. *In:* HENDERSON, R. A. & STEPHENSON, P. J. (eds) *The Geology and Geophysics of Northeastern Australia*, pp. 315–328. Geological Society of Australia Inc., Brisbane.

STOTT, D. F. 1969. Fernie and Minnes strata north of Peace River, foothills of northeastern British Columbia. *Geological Survey of Canada Paper* 67-19.

TANAKA, K. & NOZAWA, T. (eds) 1977. *Geology and Mineral Resources of Japan, Volume 1*. Geological Survey of Japan.

TATEIWA, I. 1956. *Asie—Korea*. Lexique Stratigraphique International 3.

VINOGRADOV, A. P. (ed.) 1968. *Atlas of the Lithological—Paleogeographical Maps of the U.S.S.R., Volume 3, Triassic, Jurassic and Cretaceous*. Akademia Nauk SSSR, Moscow.

WHEELER, J. O. 1961. Whitehorse map-area, Yukon Territory. *Geological Survey of Canada Memoir* 312.

XU, X., WEI, Z. S., CHEN, G. A. & JIAO, S. R. 1981. *Regional Stratigraphic Table of Tibet Plateau*. Geology Press, Beijing.

YORATH, C. J., BALKWILL, H. R. & KLASSEN, R. W. 1975. Franklin Bay and Malloch Hill map-areas, District of MacKenzie. *Geological Survey of Canada Paper* 74-36.

Evaporites

ANHUI STRATIGRAPHIC GROUP, 1978. *Regional Stratigraphic Data of Eastern China (Anhui)*. Geology Press, Beijing.

APPLIN, P. L. & APPLIN, E. R. 1965. The Comanche Series and associated rocks in the subsurface in central and south Florida. *United States Geological Survey Professional Paper* 447.

BARNES, S. U. 1976. Geology and oil prospects of

Somalia, East Africa. *American Association of Petroleum Geologists Bulletin*, **60**, 389–413.

BESAIRIE, H. 1972. *Géologie de Madagascar I. Les Terrains Sedimentaires*. Annales Géologiques de Madagascar Fascicule 35, Tananarive Imprimerie Nationale.

CENTRAL SOUTH CHINA STRATIGRAPHIC GROUP, 1974. *Regional Stratigraphic Data of Central South China (Henan, Hubei, Hunan, Guangdong, Guangxi)*. Geology Press, Beijing.

GANSU STRATIGRAPHIC GROUP, 1980. *Regional Stratigraphic Data of Northwestern China (Gansu)*. Geology Press, Beijing.

HUBER, H. (ed.) 1977. *Geological map of Iran at scale of 1:1,000,000*. National Iranian Oil Company, Geological Survey of Iran.

JIANGSU AND SHANGHAI STRATIGRAPHIC GROUP, 1978. *Regional Stratigraphic Data of Eastern China (Jiangsu and Shanghai)*. Geology Fress, Beijing.

MERLA, G., ABBATE, E., AZZAROLI, A., BRUNI, P., CANUTI, P., FAZZUOLI, M., SAGRI, M. & TACCONI, P. 1979. *A geological map of Ethiopia and Somalia at scale of 1:2,000,000*. Consiglio Nazionale delle Recerche, Italy.

MEYERHOFF, A. A. & HATTEN, C. W. 1974. Bahamas salient of North America: tectonic framework, stratigraphy and petroleum potential. *American Association of Petroleum Geologists Bulletin*, **58**, 1201–1239.

QINGHAI STRATIGRAPHIC GROUP, 1980. *Regional Stratigraphic Data of Northwestern China (Qinghai)*. Geology Press, Beijing.

ULLOA, C. & RODRIGUEZ, E. 1976. Geologia del Cuadrangulo K-12 Guateque. *Boletin Geologico INGEOMINAS*, **22**.

VINOGRADOV, A. P. (ed.) 1968. *Atlas of the Lithological—Paleogeographical Maps of the U.S.S.R., Volume 3, Triassic, Jurassic and Cretaceous*. Akademia Nauk SSSR, Moscow.

WOLFART, R. & WITTEKINDT, H. 1980. *Geologie von Afghanistan*. Gebrüder Borntraeger, Berlin.

WORKMAN, D. R. 1977. *Geology of Laos, Cambodia, South Vietnam and the eastern part of Thailand*. Great Britain Institute of Geological Sciences Overseas Geology and Mineral Resources 50.

Nubian sandstone and other non-aeolian, continental sandstones of Africa

BENDER, F. 1968. *Geologie von Jordanien*. Gebrüder Borntraeger, Berlin.

BÖTTCHER, R. 1985. Environmental model of the shallow marine Abu Ballas Formation (Aptian, Nubia Group) in south-western Egypt. *Neues Jahrbuch für Geologie und paläontologie Abhandlungen*, **169**, 261–283.

BUSSON, G. 1967. *Le Mésozoïque Saharien 1re partie: l'Extreme-Sud tunisien*. Centre de Recherches sur les Zones Arides Serie: Geologie No. 8. Editions du CNRS, Paris.

—— 1970. *Le Mésozoïque Saharien 2e partie: Essai de synthèse des données des sondages algéro-tunisiens Tome 1 & 2*. Centre de Recherches sur les Zones Arides Serie: Geologie No. 11. Editions du CNRS, Paris.

CAHEN, L., FERRAND, J. J., HAARSMA, M. J. F., LEPERSONNE, J. & VERBEEK, T. 1959. Description du sondage de Samba. *Annales du Musee Royal du Congo Belge Tervuren* 29.

——, ——, ——, —— & —— 1960. Description du sondage de Dekese. *Annales du Musee Royal du Congo Belge Tervuren* 34.

COOPER, W. G. G. & DIXEY, F. 1956. *Afrique-Nyassaland-Nyasaland*. Lexique Stratigraphique International 4.

DE SPENGLER, A. & DELTEIL, J. R. 1966. Le bassin Secondaire-Tertiaire de Côte d'Ivoire. *In:* REYRE, D. (ed.) *Bassins Sédimentaires du Littoral Africain Symposium, 1re Partie, Littoral Atlantique*, pp. 99–112. Association des Services Géologiques Africains, Paris.

DINGLE, R. V., SIESSER, W. G. & NEWTON, A. R. 1983. *Mesozoic and Tertiary Geology of Southern Africa*. Balkema, Rotterdam.

FAURE, H. 1966. Reconnaissance géologique des formations sédimentaires post-Paléozoïques du Niger oriental. *Mémoires du Bureau de Recherches Géologiques et Minières* 47.

FLORES, G. 1973. The Cretaceous and Tertiary sedimentary basin of Mozambique and Zululand. *In:* BLANT, G. (ed.) *Bassins Sédimentaires du Littoral Africain Symposium, 2eme partie, Littoral Austral et Oriental*, pp. 81–111. Association des Services Géologiques Africains, Paris.

GOUDARZI, G. H. 1970. Geology and mineral resources of Libya—a reconnaissance. *United States Geological Survey Professional Paper* 660.

GREIGERT, J. & POUGNET, R. 1967. Essai de description des formations geologiques de la Republique du Niger. *Mémoires du Bureau de Recherches Géologiques et Minières* 48.

GREKOFF, N. 1957. Ostracodes du Bassin du Congo, I. Jurassique supérieur et Crétacé inférieur du nord du bassin. *Annales du Musée Royal du Congo Belge Tervuren* 19.

GUMATI, Y. D. & KANES, W. H. 1985. Early Tertiary subsidence and sedimentary facies—northern Sirte Basin, Libya. *American Association of Petroleum Geologists Bulletin*, **69**, 39–52.

KALLENBACH, H. 1972. Beiträge zur Sedimentologie des kontinentalen Mesozoikums om Westrand des Murzukbeckens. *Geologische Rundschau*, **61**, 302–322.

KLITZSCH, E., HARMS, J. C., LEJAL-NICOL, A. & LIST, F. K. 1979. Major subdivisions and depositional environments of Nubia strata, southwestern Egypt. *American Association of Petroleum Geologists Bulletin*, **63**, 967–974.

LORENZ, J. 1980. Late Jurassic-Early Cretaceous sedimentation and tectonics of the Murzuq Basin, southwestern Libya. *In:* SALEM, M. J. & BUSREWIL, M. T. (eds) *The Geology of Libya, Volume 2*, pp. 383–392. Academic Press, London.

MOHR, P. A. 1970. The Afar triple junction and seafloor spreading. *Journal of Geophysical Research*, **75**, 7340–7352.

PETTERS, S. W. 1979. Stratigraphic history of the south-central Saharan region. *Geological Society of America Bulletin*, **90**, 753–760.

QUENNELL, A. M., MCKINLAY, A. C. M. & AITKEN, W. G. 1956. *Summary of the geology of Tanganyika*. Tanganyika Geological Survey Department Memoir 1.

SINHA, S. C. 1980. On the application of geophysical logging in the assessment of groundwater potential in Al Hamadah al Hamra Basin. *In:* SALEM, M. J. & BUSREWIL, M. T. (eds) *The Geology of Libya, Volume 2*, pp. 643–658. Academic Press, London.

VITERBO, I. 1969. Lower Cretaceous charophyta from the subsurface "Nubian Complex" of the Sirte Basin (Libya). *In:* SAID, R., BECKMANN, J. P., GHORAB, M. A., EL ANSARY, S., VIOTTI, C. & KERDANY, M. T. (eds) *Proceedings of the African Micropaleontological Colloquium, 3rd, Cairo*, pp. 393–402. NIDOC, Cairo.

WHITEMAN, A. J. 1971. *The Geology of the Sudan Republic*. Clarendon Press, Oxford.

Sand dune

MCKEE, E. D. (ed.) 1979. A study of global sand seas. *United States Geological Survey Professional Paper* 1052.

Facies studies of coal and coal-bearing strata*

Peter J. McCabe

ABSTRACT: Prior to 1960 most discussions on the depositional environment of coal-bearing strata were related to theories of the origin of cyclothems. Work during the 1960s on modern deltas, particularly the Mississippi, led to the interpretation of the classic cyclothem as deltaic in origin, with coals originating in delta-top swamps. During the 1970s and early 1980s, detailed models of depositional environments were developed for coal-bearing strata. Coals have now been described in association with virtually all types of terrestrial depositional environments.

There are surprisingly few published examples of facies analyses of coal-bearing strata. Detailed measured sections and meticulous descriptions of sedimentary structures are necessary for geologists to make independent assessments of published environmental interpretations and to make comparisons between various coal-bearing formations. The coal industry could benefit from more predictive facies models, which will require greater integration of coal and clastic facies studies.

In most mires close to areas of active clastic deposition, ash content of peat is too high for them to be precursors of coal. Most coal must have originated as peats in raised mires or in low-lying mires well removed from active clastic environments. In the latter case, a significant hiatus in deposition may be represented by the contact of coal with underlying sediments. Subsurface studies, which allow an understanding of the three-dimensional character of coal-bearing strata, may provide clues as to the type of swamp in which a coal was deposited. Recent studies in the Upper Cretaceous of Alberta, for example, provide clear evidence that coals originated in mires well removed from contemporaneous clastic depositional systems.

Keywords: sedimentology, mire, coal, Carboniferous, Cretaceous, Canada

Sediments from many types of terrestrial depositional environments have been found in association with coal. Although coals are not present in ancient desert deposits, they have been found in association with alluvial fans (Heward 1978, Hunt & Hobday 1984), sandy braided rivers (Haszeldine & Anderton 1980) and aeolian dunes (Richardson 1985). Most coals are, however, found in association with the deposits of meandering rivers, deltas and shoreline environments. The reader is referred to Galloway & Hobday (1983), McCabe (in press) and Ward (1984) for reviews of environments of deposition of coal-bearing strata. A much better understanding of these depositional environments has developed over the last 20 years. Detailed sedimentological studies carried out at mine sites have helped in the understanding of variations in roof and floor rocks. Such studies have aimed at optimizing production and lowering costs. Models have also been developed to show the areas of better coal development in regional depositional settings, and these should aid future exploration for coal resources.

This paper makes a critical assessment of the present status of facies studies applied to coal and coal-bearing strata and suggests directions for future research. A brief history of work on environments of coal deposition is presented. The concepts of facies and facies models are then reviewed with respect to coal and coal-bearing strata. The remainder of the paper deals with two recent developments in the field: (a) the attempt to understand the depositional relationships between coal and associated clastic sediment, and (b) subsurface facies studies which allow a three-dimensional understanding of facies relationships. The usefulness of subsurface studies is illustrated by examples of recent work on Cretaceous coals in western Canada.

The term mire is used in this paper to cover all non-saline wetlands, following Gore (1983). The term encompasses all ecosystems described as bog, fen, moor, muskeg, peatland and swamp (see also Moore, *this volume*). Organic material accumulates in all mires where there is net sedimentation, but the ratio of clastic to organic supply determines whether peat is formed or not.

Brief history of sedimentological studies

During the nineteenth century, geologists in many countries had recognized rhythmic cycles

* This paper should be regarded as a supplement to McCabe (1984) to which the reader is referred. Ed.

of deposition in coal-bearing strata (e.g. Phillips 1836). In 1912 Udden suggested that cycles in the Pennsylvanian of Illinois were created by repeated transgressions and regressions. The cyclothem concept, and discussions as to the causes of the transgressions and regressions, dominated studies on coal-bearing strata for the next 50 years (Hudson 1924, Weller 1930, Wanless & Weller 1932, Trueman 1946, Merriam 1964). In the 1960s attempts were made to quantify cyclicity through statistical analyses (Duff & Walton 1962, Merriam 1964, Read & Dean 1967, 1968). This was also, however, a time of scepticism about cycles (Zeller 1964) and Westoll (1968) prefaced a section of his paper on cycles in coal-bearing strata with the quote 'Use not vain repetitions, as the heathen do' (Matthew, vi, 7).

Sedimentological studies of modern environments that had started in the mid-1950s, particularly at the research laboratories of the major oil companies, led to a better understanding of ancient sediments. Moore's (1959) interpretation of Yoredale cyclothems as deltaic deposits showed the powerful potential of comparing ancient sediments with those of modern environments. Attention turned from viewing coal-bearing strata as a series of cyclothems to viewing them as a series of depositional sequences formed in a variety of environmental settings.

During the late 1960s and the 1970s sedimentologists studied both modern and ancient environments in greater detail paying particular attention to sedimentary structures and facies sequences (Reineck & Singh 1973, Reading 1986). The work of Ferm and Horne (Ferm 1974, Ferm & Horne 1979) in the Appalachians was particularly important in developing an awareness among coal geologists of the importance of sedimentological studies. The 1960s and 1970s was also a period of new initiatives in the area of coal and peat facies studies. Teichmüller (1962) related coal properties to the type of mire in which the peat had accumulated. In the USA detailed studies of the Everglades and Okefenokee Swamps (Spackman *et al.* 1966, 1976, Cohen 1974, Cohen & Spackman 1977, see also Cohen *et al., this volume*) showed the complex nature of modern mires. These studies helped considerably in understanding the origin of coal macerals. Canadian workers developed detailed facies classifications of coal seams (Hacquebard & Donaldson 1969, Cameron 1971) that aided in environmental interpretation and correlation.

The energy crisis of the late 1970s led to a considerable increase in research on coal geology. This activity has resulted in the publication of an unprecedented number of papers on environments of deposition of coal and coal-bearing strata during the 1980s. Particularly significant are the volumes edited by Ethridge & Flores (1981) and Rahmani & Flores (1984), the *Compte Rendu* of the ninth and tenth Congrès International de Stratigraphie et de Géologie du Carbonifère, and the *International Journal of Coal Geology* which began publication in 1980.

Facies

The term 'facies' has been used in many ways but is essentially 'a body of rock with specified characteristics' (Reading 1986). A sequence of sediments can, therefore, be divided into a set of facies. The type of facies defined depends on which characters are observable. Where clastic sediments are seen in the field or in core, facies are defined on the basis of nature of bedding, colour, composition, fossil content, sedimentary structures and texture. Some workers (e.g. Miall 1978) have attempted to devise facies schemes that could be applied on a universal basis to clastic sediments. However, such schemes normally involve only one or two aspects of the rock's character. It has generally been found more useful to develop a separate facies classification for the purpose of each study, dependant on the scale of the study, its purpose and the type of observations that can be made. The number of facies defined in a study must be large enough to be meaningful, but not too large to be cumbersome.

Facies analysis is carried out in a variety of ways by different scientists. One of the most rigorous approaches has been that espoused by Reading (1986) and colleagues. The essence of the method is observe, describe and interpret; in that order. A sedimentary sequence is first observed in detail to determine a plan of study. The sequence is then described in as an objective a way as possible. This leads to a classification of facies. Each facies is defined in a precise manner. Only then is the depositional setting of individual facies interpreted. Relationships between facies are examined and a final environmental interpretation can then be made.

Large mires are complex terranes with streams, lakes, relief and varying floral assemblages. Lateral shifts in environments results in the build-up of peat sequences of differing composition. The facies concept has so far been applied to peats and coals in limited ways. Most studies look at only one characteristic, defining biofacies, palynofacies or petrofacies. There are, however, several studies that have taken a more integrated approach.

Peats of the Everglades and Okefenokee Swamps (Cohen 1973, 1974, Spackman *et al.*

1976) have been studied in considerable detail (see Cohen *et al., this volume*). Microtome sections allow a determination of the major plant types in a peat. Measured sections have been made through the mires showing the major peat types and associated lithologies allowing determinations of the evolution of depositional environments. Peat profiles in the Everglades, for example, show evidence of a continuous marine transgression in the area, with mangrove peats overlying freshwater peats (see Casagrande, *this volume*, fig. 6). Barber (1981) also studied modern peats in detail, with the advantage of being able to examine trenches in a mire in northern England. He divided the peat on the basis of type of plant material and degree of humification. His cross-sections show a great degree of lateral and vertical variation in peat types. Peats from the Fraser River delta of British Columbia have been divided into a series of organic facies by Styan & Bustin (1983b) on the basis of type of plant material, degree of humification, ash content and peat texture. They were able to show the distribution of these facies in detailed cross-sections through the delta top mires.

Studies in the Everglades (Cohen & Spackman 1972) show that there is a good relationship between pollen assemblages and peat types. Studies on coal seams have also shown a correlation between petrographic composition and palynoflora. Smith (1962) examined the vertical variation in microlithotype and miospore content at 20–80 mm intervals through Carboniferous coal seams. He distinguished four miospore assemblages that were associated with coals of distinctive petrographic types. In a study of Pennsylvanian coal seams of Nova Scotia, Hacquebard & Donaldson (1969) defined facies on petrographic characteristics and related them to miospore distribution. For each facies they interpreted the types of vegetation and relative water levels at the time of deposition. Facies distributions were correlated in cores spread over a 50 km long section. Megaspores and coal facies have also been investigated in the Yorkshire Barnsley Seam by Bartram (*this volume*).

Coals have been classified on the basis of macroscopic examination (Dutcher 1978, Schopf 1960). Lithotypes of coal are normally defined on both the maceral composition and the overall texture of the coal (see Scott, *this volume*, tables 1 and 2). The lithotype concept is, therefore, similar to that of a universal facies classification as discussed above. Perhaps the most comprehensive facies classification of coal, however, is that of Ting & Spackman (1975). They define 'lithotypes' on the basis of maceral content, petrographic texture, spore content and mineral content. Coal facies, defined on this many parameters, could prove most useful in interpreting the original environment of deposition. Although Ting & Spackman propose a universal application of their facies classification, it would probably prove more useful if separate schemes are developed for different basins and coal zones.

Facies models

According to Walker (1984) a facies model can be defined as a general summary of a specific sedimentary environment, written in terms that make the summary useable to: (a) act as a norm, for purposes of comparison, (b) act as a framework and guide for future observations, (c) act as a predictor in new geological situations and (d) act as an integrated basis for interpretation of the environment or system that it represents. A model may be expressed as an idealized sequence of facies, as a block diagram or in a statistical format. Facies models should not be unique to one geological formation. On the other hand facies models, produced by distilling away all the local variability of each example, may be too simplistic really to be of use. For this reason some sedimentologists have either avoided the use of facies models or have created a plethora of end-member 'types'. There is no doubt, however, that the concept of constructing facies models has been most useful in developing our present understanding of depositional systems. Facies studies should build up a base of information that could be used to construct facies models which meet Walker's four criteria. It is worthwhile briefly examining how well past studies of coal and coal-bearing strata have done this.

Facies models have been applied to mine site problems (e.g. Horne *et al.* 1978, Guion 1984) and to exploration for mineable coal deposits (e.g. Horne *et al.* 1978, Ryer 1981, Levey 1985). However, from the point of view of an economic geologist, most models are more explanatory rather than truly predictive. Facies descriptions need to be well documented if strata from different formations are to be compared and if earlier studies are to act as guides for future observations. Some studies of coal-bearing strata (e.g. Scott 1978, Fielding 1984, 1986) have provided excellent detailed descriptions of facies and facies relationships. Unfortunately most studies have been less detailed, or at least the published data are less comprehensive. This has been particularly true in N America where there are relatively few published examples of detailed measured sections through the major coal-bearing formations. Because of excellent exposures it has been

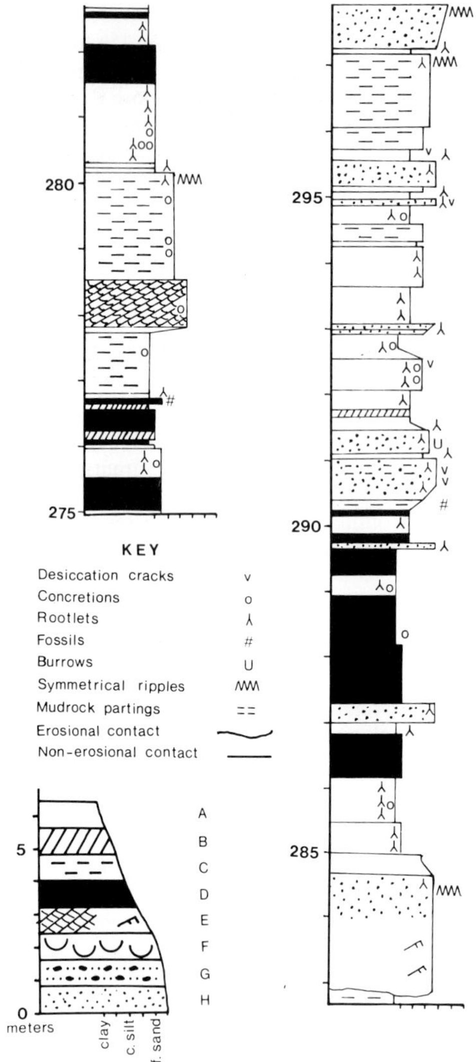

FIG. 1. Detailed log through part of the Port Hood Formation of Nova Scotia (from Gersib & McCabe 1981). Such detailed sections are necessary if comparisons are to be made between different formations.

possible to document two- or three-dimensional geometry over large areas but generally the published cross-sections show only five or six lithofacies.

Figure 1 shows a detailed measured section of a small part of the Carboniferous Port Hood Formation in Nova Scotia. The section depicts bed thickness, nature of bed contacts, average grain size, primary depositional structures and presence or organic remains. The formation was divided into eight facies that were named (e.g. Facies D = coal and carbonaceous mudrock), described and interpreted (Gersib & McCabe 1981). Such a measured section can easily be compared to other sequences and the reader is able to make an independent assessment of the environmental interpretation. Where possible, facies descriptions should also document lateral facies changes, as well as vertical transitions.

Most published facies studies of coal-bearing strata have simplistically interpreted coal beds as 'swamps' with little regard to the type of mire development (see Moore, *this volume*). On the other hand most studies on coal seams by palynologists, geochemists and coal petrographers have been done with little or no regard to surrounding clastic sediments. Cooperation between scientists from different disciplines will be necessary before facies models can be developed that form truly integrated bases for interpretation of the depositional history of sequences. As will be discussed later in the paper, many mires have developed in areas that are distinctly different from the environment in which underlying sediments were deposited. For this reason, it could be argued that separate facies models may be appropriate for clastics and coals. Although there are a few studies of in-seam facies variations, there has been no purposeful attempt at erecting a peat/coal facies model. The concept of evolution in mire types (Romanov 1968) may provide the basis for such a model (Fig. 2).

Peat-forming mires originate by the infill of lakes or poorly drained areas (Fig. 2). Accumulation of organic muds (gyttja) and the build up of floating peat mats may infill a lake and transform it into a low-lying mire. Because peat is relatively impermeable, its growth may progressively impede drainage over wider areas and low-lying mires may become very extensive (Fig. 2). These mires are able to maintain a varied and healthy flora by the throughflow of water which supplies nutrients from outside the mire. In geographical areas where annual precipitation exceeds evaporation, and where there are no long dry periods, the peat may aggrade to produce a raised mire (Fig. 2). These mires are able to build upwards because they maintain their own water table. Such conditions are found today on some tropical islands and peninsulas where daily convectional rains are prevalent, as in SE Asia (Anderson 1983), and in areas with a cool temperate maritime climate, as in NW Europe (Moore 1984). Groundwater conditions become more oligotrophic and acidic as a mire becomes more elevated. As a consequence, the flora becomes less varied and may be stunted.

The evolution of a peat-forming environment,

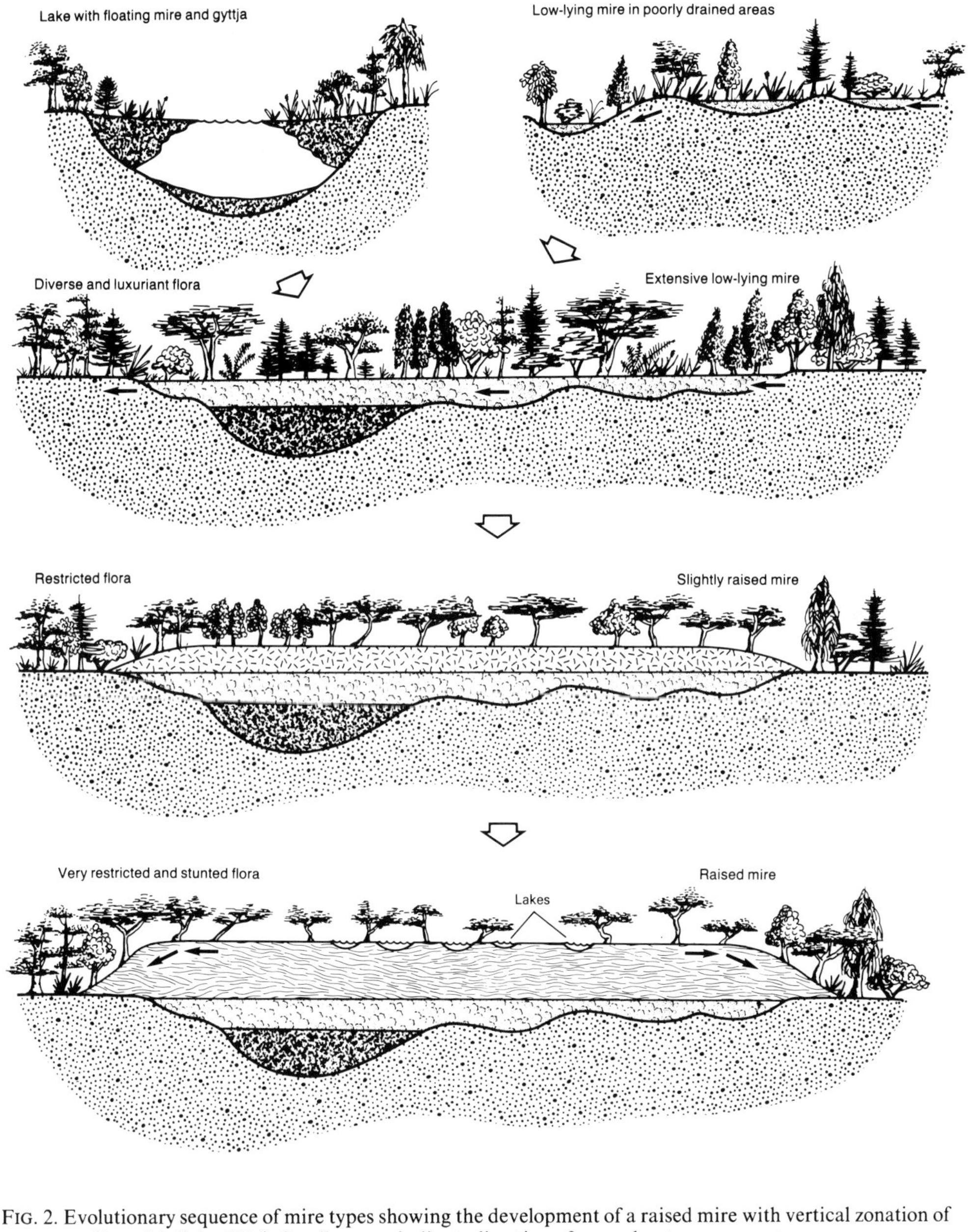

FIG. 2. Evolutionary sequence of mire types showing the development of a raised mire with vertical zonation of peat types (from McCabe 1984). Small arrows indicate direction of groundwater movement.

from a lake fill or low-lying mire to a mature raised mire, should lead to a vertical zonation of peat facies. The type of material supplied to the peat will vary from the large plant growth of the low-lying mire, to the relatively small plants of the mature raised mire or the floating mires, to the comminuted plant material of the lake bottom muds. Rates of degradation of organic material are a function of temperature, groundwater acidity and the degree of fluctuation of the groundwater table. The degree of degradation of plants should vary, therefore, through the peat profile (see Clymo, *this volume*). More studies of modern peat are needed to define the nature of such peat profiles. Large differences can be expected between peats of raised mires of the tropics, which are forested, and of temperate regions, which are not. There is, however, already some evidence from coals of evolution in mire types. For example, in a detailed study of Carboniferous coal seams, Smith (1962, 1968) observed vertical transitions in petrographic character and in miospore assemblages, which he suggested were the result of a change from wetter to drier conditions possibly as a result of raised mire development (see also Bartram, *this volume* and Fulton, *this volume*).

The relationship between coal and clastic sediments

Most facies models for coal-bearing strata show peat accumulating in close association with active clastic depositional environments such as on the floodplain of meandering rivers (e.g. Gersib & McCabe 1981), in coastal mires close to active beach-barrier systems (e.g. Snedden & Kersey 1981) in interdistributary bays and even levees of delta tops (e.g. Baganz *et al.* 1975). Such environments are, however, unlikely to be sites of peat (less than 50% ash on a moisture-free basis) accumulation unless raised mires are developed. Sediment may be introduced into low-lying mires by floods, storm surges or exceptionally high tides. This can result in very high ash contents in the sediment of mires close to active clastic environments.

It is difficult precisely to quantify rates of peat accumulation because compaction takes place within a peat in the mire environment. Calculations for Recent peats vary from 2.3 mm yr^{-1} in the tropics to 0.1 mm yr^{-1} in Arctic regions (McCabe 1984). Peat to coal compaction ratios have been calculated to be between 1.4:1 and 30:1 (Ryer & Langer 1980 (see also Collinson & Scott, *this volume*). Assuming an average ratio of 10:1, 1 mm of coal represents peat accumulation over a period of between 4 and 100 years. Low ash coals must, therefore, have formed in areas that were removed from active clastic deposition for centuries. Mudrock partings in coals probably represent events with a periodicity of thousands or even tens of thousands of years. Any more frequent events (e.g. 50 year floods) could be regarded as normal sedimentation, because the clastic sediment would become an integral part of the coal's composition.

Many sedimentologists have suggested the Mississippi delta as a modern analogue for coal-bearing strata. The first to propose this was Fisk (1960) at the International Carboniferous Congress in 1958. His classic cross-section of the Mississippi River floodplain, in the region of New Orleans (Fig. 3), shows up to 5 m of 'peat and organic muck' interfingering with levee and crevasse splay deposits. Fisk's analyses of the peats show, however, that the 'peats' have an ash content of at least 55% and average 72%, on a moisture-free basis.

Since Fisk's work, there have been more detailed studies of Mississippi delta peats (Frazier & Osanik 1969, Kosters 1983, Kosters & Bailey 1983). Recent work on peats of the abandoned lobes of the delta show that the vast majority of organic sediments have a very high ash content. Only nine of 215 samples from the Barataria and Sale–Cypremort areas, west of the modern delta had over 80% organic matter on a moisture-free basis (Kosters 1983). No peats were recorded with less than 17% ash. Peats, with less than 25% ash on a moisture-free basis, rarely exceed 1 m in thickness. A detailed examination of the sections

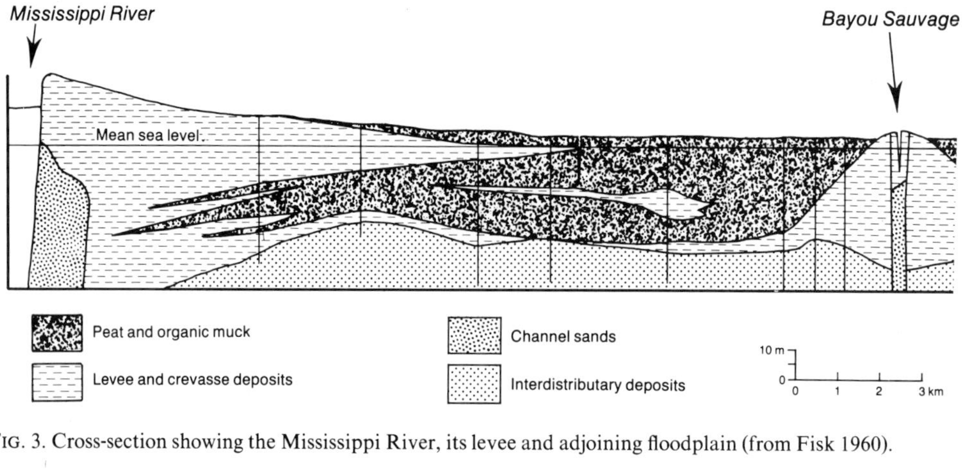

FIG. 3. Cross-section showing the Mississippi River, its levee and adjoining floodplain (from Fisk 1960).

through the peats suggests that, if preserved in the ancient record, they would form carbonaceous shales with thin coaly stringers. Leaching of the Mississippi delta peat can decrease the ash content by approximately 33% (Kosters & Bailey 1986) and Kosters & Bailey (1983) have suggested that leaching during early diagenesis could upgrade many peats to be potential precursors of commercial coal. There is, however, little evidence that such leaching processes occur. In contrast there is good evidence for epigenetic mineralization in most coals. Even if there was significant amounts of leaching, the Mississippi delta peats would still have too high an ash content to form economic coal seams.

An examination of the literature shows that significant areas of low ash peats are not present in most other deltas, and that most mires in coastal or floodplain settings are not sites of significant true peat accumulation (McCabe 1984, Breyer & McCabe 1986). The only exceptions appear to be those areas where raised mires have developed. Floating mires may also produce low ash peats (McCabe 1984) but they are usually not extensive.

Anderson (1964, 1983) described raised mires that are developed close to some shorelines and meandering river systems in SE Asia. Margins of the raised mires are relatively steep (up to 4.2 m rise over 100 m). Central parts of the mires have generally flat surfaces and are elevated between 3 and 7.5 m above flood levels. Ash content of these peats is usually less than 5%, and over large areas of mires is less than 1 or 2%. Raised mires are also present on the tide-dominated Klang-Langat delta of Malaysia (Coleman *et al.* 1970) and the Rajang delta of northern Borneo (Anderson 1964). Although thicker peats are present on adjoining coastal plains, mires on islands between the delta distributaries are raised well above storm or high tide levels. Raised mires have also been described in close association with clastic sediments in temperate climates. Peats from raised mires of the Fraser River delta, in British Columbia, have ash contents as low as 0.5–1.5% (Styan & Bustin 1983a, b). The raised mire of Borth Bog in central Wales lies immediately south of the Dovey estuary (Wilks 1979, Taylor 1983).

Rates of organic accumulation in areas of raised mires outstrip rates of sedimentation from overbank, storm or supratidal flooding. This is graphically illustrated in some of the Sarawak mires (Fig. 4) where the peat deposits typically have a biconvex structure (Anderson 1983). The deepest peats of the mires are below normal river level. This indicates that the raised mires have grown through time, progressively decreasing the width of the rivers' floodplains. Presumably an equilibrium is reached at which time sedimentation rates on the floodplain equal that of the raised mire. It is possible that this could lead to the development of stacked fluvial deposits lateral to thick peats (McCabe 1984). Although it may

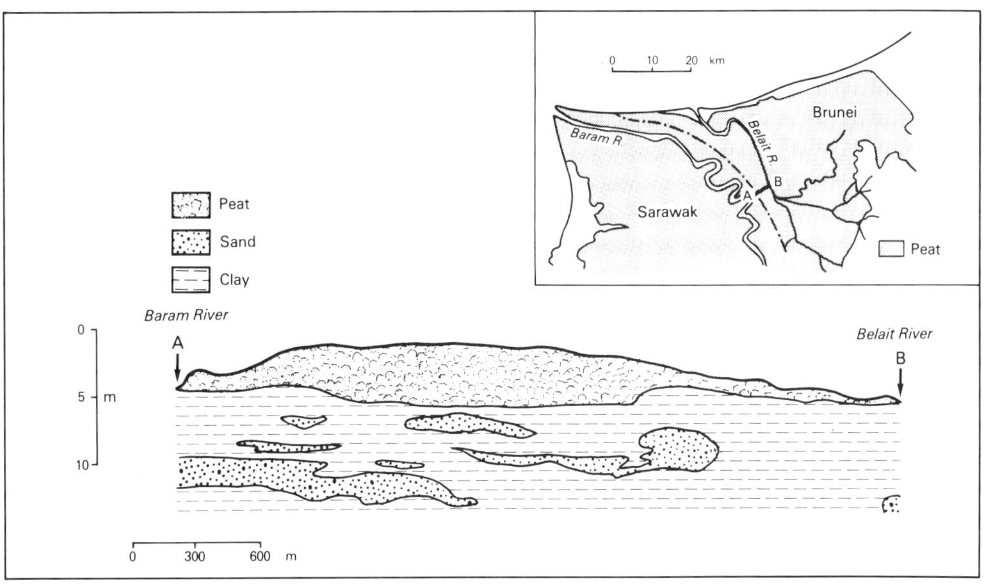

FIG. 4. Cross-section of a raised mire developed between two rivers in northern Borneo (from Wilford 1961).

be tempting to interpret all coals as having originated as peats in raised mires, many coals are thought to have formed under palaeoclimates that appear unsuitable for extensive raised mire development (e.g. subtropical or continental regimes).

Some coals are clearly contemporaneous with nearby clastics. For example, the Pennsylvanian Herrin Coal of Illinois is cut by clastic sediments of the Walshville channel but coal adjacent to the channel contains numerous splits (Treworgy & Jacobson 1985). The presence of the Walshville channel also influenced the maceral and palaeofloral content of the Herrin Coal (Harvey & Dillon 1985, Phillips *et al.* 1985). Swilleys cutting the Top Hard Coal, in the Carboniferous of central England, are also demonstrably penecontemporaneous with the coal (Elliot 1965). Cannel coals are present within the seam lateral to the swilleys. Such close juxtapositions of contemporaneous coals and clastics pose interesting questions as to the type of mire development, and the factors which confined clastic sedimentation to relatively narrow bands for long periods of time.

Although some coals and adjacent clastics may have been deposited contemporaneously, in most cases the contact between the two lithologies probably represents a considerable hiatus in deposition. Underlying sediments may have been deposited long before the mire was established; at which time the area could have been well removed from active clastic depositional environments. A considerable period of non-deposition preceded the initiation of peat accumulation in some of the mires of the southeastern USA. The peats of the Okefenokee (Cohen 1984) and Snuggedy Swamps (Staub & Cohen 1979) are developed over Pleistocene beach ridges. These mires are now well removed from the present Atlantic coast, but if preserved in the ancient record coals will be seen to overlie shoreline sediments. Care should be taken before assuming that the depositional environment of a coal is the same as underlying clastics. In some cases, overlying strata may be more genetically related to a coal than underlying strata.

Detailed studies are required to determine the genetic nature of the association between a coal seam and its enclosing clastics. Cooperative research between workers, in such fields as clastic sedimentology, palynology, palaeobotany and geochemistry, may be needed to gain a comprehensive understanding. For example, in a limited study of part of the Palaeocene Wilcox Group of Texas, Breyer & McCabe (1986) were able to discount a raised mire on palaeontological grounds. They consequently interpreted the overall environment of deposition of the coals as being much further inland than may have been suspected by looking only at the clastic sediments, which show evidence of tidal influence.

Subsurface studies

Subsurface data, from cores and well logs, or from underground mines, are essential for coal exploration and development. Detailed mapping of coal, roof- and floor-rocks in underground mines (e.g. Guion 1984, Krausse *et al.* 1979) can allow a much more comprehensive understanding of the geometry of the strata than is normally possible by studying surface exposures. Padgett & Ehrlich (1978) showed, for example, that maps of underground workings in W Virginia reflect the geometry of underlying and overlying channel systems. In the future, mine site studies may prove useful in allowing the development of better models to predict the geometry of sedimentary bodies, not just for coal mining but also for hydrocarbon and mineral exploration. Subsurface data from boreholes allow not only an understanding of the three-dimensional geometry of strata, but also permit documentation of variations in facies over long distances. Subsurface studies offer exciting possibilities in the development of future facies models for coal-bearing strata.

Ferm & Staub (1984) mapped the detailed geometry of individual coal seams and associated strata in the Carboniferous of the Appalachians. They found that the major cause of changes in seam thickness is splitting. Gradual changes in seam thickness and erosion of the coal by channels are of secondary importance. They suggested that the major factor controlling the location of thicker seams was the topography, which in turn was largely controlled by differential compaction of underlying strata. Using a similar approach, Weisenfluh & Ferm (1984) showed that differential subsidence, due to faulting contemporaneous with mire development, was an important control in determining thickness variations of the Pratt seam in the Carboniferous of Alabama.

Studies of well log information have allowed the determination of seam geometry in some Cretaceous coals of the western USA. Seam geometry has been related to the environment of deposition of underlying deltaic sediments. In a study of the coals associated with the Ferron Sandstone, Ryer (1981) showed that seams partly overlie sandstones interpreted as the regressive deposits of lobate deltas. The seams reach a maximum thickness some 10 km landward of the landward pinchout of each regressive sequence. Although they are traceable over a long distance parallel to the depositional strike, coal seams tend to be pod-like. Ryer (1981) attributes this to streams which cut across the area of mire development. In the Rock Springs Formation, coals overlie sequences interpreted as the regressive deposits of wave-dominated deltas (Levey 1985). Seams extend up to 24 km along depositional dip and up to 58 km along strike, but are thickest in areas which are only 2–3 km wide. The pod-like form to the seams is thought to be due to splitting of seams and channels cutting the coal. Unlike the Ferron Sandstone, the thickest coals overlie the regressive sequences.

Recent studies, by the Coal Geology Group of the Alberta Research Council's Geological Survey Department (McCabe *et al.* (in press), Macdonald *et al.* (in press), provide insights as to on the types of mires in which the Upper Cretaceous coals of Alberta developed. Two examples are outlined in the next section. These studies show how subsurface data can provide useful information on the environment of deposition and controls of coal deposition. This will lead to the development of better predictive models for coal exploration.

Upper Cretaceous coals of Alberta, Canada

The thickest coals in the Upper Cretaceous of Alberta are in the Horseshoe Canyon Formation, which is Maastrichtian in age. The formation averages 400 m in thickness but its lower half interfingers with, and is replaced eastwards by, marine strata of the Bearpaw Formation. Coals are present in the lower part of the Horseshoe Canyon Formation, in the zone where it interfingers with the Bearpaw Formation. The stratigraphic position may suggest that the coals formed in a delta-top or back barrier environment but, as discussed below, the regional geometry of the strata indicates that the economic coal seams originated in mires that were tens of kilometres inland.

Marine strata of the Bearpaw Formation consists of coarsening-upward sequences, which are mostly between 25 and 45 m thick. The sequences have a distinctive signature on gamma logs, with decreasing shaliness upwards (Fig. 5), and they can be correlated for distances of over 100 km. Coarsening-upward sequences begin with highly bioturbated mudstones. These pass upwards into interbedded sandstones and mudstones. Sandstone beds are up to 0.5 m thick and are generally thicker higher in the sequence. Most have sharp bases and bioturbated tops. Internally they show horizontal or hummocky cross-stratification and, in some cases, symmetrical ripples. The interbedded sandstones and mudstones are interpreted as lower shoreface deposits, having formed below fair weather wave base; an area normally of mud deposition but where sands were deposited during storm activity. Mudstones die out upward, and the upper parts of the coarsening-upward sequences consist of sandstones with hummocky cross-stratification, low angle cross-stratification and some trough cross-bedding. These sandstones are interpreted as upper shoreface deposits. The top 1 or 2 m may show well developed low angle cross-stratification, interpreted as beach deposits, but channel sandstones are cut into the upper part of many coarsening-upward sequences. Marine trace fossils occur throughout the coarsening-upward sequence and the channel sandstones show evidence of tidal current activity. Rahmani (1983), in a detailed study of the sedimentary sequence at Drumheller, interpreted the sediments as the deposits of a tide-dominated delta. However, the type of shoreline probably varied both geographically and with time and is, in any case, not critical to the present discussion.

The Horseshoe Canyon Formation is predominantly terrestrial in nature. Fining-upward sandstone sequences, up to 6 m thick, are interpreted as meandering river deposits. Interbedded mudstone sequences are generally rooted extensively, suggesting deposition on a floodplain. Thin lacustrine mudstones and crevasse-splay sandstones are also present. Virtually all of the coal is within 50 m of the marine strata of the Bearpaw Formation. Rahmani (1983) reports some brackish influence in strata overlying some coals. Individual seams are up to 5 m thick. Some coals can be traced for tens of kilometres but splitting of seams is common.

Figure 5 shows a typical cross-section of wells through the Bearpaw/Horseshoe Canyon Formation transition. The datum is a thin coal in a lower tongue of the Horseshoe Canyon Formation. It is parallel to the tops of the two overlying coarsening-upward sequences. The third coarsening-upward sequence thickens markedly from NW to SE (predominantly from west to east based on other wells). Its westward pinchout reflects the landward extent of a marine transgression which followed the formation of Coal A. The thick, predominantly barren strata of the main part of the Horseshoe Canyon Formation overlie Coal C. Figure 6 is an isopach of the total amount of coal in between the second and third coarsening-upward sequences (Coal A and associated seams). In the western part of the map, cumulative coal exceeds 7 m in places. Individual

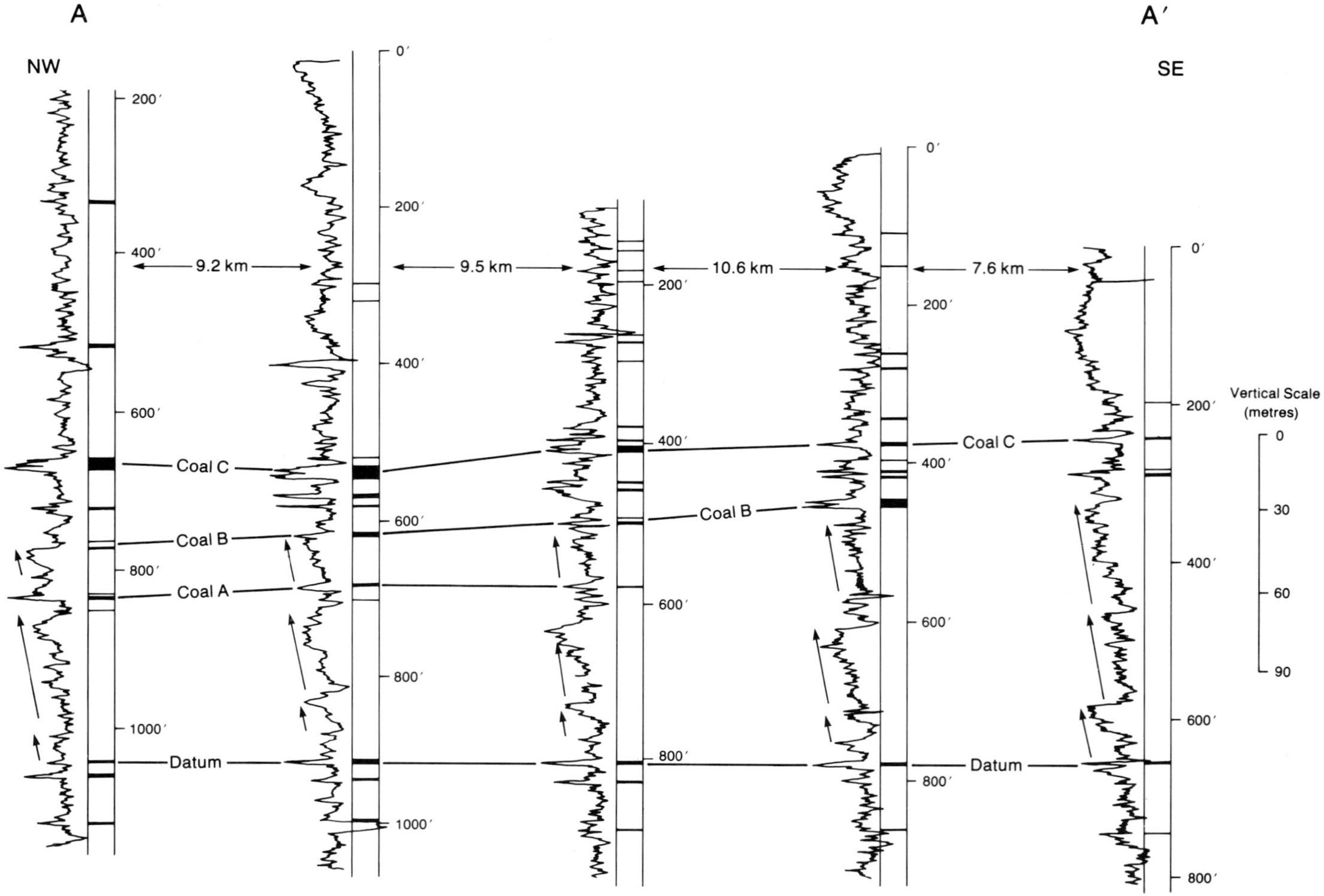

FIG. 5. Cross-section showing gamma logs from wells drilled for coal exploration along the Red Deer River valley, Alberta. Position of coal seams, determined from examination of density and resistivity logs, is shown in each column. Arrows indicate coarsening-upward sequences of the Bearpaw Formation. The thickest coals are present at the base of the overlying Horseshoe Canyon Formation. The boundary between the two formations is diachronous and is not well defined. Line of sections shown in Fig. 6.

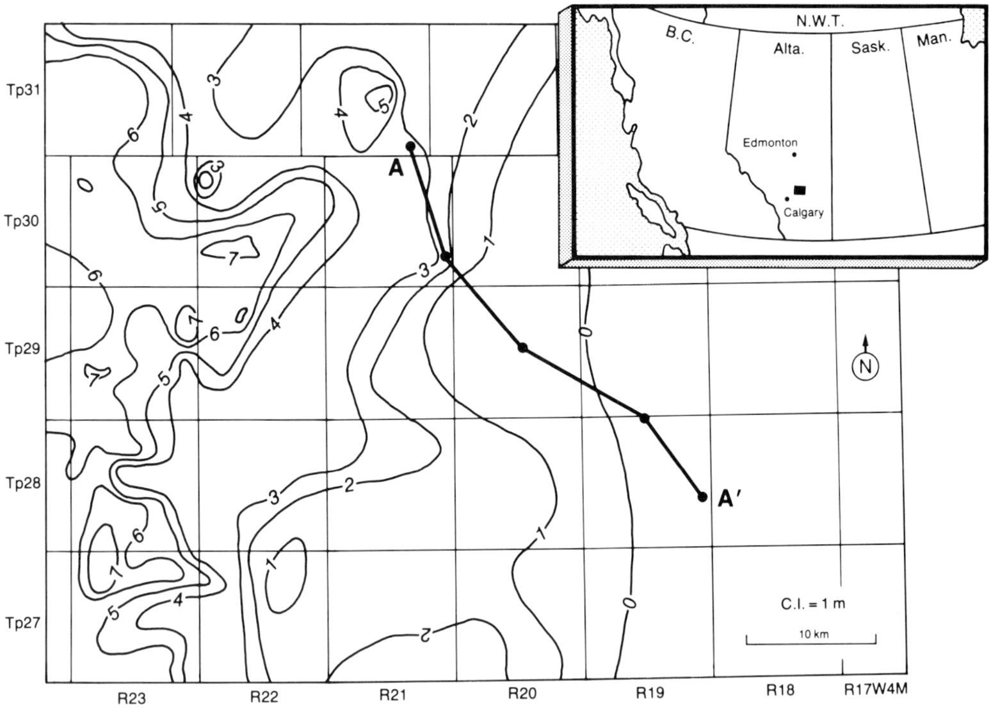

FIG. 6. Isopach of coal within a coal zone (Coal A and adjacent seams in Fig. 5) of the Horseshoe Canyon in part of central Alberta. Contours are in metres. Map is based on approximately five wells per township square in the west and two wells per township in the east. Drumheller is at Tp29, R20.

seams may exceed 3 m in thickness. Examination of core from the well at A′ shows that, although there is no coal, there are plant roots in the sandstones at the top of the second coarsening-upward sequence. However, in exposures in Township 27, Range 17 and further east, there are no roots and the highest sandstone of the sequence appears to be upper shoreface deposits, with hummocky cross-stratification and trough cross-beds (the 'nearshore bar facies' of Rahmani's (1983) section EC3). The regressive episode that formed the second coarsening-upward sequence, therefore, established terrestrial conditions only as far east as Range 18. The position of the thick coals (Fig. 6) suggests that mire conditions were favourable for the accumulation of thick peats 40–60 km landward of the shoreline at its maximum extent.

In the Campanian, coals are thinner but have been locally mined. The sedimentary facies are very similar to that of the Maastrichtian. The terrestrial Judith River Formation is underlain and interfingers with the marine Pakowki Formation in similar fashion to the relationship between the Horseshoe Canyon and Bearpaw Formations. Figure 7 is an isopach of the Pakowki Formation in the Coronation/Hanna region of Alberta. The thickness is fairly uniform in the western and easternmost parts of the map, but increases rapidly around Range 5. West–east cross-sections through the eastern half of the area resemble that of Fig. 5, but show the pinchout of several coarsening-upward sequences toward the west. For a substantial period of time, alternate transgressions and regressions moved the coastline back and forth within only a narrow band. From Fig. 7 the palaeoshoreline can be determined to have been oriented north–south, approximately at Range 5. Figure 8 is an isopach of the cumulative coal of the McKay coal zone (usually one or two seams). The coals lie at the base of the Judith River Formation and are significantly older than the younger sediments of the Pakowki Formation in the east. The coal accumulated, therefore, at the same time as marine strata was deposited in the eastern parts of the area. By comparing the position of the coals and the area of the palaeoshoreline, the thickest peats can be shown to have accumulated in mires between 40 and 80 km inland.

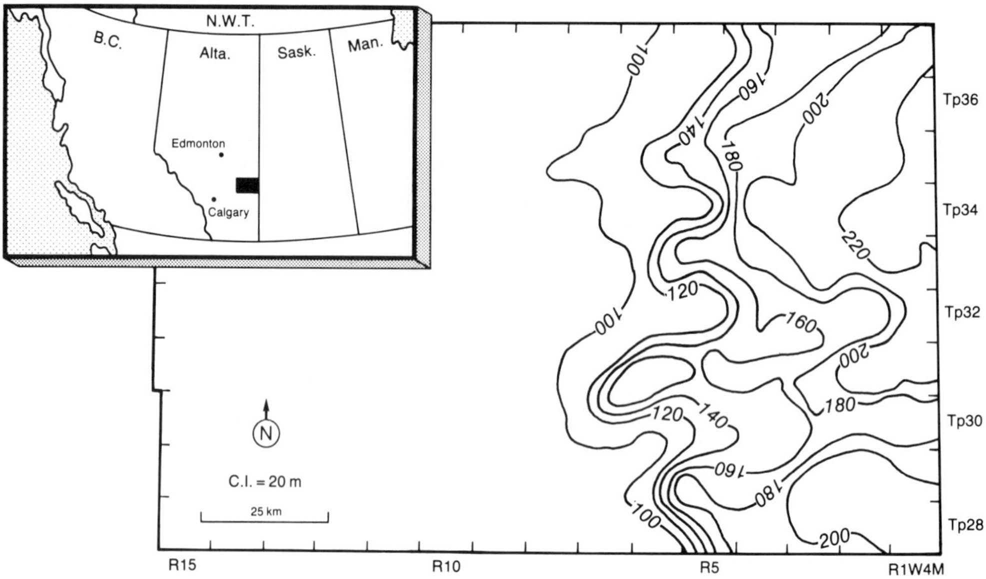

FIG. 7. Isopach of the Pakowki Formation in part of southern Alberta. Map is based on over 100 data points. See text for discussion.

A modern analogue to the Upper Cretaceous coal of Alberta may be provided by the Okefenokee Swamp of Georgia. The Okefenokee Swamp is a low-lying mire in which peats generally have less than 15% ash (Cohen *et al.* 1984). The mire lies 70–100 km inland from the present-day Atlantic shoreline and is not influenced by coastal processes. Although there is an inflow of water into the mire from surrounding areas, it is well removed from river systems with clastic sediment. Major rivers in the area flow out of the mire. As discussed earlier, even though the Okefenokee Swamp is well inland, its peats sit directly on Pleistocene coastal sediment (Cohen 1974, 1984).

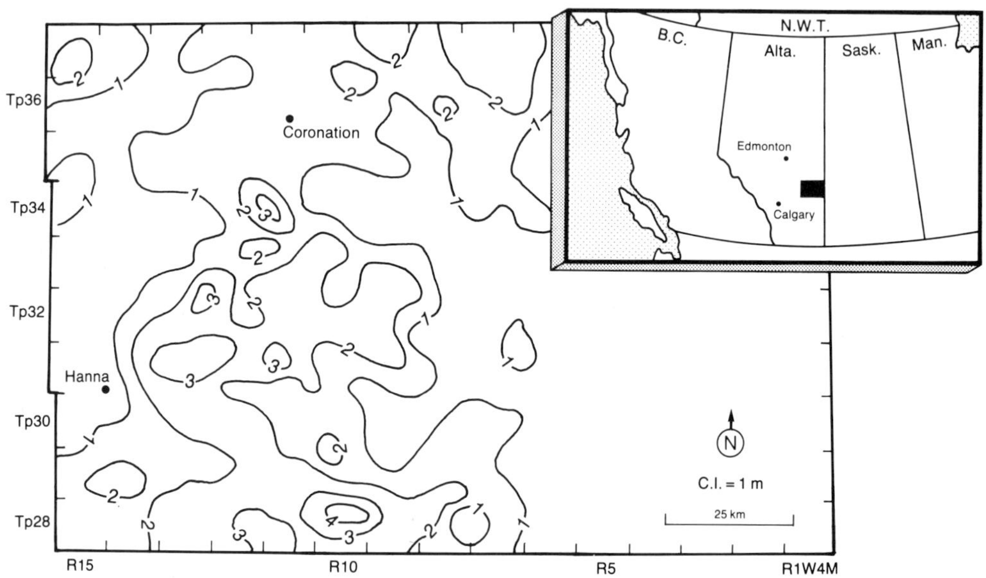

FIG. 8. Isopach of cumulative coal in the McKay coal zone at the base of the Judith River Formation. Map is based on over 100 data points. See text for discussion.

There may have been a similarly significant hiatus in deposition between the Cretaceous coal and underlying sediment. The subsurface studies on the Alberta coals are important because they demonstrate that it is not necessary to invoke raised mires to explain the juxtaposition of coastal sediments and coal. Indeed the paucity of coal close to the shoreline may well imply that the mires were of the low-lying type.

Conclusions

Facies studies are of importance from both an academic understanding of coal deposits and from an economic viewpoint for coal exploration and development. Although great advances have been made in the past, there is no doubt that studies will become significantly more sophisticated in the future. The following approaches are suggested to help solve some of the major questions in the discipline and to ensure that future facies models have greater economic applicability.

(a) More detailed facies studies need to be carried out. Published descriptions should facilitate comparisons between formations.

(b) Cooperative work between scientists of different disciplines is essential.

(c) The relationships between modern mires and active clastic depositional systems need to be thoroughly investigated.

(d) An understanding of the geometry of coals and coal-bearing strata, through subsurface studies, may prove helpful in determining the nature of ancient mires.

ACKNOWLEDGEMENTS: I wish to thank other members of the Coal Geology Group at the Alberta Research Council for use of information in this paper. I also thank Andrew Scott for his patience while I completed the paper. Alberta Research Council's Graphics Department drafted the figures and Maureen Fitz-Gerald typed the manuscript.

References

ANDERSON, J. A. R. 1964. The structure and development of the peat swamps of Sarawak and Brunei. *Journal of Tropical Geography*, **18**, 7–16.

—— 1983. The tropical peat swamps of western Malesia. *In:* GORE, A. J. P. (ed.) *Mires: Swamp, Bog, Fen and Moor*. Ecosystems of the World. Vol. 4B, *Regional Studies*, pp. 181–199. Elsevier, Amsterdam.

BAGANZ, B. P., HORNE, J. C. & FERM, J. C. 1975. Carboniferous and recent Mississippi lower delta plains; a comparison. *Transactions of the Gulf Coast Association of Geological Societies*, **25**, 183–191.

BARBER, K. E. 1981. *Peat Stratigraphy and Climatic Change*. Balkema, Rotterdam.

BREYER, J. A. & MCCABE, P. J. 1986. Coals associated with tidal sediments in the Wilcox Group (Paleogene), south Texas. *Journal of Sedimentary Petrology*, **56**, 510–519.

CAMERON, A. R. 1971. Some petrological aspects of the Harbour coal seam, Sydney Coalfield, Nova Scotia. *Bulletin of the Geological Survey of Canada*, 175.

COHEN, A. D. 1973. The petrology of some Holocene peat sediments from the Okefenokee Swamp-Marsh complex of southern Georgia. *Geological Society of American Bulletin*, **84**, 3867–3878.

—— 1974. Petrography and paleoecology of Holocene peats from the Okefenokee swamp-marsh complex of Georgia. *Journal of Sedimentary Petrology*, **44**, 716–726.

—— 1984. The Okefenokee Swamp: a low sulfur end-member of a shoreline-related depositional model for coastal plain coals. *In:* RAHMANI, R. A. & FLORES, R. M. (eds) *Sedimentology of Coal and Coal-bearing Sequences*, pp. 231–240. Special Publication of the International Association of Sedimentologists 7. Blackwell Scientific Publications, Oxford.

——, ANDREJKO, M. J., SPACKMAN, W. & CORVINUS, D., 1984. Peat deposits of the Okefenokee Swamp. *In:* COHEN, A. D., CASAGRANDE, D. J., ANDREJKO, M. J. & BEST, G. R. (eds) *The Okefenokee Swamp: its Natural History, Geology, and Geochemistry*, pp. 493–553. Wetland Surveys, Los Alamos.

—— & SPACKMAN, W. 1972. Methods in peat petrology and their application to reconstruction of paleoenvironments. *Geological Society of America Bulletin*, **83**, 129–142.

—— & —— 1977. Phytogenic organic sediments and sedimentary environments in the Everglades-Mangrove Complex. Part II. The origin, description and classification of the peats of southern Florida. *Palaeontographica B*, **162**, 71–114.

COLEMAN, J. M., GAGLIANO, S. M. & SMITH, W. G. 1970. Sedimentation in a Malaysian high tide tropical delta. *In:* MORGAN, J. P. (ed.) *Deltaic Sedimentation, Modern and Ancient*, pp. 185–197. Special Publication of the Society of Economic Paleontologists and Mineralogists 15.

DUFF, P. MCL. D. & WALTON, E. K. 1962. Statistical basis for cyclothems. A quantitative study of the sedimentary succession in the East Pennine Coalfield. *Sedimentology*, **1**, 235–255.

DUTCHER, R. R. 1978. *Field description of coal*. American Society for Testing Materials, Special Technical Publication 661.

ELLIOTT, R. E. 1965. Swilleys in the Coal Measures of Nottinghamshire interpreted as palaeo-river courses. *Mercian Geologist*, **2**, 351–371.

ETHRIDGE, F. G. & FLORES, R. M. (eds) 1981. *Recent and ancient nonmarine depositional environments:*

models for exploration. Society of Economic Paleontologists and Mineralogists, Special Publication 31.

FERM, J. C. 1974. Carboniferous environmental models in eastern United States and their significance. *In:* BRIGGS, G. (ed.) *Carboniferous of the Southeastern United States*, pp. 79–95. Geological Society of America Special Paper 148.

—— & HORNE, J. C. 1979. *Carboniferous depositional environments in the Appalachian region*. Carolina Coal Group, Department of Geology, University of South Carolina, Columbia.

—— & STAUB, J. R. 1984. Depositional controls on mineable coal bodies. *In:* RAHMANI, R. A. & FLORES, R. M. (eds) *Sedimentology of Coal and Coal-bearing Sequences*, pp. 275–289. Special Publication of the International Association of Sedimentologists 7. Blackwell Scientific Publications, Oxford.

FIELDING, C. R. 1984. Upper delta plain lacustrine and fluviolacustrine facies from the Westphalian of the Durham coalfield, NE England. *Sedimentology*, **31**, 547–567.

—— 1986. Fluvial channel and overbank deposits from the Westphalian of the Durham coalfield, NE England. *Sedimentology*, **33**, 119–140.

FISK, H. N. 1960. Recent Mississippi River sedimentation and peat accumulation. *Compte rendu, 4 Congrès l'avancement des études de stratigraphie et de géologie du Carbonifère, Heerlen 1958*, **1**, 187–199.

FRAZIER, D. E. & OSANIK, A. 1969. Recent peat deposits—Louisiana coastal plain. *In:* DAPPLES, E. C. & HOPKINS, M. E. (eds) *Environments of Coal Deposition*, pp. 63–85. Geological Society of America Special Paper 114.

GALLOWAY, W. E. & HOBDAY, D. K. 1983. *Terrigenous Clastic Depositional Systems*. Springer-Verlag, New York.

GERSIB, G. A. & MCCABE, P. J. 1981. Continental coal-bearing sediments of the Port Hood Formation (Carboniferous), Cape Linzee, Nova Scotia, Canada. *In:* ETHRIDGE, F. G. & FLORES, R. M. (eds) *Recent and Ancient Nonmarine Depositional Environments: Models for Exploration*, pp. 95–108. Society of Economic Paleontologists and Mineralogists, Special Publication 31.

GORE, A. J. P. 1983. Introduction. *In:* GORE, A. J. P. (ed.) *Ecosystems of the World. Vol. 4A, Mires: Swamp, Bog, Fen and Moor, General Studies*, pp. 1–34. Elsevier, Amsterdam.

GUION, P. D. 1984. Crevasse splay deposits and roof-rock quality in the Threequarters Seam (Carboniferous) in the East Midlands Coalfield, U.K. *In:* RAHMANI, R. A. & FLORES, R. M. (eds) *Sedimentology of Coal and Coal-bearing Strata*, pp. 291–308. Special Publication of the International Association of Sedimentologists 7. Blackwell Scientific Publications, Oxford.

HACQUEBARD, P. A. & DONALDSON, J. R. 1969. Carboniferous coal deposition association with flood-plain and limnic environments in Nova Scotia. *In:* DAPPLES, E. C. & HOPKINS, M. E. (eds) *Environments of Coal Deposition*, pp. 143–191. Geological Society of America Special Paper 114.

HARVEY, R. D. & DILLON, J. W. 1985. Maceral distributions in Illinois coals and their paleoenvironmental implications. *International Journal of Coal Geology*, **5**, 141–165.

HASZELDINE, R. S. & ANDERTON, R. 1980. A braidplain facies model for the Westphalian B Coal Measures of northeast England. *Nature*, **284**, 51–53.

HEWARD, A. P. 1978. Alluvial fan and lacustrine sediments from the Stephanian A and B (La Magdalena, Cinera-Matallana and Sabero) coalfields, northern Spain. *Sedimentology*, **25**, 451–488.

HORNE, J. C., FERM, J. C., CARUCCIO, F. T. & BAGANZ, B. P. 1978. Depositional models in coal exploration and mine planning in Appalachian region. *American Association of Petroleum Geologists Bulletin*, **62**, 2379–2411.

HUDSON, R. G. S. 1924. On the rhythmic sucession of the Yoredale Series in Wensleydale. *Proceedings of the Yorkshire Geological Society*, **20**, 125–135.

HUNT, J. W. & HOBDAY. D. K. 1984. Petrographic composition and sulphur content of coals associated with alluvial fans in the Permian Sydney and Gunnedah Basins, eastern Australia. *In:* RAHMANI, R. A. & FLORES, R. M. (eds). *Sedimentology of Coal and Coal-bearing Sequences*, pp. 43–60. Special Publication of the International Association of Sedimentologists 7. Blackwell Scientific Publications, Oxford.

KOSTERS, E. C. 1983. *Louisiana peat resources*. Technical Report, Louisiana Geological Survey, DOE/FE/05113.

—— & BAILEY, A. 1983. Characteristics of peat deposits in the Mississippi River deltaic plain. *Transactions of the Gulf Coast Association of Geological Societies*, **33**, 311–325.

—— & —— 1986. *A reassessment of Louisiana peat resources based on leaching experiments*. Coastal Geology Technical Report No. 2, Louisiana Geological Survey.

KRAUSSE, H.-F., DAMBERGER, H. H., NELSON, W. J. *et al.* 1979. *Engineering study of structural geologic features of the Herrin (No. 6) Coal and associated rock in Illinois, Volume 2—detailed report*. Final report, United States Department of the Interior, Bureau of Mines, Contract No. H0242017.

LEVEY, R. A. 1985. Depositional model for understanding geometry of Cretaceous coals: major coal seams, Rock Springs Formation, Green River Basin, Wyoming. *American Association of Petroleum Geologists Bulletin*, **69**, 1359–1380.

MACDONALD, D. E., MCCABE, P. J. & BOSMAN, A. (in press). *An evaluation of the coal resources of the Belly River Group, to a depth of 400 m in the Alberta plains*. Open file report, Alberta Research Council, Edmonton.

MCCABE, P. J. 1984. Depositional environments of coal and coal-bearing strata. *In:* RAHMANI, R. A. & FLORES, R. M. (eds) *Sedimentology of Coal and Coal-bearing Sequences*, pp. 13–42. Special Publication of the International Association of Sedimentologists 7. Blackwell Scientific Publications, Oxford.

—— (in press). Geology of coal: environments of deposition. *In:* GLUSKOTER, H. J., RICE, D. D. & TAYLOR, R. B. (eds) *Economic Geology, U.S. The Geology of North America*, P-2. Geological Society of America.

——, STROBL, R. S., MACDONALD, D. E., NURKOWSKI, J. R. & BOSMAN, A. (in press). *An evaluation of the coal resources of the Horseshoe Canyon Formation and laterally equivalent strata, to a depth of 400 m, in the Alberta plains area.* Open file report, Alberta Research Council, Edmonton.

MERRIAM, D. F. (ed.) 1964. Symposium on cyclic sedimentation. *Bulletin of the Kansas Geological Survey*, 169, Vols 1 & 2.

MIALL, A. D. 1978. Lithofacies types and vertical profile models in braided river deposits: a summary. *In:* MIALL, A. D. (ed.) *Fluvial Sedimentology*, pp. 597–604. Canadian Society of Petroleum Geologists, Memoir 5.

MOORE, D 1959. Role of deltas in the formation of some British Lower Carboniferous cyclothems. *Journal of Geology*, **67**, 522–539.

MOORE, P. D. 1984. *European Mires.* Academic Press, London.

PADGETT, G. & EHRLICH, R. 1978. An analysis of two tectonically controlled integrated drainage nets of mid-Carboniferous age in southern West Virginia. *In:* MIALL, A. D. (ed.) *Fluvial Sedimentology*, pp. 789–799. Canadian Society of Petroleum Geologists, Memoir 5.

PHILLIPS, J. 1836. *Illustrations of the Geology of Yorkshire, Part II. The Mountain Limestone District.* London.

PHILLIPS, T. L., PEPPERS, R. A. & DIMICHELE, W. A. 1985. Stratigraphic and interregional changes in Pennsylvanian coal-swamp vegetation: environmental inferences. *International Journal of Coal Geology*, **5**, 43–109.

RAHMANI, R. A. 1983. *Facies relationships and paleoenvironments of a late Cretaceous tide-dominated delta, Drumheller, Alberta.* The Mesozoic of middle North America, Field trip guidebook 2, Canadian Society of Petroleum Geologists, Calgary.

—— & FLORES, R. M. (eds) 1984. *Sedimentology of Coal and Coal-bearing Sequences.* Special Publication of the International Association of Sedimentologists 7. Blackwell Scientific Publications, Oxford.

READ, W. A. & DEAN, J. M. 1967. A quantitative study of a sequence of coal-bearing cycles in the Namurian of central Scotland, 1. *Sedimentology*, **9**, 137–156.

—— & —— 1968. A quantitative study of a sequence of coal-bearing cycles in the Namurian of central Scotland, 2. *Sedimentology*, **10**, 121–136.

READING, H. G. (ed.) 1986. *Sedimentary Environments and Facies.* 2nd edn. Blackwell Scientific Publications, Oxford.

REINECK, H.-E. & SINGH, I. B. 1973. *Depositional Sedimentary Environments.* Springer-Verlag, Berlin.

RICHARDSON, R. J. H. 1985. Quaternary geology of the North Kaipara Barrier, Northland, New Zealand. *New Zealand Journal of Geology and Geophysics*, **28**, 111–127.

ROMANOV, V. V. 1968. *Hydrophysics of Bogs.* Israel Program for Scientific Translations, Jerusalem.

RYER, T. A. 1981. Deltaic coals of Ferron Sandstone Member of Mancos Shale: predictive model for Cretaceous coal-bearing strata of Western Interior. *American Association of Petroleum Geologists Bulletin*, **65**, 2323–2340.

—— & LANGER, A. W. 1980. Thickness change involved in the peat-to-coal transformation for bituminous coal of Cretaceous age in central Utah. *Journal of Sedimentary Petrology*, **50**, 987–992.

SCHOPF, J. M. 1960. Field description and sampling of coal beds. *U.S. Geological Survey Bulletin* 1111-B.

SCOTT, A. C. 1978. Sedimentological and ecological control of Westphalian B plant assemblages from west Yorkshire. *Proceedings of the Yorkshire Geological Society*, **41**, 461–508.

SMITH, A. H. V. 1962. The palaeoecology of Carboniferous peats based on the miospores and petrography of bituminous coals. *Proceedings of Yorkshire Geological Society*, **33**, 423–474.

—— 1968. Seam profiles and seam characters. *In:* MURCHISON, D. G. & WESTOLL, T. S. (eds) *Coal and Coal-bearing Strata*, 31–40. Oliver & Boyd, Edinburgh.

SNEDDEN, J. W. & KERSEY, D. G. 1981. Origin of San Miguel lignite deposit and associated lithofacies, Jackson Group, south Texas. *American Association of Petroleum Geologists Bulletin*, **65**, 1099–1109.

SPACKMAN, W., COHEN, A. D., GIVEN, P. H. & CASAGRANDE, D. J. 1976. *The comparative study of the Okefenokee Swamp and the Everglades-mangrove swamp-marsh complex of southern Florida.* A short course presentation of the Pennsylvania State University.

SPACKMAN, W., DOLSEN, C. P. & REIGEL, W. 1966. Phytogenic organic sediments and sedimentary environments in the Everglades-mangrove complex, Part I: evidence of a transgressing sea and its effects on environments of the Shark River area of southwestern Florida. *Palaeontographica B*, **117**, 135–152.

STAUB, J. R. & COHEN, A. D. 1979. The Snuggedy Swamp of South Carolina: a back barrier estuarine coal forming environment. *Journal of Sedimentary Petrology*, **48**, 203–210.

STYAN, W. B. & BUSTIN, R. M. 1983a. Petrography of some Fraser River Delta peat deposits: coal maceral and microlithotype percursors in temperate-climate peats. *International Journal of Coal Geology*, **2**, 321–370.

—— & —— 1983b. Sedimentology of Fraser River Delta peat deposits: a modern analogue for some deltaic coals. *International Journal of Coal Geology*, **3**, 101–143.

TAYLOR, J. A. 1983. The peatlands of Great Britain and Ireland. *In:* GORE, A. J. P. (ed.) Ecosystems of the World, Vol. 4B. Mires: *Swamp, Bog, Fen and Moor, Regional Studies*, pp. 1–46. Elsevier, Amsterdam.

TEICHMÜLLER, M. 1962. Die Genese der Kohle. *Compte Rendu Quatrième Congrès International de Stratigraphie et de Géologie du Carbonifère, Heerlen, 1958*, **3**, 699–722.

TING, F. T. C. & SPACKMAN, W. 1975. The coal lithotype concept and seam profile. *Compte Rendu Septième Congrès International de Stratigraphie et de Géologie du Carbonifère, Krefeld, 1971*, **4**, 307–311.

TREWORGY, C. G. & JACOBSON, R. J. 1985. Paleoenvironments and distribution of low-sulfur coal in Illinois. *Compte Rendu, Neuvième Congrès International de Stratigraphie et de Géologie du Carbonifère, Washington and Champaign-Urbana, 1979*, **4**, 349–359.

TRUEMAN, A. E. 1946. Stratigraphical problems in the coal measures of Europe and North America. *Quarterly Journal of the Geological Society of London*, **102**, 49–92.

UDDEN, J. A. 1912. Geology and mineral resources of the Peoria quadrangle, Illinois. *U.S. Geological Survey Bulletin* 506.

WALKER, R. G. 1984. General introduction: facies, facies sequences and facies models. *In:* WALKER, R. G. (ed.) *Facies Models*, pp. 1–9. Geoscience Canada Reprint Series 1, Geological Association of Canada.

WANLESS, H. R. & WELLER, J. M. 1932. Correlation and extent of Pennsylvanian cyclothems. *Geological Society of America Bulletin*, **43**, 1003–1016.

WARD, C. R. 1984. *Coal Geology and Coal Technology*. Blackwell Scientific Publications, Oxford.

WEISENFLUH, G. A. & FERM, J. C. 1984. Geologic controls on deposition of the Pratt seam, Black Warrior Basin, Alabama, U.S.A. *In:* RAHMANI, R. A. & FLORES, R. M. (eds) *Sedimentology of Coal and Coal-bearing Sequences*, pp. 317–330. Special Publication of the International Association of Sedimentologists 7. Blackwell Scientific Publications, Oxford.

WELLER, J. M. 1930. Cyclic sedimentation in the Pennsylvanian and its significance. *Journal of Geology*, **38**, 97–135.

WESTOLL, T. S. 1968. Sedimentary rhythms in coal-bearing strata. *In:* MURCHISON, D. & WESTOLL, T. S. (eds) *Coal and Coal-bearing Strata*, pp. 71–103. Oliver & Boyd, Edinburgh.

WILFORD, G. E. 1961. *The geology and mineral resources of Brunei and adjacent parts of Sarawak with descriptions of Seria and Miri oilfields*. Geological Survey Department, British Territories in Borneo, Memoir 10.

WILKS, P. J. 1979. Mid-Holocene sea-level and sedimentation interactions in the Dovey Estuary area, Wales. *Palaeogeography, Palaeoclimatology, Palaeoecology*, **26**, 17–26.

ZELLER, E. J. 1964. Cycles and psychology. *In:* MERRIAM, D. F. (ed.) Symposium of Cyclic Sedimentation. *Bulletin of the Kansas Geological Survey*, **169**, 631–636.

PETER J. MCCABE, Alberta Research Council, PO Box 8330. Postal Station F, Edmonton, Alberta T6H 5X2, Canada. Present address: Branch of Coal Geology, US Geological Survey, Federal Center M.S. 972, Denver, Colorado 80225, USA.

Implications of vegetational change through the geological record on models for coal-forming environments

Margaret E. Collinson & Andrew C. Scott

ABSTRACT: Two broad categories of peat-forming environments, those dominated by arborescent lycopsids (Carboniferous) and those dominated by taxodiaceous conifers (Cretaceous to Recent) are used to compare some features of plant biology and plant association which would influence peat formation. Significant differences between the two categories indicate that modern analogues for ancient coal-forming environments should always take account of such plant features. Some of the more obvious implications for interpretation of coal-forming environments are discussed. A fuller appreciation of these implications requires greater understanding of the ancient plants and plant communities involved. In particular, both the individual plants and plant assemblages should be taken into account when making environmental interpretations of coal petrological studies and in the calculation of peat to coal compression ratios.

Keywords: peat, coal, Carboniferous, Tertiary, petrology, palaeobotany, palynology, palaeoenvironment, conifers, lycopsids

Modern analogues for ancient peat-forming environments undoubtedly provide many helpful and informative data concerning coal-formation (see Cohen & Spackman 1977, 1980, Cohen 1984, Spackman & Raymond, *this volume*, McCabe, *this volume*). However, they cannot have full worth unless the significant differences between modern and ancient plants and plant communities are incorporated into the relevant models.

A general outline of floras from coals is given in the Appendix. In this paper a comparison is made between Carboniferous peat-forming environments dominated by arborescent lycopsids and those of the Cretaceous to Recent dominated by taxodiaceous conifers. These two categories are themselves broad aggregates (especially the latter) and encompass many variations. Nevertheless, they serve to emphasize some general points concerning the relevance of features of plant biology and plant association which would affect peat formation. These include: nature and timing of leaf fall; arborescent versus herbaceous habit; absence/presence of secondary growth and its nature; spore versus seed habit; nature of rooting system; proportions of lignin and cellulose; presence/absence of open water and marginal aquatic plants; overall community diversity and the interactions between members of the community. Such features will affect exploitation of ground water and influence of water level variations; soil profiles; dispersal, spread and colonization of variable environments and hence stabilization of newly exposed sediments; persistence and re-establishment of peat-forming conditions; accumulation (amount and rate) of organic matter added to the peat; the composition of this organic matter and its degree of compressibility and the relative influence of variations in rainfall, temperature and a variety of other physical factors.

We attempt here, in the context of this volume, to provide some general comments on these aspects of peat-forming situations. Their implications for coal geologists are briefly considered in the interpretation of allochthonous versus autochthonous coals; compression of peat to coal; coal composition; controls on sedimentation and the development and persistence of peat-forming environments. These, often speculative, suggestions should prove provocative, fulfilling our aim of enhancing awareness of the significance of vegetational change amongst those working with all aspects of peat (coal)-forming environments.

Palaeobotanical approaches

In order to achieve a full understanding of the evolution of peat- and coal-forming environments, a thorough knowledge of ancient vegetation is essential. It is necessary to know the nature of the component plants, their functional biology and their association in ancient plant communities.

Component plants

To reconstruct ancient plant communities it is first necessary to reconstruct the plants themselves. Plants are usually represented in the fossil record by a variety of dispersed organs (pollen, spores, foliage, trunks etc.) and painstaking study is required to link these into entire plants. In some cases consistent, recurrent, association or combined occurrence of unique anatomical features may be used as evidence but it is preferable to find an organic connection between organs. Considerable progress has been made recently in this area (e.g. Pigg & Rothwell 1983, Basinger 1984, p. 282, Rothwell & Warner 1984, Wnuk & Pfefferkorn 1984, Crane & Stockey 1985, DiMichele & Phillips 1985, Wnuk 1985). The necessity to understand the whole plant is exemplified by palynological investigation of coals (Smith 1962, Hacquebard & Donaldson 1969, Scott & King 1981, Bartram, *this volume*). Unfortunately, we know few of the parent plants of the various pollen and spores found in coal, thus limiting the exploitation of this abundant evidence in interpretation of vegetational change.

Functional biology

Some insight into plant function can be derived from morphology via the uniformitarian approach. Dispersal biology (affecting spread and colonization) of seeds and spores may be inferred from their ornamentation, size, appendages, etc. (Tiffney 1985, Collinson *et al.* 1985, p. 384, Crane & Stockey 1985, Collinson 1987). Modifications for dry or wet environments may be deduced from cuticles and from internal anatomy (Wolfe & Upchurch 1986, Fahn 1974, Collinson 1987). Secondary growth, which may enable a plant to reach tree stature, is reflected in internal cellular structure (Fahn 1974, Cichan 1985, *in press*). Scars of former attachment may indicate abscission of former branches and leaves (Wnuk 1985) and other data may be used to infer deciduousness (see p. 72). A climbing, scrambling or sprawling habit may be indicated by stem form and also reflected in cellular structure (Fahn 1974, p. 410, Rothwell 1975, Rothwell & Warner 1984, Wnuk & Pfefferkorn 1984). Permineralized plant material provides the greatest detail of plant cellular structure. There are, therefore, examples of exceptional levels of information where permineralized material is abundant and has been studied in detail. The most obvious of these is the Upper Carboniferous of Euramerica where coal balls (peat permineralized by calcium carbonate) have long been studied (e.g. Phillips 1981, DiMichele & Phillips 1985, Scott & Rex 1985).

Plant communities

Once plants have been reconstructed it is necessary to interpret their distribution in ancient plant communities. Frequently, in older literature, ancient plant communities were reconstructed merely by incorporating examples of all known plants in an artistically pleasing illustration. Also, in the younger part of the geological column, reliance has been placed on ecological tolerances of nearest living relatives to deduce communities within ancient floras. More recently these approaches have been superseded by analyses of fossil occurrence in a sedimentological context combined with evidence from taphonomic studies of modern communities. As a result of this some relatively well-founded reconstructions of plant communities from ancient peat-forming environments have been made. In view of the abundance of plant compression fossils in the world's coal measures it is perhaps understandable that these have often been seen as representing the coal-forming plants whose remains have also been incorporated into associated clastic sediments (e.g. Gould 1981), but it is clear from studies involving floras from a variety of sediments that the different depositional environments preserve floras from different original communities (Scott 1977, 1978, 1979, Phillips 1981). This conclusion is also supported by taphonomic studies (Scheihing & Pfefferkorn 1984, Burnham 1985, Ferguson 1985, Gastaldo *et al.* 1985). Combining sedimentological and taphonomic evidence enables the elements of the peat-forming communities to be distinguished from those of associated vegetation (Scott 1977, 1978, 1979, Phillips 1981). The assumption that a roof shale flora may represent an example of an original peat-forming plant community is actually unjustified (Scott & Collinson 1983, fig. 22). Just as permineralized peats provide much data on plant function, they also provide evidence of peat-forming communities. Crystallization of carbonate, silica or pyrite early in the diagenetic history of the peat may prevent further decomposition and compression, structurally preserving much of the component plant material. Quantitative studies may, therefore, be applied to the assessment of changing vegetation through a single coal seam (Phillips & Peppers 1984). A great amount of detail concerning ancient plant communities, their ecology, the biomass of particular constituents and palaeoclimatic implications, may be built up (Phillips 1979, DiMichele & Phillips 1985, Phillips *et al.* 1985, Scott & Rex 1985). It is, however, important to realize that permineralized peats may not represent all elements of the coal-forming vegetation as the conditions for their formation may selectively preserve certain material (Scott & Rex 1985, see also Bartram, *this volume*). Despite some successes in this work, there remains the overall problem of representation of ancient vegetation in the fossil record. It would be naive to suggest that any ancient plant community will ever be known in complete detail. Undoubtedly certain ancient plants will never be fully reconstructed, perhaps being known only from isolated spores, while others may leave no fossil record. In the context of this paper it is fortunate that wetland communities are amongst those which suffer the least taphonomic bias in their fossil record (Scheihing & Pfefferkorn 1984, Ferguson 1985).

Significant features of peat-forming plants and plant communities

Features of plants which would influence peat formation may be grouped into four major areas of plant form and function: anchoring systems, reproductive biology, leaf and shoot biology and the detailed structure of woody axes. Overall diversity within each of these categories must also be considered along with some aspects of microhabitat occupation. In the following section a comparative assessment of these features is made for two broad categories of peat-forming environment, those dominated by arborescent lycopsids (Carboniferous) and those dominated by taxodiaceous conifers (Cretaceous to Recent). These two categories have been chosen as they are the two periods of coal formation about which most is known concerning the plants and plant communities involved and are the most important as regards world coal reserves. Each category itself includes considerable variation so that the following text must be considered merely as a general overview. It is not our intention to imply that all coals dominated by taxodiaceous conifers formed from the Cretaceous to Recent are the same; they merely show a gross similarity and collectively show striking differences from those of the Upper Carboniferous. The two categories chosen may be placed into the overall context of variation of floras from coals through time by consulting the Appendix.

Anchoring systems

Anchoring systems of the arborescent lycopsids had a horizontally spreading form. They consisted of large dichotomizing stigmarian axes and smaller, spirally arranged, rootlets which also dichotomized (Frankenberg & Eggert 1969, Stewart 1983). On stump casts the larger axes have been seen to extend some 2–4 m laterally in a more or less horizontal plane (Frankenberg & Eggert 1969, pl. 1, Gastaldo 1985, 1986, Wnuk 1985), but a complete anchoring system has never been recorded. The axes are usually seen compressed in very fine grained shales (e.g. Frankenberg & Eggert 1969, pl. 1). W. G. Chaloner (personal communication, 1986) has commented that this might well give a false impression of their horizontal attitude due to the predominance of vertical compaction of the original muds. Wnuk (1985) and Gastaldo (1985, 1986) describe the stump casts from a variety of lithologies ranging from claystone to mudstones and siltstones. Gastaldo (1986) observed that in most cases the stigmarian axes were rather shallowly horizontally spreading and that higher placed systems did not penetrate those below. Wnuk (1985) invoked a scene where 'the swamp forest was criss-crossed over, and just below, its surface by an extensive network of stigmarian roots'. The overall impression, therefore, is one of a very shallowly spreading, horizontal anchoring system. Detailed sedimentological evidence from *in situ* specimens is needed in order to substantiate this interpretation.

Rooting systems of all modern wetland plants are generally shallow due to the anaerobic nature of the rooting substrate (Hook & Crawford 1978, Kozlowski 1984). Gross morphological details of entire rooting systems are scarce in the literature. Rather more seems to be known about the stigmarian systems that about rooting systems of modern or fossil taxodiaceous conifers. Fowler *et al.* (1973) provided a detailed account of *in situ* taxodiaceous stumps from the Eocene of southern England. These had a horizontally spreading rooting system extending for at least 3 m (the apices were not found). The laterally spreading roots formed interwoven mats at certain horizons. However, at both sites where they occurred, the stumps included downward penetrating roots, extending at least 1 m into the strata below. Furthermore, at one site, two levels of stumps are preserved and the rooting systems overlapped. Harrar & Harrar (1962) and Stone & Vasey (1968) both refer to horizontally spreading shallow root systems in modern Taxodiaceae but an illustration (Henry & McIntyre 1926), albeit of specimens planted away from their natural environment, shows a downward penetrating habit.

Hook (1984) and Stone & Vasey (1968) emphasize the importance of adventitious roots in modern Taxodiaceae, both in re-establishing an actively absorbing root system after inundation and in trapping organic debris around the stumps thus raising substrate level. Adventitious roots are not known for arborescent lycopsids (Rothwell & Erwin 1985).

Whereas certain reconstructed fossil taxodiaceous conifers combine features now segregated in several modern genera (e.g. Fowler *et al.* 1973) others show striking similarity to a single modern species (Basinger 1984). Pollen and seed cones, foliage and wood are all well known but details of roots are scarce and of root apices totally lacking. There seems no reason to suppose that roots of the ancient taxodiaceous conifers should have differed fundamentally from those of the family today when all of the other organs are so very similar. If this is indeed the case, then the apical system and the details of the root branching differed strikingly from that of the stigmarian axis. Modern coniferous apices are small compared to the massive structure (up to 1.3 cm across the truncate tip) recently described for *Stigmaria* (Rothwell 1984). Furthermore, the branching produced by the coniferous system includes a complex series of lateral roots with a continuum of root and rootlet size compared with the 'large' and 'small' organs of the *Stigmaria*. In combination these differences would influence the range of substrates penetrated by the rooting systems, with limitations imposed on the stigmarian axes in clastic substrates.

The underground portions of the herbaceous lycopsids associated with the arborescent forms in the Carboniferous coal-forming environments (DiMichele & Phillips 1985, Phillips *et al.* 1985) are poorly known. Some were probably cormose with numerous fine rootlets (Pigg & Taylor 1985), others may have been rhizomatous, like some of their living relatives (see also Marques-Toigo & Correa da Silva 1984).

The rooting systems of other associates of the arborescent lycopsids, the tree ferns and pteridosperms, are also poorly known. On the basis of their aerial portions they may well have been similar to those of modern tree ferns, i.e. with a mantle of small adventitious roots produced by the stem (see, e.g. Wnuk & Pfefferkorn 1984). Cordaite root systems have recently been described by Costanza (1985). These plants were subdominants at certain periods of Carboniferous coal-formation (DiMichele *et al.* 1985, Phillips *et al.* 1985). Their roots were also adventitious but resembled the stilt rooting systems of some modern mangrove flowering plants (Costanza 1985).

In the peat-forming environments dominated by taxodiaceous conifers the associated plants included ferns and tree ferns and, by the late Cretaceous, an increasing diversity of flowering plant trees, shrubs and herbs (see references in the Appendix). Direct evidence concerning their rooting systems is almost entirely lacking although forms showing anatomical modifications for a wetland existence have been described (summarized by Collinson 1987). In view of these, and of the range of fossil flowering plant organs (fruits, seeds, leaves, pollen, flowers, wood) similar to those of modern angiosperms by at least the early Tertiary (see p. 71), it seems reasonable to assume that rooting strategies included a range similar to that of the present-day flowering plants.

Peat-forming environments dominated by arborescent lycopsids thus differed from those dominated by taxodiaceous conifers in several aspects of their anchoring systems. Trees anchored by stigmarian axes were less able to exploit clastic substrates and constraints on their branching may have resulted in a less interwoven anchoring network; the anchoring system was shallowly horizontal, lacking vertically descending portions. The plant association along with the aborescent lycopsids was of low diversity (DiMichele & Phillips 1985) even allowing for gaps in the fossil record. There were apparently few herbs and anchoring systems of the tree ferns, pteridosperms and cordiates differed fundamentally from those likely to have existed on the angiosperm sub-dominants within the taxodiaceous conifer peat-forming environments. The latter also included a far more diverse flora. It is probable that the sediment stabilizing capacity of the arborescent lycopsid communities was limited in comparison to that of the taxodiaceous communities. The ability of the former to tolerate clastic inundation, to spread into adjoining non-peat-forming environments and to tolerate fluctuation in the level of the water table was similarly reduced. The stability of the component plants and resultant soil profiles would also have differed between the two categories of peat-forming environments.

Reproductive biology

Arborescent lycopsids reproduced using the heterosporous strategy. This requires free water for completion of the life cycle and relies on a vulnerable independent gametophyte generation. Some protection and food reserve are available to assist sporophyte establishment, especially in *Lepidocarpon* (see Phillips 1979, Stewart 1983) which, according to DiMichele & Phillips (1985), is one of the commonest and most uniformly distributed of all Carboniferous 'coal swamp' fossils. Nevertheless, these authors suggest that the requirement for water mediated fertilization would have eliminated all non-apomictic heterosporous forms from non-aquatic environments because of the crucial nature of the timing of spore release with favourable conditions (see also Phillips 1979). In addition, even in *Lepidocarpon*, the embryo is not fully protected in the manner found in the seed-bearing strategy.

Vegetative propogation could circumvent the critical timing problem of the heterosporous strategy. However, there is no evidence that any arborescent lycopsids were able to spread in this manner (Phillips 1979, DiMichele & Phillips 1985, Rothwell & Erwin 1985). Homosporous plants such as ferns, sphenopsids and herbaceous lycopsids have an even more vulnerable sexual reproductive strategy with little or no food reserve or protection to assist sporophyte establishment. Some of these may well have been rhizomatous such as the sphenopsids (DiMichele *et al.* 1985). Tiffney & Niklas (1985) refer to rhizomatous growth and adventitious buds in coenopterid ferns. However, neither of these groups were significant components of the Carboniferous peat-forming vegetation (DiMichele & Phillips 1985, DiMichele *et al.* 1985). Adventitious buds have been recorded in cordaites (Constanza 1985) though their role in the propagation of this group is unclear. Apomixis may well have been involved in enabling some heterosporous arborescent lycopsids, such as *Sigillaria*, to spread into drier areas (DiMichele & Phillips 1985).

The ability of spore-bearing plants to persist as dominants of drier environments is limited to those with effective perennating organs (e.g. rhizomes). The short spore longevity and the absence of embryo protection or a period of dormancy make survival as propagules through unfavourable period improbable. Recolonization, following loss of the peat-forming wetland habitat, would be restricted. Small-scale environmental fluctuations might, therefore, result in vegetational change. In contrast, persistence as dormant seeds would improve the resistance of seed-plant dominated communities to such fluctuations.

Some associates of the heterosporous arborescent lycopsids were seed-bearing plants (cordaites and pteridosperms) with the attendant advantages of this strategy. However they never approached the abundance or diversity of the seed-bearing plants in the taxodiaceous conifer dominated peat-forming environments. Taxodiaceous conifers are themselves seed-bearers and many of their associates in the peat-forming environments also reproduced using seeds. Homosporous ferns generally were subdominants as understory elements and were largely ousted from this role by flowering plants by the mid-Cretaceous (see the Appendix). The elaborate reproductive biology of the flowering plants involves pollen recognition systems and outbreeding mechanisms, a range of specific pollination and dispersal vectors and a variety of dormancy mechanisms (Stebbins 1974, 1981, Crepet 1983, 1985). The diversity of angiosperm pollen (Muller 1985), fruits and seeds (Tiffney 1985), flowers (Crepet 1985, Friis 1985) and leaves (Hickey & Doyle 1977) suggests that these mechanisms were already dominant in plant communities from the mid-Cretaceous onwards and certainly by the early Tertiary. Mai (1985) and Collinson (1987) review the fossil record of wetland flowering plants. The special reproductive biology of the angiosperms has resulted in their great diversity today (Stebbins 1974, 1981, Crepet 1983, 1985). This is also expressed in an assessment of diversity within ancient plant communities (Knoll 1986) which shows an increase from the time of angiosperm origin. This diversity of flowering plant communities further improves their resistance to environmental fluctuation in that any single change is less likely to affect all members of the community. Dormancy mechanisms and seed dispersal vectors also enhance opportunity for spread into, and colonization of, a variety of microhabitats (Collinson & Scott 1987).

The angiosperms also include numerous herbaceous forms whereas, as far as is known, all herbaceous plants in the Carboniferous communities were spore-bearing plants. Thus, where a diversity of seed-bearing angiosperms (many with likely rhizomatous and stoloniferous growth, Tiffney & Niklas 1985) could spread to form a ground cover vegetation at drier margins of peat-forming environments, only rhizomatous spore-bearers could have played this role in the Carboniferous. Few spore-bearing herbs are recorded as sub-dominants of the Carboniferous arborescent lycopsid communities (DiMichele & Phillips 1985, Phillips *et al.* 1985—but see also Bartram, *this volume*). The rhizomatous habit is lacking in many of these (Pigg & Taylor 1985). The sphenopsids are likely to have been rhizomatous but these were shrubby or vine-like plants (DiMichele *et al.* 1985). Tetrapod assemblages also indicate a low abundance of ground level herbage in the Carboniferous (Milner 1980). It seems likely, therefore, that the spread of peat-forming environments over adjacent sites (see also microhabitat occupation p. 73) and increased margin stability was greatly enhanced where angiosperms formed the dominant herbaceous component.

Phases of spore liberation (especially critical in the heterosporous system), the presence of homosporous (e.g. tree ferns) and heterosporous (e.g. arborescent lycopsids) strategies, the numbers of spores produced and the presence or absence of seed-bearing plants all affect the composition of peats and of spore and pollen assemblages as examined in palynological studies. The free sporing versus retained megaspore systems (Phillips 1979) and the monocarpic (with a single period of propagule production at maturity) reproduction of some forms (DiMichele & Phillips 1985, Wnuk 1985) would also influence the composition of peats and palynological assemblages. Herbaceous plants (consisting mainly of soft, easily decomposed tissues and with few organs abscissed) may be represented in the fossil record only by their propagules such as spores and seeds. As Bartram (*this volume*) has shown for a Carboniferous example, the herbaceous elements or phases in coal-forming communities may be underestimated unless the propagules are examined.

Leaf and shoot biology

Several aspects of leaf and shoot biology may affect the peat-forming environment. Perhaps the most important is the deciduous or persistent nature of the foliage and branches as this will have considerable influence over the composition and accumulation of the organic material in the peat. Frequent, non-synchronous, leaf fall might permit decomposition to keep pace, whereas a sudden high leaf influx (as in temperate deciduous systems) might saturate the decomposer niche or at least prevent further decomposition in lower layers of litter. Plant material which decomposes on the plant, e.g. a fern frond or a small entire herb, unless it has the specific role of a *Sphagnum*-like plant in a continuously peat-building situation, is unlikely to contribute recognizable material to the peat (see Scheihing & Pfefferkorn 1984, Ferguson 1985), although its existence may be vital for the persistence of peat-forming conditions. The proportions of leafy versus woody litter making up the peat will also control peat compaction and ultimate coal composition. In addition the deciduous versus persistent nature of the canopy foliage must influence the nature of lower levels in the canopy and of herbaceous ground cover.

In Upper Carboniferous coal-forming communities dominated by arborescent lycopsids, leaf-shedding was not the general rule (DiMichele & Phillips 1985, Wnuk 1985). A variety of strategies typified several subcommunities. Some trees were clothed with leaves for much of their lives, others may have shed some leaves or perhaps these decomposed on the parent plant. Some trees had deciduous lateral branches, others simply a determinate crown of branches developing at maturity. In the latter cases the trees were often unbranched for much of their life and the crown itself was not densely branching. Most lycopsid forests were therefore open in nature (DiMichele & Phillips 1985). Even leaf-shedding from these trees would have produced relatively little leaf-litter compared with that produced by an extensively branching coniferous or angiospermous tree. Communities with tree-ferns as dominants or subdominants probably received no regular litterfall from these largely unbranched plants, the fronds of which decomposed on the parent plant. Scheihing & Pfefferkorn (1984) provide taphonomic evidence supporting this suggestion. This would also apply to medullosan pteridosperms (Wnuk & Pfefferkorn 1984) which formed the understory in some lycopsid forests (DiMichele & Phillips 1985, Wnuk 1985). Wnuk & Pfefferkorn (1984) noted that even small details like the basal attachment of pinnules (along their full length or at the midvein only) may influence the occurrence of such foliage in sediments.

The shedding of leafy shoots in the taxodiaceous conifers is well documented from their first occurrences in peat-forming floras (Fowler *et al.* 1973, Krumbiegel *et al.* 1983, Basinger 1984, Lapasha & Miller 1984, Miller & Lapasha 1984, Wing 1984, Smiley & Rember 1985, Spicer & Parrish 1986). Leaf shedding in many of the associated angiosperm trees and shrubs may be inferred in a variety of ways: from leaf form, anatomy, cuticles etc. (Wolfe & Upchurch 1986); from nearest living relatives; from palaeobiochemical analyses of leaf tissues and their preservation states (Niklas *et al.* 1985); from taphonomic evidence (compare Spicer 1980 with Scheihing & Pfefferkorn 1984, Ferguson 1985) where large numbers of leaves occur in sediments resulting from accumulations in the absence of storms (e.g. Smiley & Rember 1985) or from possible abscission scars on woody axes (Crane & Stockey 1985, Dilcher & Crane 1985). Timing of leaf-shedding (e.g. synchronous leaf fall typical of modern temperate climates versus the 'semi-deciduous' examples in some tropical forests) is very difficult to deduce from the fossil record and even evergreen trees sometimes shed senescent leaves rather than these decomposing on the parent plants. Such differences may be implied from nearest living relatives or deduced via features of leaf form and anatomy (Wolfe & Upchurch 1986). The abundance of leaf- and leafy shoot-shedding in the peat-forming environments dominated by the taxodiaceous conifers is therefore in marked contrast to those dominated by arborescent lycopsids.

Biology of woody axes

The overall height of the tree-sized members of communities influences canopy structure. Light interception is controlled by the nature of branching of the successive canopy members. The Upper Carboniferous lycopsid-dominated forests exhibited several canopy layers (DiMichele & Phillips 1985, Wnuk 1985) comparable to some modern communities. However, their branching patterns, with determinate growth, were very different and resulted in an open canopy in most lycopsid dominated forests (DiMichele & Phillips 1985). Pteridosperms probably formed the lower canopy layer at least in some clastic swamp communities (Wnuk 1985). As lycopsids make up 69–75% of identifiable material in coal ball peats (DiMichele & Phillips 1985) a pteridosperm understory may not have been a feature of peat-forming communities. A lower canopy composed of either pteridosperms, or leaf-retaining arborescent lycopsids, would restrict the light levels at the ground thus partially explaining the small numbers of herbs apparently associated with these communities (see p. 71). Pteridosperms, with their huge leaves retained on the plant, would certainly have formed a very dense lower canopy layer (Wnuk & Pfefferkorn 1984, Wnuk 1985). In this respect the modern analogue used by Scheihing & Pfefferkorn (1984) of palm-dominated wetlands for pteridosperm and tree fern associations of the past becomes less valid, although the analogy in terms of overall tree architecture and leaf decomposition on the parent plant is helpful.

In contrast to the arborescent lycopsid dominated forests, those formed from taxodiaceous conifers today have extensively branched canopies and form closed forests. However, the deciduous nature of taxodiaceous shoots produces the very special 'bald' cypress situation (Wade *et al.* 1980) and may well have done so in the past, providing a seasonality of light access to be exploited by suitably adapted angiosperms as herbaceous ground cover. The relative height and spacing of the conifers and the angiosperm subdominants of ancient forests (and hence the canopy structure) are largely unknown.

At a more detailed level, both the lycopsid trees and tree ferns have very distinctive internal anatomy different from that of taxodiaceous conifers and angiospermous trees. These differences are expressed in the relative amounts of lignified tissues within the axes and hence the relative compressibility and rapidity of decomposition. The arborescent lycopsid axis may be considered as a set of concentric cylinders; the central one of soft tissued surrounded by a narrow lignified zone (secondary xylem); this surrounded by a wide soft zone succeeded by an often thick cylinder of lignified secondary cortex ('periderm'). The various arborescent lycopsids differ in the precise amounts of each tissue (DiMichele & Phillips 1985) according to the subcommunities in which they grew. However, the overall pattern differs greatly from the essentially thick, solid cylinder of lignified secondary xylem which makes up the coniferous or angiospermous trunk. Compressibility, access for decomposers and lignin proportions all differ immensely between the two systems. Tree ferns possess another very different structure, essentially a 'central cylinder' of softer tissues within which are very narrow, often convoluted, more lignified zones (meristeles). The 'central cylinder' is surrounded by a thick lignified sheath and this by a mantle of adventitious roots and persistent leaf bases of mixed composition.

The precise chemical nature of these various tissues is a scarcely explored aspect although it seems reasonable to predict that lycopsid lignin may differ from angiospermous lignin (see, e.g. Logan & Thomas 1985). These chemical differences may affect decomposition rates. According to Stubblefield *et al.* (1985) the decomposer mechanisms of wood decay have undergone little change since the first recorded instance (in the Devonian) of the preservation of the micro-organisms involved.

Microhabitats

Within the Carboniferous the existence of certain plant communities, e.g. lycopsid forest, pteridosperm forest, tree-fern forest and their various sub-communities, has been established (Phillips 1981, Wnuk & Pfefferkorn 1984, DiMichele & Phillips 1985, DiMichele *et al.* 1985, Phillips *et al.* 1985, Wnuk 1985, Gastaldo 1986). Herbaceous phases in the peat-forming environment have also been recognized (Phillips *et al.* 1985, Bartram, *this volume*). However, there are no records of any open water aquatic vascular plants in these coal-forming environments. Calamites are likely to have been marginal aquatic small trees or shrubs (Scott 1978) and some lycopsid trees may have lived in standing water (DiMichele & Phillips 1985). There is no evidence for obligate aquatics; no clear evidence that any herbaceous forms occupied marginal aquatic sites and certainly no evidence of free-floating open water aquatic vascular plants. Aquatics and marginal aquatics play an important role in the initial stages of wetland seral succession (in the encroachment of vegetation over water, producing organic debris and trapping inorganic sediment) and generally buffer the junction between open water and mire (Collinson 1987). The absence of such plants must have greatly modified the peat-forming environment.

Highly complex inter-relationships where peat formation can be caused or prevented by small-scale vegetational variations have been documented in the highly diverse vegetation of modern wetlands dominated by taxodiaceous conifers and angiosperms (Davis 1946, Spackman *et al.* 1969). When almost all of the plants causing these fluctuations (and their reproductive and other strategies) are absent it is difficult even to speculate upon the implications. Perhaps variations in peat formation were under much less subtle control in the Carboniferous than they are today. It is obviously important, where possible, to search for modern peat-forming analogues with low diversity floras. In the Carboniferous peat-forming environments one might predict a much sharper boundary between mire and clastic sedimentation and sudden fluctuations in peat formation rather than gradual encroachment, compared with those peat-forming environments dominated by taxodiaceous conifers.

The suggested absence of aquatics may be more apparent than real due to inadequacy of the fossil record. However, a simple test of searching for any plant fossil (leaf, spore, seed etc.) consistently associated with lacustrine sediments (or similar) would greatly help in answering this question. We are unaware of any such occurrences in the Carboniferous. A diversity of megaspores whose modifications suggest affinity with water ferns occur in the Mesozoic and macrofossils of water ferns are found in lacustrine sediments (Collinson 1987). It seems unlikely that these have gone unrecognized in the Carboniferous. Other free-floating aquatic plants are angiosperms, a group absent from Carboniferous coal-forming environments but which, as already noted, were present by the mid-Cretaceous. Abundant taxa whose nearest living relatives are aquatics, and whose facies association supports this interpretation for their ancient ecology, are known from at least the early Tertiary (Collinson 1987, Mai 1985).

Implications for the geology of coal-forming environments

Allochthonous versus autochthonous coals

Use of modern models in interpreting allochthonous versus autochthonous circumstances on the basis of soil profiles or rooting systems may not always be appropriate. Horizontal stigmarian anchoring systems along with fine roots from cormose bases and adventitious roots from tree ferns are likely to have produced distinctive soil profiles with no close modern analogues. Some rooting systems of ancient plants are unknown, others may be difficult to recognize in field observations. Whereas the stigmarian axis may be easily documented other *in situ* rooting systems may pass unnoticed, especially if they differ from known modern analogues. An absence of *in situ* roots may, therefore, be more apparent than real.

Compression ratios

Peats formed from different plant types can be expected to behave differently in terms of compression. In view of the considerable variation in plant types dominating various peat-forming environments (e.g. Carboniferous arborescent lycopsids, Tertiary taxodiaceous conifers) it may be inappropriate to compare compression ratios between different times or to expect any kind of 'standard series' for compression data. The situation is further complicated by the compression history of the peat. Clearly a peat, even if formed from the same plant material, could experience different compression under varying geological regimes and over different periods of time. A striking difference in compression potential is illustrated by a coniferous tree trunk versus a lycopsid tree trunk (especially forms of the latter with limited periderm: see DiMichele & Phillips 1985 and compare with White 1986, see also Rex 1986). More subtle differences, however, would be due to proportions of spores to foliar and woody material, dependent, for example, on phasing of spore release and leaf fall.

There need be no simple relationship between coal rank (e.g. peat to brown coal to bituminous coal) and the peat to coal compression ratio (e.g. 6:3:1, Stach *et al.* 1982). Differences can be predicted within a coal of a single rank dependent on botanical composition. Winston (1986) recorded compression ratios for different tissues between permineralized peats and passing into adjacent coals, e.g. lycopsid periderm 7:1, stigmarian rootlet 30:1, cordaitean wood 3.3:1. If taken to extremes these imply that 1 m of coal could be derived from 3.3 m of peat composed of cordaite wood or 30 m of stigmarian rootlets. Whereas such peat compositions are highly improbable Winston's study clearly shows that compression ratios are dependent on the tissues involved.

Individual figures for compression ratios for various coals are often quoted (Ryer & Langer 1980, Baird & Woodland 1982, Stach *et al.* 1982, Courel 1987). These, e.g. 16:1 for a Cretaceous coal (Ryer & Langer 1980), may reflect vegetational composition as much as they reflect coal rank (but see also White 1986).

Composition and coal petrology

The nature of plant material forming the peat (both in terms of tissues involved and their chemical composition) will affect its structure and hence that of the coal as revealed by coal petrology, expressed as macerals and lithotypes (Cohen & Spackman 1977, 1980, Cohen *et al.*, *this volume*). In past peat-forming environments an obvious example is the differential proportions of lignified tissues in lycopsid trunks versus coniferous trunks. More subtle differences may relate to access for decomposers via softer tissues in between lignified tissues in the lycopsid axis, potentially altering the composition of the tissue prior to incorporation in the peat. Proportions of wood, cuticle and sporopollenin will be influenced by phasing of branch and leaf fall and spore or pollen liberation.

It would be expected that comparable coal lithotypes could be produced by very different plant communities, e.g. exinite-rich coals by either herbaceous fern or lycopsid communities or arborescent fern or lycopsid communities. Equally, one community might also produce very different coal lithotypes if sampled at different phases in community behaviour (phenological phases). Decompositional and compression history presumably will also influence the ultimate coal composition. There need be no simple relationship between coal maceral or lithotype and an ancient plant community.

A common assumption in coal petrology is that inertinite-rich coals are formed above ground water level; vitrinite-rich coals in 'forest swamps' and sporinite-rich coals in reed swamps and subaquatic conditions (Stach *et al.* 1982). This need not be the case. For example, in an arborescent lycopsid-dominated community, abscissed branches might give rise to a vitrinite-rich coal whereas high spore production from the monocarpic species could give rise to a sporinite-rich coal. Equally at any time, fire, producing charcoal, might sweep across all environments resulting in an inertinite-rich coal.

The persistence of a petrographic feature need not mean a constant environment. Subtle variations in vegetation need not be reflected in coal petrology (Bartram, *this volume*). Furthermore, considerable variation in vegetation would be encompassed within a single coal maceral. The 'woody tissues' of vitrinite might be derived from arborescent lycopsids, pteridosperms, cordaites or conifers and from 'bark' or secondary xylem of these plants. Variations within the maceral vitrinite are to be predicted on this basis. Differences within a single maceral are sometimes interpreted in terms of thermal history, tectonic setting etc. (Jones *et al.* 1984, Murchison *et al.* 1985) but the possible role of botanical composition is usually ignored.

It is therefore clear that simple palaeoenvironmental models based only on coal petrology (e.g. Diessel 1980, 1985, Gould & Shibaoka 1980, Marchioni 1980, Stach *et al.* 1982) may lead to erroneous interpretations. Such studies need to be integrated with analyses of coal forming plant parts. An interesting technique of ion probe etching of polished blocks has shown that palaeobotanical data are available even in high rank coals (anthracites: Kizilshtein & Shpitzgluz 1982). Palaeobotanical evidence, interpreted in terms of source vegetation, should be combined with petrological studies in interpretation of coal-forming environments.

Other variables in coal composition, e.g. of ash and sulphur (Casagrande, *this volume*), may also be influenced by the composition of the coal-forming vegetation as well as by the overall community structure and its controls over peat forming conditions (see following sections) (Cohen *et al.*, *this volume*).

Sedimentological models

The framework produced by anchoring systems and the growth of plants in the peat-forming community will influence resistance to erosion at the margins of the peat-forming environment (McCabe 1984). The horizontal systems of Carboniferous arborescent lycopsids may have aided stability of the plants in soft peats (as suggested for soft clastic sediments by Wnuk 1985). They differed, however, from those of the taxodiaceous conifers perhaps producing a less stable peat framework and with vulnerability to undercutting. In combination with the absence of aquatic vascular plants and herbaceous r-strategy seed-bearing colonizers of exposed sediments (i.e. with no marginal vegetation buffering the peat) this suggests that the boundary between peat-forming circumstance and clastic sedimentation would have been sharper. The mire vegetation itself may have controlled aspects of sedimentology such as channel migration. Guion (1987) has noted that many channel systems do not cut into Carboniferous peats (see also McCabe 1984).

Schumm (1968) suggested that vegetation has played a controlling role in the development of sedimentary systems which did not reach their modern form until after the evolution of the grasses. Vegetational control has been clearly demonstrated in the Florida wetlands with particular reference to the occurrence of peat versus clastic sedimentation (Spackman *et al.* 1969, see also Collinson 1983). Wing (1984) has demonstrated vegetational control on sedimentation on an Eocene floodplain with wetland plant communities including taxodiaceous conifers. The relationship between plants and sedimentation has been examined in several publications (Sigafoos 1964, Scheihing & Pfefferkorn 1984, Collinson 1986).

Thiadens & Haites (1944) noted a higher proportion of seam splits and washouts in Netherlands Carboniferous versus Tertiary coals. This they attributed to higher sediment and water carrying capacity of channels due to higher precipitation. However, vegetational control might equally be responsible. In the Carboniferous coal-forming environment, in the absence of marginal vegetation, flood waters would easily carry clastic debris into the peat-forming environment. In the Tertiary situation much of this debris would be filtered by the marginal plants. Furthermore many Tertiary peat-forming plants could tolerate or re-colonize the slight rise in the substrate level caused by a small clastic input. Peat formation might therefore continue more or less unchecked. In contrast, the specifically adapted, low diversity, lycopsid communities might be destroyed by a similar clastic input.

Gastaldo (1985, 1986) has noted that Carboniferous coal seam splits are often related to vegetational change, with pteridosperm-dominated communities replacing those dominated by lycopsids. Phillips *et al.* (1985) further discuss the role of vegetational and climatic change in relation to Carboniferous coal resources. It is, however, necessary to distinguish between the raised ombrotrophic mires (such as that in Sarawak—Anderson 1973), where rainfall is high, versus swamp or rheotrophic mires where water level is maintained by groundwater (Moore, *this volume*). Interpretations of climate, and rainfall in particular, depend on accurate identification of the type of mire system being considered. Modern wetland sedimentation (e.g. invasion of mire vegetation over open water) may be dependent on particularly adapted plants or plant communities some of which may have evolved quite recently. This should also always be considered when modern analogues are applied to ancient peat-forming environments.

Persistence and re-establishment of peat-forming conditions

Spore-bearing plants and those with a horizontal anchoring (and hence absorptive) system, such as that of the Carboniferous arborescent lycopsids, were particularly susceptible to lowering of ground water. Comparatively small variations in local environmental conditions might have prevented peat formation resulting in clastic sedimentation. This might result in incorporation of dirt bands or tonsteins in otherwise autochthonous peat-forming sequences (see also previous section). Re-establishment of peat formation would have required lateral migration from surrounding peat-forming communities and might only have been possible if exactly the original conditions were repeated.

The relative diversity of the peat-forming communities would greatly influence re-establishment of peat formation given environmental fluctuation. In modern peat-forming environments, e.g. the Everglades (Davis 1946, Spackman *et al.* 1969), the diversity of angiosperms provides the basis for a huge number of peat-forming sub-communities which are controlled by such variables as substrate and water table. Comparable diversity is lacking in Carboniferous communities even at their most diverse (DiMichele & Phillips 1985) and even allowing for inadequacy of the fossil record (Gastaldo 1986). It is noteworthy that peat formation by these low diversity, arborescent lycopsid-dominated communities persisted for such a prolonged period. These peat-forming communities have been 'self-maintaining' due to their specifically adapted biology, only ousted by major environmental change such as the later Palaeozoic aridity (DiMichele *et al.* 1985).

Other aspects of the environment of peat formation, such as the effects and significance of forest fires, are often largely ignored. Wade *et al.* (1980) concluded that fires may exercise a controlling influence in maintaining the taxodiaceous conifer communities in Florida today. These authors also consider the role of fire throughout the Florida wetlands. Komarek (1972), Cypert (1972) and Cohen & Spackman (1977, Cohen *et al.*, *this volume*) also discuss the role of fire in modern peat-forming ecosystems.

Fire may also have been involved in the catastrophic event at the Cretaceous–Tertiary boundary after which a peat-forming vegetation dominated by ferns became established (Tschudy *et al.* 1984, Wolbach *et al.* 1985). Phillips *et al.* (1985) comment upon high proportions of fusain (fossil charcoal) associated with medullosan pteridosperm dominance in coal balls. Recognition that fire might have played a role in peat-forming communities draws particular attention to biological features of fire adaptation in plants (e.g. Daubenmire 1959). These features, e.g. thick resistant bark, seeds germinating or cones opening after fire, might, in the future, be compared with ancient plant communities in order to consider their relative fire adaptation. For further comment on the importance of fires in the fossil record see Scott & Collinson (1978) and Cope & Chaloner (1985).

Discussion

Peat formation and all aspects of peat-forming environments are influenced, in some cases controlled, by the plants and plant communities which contribute to the peat. We have attempted to provide general, somewhat speculative, comment as to the distinctive features of different peat-forming plant communities and the variable influence which these have exerted over peat and coal formation.

The extensive data from Euramerican Upper Carboniferous coal balls provide our only adequate insight into ancient plant communities and vegetational changes through a coal-forming sequence (Phillips 1979, DiMichele & Phillips 1985, Phillips *et al.* 1985). In spite of the excellence of this work it must be recognized that the circumstances of formation of these permineralized peats may not include all elements of the coal-forming vegetation. Additional work such as that of Winston (1986) tracing plant components into coal from the permineralized peats will help to enhance our understanding of the representative nature of coal balls. Studies such as those of Willard (1986) comparing megaspores and microspores in coals with associated coal ball macro-floras are of considerable importance. As remarked by Willard, the inclusion of megaspores (+ seeds) in any investigation is vital as this provides data on the heterosporous and seed-bearing members of the communities. As demonstrated by Bartram (*this volume*), herbaceous phases may pass unnoticed if megaspores are ignored. A further study we would advocate is a comparative investigation of megaspores dispersed within coal balls with those from adjacent coals as a possible means of evaluating the representative nature of the coal ball flora or at least recognizing the bias involved. A plant with a very large spore output could dominate a palynological sample at the expense of one producing very few spores, even if the latter were the dominant element in the vegetation. Again, as remarked by Willard (1986) this aspect may be evaluated using a combination of reproductive organs containing spores from permineralized peats and adjacent coal palynology. All palynological analyses will benefit from any increased data on parent plants and some recent attempts to interpret coal palynofloras in terms of source vegetation via parent plants may be noted (Scott 1978, Scott & King 1981, Bartram, *this volume*). Smith's (1962, 1968) series of classic studies related spore assemblages as spore phases to coal petrology and hence to depositional environment. Integration of these two approaches to palynological analysis of coals would provide a much enhanced data base concerning the relationship of palynofloras to vegetational change.

Ideally, data from palynological studies, including both mega- and microspore analyses along with coal petrological data, should be integrated with information from permineralized peats. Knowledge of associated compression floras from clastic sediments and taphonomic evidence would compliment these data resulting in a very well-founded reconstruction of ancient coal-forming vegetation and component communities. Whereas this may seem an unattainable aim much progress has already been made towards this in the Upper Carboniferous of Euramerica. Nevertheless there are many other coal-forming environments where next to nothing is known concerning ancient vegetation. Studies undertaken on many permineralized peats often have been concerned with floristic data, including aspects of plant function. These studies should be undertaken along with an ecological assessment taking into account their field and sedimentological context and if possible the sampling of adjacent coals. Negative evidence, especially where usually peat-forming plants or plant communities are discovered in non-peat-forming circumstances, must also be evaluated.

In using modern analogues to interpret ancient peat-forming environments it is clearly necessary to allow for the variation in ancient and modern plant biology and plant association which we have described. In some cases it is possible to find modern analogues for some aspects of plant form and this results in some rather unexpected comparisons. Scheihing & Pfefferkorn (1984), in their study of the taphonomy of the Orinoco delta, placed great emphasis on the presence of architectural analogues for Carboniferous plants. They drew analogies between certain modern palms (monocotyledonous flowering plant trees), the fossil tree fern *Psaronius* and the medullosan pteridosperms. Modern flowering plant mangroves were used as analogues for fossil cordaites and a monocotyledonous aroid reaching 2–3 m in height (*Montricardia arborescens*) for fossil calamites. Scheihing & Pfefferkorn (1984) were dealing with taphonomy of clastic sediments and did not concern themselves with finding modern analogues for various arborescent lycopsids.

Some aspects of these analogies extend beyond mere architecture ('tree shape'), e.g. the decomposition on the plant of palm, tree-fern and pteridosperm foliage. Other aspects are less valid, e.g. using seed plants (palms and aroids) to compare with spore-bearing plants (ferns and calamites). Also the pteridosperm fronds (see Wnuk & Pfefferkorn 1984) were generally much larger than palm leaves and certainly were many more times pinnate. In addition, the associated modern plant communities differ from the ancient ones, particularly in their overall diversity, the presence of flowering plant herbs and of marginal aquatic communities, as we have already discussed. Nevertheless, this principle of searching for appropriate modern analogues not merely at the level of peat-forming environments, but in aspects of the component plants themselves, is an excellent one. Using architectural analogues at least removes one of the many variables. Some taphonomic work in tropical peat-forming environments such as those of Sarawak (Anderson 1973) or those referred to by Scheihing & Pfefferkorn (1984) and Richards (1952, p. 292) in E Asia and tropical America, is clearly needed.

Conclusions

Peat-forming vegetation has undergone major changes through the geological record. These changes include significant variations in many features of plant form and function which influence peat formation. Diversity within plant communities has increased and various new habitats have been occupied, particularly following the spread of the angiosperms. Some specially adapted wetland plants have a comparatively recent origin, whereas others, e.g. the arborescent heterosporous lycopsids, have no close modern analogues.

Models for ancient peat-forming environments based on modern analogues cannot have full worth unless these differences between modern and ancient plants and plant communities are taken into account. Several aspects of coal geology, such as the interpretation of macerals or lithotypes in terms of environmental events and the relationship between coal rank and compression ratios, must also take account of likely vegetational control.

Future studies should ideally integrate data from permineralized peats with that from adjacent coals. Coal analysis should involve not only conventional palynology and petrology but also extraction of megaspores, seeds and any other determinate plant parts along with any data available from thin or polished sections of orientated blocks of coal. Investigation of plant distribution in associated clastic sediments will help to produce a regional model. Taphonomic studies, especially in areas where some features (e.g. plant architecture; leaf behaviour of dominant trees, community diversity, mire category) are analogous with ancient communities, will also be very valuable.

In combination, these data potentially provide the basis for a thorough evaluation of factors controlling peat and hence coal formation. The influence of abiotic features such as climatic variation, sedimentary regimes, fire, ground water fluctuation etc. can be considered and the role of vegetational change may be more fully explored.

ACKNOWLEDGEMENTS: This paper has been written as a result of ideas crystallized while attending the 'Coal and Coal-bearing Strata' symposium and we would like to thank the conference participants for producing this stimulating environment. In particular K. M. Bartram, C. Cleal and B. Thomas provided helpful discussion. W. G. Chaloner and L. E. Frostick are thanked for their valued comments on an earlier draft of this paper.

This work was undertaken by M.E.C. while in receipt of a Royal Society 1983 University Research Fellowship which is gratefully acknowledged.

Appendix

Examples of floras from coals

In the main body of this paper two broad categories of ancient peat-forming environment have been compared and significant differences noted. For many of the world's coals there is only a very limited knowledge of the peat- (coal-) forming floras. The general outline given below should however serve to demonstrate the variation in peat-forming vegetation through time. This further emphasizes the caution with which knowledge of any one peat-forming environment should be applied in interpretation of others.

The only data concerning Devonian peat-forming floras come from the work of Scheckler (1986) on peats permineralized with pyrite from W Virginia. This reveals that the peat was dominated by the enigmatic 'fern' *Rhacophyton*. Rex & Scott (1987) have shown that lycopsids were an important element in some Scottish Lower Carboniferous peats (Fig. 1a) but ferns were also a significant component of Lower Carboniferous peats (Scott *et al.* 1984).

Upper Carboniferous peats from the tropical zone of Euramerica are very well known through the classic studies of Phillips *et al.* (1985 and references cited therein). Westphalian coals are formed predominantly of arborescent lycopsids

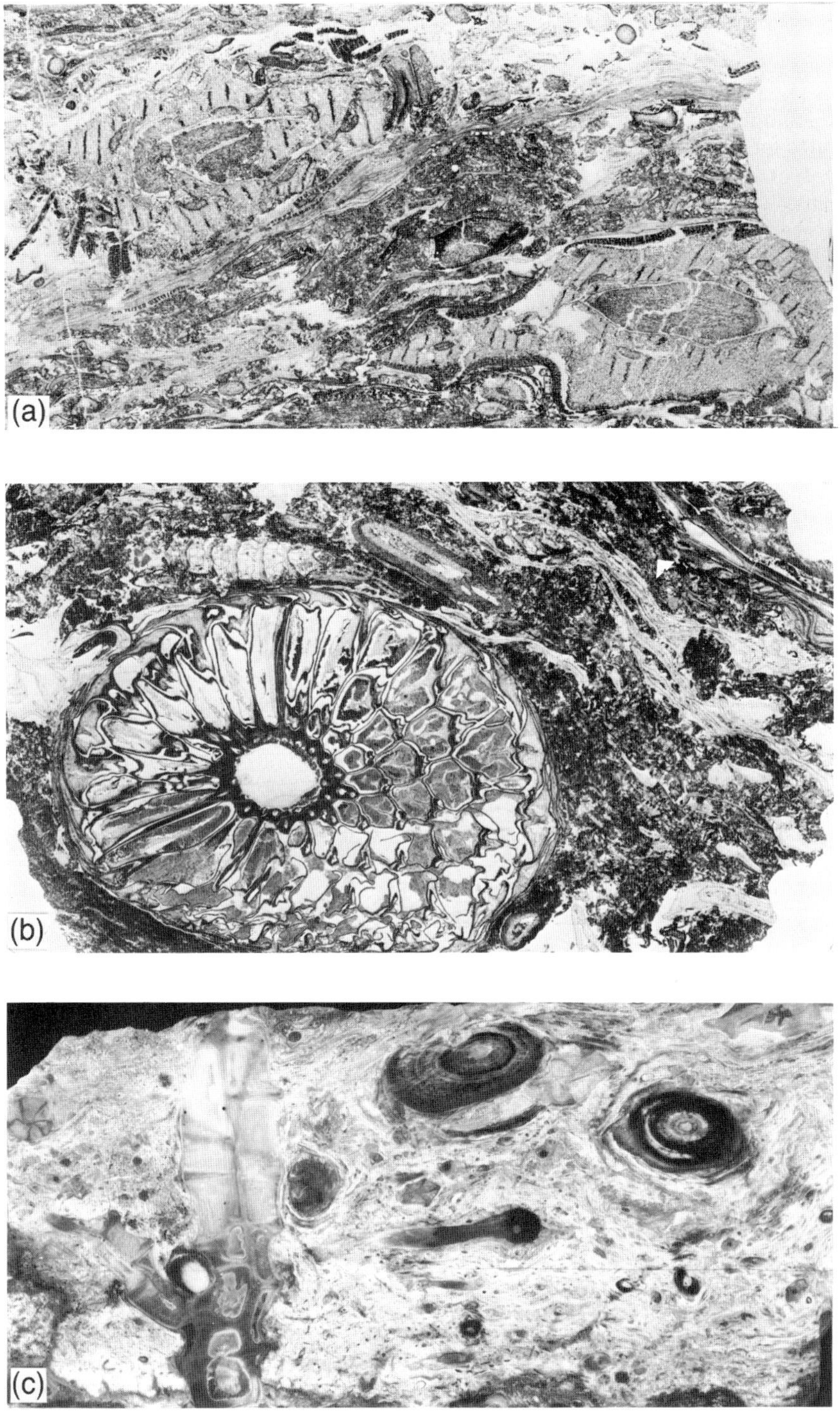

FIG. 1. Palaeozoic permineralized peats.
(a) 'Pettycur Limestone' (× 2): a calcareous permineralized peat from a volcanic terrane showing lycopods, ferns and the pteridosperm *Heterangium* (two large stems). Lower Carboniferous. Pettycur, Fife, Scotland.
(b) Upper Carboniferous coal ball (× 3): a calcareous permineralized peat composed mainly of lycopods and some ferns showing a large lycopod cone. Lancashire, England.
(c) Siliceous permineralized peat (× 1) composed mainly of the seed plant *Glossopteris* with branches showing growth rings and its roots (*Vertebraria*). Permian, Bowen Basin, Australia.

(DiMichele & Phillips 1985) although herbaceous lycopsids, tree ferns, pteridosperms and cordaites were important sub-dominants and sometimes dominated for short periods (Phillips *et al.* 1985, Fig. 1b). Bartram (*this volume*) has shown that herbaceous lycopsids may play a more important role in these coals than has previously been thought. Stephanian coals in this area are dominated by tree ferns (Phillips & Peppers 1984, Phillips *et al.* 1985) and those of the Permian by tree ferns and calamites (Galtier 1980, Galtier & Phillips 1985).

In contrast to this late Carboniferous–Permian change to tree fern dominance in Euramerica, lycopsid trees continued to thrive in the coal-forming environments in the Cathaysian province, e.g. in China (Phillips *et al.* 1985, pp. 91–92) until the end of the Permian, in association with tree ferns and cordaites. Tian Bao-lin (1979), however, records coal-forming environments dominated by seed plants (cordaites and gymnosperms) rather than spore-bearing plants in the Permian of China. In other areas seed plants also dominated Palaeozoic coal-forming environments, e.g. glossopterids in Gondwanaland (Fig. 1c) (Foster 1975, 1979, Gould & Shibaoka 1980, Fleming 1982) although ferns were also an important element along with the seed plants. Gondwanan coal-forming vegetation is poorly understood although permineralized peats are recorded from Australia and Antarctica (Schopf 1970, 1971, Gould & Delevoryas 1977) (Fig. 1c). Marques-Toigo & Correa da Silva (1984) describe Brazilian Permian coals comprising assemblages of herbaceous lycopsids with subordinate seed plants.

Early Mesozoic coal-forming floras are even more poorly known, our knowledge being based largely on compression floras from associated clastic sediments, palaeosols and palynology (e.g. Gould 1981, Retallack 1981). Some information has come from coal petrology (e.g. Lapo 1978a, b, Smyth & Cameron 1982). Gould (1981) and Gould & Shibaoka (1980) indicate that the coals of the Middle Jurassic Walloon Coal Measures of Australia are composed mainly of conifers.

Several early, mid- and late Cretaceous coals have been shown to be dominated by taxodiaceous conifers with an understory of tree ferns or herbaceous ferns (Parker 1976, Lapasha & Miller 1984, Pelzer 1984, Spicer & Parrish 1986). In the late Cretaceous, Parker (1976) records a deciduous angiosperm as sub-dominant to the taxodiaceous conifers. Palms and swamp margin and aquatic herbaceous angiosperms are associates of this coal-forming flora. Retallack & Dilcher (1981) recorded angiosperms as dominant in a mid-Cretaceous peat-forming flora along with a fern understory. Coals at the Cretaceous–Tertiary boundary are dominated by ferns (Tschudy *et al.* 1984, Saito *et al.* 1986, Tschudy & Tschudy 1986, Wolfe & Upchurch 1986) following the devastation of angiosperm-dominated vegetation by the end Cretaceous event.

Tertiary peat-forming environments are more clearly understood, at least partly because cuticles, fruits, seeds etc. have been extracted from the less mature brown coals. Plant material has also been investigated by petrological analyses (Teichmüller & Teichmüller 1968, Klein-Reesink *et al.* 1982). Various permineralized Tertiary peats have also been studied, e.g. in India, Canada and the USA although very few (e.g. Arnold & Daugherty 1964, Basinger 1984) have yet been considered from an ecological standpoint. Tertiary coals are dominated by conifers, especially Taxodiaceae and a diverse range of angiosperms including tree, shrub and herbaceous forms (Traverse & Barghoorn 1953, Teichmüller 1958, Teichmüller & Teichmüller 1968, Gregor 1978, 1982, Gould & Shibaoka 1980, Klingebiel *et al.* 1981, Krumbeigel *et al.* 1983, Wing 1984, Smiley & Rember 1985). No equivalents to the northern hemisphere temperate peats formed by *Sphagnum* (Styan & Bustin 1983) have yet been identified in the pre-Quaternary fossil record (see comments of Clymo, *this volume*).

References

ANDERSON, J. A. R. 1973. An introduction to the ecology of the peat swamps of Sarawak and Brunei. *Proceedings of the International Peat Society Symposium on the Classification of Peat and Peatlands*, pp. 44–51.

ARNOLD, C. A. & DAUGHERTY, L. H. 1964. A fossil dennstaedtioid fern from the Eocene Clarno Formation of Oregon. *Contributions from the Museum of Paleontology of the University of Michigan*, **19**, 65–88.

BAIRD, G. C. & WOODLAND, B. G. 1982. Pennsylvanian coalified rhizomorphs in Illinois: evidence for non-compressive coalification to bituminous coal rank. *Sedimentology*, **29**, 3–15.

BASINGER, J. F. 1984. Seed cones of *Metasequoia milleri* from the Middle Eocene of Southern British Columbia. *Canadian Journal of Botany*, **62**, 281–289.

BURNHAM, R. J. 1985. Deposition and preservation of plants in a modern paratropical basin. *American Journal of Botany*, **72**, 890.

CICHAN, M. A. 1985. Vascular cambium and wood development in Carboniferous plants. I. Lepidodendrales. *American Journal of Botany*, **12**, 1163–1176.

—— 1986. Conductance of the wood of selected Carboniferous plants. *Paleobiology*, **12**, 302–310.

COHEN, A. D. 1984. The Okefenokee Swamp: a low sulphur end-member of a shoreline-related depositional model for coastal plain coals. *In:* RAHMANI, R. A. & FLORES, R. M. (eds) *Sedimentology of Coal and Coal-bearing Sequences*, pp. 231–240. *Special Publication of the International Association of Sedimentologists*, **7**. Blackwell Scientific Publications, Oxford.

—— & SPACKMAN, W. 1977. Phytogenic organic sediments and sedimentary environments in the Everglades–Mangrove Swamp-marsh complex. Part II Petrography of peats. *Palaeontographica B*, **162**, 71–114.

—— & —— 1980. Phytogenic organic sediments and sedimentary environments in the Everglades–Mangrove complex. Part III Decomposition of plant tissues and the origin of coal macerals. *Palaeontographica B*, **172**, 16–149.

COLLINSON, J. D. 1986. Alluvial sediments. *In:* READING, H. G. (ed.) *Sedimentary Environments and Facies*, 2nd edn, pp. 20–62. Blackwell Scientific Publications, Oxford.

COLLINSON, M. E. 1983. Palaeofloristic assemblages and palaeoecology of the Lower Oligocene Bembridge Marls, Hamstead Ledge, Isle of Wight. *Botanical Journal of the Linnean Society*, **86**, 177–225.

—— 1987. Freshwater macrophytes in palaeolimnology. *Palaeogeography, Palaeoclimatology, Palaeoecology*, **60**.

——, BATTEN, D. J., SCOTT, A. C. & AYONGHE, S. 1985. Palaeozoic, Mesozoic and contemporaneous megaspores from the Tertiary of Southern England: indicators of sedimentary provenance and ancient vegetation. *Journal of the Geological Society of London*, **142**, 375–395.

—— & SCOTT, A. C. 1987. Factors controlling the organization of ancient plant communities. *In: Organisation of Communities—past and present*, pp. 367–388. British Ecological Society, London.

COPE, M. J. & CHALONER, W. G. 1985. Wildfire: an interaction of biological and physical processes. *In:* TIFFNEY, B. (ed.) *Geological Factors and the Evolution of Plants*, pp. 257–277. Yale University Press, New Haven.

COSTANZA, S. H. 1985. *Pennsylvanioxylon* of Middle-Upper Pennsylvanian coals from the Illinois Basin and its comparison with *Mesoxylon*. *Palaeontographica B*, **197**, 81–121.

COUREL, L. 1987. Stages in the compaction of peat: examples from the Stephanian and Permian of the Massif Central, France. *Journal of the Geological Society of London*, **144**, 489–493.

CRANE, P. R. & STOCKEY, R. A. 1985. Growth and reproductive biology of *Joffrea speirsii* gen et sp nov., a *Cercidiphyllum*-like plant from the late Paleocene of Alberta, Canada. *Canadian Journal of Botany*, **63**, 340–364.

CREPET, W. L. 1983. The role of insect pollination in the evolution of the angiosperms. *In:* REAL, L. (ed.) *Pollination Biology*, pp. 29–50. Academic Press, Orlando, Florida.

—— 1985. Advanced (constant) insect pollination mechanisms: pattern of evolution and implications vis-a-vis angiosperm diversity. *Annals of the Missouri Botanical Garden*, **71**, 607–630.

CYPERT, E. 1972. Plant succession on burned areas in Okefenokee swamp following the fires of 1954 and 1955. *Proceedings of the Annual Tall Timbers Fire Ecological Conference*, **12**, 199–217.

DAUBENMIRE, R. F. 1959. *Plants and Environment. A Textbook of Plant Autecology*, 2nd edn, p. 422. Wiley, New York.

DAVIS, J. H. 1946. The peat deposits of Florida; their occurrence, development and uses. *Bulletin of the Florida Geological Survey*, **30**, 1–247.

DIESSEL, C. F. K. 1980. Newcastle and Tomago Coal Measures. *In:* HERBERT, C. & HELBY, R. (eds) *A Guide to the Sydney Basin*, pp.100–114. Bulletin 26, Geological Survey of New South Wales.

——1985. Macerals as coal facies indicators. *Compte Rendu Dixieme Congrès International de Stratigraphie et de Géologie du Carbonifère*, **3**, 367–373.

DILCHER, D. L. & CRANE, P. R. 1985. *Archaeanthus*: an early Angiosperm from the Cenomanian of the Western Interior of North America. *Annals of the Missouri Botanical Garden*, **71**, 351–383.

DIMICHELE, W. A. & PHILLIPS, T. L. 1985. Arborescent lycopod reproduction and paleoecology in a coal-swamp environment of late Middle Pennsylvanian age (Herrin Coal, Illinois). *Review of Palaeobotany and Palynology*, **44**, 1–26.

——, —— & PEPPERS, R. A. 1985. The influence of climate and depositional environment on the distribution and evolution of Pennsylvanian coal-swamp plants. *In:* TIFFNEY, B. H. (ed.) *Geological Factors and the Evolution of Plants*, pp. 223–256. Yale University Press, New Haven.

FAHN, A. 1974. *Plant Anatomy*, 2nd edn (1st edn translated by Broido-Altman, S.), p. 611. Pergamon Press, Oxford.

FERGUSON, D. K. 1985. The origin of leaf-assemblages—new light on an old problem. *Review of Palaeobotany and Palynology*, **46**, 117–188.

FLEMING, P. J. G. 1982. Coal-forming and associated environments, Permian Denison Trough, Queensland. *Queensland Government Mining Journal*, **83**, 284–292.

FOSTER, C. B. 1975. Permian plant microfossils from the Blair Athol Coal Measures, Central Queensland, Australia. *Palaeontographica B*, **154**, 121–171.

—— 1979. Permian plant microfossils of the Blair Athol Coal Measures, Baralaba Coal Measures, and Basal Rewan Formation of Queensland. *Geological Survey of Queensland Palaeontological Paper*, **45**, 1–244.

FOWLER, K., EDWARDS, N. & BRETT, D. W. 1973. *In situ* coniferous (Taxodiaceous) tree remains in the Upper Eocene of southern England. *Palaeontology*, **16**, 205–217.

FRANKENBERG, J. M. & EGGERT, D. A. 1969. Petrified

Stigmaria from North America: Part I. *Stigmaria ficoides*, the underground portions of Lepidodendraceae. *Palaeontographica B*, **108**, 43–92.

FRIIS, E. M. 1985. Preliminary report of Upper Cretaceous Angiosperm reproductive organs from Sweden and their level of organization. *Annals of the Missouri Botanical Garden*, **71**, 403–418.

GALTIER, J. 1980. Les végétaux silicifies du Permien d'Autun: analyse quantitative et interpretation possible. *Bulletin Société d'Histoire Naturelle d'Autun*, **95**, 35–39.

—— & PHILLIPS, T. L. 1985. Swamp vegetation from Grand Croix (Stephanian) and Autun (Autunian), France, and comparisons with coal-ball peats of the Illinois Basin. *Compte Rendu Neuvième Congrès International de Stratigraphie et de Géologie du Carbonifère (1979)*, **5**, 13–24.

GASTALDO, R. A. 1985. Upper Carboniferous paleoecological reconstructions: observations and considerations. *Compte Rendu Dixième Congrès International de Stratigraphie et de Géologie du Carbonifère*, Madrid, 1983, **2**, 281–296.

—— 1986. Implications on the paleoecology of autochthonous lycopods in clastic sedimentary environments of the early Pennsylvanian of Alabama. *Palaeogeography, Palaeoclimatology, Palaeoecology*, **53**, 191–212.

——, MCCARROLL, S. M. & DOUGLASS, D. P. 1985. Interdistributary plant accumulating environments of the lower delta plain, Mobile Delta, Alabama. *American Journal of Botany*, **72**, 893.

GOULD, R. 1981. The coal-forming flora of the Walloon Coal Measures. *Coal Geology (Australia)*, **1**, 83–105.

—— & DELEVORYAS, T. 1977. The biology of *Glossopteris*: evidence from petrified seed-bearing and pollen-bearing organs. *Alcheringa*, **1**, 387–399.

—— & SHIBAOKA, M. 1980. Some aspects of the formation and petrographic features of coal members in Australia, with special reference to the Tasman orogenic zone. *Coal Geology, Journal of the Coal Geology Group of the Geological Society of Australia*, **2**, 1–29.

GREGOR, H.-J. 1978. The leaf and fruit floras of the Schwardorf area (Eastern Bavaria) in ecological and stratigraphical correlation with other Miocene brown coal deposits. *Courier Forschungsinstitut Senckenberg*, **30**, 107–111.

—— 1982. *Die jungtertiären Floren Süddeutschlands*. Ferdinand Enke Verlag, Stuttgart.

GUION, P. D. 1987. Palaeochannels in mine workings in the High Hazel Seam (Westphalian B), Nottinghamshire Coalfield, England. *Journal of the Geological Society of London*, **144**, 471–488.

HACQUEBARD, P. A. & DONALDSON, J. R. 1969. Carboniferous coal deposition associated with flood-plain and limnic environments in Nova Scotia. *In:* DAPPLES, E. C. & HOPKINS, M. E. (eds) *Environments of Coal Deposition*, pp. 143–191. Geological Society of America Special Paper 114.

HARRAR, E. S. & HARRAR, J. E. 1962. *Guide to Southern Trees*. 2nd edn. New York, Dover.

HENRY, A. & MCINTYRE, M. 1926. The swamp cypresses, *Glyptostrobus* of China and *Taxodium* of America, with notes on allied genera. *Proceedings of the Royal Irish Academy, B*, **37**, 90–116.

HICKEY, L. J. & DOYLE, J. A. 1977. Early Cretaceous fossil evidence for angiosperm evolution. *The Botanical Review*, **43**, 3–104.

HOOK, D. D. 1984. Adaptations to flooding with fresh water. *In:* KOZLOWSKI, T. T. (ed.) *Flooding and Plant Growth*. Academic Press, Orlando, Florida.

—— & CRAWFORD, R. M. M. (eds) 1978. *Plant Life in Anaerobic Environments*. Ann Arbor, Michigan.

JONES, J. M., DAVIS, A., COOK, A. C., MURCHISON, D. G. & SCOTT, E. 1984. Provincialism and correlations between some properties of vitrinites. *International Journal of Coal Geology*, **3**, 315–331.

KIZILSHTEIN, L. YA. & SHPITZGLUZ, A. L. 1982. New method of petrographic analysis of anthracite. *Doklady Akademic Nauk SSSR*, **263**, 175–179.

KLEIN-REESINK, J., RIEGEL, W. & SCHAUB, K. 1982. Zur petrographischen konstitution alt—und jungtertiarer Braunkohlen Niederhessens. *Zeitschrift der Deutschen Geologischen Gesellschaft*, **113**, 309–337.

KLINGEBIEL, A., LEGIGAN, P., NICOLAS, G. & ROBERT, P. 1981. Le Lignite d'Arjuzanx. *Bulletin Centres Recherche Exploration-production Elf-Aquitaine*, **5**, 681–699.

KNOLL, A. H. 1986. Patterns of change in plant communities through time. *In:* DIAMOND, J. & CASE, T. J. (eds) *Community Ecology*, pp. 126–148. Harper & Row, New York.

KOMAREK, E. V. 1972. Ancient fires. *Proceedings of the Annual Tall Timbers Fire Ecological Conference*, **12**, 219–240.

KOZLOWSKI, T. T. (ed.) 1984. *Flooding and Plant Growth*. Academic Press, Orlando, Florida.

KRUMBIEGEL, G., RUFFLE, L. & HAUBOLD, H. 1983. *Das Eozane Geiseltal*. Ziemsen Verlag, Wittenberg Lutherstadt.

LAPASHA, C. A. & MILLER, C. N. 1984. Flora of early Cretaceous Kooteria Formation Montana. Paleoecology. *Palaeontographica B*, **194**, 109–130.

LAPO, A. V. 1978a. Comparative characteristics of vitrinites of Carboniferous coals of the Ukraine and Jurassic coals of Siberia. *Fuel*, **57**, 179–183.

—— 1978b. Phyterals of Jurassic coals of Tuva. *Palaeobotanist*, **25**, 205–216.

LOGAN, K. J. & THOMAS, B. A. 1985. Distribution of lignin derivatives in plants. *New Phytologist*, **99**, 571–585.

MCCABE, P. J. 1984. Depositional environments of coal and coal-bearing strata. *In:* RAHMANI, R. A. & FLORES, R. M. (eds) *Sedimentology of Coal and Coal-bearing Environments*, pp. 2–30. Special Publication of the International Association of Sedimentologists, 7. Blackwell Scientific Publications, Oxford.

MAI, D. H. 1985. Entwicklung der Wasser—und Sumpfpflanzen—Gesellschaften Europas von der Kreide bis ins Quartar. *Flora* (Jena), **176**, 449–511.

MARCHIONI, D. L. 1980. Petrography and depositional environments of the Liddell Seam, Upper Hunter Valley, New South Wales. *International Journal of Coal Geology*, **1**, 35–61.

MARQUES-TOIGO, M. & CORREA DA SILVA, Z. C. 1984. On the origin of Gondwanic South Brazilian Coal Measures. *In:* LEMOS DE SOUSA, M. J. (ed.) *Symposium on Gondwana Coals*, Lisbon, 1983. Proceedings and papers. *Comunicacoes dos servicos Geologicos de Portugal*, **70**, 151–160.

MILLER, C. N. & LAPASHA, C. A. 1984. Flora of the early Cretaceous Kooterai Formation in Montana, conifers. *Palaeontographica B*, **193**, 1–120.

MILNER, A. R. 1980. The tetrapod assemblage from Nýřany, Czechoslovakia. *In:* PANCHEN, A. L. (ed.) *The Terrestrial Environment and the Origin of Land Vertebrates*, pp. 439–496. Academic Press, London.

MULLER, J. 1985. Significance of fossil pollen for angiosperm history. *Annals of the Missouri Botanical Garden*, **71**, 419–443.

MURCHISON, D. G., COOK, A. C. & RAYMOND, A. C. 1985. Optical properties of organic matter in relation to thermal gradients and structural deformation. *Philosophical Transactions of the Royal Society of London A*, **315**, 157–186.

NIKLAS, K. J., BROWN, R. M. JR & SANTOS, R. 1985. Ultrastructural states of preservation in Clarkia Angiosperm leaf tissues: implications on models of fossilization. *In:* SMILEY, C. J., LEVITON, A. E. & BERSON, M. (eds) *Late Cenozoic History of the Pacific Northwest*, pp. 143–160. Interdisciplinary Studies on the Clarkia Fossil Beds of Northern Idaho. Pacific Division. American Association for the Advancement of Science, San Francisco.

PARKER, L. R. 1976. The palaeoecology of the fluvial coal-forming swamps and associated floodplain sediments in the Blockhawk Formation (Upper Cretaceous) of central Utah. *Brigham Young University General Studies*, **22**, 99–116.

PELZER, G. 1984. Cross section through a fluvial environment in the Wealden of Northwest Germany. *In:* REIF, W.-E. & WESTPHAL, F. (eds) *Third Symposium on Mesozoic Terrestrial Ecosystems, Short Papers*, pp. 181–186. Attempto Verlag, Tubigen.

PHILLIPS, T. L. 1979. Reproduction of heterosporous arborescent lycopods in the Mississippian-Pennsylvanian of Euramerica. *Review of Palaeobotany and Palynology*, **27**, 239–289.

—— 1981. Stratigraphic occurrences and vegetational patterns of Pennsylvanian pteridosperms in Euramerican coal swamps. *Review of Palaeobotany and Palynology*, **32**, 5–26.

—— & PEPPERS, R. A. 1984. Changing patterns of Pennsylvanian coal-swamp vegetation and implications of climatic control on coal occurrence. *International Journal of Coal Geology*, **3**, 205–255.

——, —— & DIMICHELE, W. A. 1985. Stratigraphic and interregional changes in Pennsylvanian coal-swamp vegetation: environmental inferences. *International Journal of Coal Geology*, **5**, 43–109.

PIGG, K. B. & ROTHWELL, G. W. 1983. *Chaloneria* gen. nov.; heterosporous lycophytes from the Pennsylvanian of North America. *Botanical Gazette*, **144**, 132–147.

—— & TAYLOR, T. N. 1985. *Cormophyton* gen.nov. A cormose lycopod from the Middle Pennsylvanian Mazon Creek Flora. *Review of Palaeobotany and Palynology*, **44**, 165–181.

RETALLACK, G. J. 1981. Two new approaches for reconstructing fossil vegetation with examples from the Triassic of Eastern Australia. *In:* GRAY, J., BOUCOT, A. J. & BERRY, W. B. N. (eds) *Communities of the Past*, pp. 271–295. Hutchinson Ross, Stroudsburg, Pennsylvania.

—— & DILCHER, D. L. 1981. Early angiosperm reproduction: *Prisca reynoldsii*, gen. et sp. nov. from mid-Cretaceous coastal deposits in Kansas, USA. *Palaeontographica B*, **179**, 103–137.

REX, G. 1986. Further experimental modelling of the formation of plant compression fossils. *Lethaia*, **19**, 143–159.

—— & SCOTT, A. C. 1987. The sedimentology, palaeoecology and preservation of the Lower Carboniferous plant deposits at Pettycur, Fife, Scotland. *Geological Magazine*, **124**, 43–66.

RICHARDS, P. W. 1952. *The Tropical Rain Forest. An Ecological Study*. Cambridge University Press.

ROTHWELL, G. W. 1975. The Callistophytaceae (Pteridospermopsida) 1. Vegetative structures. *Palaeontographica B*, **151**, 171–196.

—— 1984. The apex of *Stigmaria* (Lycopsida), rooting organ of Lepidodendrales. *American Journal of Botany*, **71**, 1031–1034.

—— & ERWIN, D. M. 1985. The rhizomorph apex of *Paurodendron*; implications for homologies among the rooting organs of Lycopsida. *American Journal of Botany*, **72**, 86–98.

—— & WARNER, S. 1984. *Cordaixylon dumusum* n.sp. (Cordaites). 1. Vegetative structures. *Botanical Gazette*, **145**, 275–291.

RYER, T. A. & LANGER, A. W. 1980. Thickness change involved in the peat-to-coal transformation for a bituminous coal of Cretaceous age in central Utah. *Journal of Sedimentary Petrology*, **50**, 987–992.

SAITO, T., YAMANOI, T. & KAIHO, K. 1986. End-Cretaceous devastation of terrestrial flora in the boreal Far East. *Nature, London*, **323**, 253–256.

SCHECKLER, S. E. 1986. Geology, floristics and paleoecology of Late Devonian coal swamps from Appalachian Laurentia (USA). *Annales de la Societe geologique de Belgique*, **109**, 209–222.

SCHEIHING, M. H. & PFEFFERKORN, H. W. 1984. The taphonomy of land plants in the Orinoco Delta: a model for the incorporation of plant parts in clastic sediments of Late Carboniferous age of Euramerica. *Review of Palaeobotany and Palynology*, **41**, 205–240.

SCHOPF, J. M. 1970. Petrified peat from a Permian coal bed in Antarctica. *Science*, **169**, 274–277.

—— 1971. Notes on plant tissue preservation and mineralization in a Permian Deposit of peat from Antarctica. *American Journal of Science*, **271**, 522–543.

SCHUMM, S. A. 1968. Speculations concerning the paleohydrologic control of terrestrial sedimentation. *Bulletin of the Geological Society of America*, **79**, 1573–1588.

SCOTT, A. C. 1977. A review of the ecology of Upper Carboniferous plant assemblages with new data from Strathclyde. *Palaeontology*, **20**, 447–473.

—— 1978. Sedimentological and ecological control of Westphalian B plant assemblages from West Yorkshire. *Proceedings of the Yorkshire Geological Society*, **41**, 461–508.

—— 1979. The ecology of Coal Measure floras from northern Britain. *Proceedings of the Geologists Association*, **90**, 97–116.

—— & Collinson, M. E. 1978. Organic sedimentary particles: results from Scanning Electron Microscope studies of fragmentary plant material. *In:* Whalley, W. B. (ed.) *Scanning Electron Microscopy in the Study of Sediments*, pp. 137–167. Geoabstracts, Norwich.

—— & —— 1983. Investigating fossil plant beds. *Geology Teaching*, **7**, 114–122 and **8**, 12–26.

——, Galtier, J. & Clayton, G. 1984. The distribution of Lower Carboniferous anatomically preserved floras in Western Europe. *Transactions of the Royal Society of Edinburgh: Earth Sciences*, **75**, 311–340.

—— & King, G. R. 1981. Megaspores and coal facies: an example from the Wesphalian A of Leicestershire, England. *Review of Palaeobotany and Palynology*, **34**, 107–113.

—— & Rex, G. 1985. The formation and significance of Carboniferous coal balls. *Philosophical Transactions of the Royal Society of London B*, **311**, 123–137.

Sigafoos, R. S. 1964. Botanical evidence of floods and flood-plain deposition. *United States Geological Survey Professional Paper A*, 485, 1–35.

Smiley, C. J. & Rember, W. C. 1985. Composition of the Miocene Clarkia Flora. *In:* Smiley, C. J., Leviton, A. E. & Berson, M. (eds) *Late Cenozoic History of the Pacific Northwest. Interdisciplinary studies on the Clarkia Fossil Beds of Northern Idaho*, pp. 95–112. Pacific Division. American Association for the Advancement of Science, San Francisco.

Smith, A. H. V. 1962. The palaeoecology of Carboniferous peats based on the miospores and petrography of bituminous coals. *Proceedings of the Yorkshire Geological Society*, **33**, 423–474.

—— 1968. Seam profiles and seam characters. *In:* Murchison, D. G. & Westoll, T. S. (eds) *Coal and Coal-bearing Strata*, pp. 31–40. Oliver & Boyd, Edinburgh.

Smyth, M. & Cameron, M. 1982. Organic petrology and source rock potential of sediments in the Eromanga Basin, South Australia. *International Journal of Coal Geology*, **1**, 263–281.

Spackman, W., Riegel, W. L. & Dolsen, C. P. 1969. Geological and biological interactions in the swamp-marsh complex of Southern Florida. *In:* Dapples, E. C. & Hopkins, M. E. (eds) *Environments of Coal Deposition*, pp. 1–35. Geological Society of America Special Paper 114.

Spicer, R. A. 1980. The importance of depositional sorting to the biostratigraphy of plant megafossils. *In:* Dilcher, D. L. & Taylor, T. N. (eds) *Biostratigraphy of Fossil Plants*, pp. 171–183. Dowden, Hutchinson & Ross, Stroudsburg, Pennsylvania.

—— & Parrish, J. T. 1986. Paleobotanical evidence for cool north polar climates in middle Cretaceous (Albian–Cenomanian) time. *Geology*, **14**, 703–706.

Stach, E., Mackowsky, M. T., Teichmüller, M., Taylor, G. H., Chandra, D. & Teichmüller, R. 1982. *Stach's Textbook of Coal Petrology*, 3rd edn. Gebrüder Borntraeger, Berlin.

Stebbins, G. L. 1974. *Flowering Plants, Evolution above the Species Level.* Belknap Press of the Harvard University Press, Cambridge, Massachusetts.

—— 1981. Why are there so many species of angiosperms? *Bioscience*, **31**, 573–576.

Stewart, W. N. 1983. *Paleobotany and the Evolution of Plants.* Cambridge University Press.

Stone, E. C. & Vasey, R. B. 1968. Preservation of coast Redwood on alluvial flats. *Science*, **159**, 157–161.

Stubblefield, S. P., Taylor, T. N. & Beck, C. B. 1985. Studies of Paleozoic fungi. V. wood-decaying fungi in *Callixylon newberryi* from the Upper Devonian. *American Journal of Botany*, **72**, 1765–1774.

Styan, W. B. & Bustin, R. M. 1983. Petrography of some Fraser River Delta peat deposits: Coal maceral and microlithotype precursors in temperate-climate peats. *International Journal of Coal Geology*, **2**, 321–370.

Teichmüller, M. 1958. Rekonstruktion vershiedener Moortypen des Hauptflozes der niederrheinischen Braunkohle. *Fortschritte in der Geologie von Rheinland und Westfalen*, **2**, 599–612.

—— & Teichmüller, R. 1968. Cainozoic and Mesozoic deposits in Germany. *In:* Murchison, D. & Westoll, T. S. (eds) *Coal and Coal-bearing Strata*, pp. 347–379. Oliver & Boyd, Edinburgh.

Thiadens, A. A. & Haites, T. B. 1944. Splits and washouts in the Netherlands Coal Measures. *Mededelingen Geologische Stichting C-11-1*, No. 1, 1–51.

Tian Bao-Lin 1979. Coal balls in the coal seams in China. *Abstracts of the ninth International Congress of Carboniferous Stratigraphy and Geology*, Urbana, 1979, p. 214.

Tiffney, B. H. 1985. Seed size, dispersal syndromes and the rise of the Angiosperms: evidence and hypothesis. *Annals of the Missouri Botanical Garden*, **71**, 551–576.

—— & Niklas, K. J. 1985. Clonal growth in land plants: a paleobotanical perspective. *In:* Jackson, J. B. C., Buss, L. W. & Cook, R. E. (eds) *Population Biology and Evolution of Clonal Organisms*, pp. 35–66. Yale University Press, New Haven,

Traverse, A. & Barghoorn, E. S. 1953. Micropaleontology of the Brandon lignite, an early Tertiary coal in central Vermont: preliminary note. *Journal of Paleontology*, **27**, 289–293.

Tschudy, R. H., Pilmore, C. L., Orth, C. J., Gilmore, J. S. & Knight, J. D. 1984. Disruption of the Terrestrial plant ecosystem at the Cretaceous-Tertiary boundary, Western Interior. *Science*, **225**, 1030–1032.

—— & Tschudy, B. 1986. Extinction and survival of plant life following the Cretaceous/Tertiary boundary event, Western Interior, North America. *Geology*, **14**, 667–670.

WADE, D., EWEL, J. & HOFSTETTER, R. 1980. *Fire in South Florida Ecosystems*. U.S. Department of Agriculture Forest Service General Technical Report SE-17, Southeast. Forest Experimental Station, Asheville (N.C).

WHITE, J. M. 1986. Compaction of Wyodac Coal, Powder River Basin, Wyoming, U.S.A. *International Journal of Coal Geology*, **6**, 139–147.

WILLARD, D. A. 1986. Comparison of megaspores, miospores and coal balls in analysis of coal-swamp vegetation. *American Journal of Botany*, **73**, 715 (Abstract 309).

WING, S. L. 1984. Relation of paleovegetation to geometry and cyclicity of some fluvial carbonaceous deposits. *Journal of Sedimentary Petrology*, **54**, 52–66.

WINSTON, R. B. 1986. Characteristic features and compaction of plant tissues traced from permineralized peat to coal in Pennsylvanian coals (Desmoinsian) from the Illinois Basin. *International Journal of Coal Geology*, **6**, 21–41.

WNUK, C. 1985. The ontogeny and paleoecology of *Lepidodendron rimosum* and *L. bretonense* trees from the Middle Pennsylvanian of the Bernice Basin (Sullivan County, Pennsylvania). *Palaeontographica B*, **195**, 153–181.

—— & PFEFFERKORN, H. W. 1984. The life habits and paleoecology of middle Pennsylvanian medullosan pteridosperms based on an *in situ* assemblage from the Bernice Basin (Sullivan County, Pennsylvania, U.S.A.) *Review of Palaeobotany and Palynology*, **41**, 329–351.

WOLBACH, W. S., LEWIS, R. S. & ANDERS, E. 1985. Cretaceous extinctions: evidence for wildfires and search for meteoritic material. *Science*, **230**, 167–170.

WOLFE, J. A. & UPCHURCH, G. R. 1986. Vegetational, climatic and floral changes at the Cretaceous–Tertiary boundary. *Nature, London*, **324**, 148–152.

MARGARET E. COLLINSON, Department of Biology, King's College London (KQC), University of London, Kensington Campus, Campden Hill Road, London W8 7AH, England.

ANDREW C. SCOTT, Department of Geology, Royal Holloway and Bedford New College, University of London, Egham Hill, Egham, Surrey TW20 0EX, England.

Sulphur in peat and coal

D. J. Casagrande

ABSTRACT: Coal sulphur content is a major consideration in coal marketing. To predict sulphur content and distribution in a coal basin, a sulphur geochemical model needs to be developed. A series of questions are posed regarding coal sulphur distribution, forms and origins. Salient aspects of literature on coal sulphur content and distribution are reviewed to establish a framework for developing a sulphur geochemical model.

Some coal sulphur originates from peat-forming processes. Sulphate, ferrous iron, and microorganisms are key ingredients for the origins of coal pyrite and organic sulphur. Hydrogen sulphide is an important intermediate for pyrite and organic sulphur formation in peats. Sulphur isotope values for peat sulphate, plant sulphur, peat pyrite and peat organic sulphur corroborate the hypothesis that microorganisms reduce sulphate to hydrogen sulphide, which in turn reacts with available ferrous iron or organic matter to produce pyrite and organic sulphur respectively.

Keywords: peat, coal, sulphur, pyrite, sulphur dioxide, acid rain

Due to the volume of information on sulphur in coal, this paper will focus on components of the following subject areas:

(1) The importance of sulphur in coal.
(2) Facts about sulphur forms (pyrite and organic sulphur) in US coals.
(3) The origins of sulphur in coal.

There will be little narrative on detailed experimental methods. Descriptions of all methodologies are available in the cited literature.

The importance of sulphur in coal

Sulphur is probably the one constituent in coal that most affects coal marketing. Virtually every coal contract includes specifications for coal sulphur and penalties if the total sulphur, or sulphur form limit is exceeded.

The environmental laws in the US are becoming increasingly stringent with respect to sulphur dioxide emissions. The acid rain problem will amost certainly generate further US legislation on SO_2 emissions. This legislation may cause the shut-down of some older boilers and increase sulphur removal costs for utilities having long-term contracts for high sulphur coal.

In addition to SO_2 related problems, there is the issue of acid water generation from pyrite in waste piles. This environmental problem has come to the forefront in recent years. It is of economic concern wherever there is revegetation, reclamation and backfilling with (high sulphur) pyrite-bearing sediments.

US regulations tie SO_2 to coal BTU content—namely 1.2 lb of SO_2/million BTU. Therefore, some western USA coals having low sulphur and low BTU could have problems meeting environmental regulations. Generally, lignitic rank coals with less than 0.6% sulphur, and bituminous rank coals with less than 0.8% sulphur can meet emission standards. All values are on a dry weight basis.

Predicting sulphur distribution

With respect to coal exploration, there is strong incentive to find the lowest sulphur coal in a basin. Some basins are hundreds of miles in extent. How does one predict the quantity and distribution of sulphur in a basin in a way that is cost effective and efficient?

In order to predict the quantity and distribution of sulphur in a coal basin, one needs to develop a geochemical model for coal sulphur distribution. In the course of developing the geochemical model, a series of questions should be answered: (1) What are the forms of sulphur and what is their distribution in coal? (2) Why does the form and content of sulphur vary within a coal and what caused this variation? (3) What is the source of sulphur in coal? (4) How does sulphur become incorporated into coal? (5) What is the timing of sulphur incorporation in coal? (6) What is the influence of the surrounding confining strata on coal sulphur content?

Two integrated approaches to answering these questions include:

(1) Use a target coal seam or seams to carry out geochemical analysis and develop a geochem-

ical model for sulphur distribution that is specific to the target seam.

(2) Develop geochemical investigations in the coal-forming sequence of plant—peat—lignite—bituminous coal to understand the fundamental processes associated with sulphur origins in coal.

For the remainder of this paper, research results are discussed which may provide answers to some of these questions.

Both approaches have been used in an attempt to understand the forms and distribution of sulphur in coal and to develop a sulphur geochemical model for coal. Initially, there will be a focus on the distribution of sulphur in coal and then there will be a discussion of coal sulphur origins.

Sulphur in coal

General comments

There are considerable data on the distribution of sulphur in US coals. The range of sulphur is from <0.1% to over 10% on a dry weight basis. Typically, total sulphur falls in the range of 0.5–5% for US coals (Walker & Hartner 1966, *USBM Circ.* 8680, 1966, Ruch *et al.* 1974, Davis 1982).

The terms 'high' or 'low' sulphur coal are relative to the coal basin. In the Illinois Basin, high sulphur coal will often contain greater than 4% sulphur, whereas at El Cerrejon, in Colombia, high sulphur coal is considered to be greater than 0.7% sulphur.

Within the six coal provinces of the US, the Carboniferous coals from the Eastern and Interior Provinces generally have higher sulphur values than coals of the Western or Gulf Coast Provinces (Table 1).

However considerable variation can be seen even within a province and in some cases even within the same seam. An excellent example of this variability is the Herrin No. 6 seam in the Illinois Basin. There are locations in the Herrin seam having over 8% sulphur. There are other locations in what is called the quality circle area in southern Illinois in which the Herrin seam has about 1% coal sulphur (Gluskoter & Simon 1968, Ruch *et al.* 1974).

Williams & Keith (1963) related roof lithology to coal quality in Appalachian coals and Reidenouer *et al.* (1967) showed similar relations for Western Pennsylvania coals. Gluskoter & Simon (1968) present evidence that Illinois Basin roof lithology and thickness is related to sulphur content of the associated Herrin No. 6 coal seam. Grey, freshwater-derived shale is associated with coals having less than 3% sulphur; whereas black, marine-derived shale is associated with coals having more than 3% sulphur. Furthermore, in areas where grey shale is immediately associated with coal and the grey shale is overlain with black shale, a grey shale thickness of 6 m is required to eliminate the sulphur influence of the overlying marine black shale. McCabe (1984) provides an excellent summary on coal versus geological environment.

There is no 'typical' distribution of sulphur or sulphur forms in US coals. However, there are six observations that appear pertinent in developing a sulphur geochemical model:

TABLE 1. *Sulphur values of US coal (from Davis 1982)*

Province/region	States	Age	Sulphur (% dry basis) Total	Pyrite	Organic	Number of samples
Eastern						
Appalachian	AL, E, KY, MD, OH, PA, TN, VA, WA, VA	Pennsylvanian	0.9	0.3	0.6	12
Anthracite	PA	Pennsylvanian	2.2	1.3	0.8	175
Interior						
Eastern	IL, IN, WA, KY	Pennsylvanian	3.9	2.2	1.5	48
Western	AR, IA, KS, MO, OK	Pennsylvanian	4.3	2.3	1.6	43
			4.1	2.2	1.6	91
Northern Great Plains	CO, MT, ND, NM, WY	Upper Cretaceous–Eocene	0.6	0.2	0.5	22
Rocky Mountain	AZ, CO, NM, UT, WY	Upper Cretaceous–Eocene	0.7	0.2	0.5	53
Pacific	WA	Eocene	0.9	0.3	0.6	3
Gulf	AR, LA, TX	Palaeocene–Eocene	1.3	0.3	1.0	11

TABLE 2. *Sulphate minerals identified in coal*

Aluminocopiapite	$AlFe_4(SO_4)6 \cdot O(O) \cdot 2OH_2O$	Kieserite	$MgSO_4 \cdot H_2O$
Alunite	$K_2O \cdot 3Al_2O_3 \cdot 4SO_3 \cdot 6H_2O$	Melanterite	$FeSO_4 \cdot 7H_2O$
Alunogen	$Sl_2(SO_4)_3 \cdot 18H_2$	Mirabilite	$Na_2SO_4 \cdot 10H_2O$
Anhydrite	$CaSO_4$	Natroalunite	$Na_2O \cdot 3Al_2O_3 \cdot 4SO_3 \cdot 6H_2O$
Barite	$BaSO_4$	Natrojarosite	$Na_2O \cdot 3Fe_2O_3 \cdot 4SO_3 \cdot 6H_2O$
Bassanite	$CaSO_4 \cdot 1/2H_2O$	Pickeringite	$MgSO_4 \cdot Al_2(SO_4)_3 \cdot 22H_2O$
Celestite	$SrSO_4$	Potassium alum	$K_2O \cdot Al_2O_3 \cdot 4SO_4 \cdot 24H_2O$
Coquimbite	$Fe_2(SO_4)_3 \cdot 9H_2O$	Roemerite	$FeSO_4 \cdot Fe_2(SO_4)_3 \cdot 12H_2O$
Epsomite	$MgSO_4 \cdot 7H_2O$	Sideronatrite	$2Na_2O \cdot Fe_2O_3 \cdot 4SO_3 \cdot 7H_2O$
Gypsum	$CaSO_4 \cdot 2H_2O$	Siderotil	$FeSO_4 \cdot SH_2O$
Halotrichite	$FeSO_4 \cdot Al_2(SO_4)_3 \cdot 22H_2O$	Szomolnokite	$FeSO_4 \cdot H_2O$
Jarosite	$K_2O \cdot 3Fe_2O_3 \cdot 4SO_3 \cdot 6H_2O$		

(1) The sulphate sulphur content of US coals is generally less than 0.3% and in most cases less than 0.01%. Table 2 shows sulphate minerals observed in coal (Davis 1982).

(2) In general, low sulphur coals contain more organic sulphur than pyritic sulphur.

(3) High sulphur coals often contain similar proportions of pyritic and organic sulphur.

(4) Some coals have greater than 5% organic sulphur but have less than 0.5% pyritic sulphur.

(5) High sulphur coals are generally associated with marine roof.

(6) Total sulphur and sulphur forms distribution is often variable. Higher sulphur contents are often observed in bench samples near the top and bottom of a seam.

Pyritic sulphur

While there has been an occasional report in which elemental sulphur has been identified in coal (Berteloot 1947), the rest of this section will focus on coal pyritic and organic sulphur content.

With respect to coal pyrite, there are some observations that are important when developing a sulphur geochemical model:

(1) Macroscopically, pyrite occurs as veins, lenses, nodules and disseminated crystals (Thiesen 1920, 1945, Cady 1935, Wanless 1959, Yurovskii 1959, 1960, Neavel 1966, Reyes-Navarro & Davis 1976).

(2) Microscopically, pyrite occurs as blebs, fine veinlets, dendrites, euhedral crystals and replacement of plant material (references as above).

(3) The various pyrite forms or modes of occurrence can be found within centimetres of one another in a coal seam (references as above).

(4) Elements such as arsenic, bismuth, cadmium, copper, lead, mercury, molybdenum, selenium, silver, tin and zinc are components of sulphide minerals in coal (Finkelman 1980). Table 3 is a list of some sulphides identified in coals (Davis 1982).

(5) Because of the intimacy of the pyrite-coal particle matrix, it appears that some pyrite could only have been formed syngenetically, that is at the time of deposition or before consolidation of the sediment; finely disseminated pyrite is an example (Neavel 1966).

(6) Some pyrite forms appear to form epigenetically, that is, after the sediment has consolidated. Pyrite found in cleat could not have been formed in the original peat swamp, because cleat is only known to occur in coal with a rank of lignite or higher. Smith & Batts (1974) and Price & Shieh (1979) have documented the epigenetic occurrence of pyrite using sulphur isotopes.

TABLE 3. *Sulphides in coal*

Argentite	Ag_2S	Marcasite	FeS_2
Arsenopyrite	$FeAsS$	Millerite	NiS
Bismuthinite	Bi_2S_3	Orpiment	As_2S_3
Bornite	Cu_5FeS_3	Polydymite	Ni_3S_4
Chalcopyrite	$CuFeS_2$	Pyrite	FeS_2
Cinnabar	HgS	Pyrrhotite	$Fe_{1-x}S$
Covellite	CuS	Realgar	AsS
Galena	PbS	Sphalerite	ZnS
Greigite	Fe_3S_4	Ulmannite	$NiSbS$
Linnaeite	Co_3S_4		

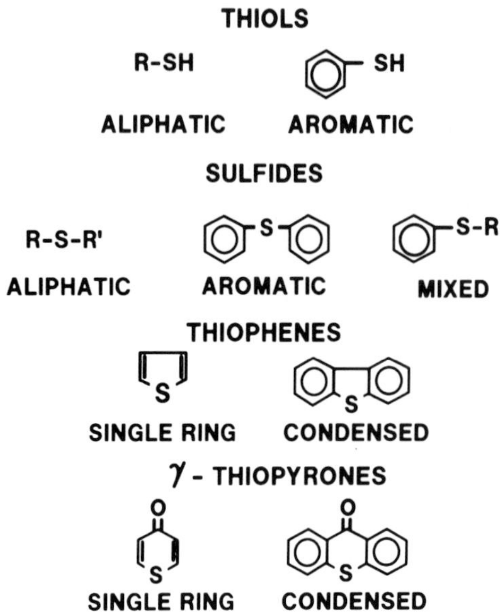

FIG. 1. Proposed organic sulphur forms in coal (from Given & Wyss 1961).

(7) Pyritic sulphur content is often higher in upper and lower portions of a seam than in the middle—perhaps implying coal sulphur content is influenced by roof or floor strata (Williams & Keith 1963, Gluskoter & Simon 1968, Gluskoter 1977).

Organic sulphur

Knowledge about the forms of organic sulphur in coal is minimal and virtually all the information is by inference or obtained by indirect means (Brooks 1956, Given & Wyss 1961, Attar 1979). Given & Wyss (1961) suggest coal organic sulphur forms or functionalites, including thiols, sulphides, thiophenes and thiopyrones; these are shown in Fig. 1.

Raymond (1979) has used the electron microprobe to establish the distribution of organic sulphur in the various coal macerals. He showed organic sulphur distribution in macerals of US coals as follows:

sporinite > vitrinite > resinite > fusinite

Decomposition, depolymerization, spectroscopic and solvent extraction methods (Speight 1978, Attar 1979, Rodionova & Baranskii 1970, Miura & Yanagi 1963) are not adequate for establishing the forms (or organic functionalities) of organic sulphur in coal. All methods suffer from the same analytical problems: (1) poor mass transfer of solvent into the coal matrix, and (2) a small number of organic sulphur atoms in a large coal matrix.

One technique, thermal reduction, which may provide some information on coal organic sulphur functionalities is discussed by Attar & Dupuis (1979). The outline for the technique (Fig. 2) shows that under reducing conditions, each form of organic sulphur will reduce to H_2S at a particular temperature. Using this technique on Illinois Basin, Herrin No. 6 coal, Attar & Dupuis (1979) observed thiophene functionalities constituted over 50% of the total sulphur.

There are experimental problems with the Attar & Dupuis (1979) technique:

(1) Thiphenes are generally calculated by difference because thiophene reduction is not complete and does not occur until reactor temperatures exceed 450°C. At this temperature and above

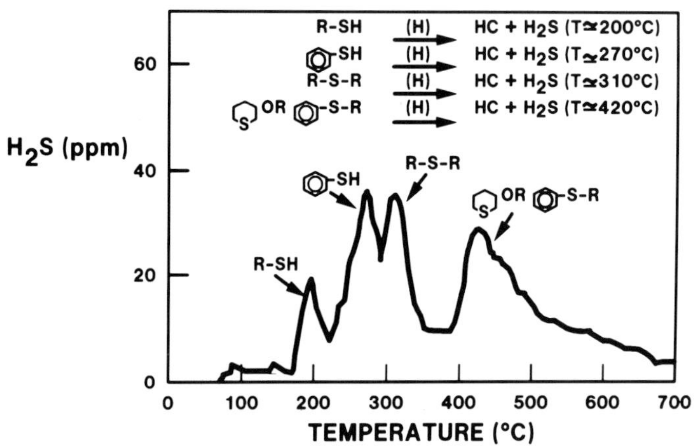

FIG. 2. Kinetograms using the thermal reduction technique.

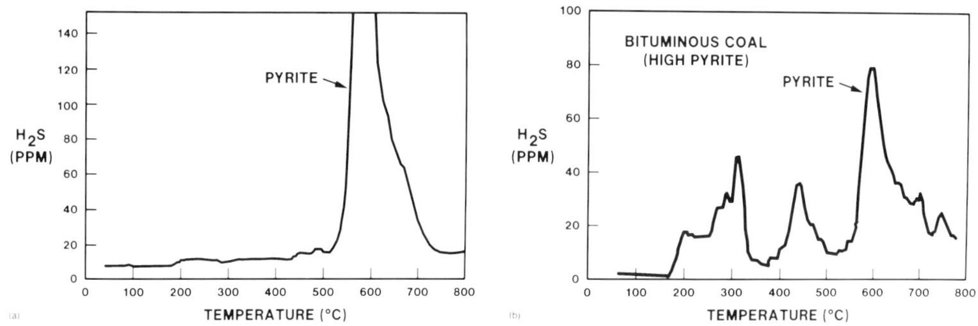

FIG. 3. Kinetogram of pyrite (a) versus high pyrite coal (b).

there is a question as to what reactants are still present in the reactor and what the generated H_2S is related to.

(2) Pyrite is reported to be a non-participant in the thermal reduction reactions (Attar & Dupuis 1979). However, Fig. 3 shows a kinetogram of pure pyrite (a) and a kinetogram of high pyrite coal (b). Pyrite appears to participate in the thermal reduction process and has a negative influence on the quantification of organic sulphur functionalities such as thiophene, which also reduce to H_2S at similar temperatures.

(3) There is a lack of standard sulphur-containing polymers that adequately mimic coal structures. Thus, standard kinetograms are difficult to obtain and there are serious questions regarding the relationship of standard kinetograms to sample kinetograms. Calibration of the kinetograms is therefore very difficult.

(4) Mass transfer of reducing agent into the coal matrix appears to be a problem (this may account for poorly resolved peaks).

The thermal reduction technique is an improvement over available techniques, but analytical results from thermal reduction must be evaluated in the light of the experimental difficulties.

Even though important progress has been made in establishing the form and structure of organic sulphur in coal, organic sulphur remains the least understood form of sulphur in coal.

Origins of sulphur in coal

General comments

This section summarizes results on the origins of sulphur in coal with particular emphasis to processes occurring during the peat-forming stage. In all cases, peat-forming systems from the Okefenokee Swamp, Georgia and the Everglades Swamp, Florida were used for these investigations. The use of these contemporary peat-forming systems as suitable coal-forming models has been discussed elsewhere (Spackman *et al.* 1976). Listed below are a series of experimental results that have been integral to the development of a coal sulphur geochemical model.

Figure 4 is an overview of the fundamental geochemistry of the peat-forming environment of the coal-forming system. Coal formation starts with plant polymers (macromolecules) and ends with coal macerals. In some cases, plant anatomical structure is maintained (see Cohen *et al.*, *this volume*). In other cases, considerable degradation of plant polymer material occurs. Plant polymer degradation includes the synthesis of geopolymers such as humic acids, fulvic acids, etc. These geopolymers along with some primary plant polymers subsequently yield the coal macerals.

Sulphur content of peat

Peats contain all the forms of sulphur found in coal including pyritic, sulphate and organic sulphur (Casagrande *et al.* 1977, Altschuler *et al.* 1983, Cohen *et al.* 1983, Casagrande 1985, Given & Miller 1985). The sulphur content of contemporary peats can account for all the sulphur found in coal. Peats in the Florida Everglades contain 0.1–10% total sulphur on a dry basis (Casagrande *et al.* 1977, Altschuler *et al.* 1983, Cohen *et al.* 1983, Casagrande 1985, Given & Miller 1985).

We have studied the distribution of sulphur in the operationally defined organic fractions of peat; humic acid, fulvic acid, humin, water soluble and benzene/methanol soluble (Casagrande *et al.* 1978, 1980). Table 4 shows the relative abundances of these fractions and their sulphur contents in peats from the Okefenokee and Everglades Swamps. The principal conclusions are: (a) humin represents most of the sediment, (b) humic acids are quantitatively more important in the marine environment and (c) the

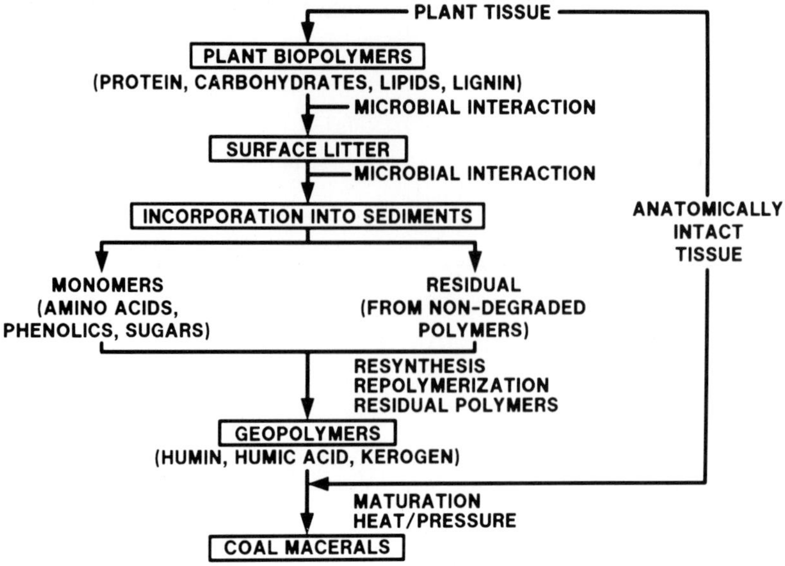

FIG. 4. Peat-forming geochemistry.

humic and humin fractions have sulphur contents similar to the whole peat sulphur content. Given the inherent chemical stability of humic acids and humin, sulphur associated with these organic fractions is presumed to be largely geochemically stable and therefore is likely to be an important part of the sulphur eventually found in coal.

Marine-influenced peats generally have a higher sulphur content than freshwater-influenced peats (Casagrande *et al.* 1977, Casagrande 1985). The Minnie's Lake site in the Okefenokee Swamp was not influenced by marine waters. The peat petrography associated with this site was shown (Cohen 1968) to be a cypress (*Taxodium*) peat overlying peat derived from water lily (*Nymphaea*). Figure 5 shows the sulphur content of Minnie's Lake peat as a function of depth. Sulphur content is low and changes little as a function of depth. Table 5 gives the mean, and range of sulphur forms values. Most of the sulphur

TABLE 4. *Distribution of sulphur in peat fractions*

Sample	% of total organic matter	% of total sulphur (DWB)	% of total sulphur in fraction	% of total sulphur as	
				Organic	Inorganic
Okefenokee[a] (Grand Prairie—freshwater)					
Total	—	0.20	—	79.7	20.3
Water sol.	4.12	0.38	12.5	66.3	33.7
Ben/met sol.	4.78	0.15	1.56	60.2	39.8
Humin	69.2	0.16	67.8	78.3	21.7
Humic acid	5.83	0.33	13.2	70.5	29.5
Fulvic acid	16.1	0.048	4.47	35.2	64.8
Everglades[b] (Little Shark—marine)					
Total	—	1.66	—	59.4	40.6
Water sol.	2.05	5.92	7.92	0.83	99.1
Ben/Met sol.	3.69	5.35	11.4	49.9	50.1
Humin	48.1	1.78	51.6	76.3	23.7
Humic acid	23.5	1.25	17.5	62.2	27.8
Fulvic acid	22.6	0.91	12.2	12.9	87.1

[a] Mean of five depths of peat in Grand Prairie freshwater peat core.
[b] Mean of five depths of peat in Little Shark marine water peat core.

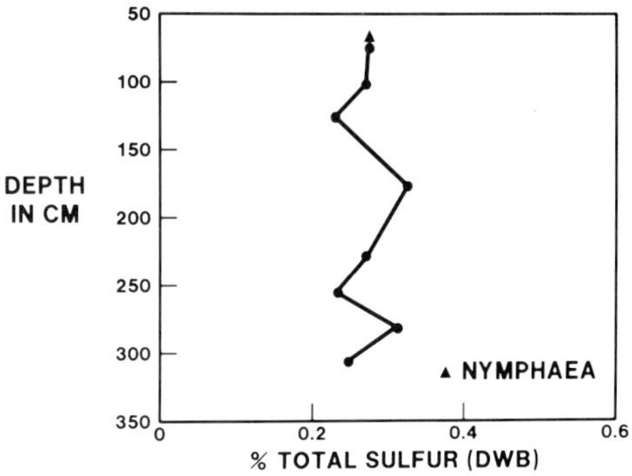

FIG. 5. Sulphur distribution in Minnie's Lake peat.

is in an organic combination. There is very little pyritic sulphur in these sediments.

On the other hand, the marine-influenced Little Shark River area in the Florida Everglades shows some remarkable differences from freshwater peats with respect to sulphur. A reconstructed environmental setting is shown in Fig. 6 (Spackman *et al.* 1976). Freshwater peats developed in a water lily (*Nymphaea*) environment are overlain by peats derived from halophytic plants such as the Red Mangrove (*Rhizophora mangle* L.).

Sulphur content in the Little Shark peat profile is shown in Fig. 7 and the sulphur forms distribution is shown in Table 6. Total sulphur content ranges from 1.29 to 10.11% on a dry peat basis. Although organic sulphur is dominant, pyritic sulphur has increased importance relative to freshwater peats.

From the comparison of sulphur content of freshwater- and marine-influenced peats, sulphate content in the water is the key difference between the two systems. As will be shown, sulphate is the principal reactant for establishing pyritic and organic sulphur levels in peat. Plant sulphur plays a major contributing role to sulphur in freshwater peat, but is not the principal sulphur source for marine peats.

Sulphur forms and reaction

Using techniques developed for soils (Freney 1967), we found that ester-sulphate comprises 25% of the total sulphur in peat (Casagrande *et al.* 1977 and Casagrande & Siefert 1977). Figure 8 shows a plot of ester-sulphate contents versus peat depth in the Little Shark peat profile. Ester-sulphate bonding includes a carbon–oxygen–sulphur linkage as opposed to the more common carbon–sulphur linkage. Ester-sulphate may be an important, relatively stable source of sulphate for microbial reduction to hydrogen sulphide. The hydrogen sulphide could react with peat components, thus incorporating sulphur in the organic material (Casagrande & Siefert 1977). Altschuler *et al.*'s (1983) data indicate that sulphur associated with ester-sulphate may be the source of pyritic sulphur in low sulphur freshwater peats. This would add credance to the above work on the importance of ester-sulphate. However, there is disagreement about the source of ester-sulphate. Given & Miller (1985) indicate that very little ester-sulphate is associated with plants. However, studies indicate that 17–30% of the plant sulphur is in ester-sulphate combination (Casagrande *et al.* 1977) Also, bacteria may synthesize ester-sulphate functionalities and, through bacterial reworking, bacterial ester-sulphate becomes an intimate part of the sediment.

Hydrogen sulphide generation in peat

There is an active microbial community in peat (Given *et al.* 1983). Those who have walked in a marine peat environment know hydrogen sulphide is being formed.

TABLE 5. *Distribution of sulphur in low sulphur peat*

	Mean	Range	% of total sulphur (mean)
Total sulphur (peat)	0.187	0.17–0.21	100
Sulphur form (peat)			
Sulphate	0.0035	0.015–0.073	1.87
H_2S	0.0008	0.0007–0.0014	0.42
Elemental	0.0046	0.0031–0.0064	2.45
Ester sulphate	0.047	0.041–0.052	25.13
Pyrite	0.0017	0.0014–0.0022	0.90
Organic[a]	0.129	0.10–0.19	69.23

Synthesis of data from Casagrande *et al.* (1977).
[a] Organic sulphur obtained by difference.

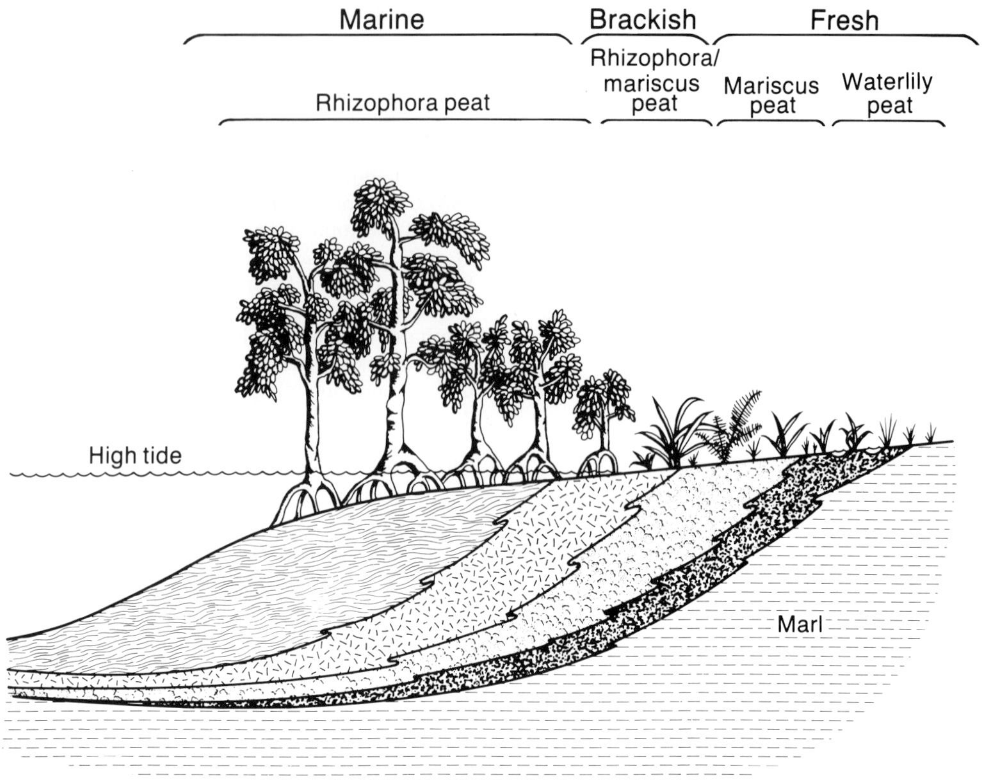

FIG. 6. Reconstructed environmental setting at Little Shark Site (from McCabe 1984, after Spackman *et al.* 1976).

The author agrees with the conventional theory that sulphate is bacterially reduced to form hydrogen sulphide (Harrison & Thode 1958, Zobell 1963, Goldhaber & Kaplan 1975, 1980, Jorgensen 1982, Howarth & Morgensen 1984).

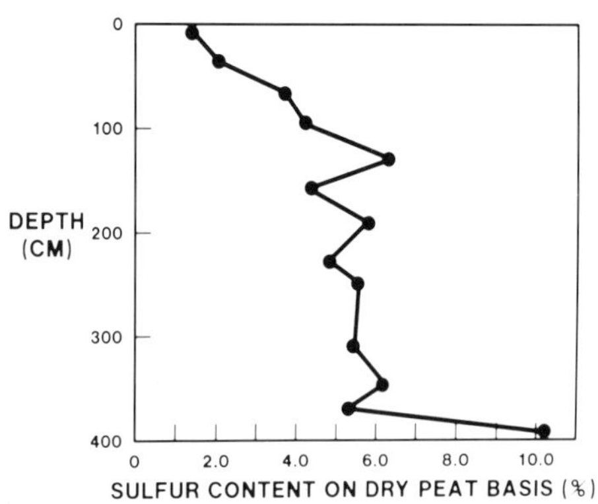

FIG. 7. Distribution of sulphur in Little Shark peat profile.

However, sulphate reducing bacteria require specific pH and anaerobic conditions that are not ubiquitous in peats and thus hydrogen sulphide generation may readily occur in some sulphate-rich environments and not in others.

Several attempts have been made to quantify H_2S production in peat (Casagrande *et al.* 1977). Some of these data are shown in Fig. 9. Given & Miller (1985) show lower levels of H_2S in peat. The exact quantities are not as relevant as the large differences (about an order of magnitude) between the freshwater- and marine-influenced sediments.

We feel that sulphate availability is important in the peat-forming system and, as will be shown in subsequent sections of this paper, hydrogen sulphide plays a key role in the origins of sulphur in coal.

Pyrite formation in peats

There has been considerable work by Yurovskii (1960), Kaplan *et al.* (1963), Berner (1970, 1984,

TABLE 6. *Distribution of sulphur in high sulphur peat*

	Mean	Range	% of total sulphur (mean)
Total sulphur	5.16	1.19–10.11	100
Sulphur form			
Sulphate	0.36	0.24–0.53	6.97
H_2S	0.015	0.003–0.032	0.29
Elemental	0.39	0.054–0.63	7.56
Ester sulphate	1.22	0.42–2.43	23.64
Pyrite	0.74	0.073–5.34	14.34
Organic[a]	2.44	1.30–4.56	47.20

Synthesis of data from Casagrande *et al.* (1977).
[a] Organic sulphur obtained by difference.

1985), Rickard (1975), Goldhaber & Kaplan (1980). Love *et al.* (1983) and others that document reduction of sulphate to H_2S and its subsequent reaction with ferrous iron to produce pyrite. We accept this explanation for the syngenetic incorporation of sulphur as pyrite in the peat-forming stage. However, it should be noted that in addition to hydrogen sulphide, ferrous ions need to be available for pyrite formation. Ferrous ions are not always readily available. The quantity of ferrous iron brought into the system may depend upon the dispersal of clays transported via freshwater streams to the marine environment. Also, pH will determine the form of iron available (Fe^{++} versus Fe^{+++}). Coals high in organic sulphur (>4%) and low in pyritic sulphur (<1%) may reflect the non-availability of Fe^{++} ions during peat formation and diagenesis. Whereas the chemistry of pyrite formation appears to be established, Cohen (1968) has shown that some modes of occurrence of peat pyrite indicate a strong bacterial or plant tissue influence on pyrite crystal growth.

However, sight should not be lost of the formation of epigenetic pyrite (in cleats, for example) as well as the above mentioned syngenetic coal pyrite.

An example of pyrite distribution in low and high sulphur peats is shown in Fig. 10. There is very little pyrite in freshwater-influenced peats, but considerable pyrite content in marine-influenced peats.

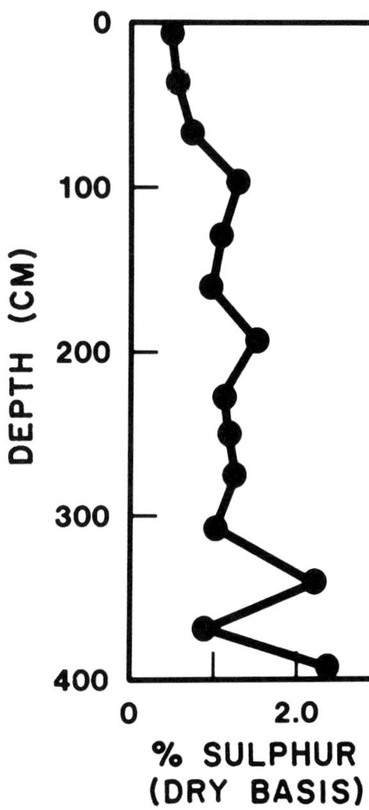

FIG. 8. Ester-sulphate distribution in Little Shark peat profile.

Derivation of organic sulphur

Plants and microbial cells have sulphur-containing amino acids, such as methionine (Freney 1967). Much of the organic sulphur associated with plants and microorganisms may be in this form. To test the hypothesis that sulphur-containing amino acids will contribute to peat sulphur, labelled ^{35}S-methionine was introduced into a peat (Casagrande 1984). The peat was fractionated and ^{35}S-methionine was found to be readily incorporated into the humic acid fraction (Fig. 11). Thus plant or microbial sulphur-containing amino acid can contribute to peat sulphur.

Since sulphate availability seems important for

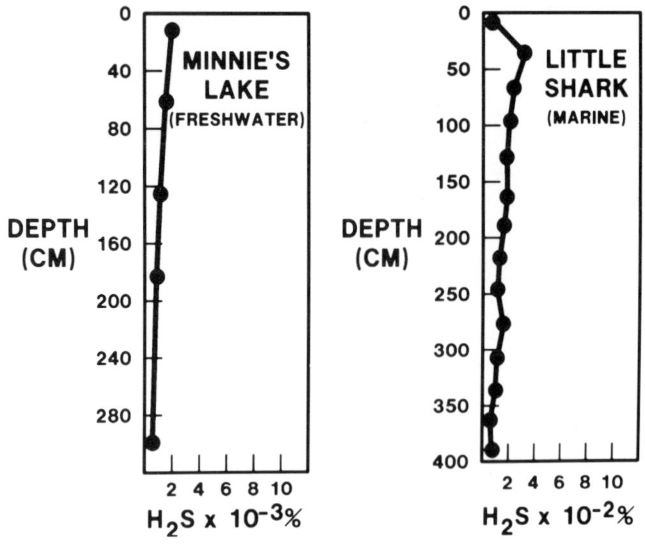

FIG. 9. Hydrogen sulphide distribution in peat profiles.

hydrogen sulphide formation and hydrogen sulphide is thought to be a key intermediary in the formation of pyrite, the author felt that hydrogen sulphide may also be an important reactant leading to peat organic sulphur formation.

In one set of experiments, Casagrande *et al.* (1979) demonstrated that not only is $H_2{}^{35}S$ incorporated in peat as organic sulphur, but as seen in Fig. 12 the reaction is accomplished quite rapidly. The mechanisms for the incorporation are not known. However, as early as 1913, Shorey, in his pioneering work on sulphur in soils, speculated that H_2S reacted with benzaldehyde (from lignin) to produce trithiobenzaldehyde. Schwab *et al.* (1976) demonstrated that saturated aldehydes react with H_2S to produce a variety of sulphur-containing compounds at ambient temperature and pressure. More recently Mango

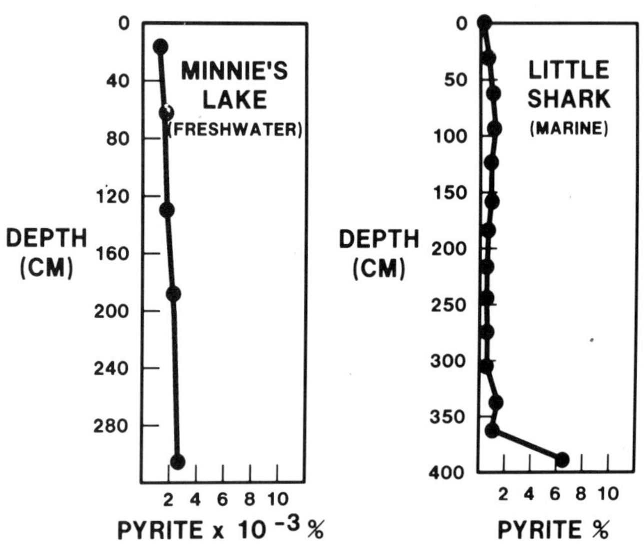

FIG. 10. Pyrite content in low and high sulphur peat.

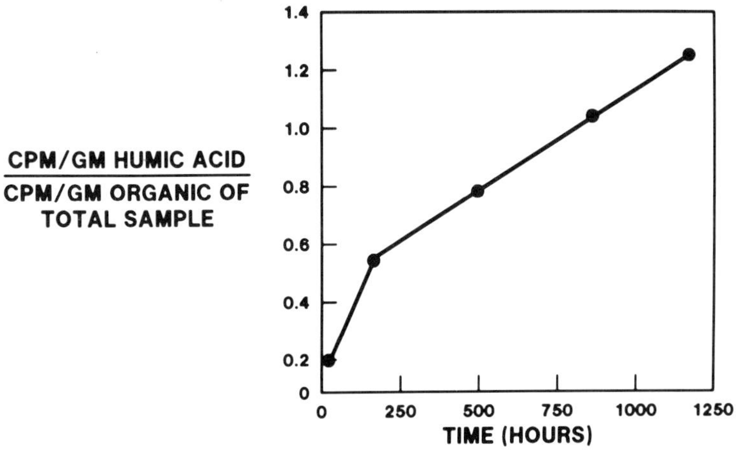

FIG. 11. Incorporation of ^{35}S-methionine in peat.

(1983) showed that H_2S reacts with certain carbohydrate moeites to yield organic sulphur.

Another set of experiments showed the potential role of elemental sulphur as a source of organic sulphur in peat (Casagrande & Ng 1979). Elemental sulphur is a key intermediary in the sulphur cycle. Furthermore, elemental sulphur is found in peats, especially those having high total sulphur content (Casagrande *et al.* 1977, Given & Miller 1985). ^{35}S-labelled elemental sulphur was reacted with peat humic acid. As shown in Fig. 13 some of the elemental sulphur was taken up in the humic acid as organic sulphur. The conditions for this reaction included a gentle reflux of elemental sulphur (in chloroform) with peat humic acid. Chemical reactions have not been ascertained for this reaction but, given the relatively mild conditions of the experiment, and the availability of elemental sulphur and humic acids in peat, it appears to be another plausible mechanism for organic sulphur incorporation in coal formation.

Sulphur isotope distribution in peat

Sulphur isotope techniques have been used to build on the idea that sulphate is the ultimate

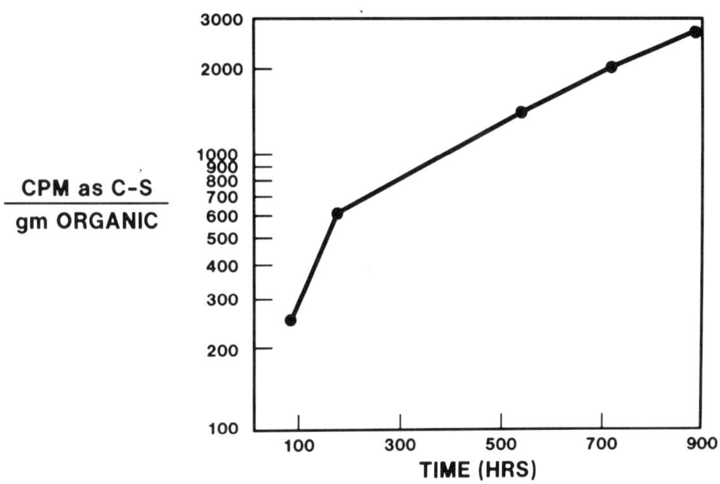

FIG. 12. Incorporation of $H_2{}^{35}S$ in peat.

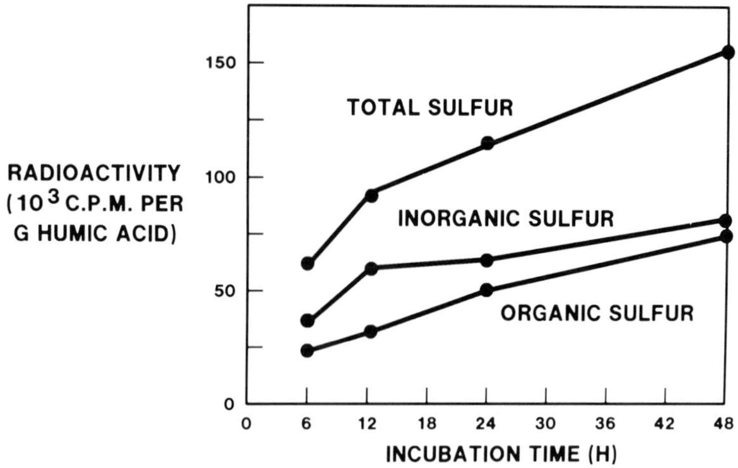

FIG. 13. Incorporation of ^{35}S in peat.

syngenetic coal sulphur source and that bacterially-derived hydrogen sulphide is a key intermediary in the process of sulphur incorporation at the peat-forming stage of coalification (Casagrande & Price 1981).

Since sulphur isotope data will be presented, it is important to define sulphur isotope terminology. Sulphur occurs in nature in four stable isotopic species. Isotope geochemists measure isotope ratios and, in the case of sulphur, the ratio of $^{34}S/^{32}S$ is reported. Because of the limitations of the mass spectrometer, one does not measure the absolute ratio $^{34}S/^{32}S$, but reports data in terms of a relative abundance ratio:

$$\delta^{34}S = \left[\frac{(^{34}S/^{32}S)\ \text{sample}}{(^{34}S/^{32}S)\ \text{standard}} - 1\right] \times 1000$$

The relative abundance ratio is defined as the $\delta^{34}S$ value which is equal to the ratio of the sample minus the ratio of the standard, divided by the standard times 1000—where the standard is sulphur from the troilite phase from the Canyon Diablo Meteorite. Thus, a $\delta^{34}S$ value >0 indicates a relative enrichment of the $\delta^{34}S$ isotope and $\delta^{34}S$ 0 would indicate relatively larger amounts of the ^{32}S isotope relative to the standard.

There are a number of chemical and physical processes that influence this ratio (Harrison & Thode 1958, Nakai & Jensen 1964). It is well known that when bacteria reduce readily available sulphate to hydrogen sulphide, the resulting hydrogen sulphide is significantly enriched in the lighter isotope (Kaplan *et al.* 1963, Kaplan & Rittenburg 1964).

Sulphur isotope work on coals by Smith & Batts (1974) and Price & Shieh (1979), demonstrated that low organic sulphur ($<1\%$) coals have a rather narrow isotopic variation ($+4.6$ to $+7.3$ per mil); but, high organic sulphur ($>1\%$) coals have a large isotopic variation ($+2.9$ to $+24.4$ per mil). Information on coal sulphur isotopes, coupled with ideas on peat sulphur incorporation via hydrogen sulphide, led the author to analyse a selected number of peat samples from different salinity regimes and different source vegetation.

In low sulphur, freshwater-derived peats, from the Okefenokee Swamp, sulphur isotope ratios vary from $+9$ to $+13$ per mil. Sulphur isotope distribution in a peat profile is presented in Fig. 14. The isotope ratios represent a relatively close grouping of values with a $\delta^{34}S$ range of 5.2.

From this information and other data on low sulphur peats we concluded that:

(1) The distribution of $\delta^{34}S$ was remarkably uniform in terms of: (a) geographical area in the same swamp, (b) source plant materials, (c) peats from different depths of the same core and (d) peat fractions.

(2) The $\delta^{34}S$ of the peat is similar to that of sulphate sulphur in associated freshwater ($\delta^{34}S = +9$). Thus, little isotopic fractionation has taken place.

(3) Isotope values from the plants ($+10$ per mil) were similar to isotope values from associated peats and waters substantiating the findings of Chukrov *et al.* (1980) that plants do not fractionate sulphur derived from sulphate.

We carried out a similar sulphur isotope investigation in the marine influenced peats of

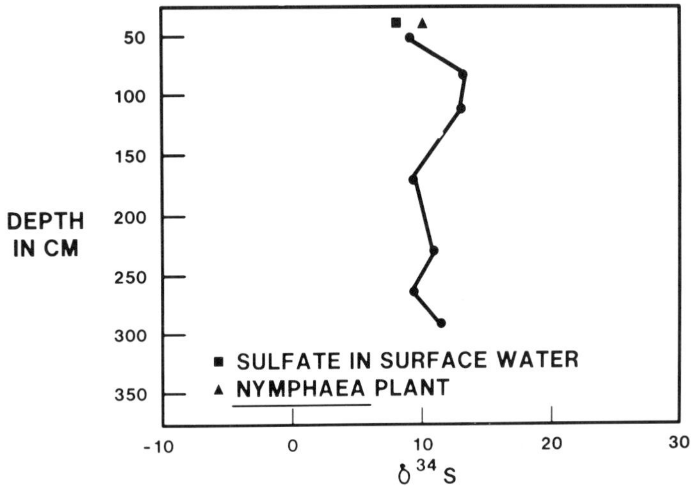

FIG. 14. Sulphur isotope ratios in Okefenokee peat.

the Florida Everglades. As indicated previously, there are substantial amounts of sulphur (1–10%) associated with this sediment.

The distribution of $\delta^{34}S$ in peats from a marine-influenced system (Fig. 15) shows:

(1) Negative isotope values in the peats (−8 to −30 per mil).

(2) There is substantial isotopic fractionation between seawater sulphate (+20 per mil) and peats (−8 to −30 per mil), but not as much fractionation between seawater sulphate and plant sulphur (+6 to +13 per mil).

In a continuation of the basic sulphur isotope study on whole peat samples, we carried out sulphur isotope analysis on sulphur forms in a peat profile. The peat profile studied was collected from the Little Shark River area of the Florida Everglades. Figure 16 shows the variation in plant type and environmental interpretation in the core profile. The marine-influenced peat overlies brackish- and freshwater-influenced peat. Methodology for petrographic analysis can be found in Cohen (1968).

Figure 17 shows sulphur isotope ratios of sulphate, pyrite and organic sulphur versus peat depth. The peat-forming environment, as interpreted from petrographic data, is indicated by wavy lines. The $\delta^{34}S$ of sulphate is very similar to current seawater sulphate. The $\delta^{34}S$ of pyrite sulphur and organic sulphur are considerably more negative than the $\delta^{34}S$ of associated sulphate. These data indicate considerable bacterially-mediated sulphate reduction to hydrogen sulphide. The resulting hydrogen sulphide then participates in reactions that yield pyrite and organic sulphur.

Because the $\delta^{34}S$ values for organic and pyritic sulphur are so negative, we suggest a virtually limitless supply of sulphate is available from associated waters. If sulphate ions were limited, one would anticipate that pyrite and/or organic sulphur isotope ratios would approach the original $\delta^{34}S$ of seawater (+20 per mil). It is anticipated that as the peat is consolidated and less sulphate is available, more positive $\delta^{34}S$ values may result.

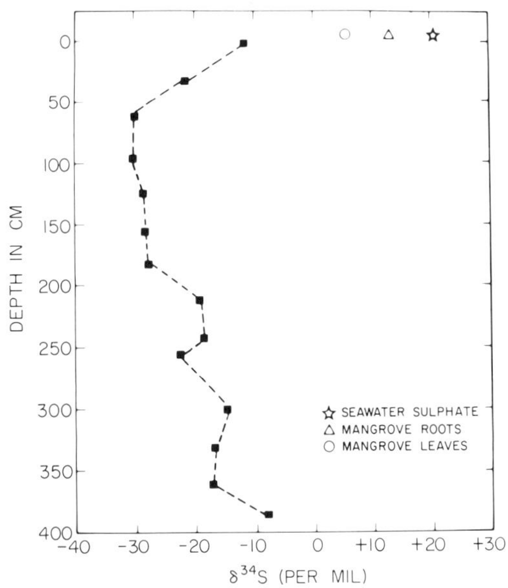

FIG. 15. Sulphur isotope ratios in Little Shark peat.

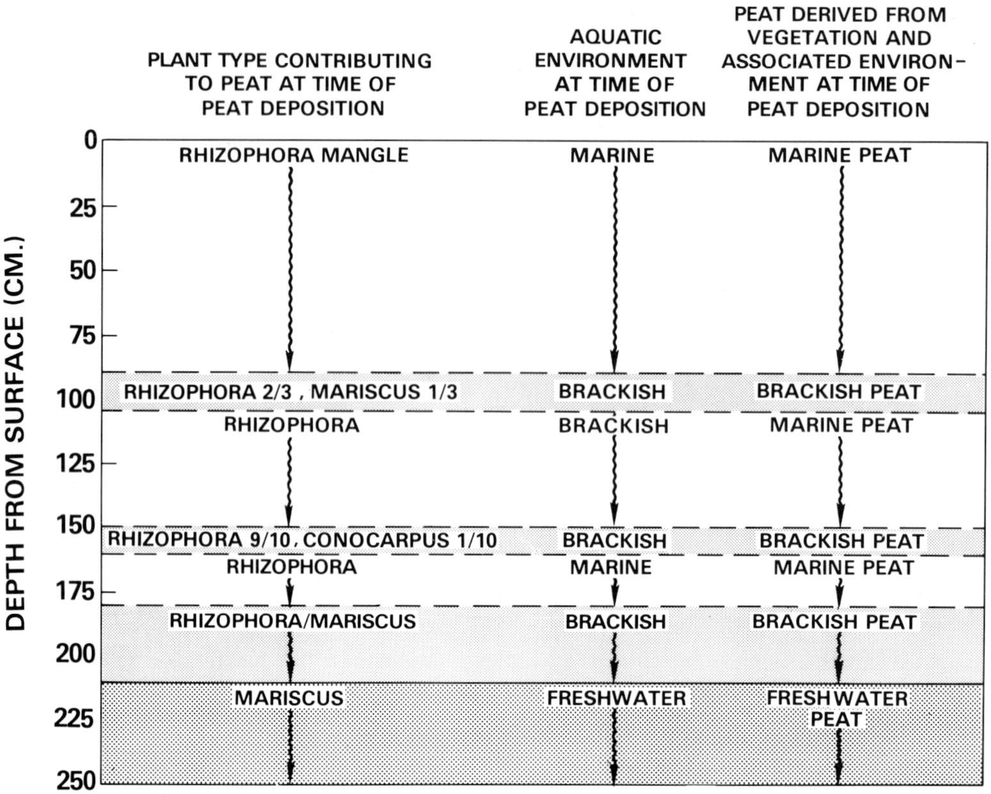

FIG. 16. Peat petrology of Little Shark site.

When a comparison is made of the $\delta^{34}S$ of organic sulphur and pyritic sulphur from marine inundated freshwater peat with $\delta^{34}S$ of non-marine influenced freshwater peats (Fig. 7) the impact of marine water derived sulphate is evident. Thus marine waters can easily mask a previous freshwater isotope signature. This could have implications on what is eventually found in coal and how environments of coal deposition are reconstructed.

Results from the investigations discussed above point to a relatively complex set of processes that will control the quantity and distribution of coal sulphur.

Figure 18 is a crossplot of sulphur isotope values from low and high organic sulphur coals (Price & Shieh 1979) and low and high organic sulphur peats. Both low sulphur peats and coals have relatively narrow ranges of isotope ratios associated with them, whereas high organic sulphur peat and coal have a wide range of isotope ratio variation. It should be noted that the peats have not been influenced by epigenetic sulphur as have the coals. More positive $\delta^{34}S$ values for hydrogen sulphide (and hence pyritic and organic sulphur) may be expected for higher sulphur peats, with diminishing sulphate content.

Sulphur geochemistry model

Figure 19 is a schematic diagram that shows the fate of sulphur in peat-forming systems. The central role of sulphate and its impact on eventual coal sulphur content is evident. However, other factors are equally important; bacteria must reduce sulphate to elemental sulphur and/or hydrogen sulphide before the diagenetic sulphur is incorporated. Sulphate-reducing bacteria have anaerobic, nutrient and pH requirements that may not always be met in all areas of a peat-forming environment.

Syngenetic coal pyrite content and morphology hinges on the availability of sulphate, functioning bacteria (pH and nutrients) and ferrous iron. Those areas in a coal-forming environment influenced by higher sulphate content will yield lignites having higher sulphur content than areas

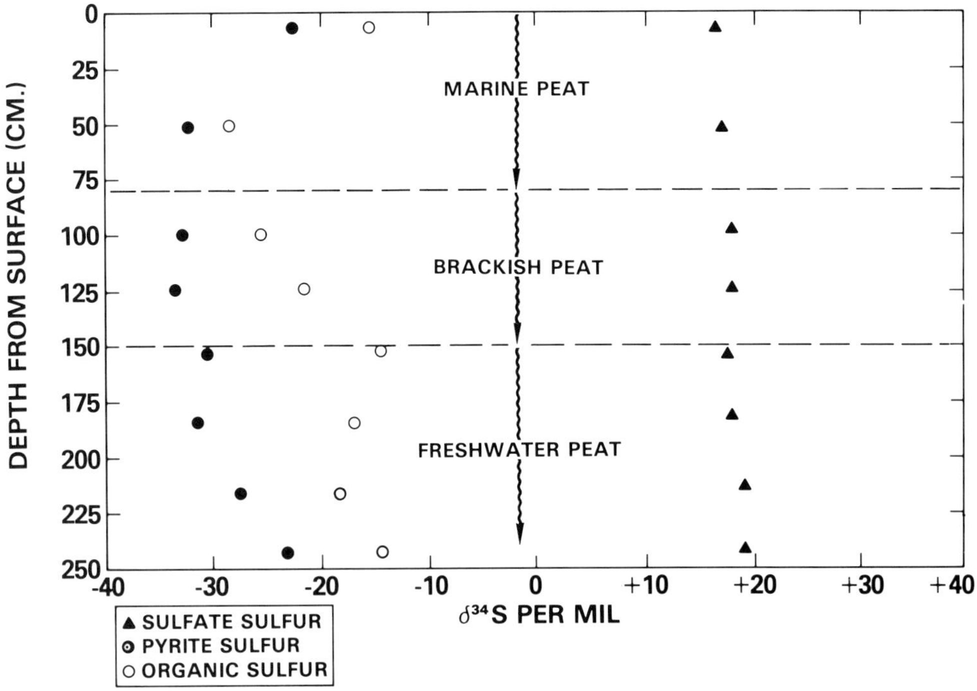

FIG. 17. Sulphur isotope ratios of sulphate, organic and pyrite sulphur in peat.

of the same seam that were not marine-influenced. Ferrous iron may only be available through the deposition of clay particles coming into the system from fluvial sources. Ferrous ions could easily be in limited supply in some areas of the swamp. These processes coupled with reactant availabilities yield the sometimes large variability of total sulphur and sulphur forms isotopic values in both vertical and horizontal seam profiles.

The origins of epigenetic pyrite are less well understood. It may be difficult to imagine similarities between pyrite-forming processes occurring at the cleat-forming stage (or beyond the lignite stage) and the unconsolidated peat-forming stage.

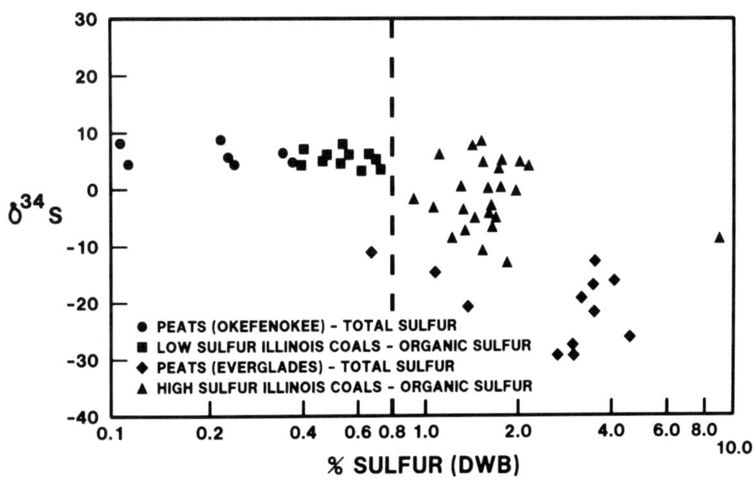

FIG. 18. Crossplot of sulphur isotope ratios versus sulphur content in Illinois No. 6 coal versus peat.

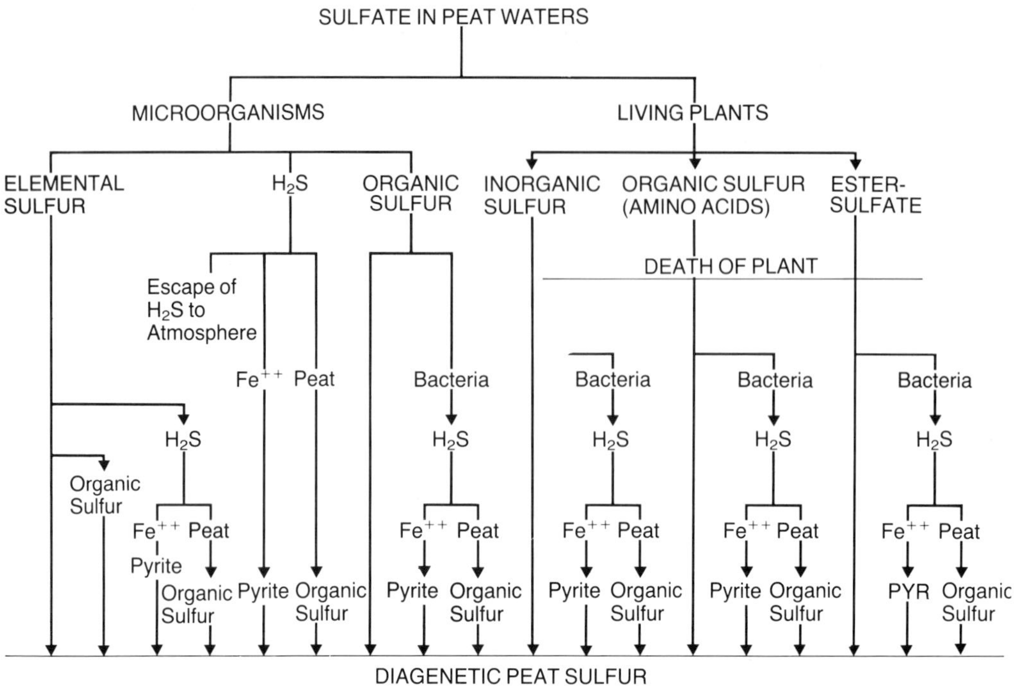

FIG. 19. Sulphur diagenesis in peat.

However, the physical evidence for epigenetic pyrite formation (cleat-filling and fusian-filling) appears overwhelming.

Organic sulphur in low sulphur coals comes mainly from plant material organic sulphur. It is likely that little organic sulphur in low sulphur coals originates from sulphate through hydrogen sulphide formation. Similarities of organic sulphur contents, low pyrite contents and sulphur isotope composition of low sulphur peats and coals are too striking to imagine any other dominant sulphur-incorporating process.

Organic sulphur in high sulphur coals undoubtedly originates from sulphate through hydrogen sulphide. The question is one of timing; when does the organic sulphur get incorporated? We have shown that all the organic sulphur in coal can be accounted for in the peat-forming stage. However, were all high sulphur coals laid down under marine conditions? The author feels that this is most unlikely. It is more likely that most high sulphur coals were originally laid down under freshwater conditions and subsequently influenced by high sulphate waters. Did organic sulphur enrichment occur at the time of pyrite content enrichment (after cleat formation)? If this is true one would expect organic sulphur enrichment near cleats. The author is not aware of research in this area. A more likely scenario is that ancient peats were laid down under freshwater conditions and were inundated by marine waters during marine transgressions. This would allow for a marine sulphur signature to be impressed on a freshwater sulphur signature, much like that observed in the Everglades Little Shark site.

One troubling component of the isotope investigation is the very negative isotope values in high organic sulphur peats versus the more positive values in Illinois high organic sulphur coals. Figure 18 compares organic sulphur in the coals and total sulphur including pyrite in the peat. It is possible that as sulphate content became limited in high sulphur environments, the isotope ratio of produced H_2S became more positive (closer to seawater sulphate) yielding more positive organic and pyritic sulphur isotope values. This, however, is a minor quantitative addition to the bulk of earlier available sulphur.

Figure 20 is a schematic showing the incorporation of epigenetic sulphur into coal. It is clear that there are numerous timing possibilities for sulphur incorporation. It is little wonder that the result is a wide range of sulphur contents and sulphur forms and isotopic values in coals.

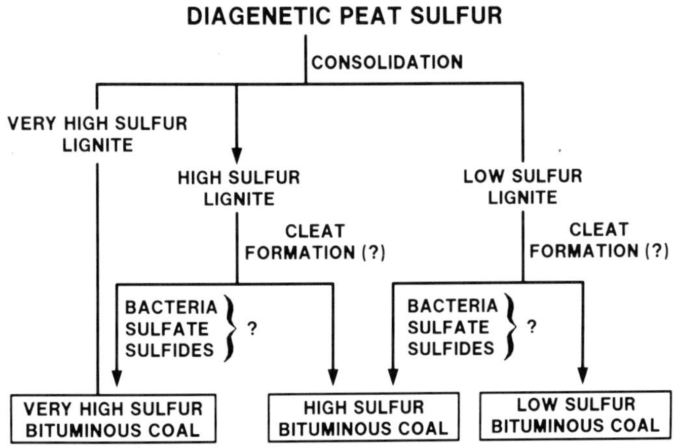

FIG. 20. Epigenetic sulphur: origin of high sulphur bituminous coals.

Conclusions

The salient conclusions derived from this study are as follows:

(1) With increased environmental regulations, there are important economic incentives to explore for low sulphur coals.

(2) To explore for low sulphur coals in a province, basin or prospect, a suitable sulphur geochemical model needs to be developed.

(3) Sulphur content of US coals is highly variable with a range of 0.1 to over 10%; pyritic sulphur and organic sulphur are the dominant sulphur forms.

(4) Virtually all the sulphur in coal can be accounted for in the peat-forming stage, where syngenetic sulphur (pyrite and organic sulphur forms) are incorporated.

(5) Sulphate content, an anaerobic environment suitable for microbial sulphate reduction, ferrous iron and peat organic matter are the principal ingredients for the formation of syngenetic pyrite and organic sulphur.

(6) Hydrogen sulphide formed from microbial reduction of sulphate (1) reacts with organic matter to form organic sulphur and (2) reacts with ferrous iron to produce pyrite.

(7) The availability of appropriate reducing environments and ferrous iron will dictate the formation of pyrite.

(8) Sulphur isotope values of modern peat pyrite and organic sulphur are similar to values found in high and low sulphur coals.

ACKNOWLEDGEMENTS: I thank the authors in the various investigations that were cited. William Spackman and Arthur Cohen are thanked for their help in the field and for providing petrography on cores.

I thank Fred Price for his contribution to the sulphur isotope component of this work and I thank Fred Price, Robert Finkelman and Jesse Yeakel for their review of this manuscript.

References

ALTSCHULER, Z. S., SSHNOPFE, M. M., SILBER, C. C. & SIMON, F. O. 1983. Sulfur diagenesis in Everglades peat and the origin of pyrite in coal. *Science*, **221**, 221–227.

ATTAR, A. 1979. Sulfur groups in coal and their determination. *In: Analytical Methods for Coal and Coal Products, Vol. 111*, pp. 353–357. Academic Press, New York.

—— & DUPUIS, F. 1979. Data on the distribution of organic sulfur functional groups in coal. *Preprints for Division of Fuel Chemistry, American Chemical Society* 24(1), 166–177.

BERNER, R. A. 1970. Sedimentary pyrite formation. *American Journal of Science*, **268**, 1–23.

—— 1984. Sedimentary pyrite formation: an update. *Geochimica et Cosmochimica Acta*, **48**, 605–615.

—— 1985. Sulphate reduction, organic matter decomposition and pyrite formation. *Philosophical Transactions of the Royal Society of London A*, **315**, 25–38.

BERTELOOT, J. 1947. Presence of native sulfur in coal. Variations in the sulfur content from top to bottom of a coal seam. *Annals of the Geology Society of Norway*, pp. 195–206; *Chemical Abstracts*, **44**, 818.

BROOKS, J. P. 1956. Organic sulfur in coal. *Journal of the Institute of Fuels*, **29**, 82–90.

CADY, G. H. 1935. Distribution of sulfur in Illinois coals and its geological implications. *Illinois Geological Survey Report of Investigations*, 35.

CASAGRANDE, D. J. 1984. Organic geochemistry of Okefenokee peats. *In:* COHEN, A., CASAGRANDE, D., ANDREJKO, M. & BEST, R. (eds) *The Okefenokee Swamp: its Natural History, Geology and Geochemistry*. Wetland Surveys, Los Alamos.

—— 1985. The distribution of sulfur in progenitors of low sulfur coal: the origins of organic sulfur. *In:* CROSS, A. T. (ed.) *Neuvième Congrès International de Stratigraphie et de Géologie du Carbonifere*, **4**, pp. 299–307, May 1979.

——, GRONLI, K. & PATEL, V. 1978. The distribution of sulfur in organic fractions of peat: origins of sulfur in coal. *In:* KRUMBEIN, W. E. (ed.) *Environmental Biogeochemistry and Geomicrobiology*, pp. 515–521. Ann Arbor Science, Ann Arbor.

——, —— & SUTTON, N. 1980. The distribution of sulfur and organic matter in various fractions of peat: origins of sulfur in coal. *Geochimica et Cosmochimica Acta*, **44**, 25–32.

——, IDOSU, G., FRIEDMAN, A., RICKERT, P., SIEFERT, K. & SCHLENZ, D. 1979. H_2S incorporation in coal precursors: origins of organic sulphur in coal. *Nature, London*, **282**, 599–600.

—— & NG, L. 1979. Incorporation of elemental sulphur in coal as organic sulphur. *Nature, London*, **282**, 598–599.

—— & PRICE, F. 1981. Sulfur isotope distribution in coal precursors. *GSA Annual Meeting, Cincinnati Abstract.*

—— & SIEFERT, K. 1977. Origins of sulfur in coal: Importance of ester-sulfate content. *Science*, **195**, 675–676.

——, ——, BERSCHINSKI, C. & SUTTON, N. 1977. Sulfur in peat-forming systems of the Okefenokee Swamp and Florida Everglades: origin of sulfur in coal. *Geochim et Cosmochim Acta*, **41**, 161–167.

CHUKROV, F. S., ERMILOVA, L. P., CHURIKOV, V. S. & NOSIK, L. P. 1980. The isotope composition of plant sulfur. *Organic Geochemistry*, **2**, 69–75.

COHEN, A. D. 1968. The petrology of some peats of Southern Florida. *PhD thesis.* Pennsylvania State University.

——, SPACKMAN, W. & DOLSEN, C. P. 1983. Occurrence and distribution of sulfur in peat-forming environments of southern Florida. *In:* RAYMOND, R. & ANDREJKO, M. J. (eds) *Workshop on Mineral Matter in Peat*, pp. 87–112. Los Alamos.

DAVIS, A. 1982. Sulfur in coal. *Earth and Mineral Science Letters*, **51**(2), 1–18.

FINKELMAN, R. B. 1980. Modes of occurrence of trace elements in coal. *PhD thesis*, University of Maryland. (Also *United States Geological Survey Open File*, 81–99.)

FRENEY, J. R. 1967. Sulfur-containing organics. *In:* MCCLAREN & PETERSON (eds) *Report Soil Biochemistry*, pp. 229–259. Marcel Dekker, New York.

GIVEN, P. H. & MILLER, R. N. 1985. Distribution of forms of sulfur in peats from saline environments in the Florida Everglades. *International Journal of Coal Geology*, **5**, 397–409.

——, SPACKMAN, W., IMBALZANO, J. R., CASAGRANDE, D. J., LUCAS, A. J., COOPER, W. J. & EXARCHOS, C. 1983. Physiochemical characteristics and levels of microbial activity in some Florida peat swamps. *International Journal of Coal Geology*, **3**, 77–99.

—— & WYSS, W. F. 1961. The chemistry of sulfur in coal. *British Coal Utilization Research Association Monthly Bulletin*, **25**, 165–179.

GLUSKOTER, H. J. 1977. Inorganic sulfur in coal. *Energy Sources*, **2**, 125–131.

—— & SIMON, J. A. 1968. Sulfur in Illinois coals. *Illinois State Geological Survey Circular* 432. Urbana, Illinois.

GOLDHABER, M. B. & KAPLAN, I. R. 1975. The sulfur cycle. *In:* GOLDBERG, E. P. (ed.) *The Sea*, pp. 569–655. Wiley, New York.

—— & —— 1980. Mechanisms of sulfur incorporation and isotope fractionation during early diagenesis in sediments of the Gulf of California. *Marine Chemistry*, **9**, 95–143.

HARRISON, A. G. & THODE, H. G. 1958. Mechanism of the bacterial reduction of sulphate from isotope fractionation studies. *Transactions of the Faraday Society*, **54**, 84–92.

HOWARTH, R. W. & JORGENSEN, B. B. 1984. Formation of ^{35}S-labelled elemental sulfur and pyrite in coastal marine sediments (Limfjorden and Kysing Fjord, Denmark) during short term $^{35}SO_4$-reduction measurements. *Geochimica et Cosmochimica Acta*, **48**, 1807–1818.

JORGENSEN, B. B. 1982. Ecology of the bacteria of the sulfur cycle with special reference to anoxic-oxic interface environments. *Transactions of the Royal Society of London B*, **298**, 543–561.

KAPLAN, I. R., EMERY, K. O. & RITTENBERG, S. C. 1963. Redistribution and isotopic abundance of sulfur in recent marine waters off Southern California. *Geochimica et Cosmochimica Acta*, **27**, 297–331.

—— & RITTENBURG, S. C. 1964. The microbiological fractionation of sulfur isotopes. *Journal of General Microbiology*, **34**, 195–212.

LOVE, L. J., COLEMAN, M. L. & CURTIS, C. D. 1983. Diagenetic pyrite formation and sulfur isotope fractionation associated with a Westphalian marine incursion, northern England. *Transactions of the Royal Society, of Edinburgh Earth Science*, **74**, 165–182.

MANGO, F. D. 1983. The diagenesis of carbohydrates by hydrogen sulfide. *Geochimica et Cosmochimica Acta*, **47**, 1433–1442.

MCCABE, P. J. 1984. Depositional environments of coal and coal-bearing strata. *In:* RAHMANI, R. A. & FLORES, R. M. (eds) *Sedimentology of Coal and Coal-bearing Sequences*, pp. 13–42. Special Publication of the International Association of Sedimentologists, **7**, Blackwell Scientific Publications, Oxford.

MIURA, Y. & YANAGI, Y. 1963. Chemical studies on the sulfur of Japanese coals (II). *Chemical Abstracts*, **61**, 10,501 (1964).

NAKAI, N. & JENSEN, M. L. 1964. The kinetic isotope effect in bacterial reduction and oxidation of sulfur. *Geochimica et Cosmochimica Acta*, **28**, 1983–1912.

NEAVEL, R. 1966. Sulfur in coal: its distribution in the seam and mine products. *PhD thesis*. The Pennsylvania State University.

PRICE, F. T. & SHIEH, Y. N. 1979. The distribution and isotopic composition of sulfur in coals from the Illinois Basin. *Economic Geology*, **74**, 1445–1461.

RAYMOND, R. 1979. Correlating organic sulfur in coal to depositional environments by electron probe microanalysis. *Ninth International Meeting of Carboniferous Stratigraphy and Geology*, Champaign, Urbana, Illinois.

REYES-NAVARRO, J. & DAVIS, A. 1976. Pyrite in coal—its forms and distribution as related to the environments of coal deposition in three selected coals from western Pennsylvania. *Special Research Report* ST-110, College of Earth and Mineral Sciences, The Pennsylvania University.

RICKARD, D. T. 1975. Kinetics and mechanism of pyrite formation at low temperatures. *American Journal of Science*, **275**, 636–652.

REIDENOUER, D., WILLIAMS, E. G. & DUTCHER, R. R. 1967. The relationship between paleotopography and sulfur distribution in some coals of western Pennsylvania. *Economic Geology*, **62**, 632–649.

RODIONOVA, L. E. & BARANSKII, A. O. 1970. Active sulfur groups of Irkutsk coals studied by a chemical reaction method. *Izvestisa Nauchno-Issled. Inst. Nefte-Uglekhim. Sint. Irkutsk, Univ.* 12, 93; *Chemistry Abstracts* **45**, 5122 (1971).

RUCH, R. H., GUSKOTER, H. J. & SHIMP, N. F. 1974. Occurrence and distribution of potentially volatile trace elements in coal. *Illinois State Geology Survey Environmental Geology Notes* 72. Urbana, Illinois.

SCHWAB, A. W., Gast, L. E. & ROHWEDDER, W. K. 1976. Nucleophilic and radical addition of H_2S to methyl linolenate and linseed oil. *Journal of the American Oil Chemical Society*, **53**, 762–766.

SHOREY, E. C. 1913. Some organic soil constituents. *Bureau of Soils Bulletin* 88. US Department of Agriculture.

SMITH, J. W. & BATTS, B. D. 1974. The distribution and isotopic composition of sulfur in coal. *Geochimica et Cosmochimica Acta*, **38**, 121–133.

SPACKMAN, W., COHEN, A. D., GIVEN, P. H. & CASAGRANDE, D. J. 1976. *The comparative study of the Okefenokee Swamp and the Everglades—mangrove swamp—marsh complex of southern Florida.* Field guidebook printed for the Geological Society of America, pre-convention field trip, 15–17 November 1974. (Subsequently published by Coal Research Section, The Pennsylvania State University.)

SPEIGHT, J. G. 1978. Assessment of structures in coal by spectroscopic techniques. *In: Analytical Methods for Coal and Coal Products*, Volume II, pp. 75–101. Academic Press, New York.

THIESSEN, R. 1920. Occurrence and origin of finely disseminated sulfur compounds in coal. *Transactions of the American Institute of Mining Engineers*, **63**, 913–926.

—— 1945. Forms of sulfur in coal. *In:* LOWRY, X. (ed.) *Chemistry of Coal Utilization*, Vol. 1, pp. 425–449. Wiley, New York.

U.S. DEPARTMENT OF INTERIOR, Washington, D.C. 1966. The reserve base of U.S. coals by sulfur content. *U.S. Bureau of Mines Information Circular* 8680.

WALKER, F. E. & HARTNER, F. E. 1966. Forms of sulfur in coals. *U.S. Bureau of Mines Information Circular* 8301.

WANLESS, A. M. 1959. The occurrence of sulfur in British coals. *Journal of the Institute of Fuels*, **32**, 258–266.

WILLIAMS, E. G. & KEITH, M. L. 1963. Relationship between sulfur in coals and occurrence of marine roof beds. *Economic Geology*, **58**, 720–729.

YUROVSKII, A. Z. 1959 Basis of the theory of the origin of sulfur in coal. *Symposium on Nature of Coal Production*, pp. 51–55. Central Fuel Research Institute, Jealgora, India.

—— 1960. *Sulfur in Coals*. USSR Academy of Sciences, Institute of Mineral Fuels. (English translation by Indian Nation Scientific Documentation Center, 1974.)

ZOBELL, C. E. 1963. Organic geochemistry of sulfur. *In:* BREGER, I. A. (ed.) *Organic Geochemistry*, pp. 543–578. Pergamon Press, New York.

D. J. CASAGRANDE, Exxon Production Research Co., Houston, Texas 77036, USA. Present address: Environmental and Coal Associates, PO Box 771785, Houston, Texas 77215, USA.

Interpreting the characteristics of coal seams from chemical, physical and petrographic studies of peat deposits

Arthur D. Cohen, William Spackman & Robert Raymond Jr

ABSTRACT: The purpose of this paper is to illustrate some of the kinds of information about coals that are at present being generated from studies of peat deposits and also to point out some possible new areas of research that might be undertaken in the future.

Some notable examples of new ideas about coal seam composition or formation that have been generated from studies of modern peat deposits include: (1) discovery of the probable progenitors of certain coal macerals and elucidation of the processes by which they may have formed; (2) evidence that some types of mineral matter may be dissolved out of peat deposits; (3) verification of the role of marine waters in emplacement of sulphur in peat; (4) discovery of the importance of 'doming' in peat deposit development; (5) discovery of a new way to form a split in a coal seam, i.e. development of a 'fire splay'; (6) elucidation of the mechanisms responsible for producing stratification in coal seams; (7) recognition of the world-wide importance of 'back-barrier' coal-forming environments.

These kinds of discoveries are important in themselves; however, they can also be shown to have many additional practical applications, especially if woven into models to predict the economic characteristics of coal seams.

Keywords: coal, peat, petrography, organic

The purpose of this paper is to review some of the ways in which studies of modern peat deposits have been used in the past or could be used in the future to help understand and predict the characteristics of coal seams. It is not our purpose to review all of the published literature on peats but rather to give examples of the kinds of studies that have been undertaken (or are at present being undertaken) that are particularly relevant to interpreting coal seam characteristics. Where applicable, we have given examples from our own research rather than that of others because we are more intimately familiar with its basis. For that reason, emphasis is on studies of the peat deposits of the southeastern United States.

General characteristics of peat deposits

Peat is a sedimentary deposit formed primarily from plant tissues (and their products) and secondarily from inorganic substances. It is, thus, an aggregate of 'macerals' (or 'pre-macerals') and minerals that, when subjected to certain (still somewhat poorly understood) processes, can be converted to coal. Given the present state of our knowledge with respect to coalification processes, it is difficult to place a precise boundary between what might be a precursor of a carbonaceous shale and what might be a precursor of a coal. However, nearly everyone agrees that the starting material must be primarily organic, and it would probably not be too controversial to say that the precursor of a good coal should consist of at least 70% (by dry weight) organic material.

In order for peat to accumulate in large quantities, three requirements must be satisfied: (1) there must be a large enough source of plant tissues; (2) organic decomposition processes must be retarded; and (3) the rate of influx of inorganic sediments must be low (see also Moore, Clymo, *this volume*). These requirements can be satisfied in a number of different ways, and in fact, these three requirements are interdependent. If one parameter happens to go up or down, another must also change in order for peat of the 'same' quality to accumulate. For example, if the rate of decomposition increases (due perhaps to a temperature increase), then the phytogenic sedimentation rate must also increase and/or the inorganic sedimentation rate must decrease to produce a coal precursor with the same 'ash' content. However, nothing is that simple in nature because changing the temperature would also undoubtedly change the source plant community, which would in turn change the organic (and possibly also the inorganic) composition of the coal.

Because peat is an aggregate of both organic and inorganic substances, its formation is very complex, being a blend of both 'normal' inorganic sedimentation processes and organic sedimentation processes. A single layer within a peat deposit contains substances that have arrived at that spot by many different mechanisms and through many different stages of transformation. For example, one leaf from a tree may fall directly into the waters of a swamp to be exposed to very little decay or shrivelling, whereas another leaf from the same tree may fall on a dry spot and be altered by oxygen, fire, aerobic microorganisms, etc., before it is eventually washed or blown into the deposit. The two leaves (or their remains) may become juxtaposed in the deposit and may later become part of a coal seam. However, even though they were derived from the same plant, they have produced somewhat different macerals in the coal because their alteration pathways were fixed before they entered the deposit. Furthermore, accidents of association can cause changes to occur within 'microenvironments' that may be as small as a few square micrometres in size. It is the rule (not the exception) that, in any one view of peat through a microscope at 100 ×, one would expect to find plant substances (even from the same species) at many different stages of alteration.

Thus, the number of factors affecting the composition of a peat deposit may be significantly greater than those affecting an inorganic sedimentary deposit. Some of these factors are discussed below.

Sources of primary ingredients

Botanical constituents

Peats generally form *in situ* (although they may rarely form entirely from transported material and always have some small transported component). Therefore most of the organic components of peats come from plants growing within wetland environments, where surface water flow is slow and where water is at or near the surface for most of the year.

This persistence of water restricts the kinds and morphologies of plants contributing to peat deposits, but still a great many different types of plants can exist under these conditions. Peats may, for example, be produced from gymnospermous woody plants (e.g. *Picea*, *Taxodium*) or angiospermous woody plants (e.g. *Persea*), from floating herbaceous aquatics (e.g. *Nymphaea*), or emergent herbaceous aquatics (e.g. grasses and sedges). They may also be formed predominantly from lower plants, such as mosses (*Sphagnum*) or algae (Cohen *et al.* 1984). Peat deposits are most common in freshwater settings but they may form in brackish or marine environments (e.g. mangrove swamps or salt marshes).

Peats occur in almost every climatic zone and at almost any altitude where source plants and hydrologic conditions are suitable. For example, peat deposits of considerable thickness and geographical extent occur in lowland tropical regions of Costa Rica and Panama (Cohen *et al.* 1985a, Cohen & Raymond, unpublished report), in temperate regions such as New York and Pennsylvania (Cameron 1970a, b), or in cold climates such as northern Minnesota (Malterer 1979) and Canada (Radforth & Browner 1977). Peats of considerable thickness have been found in the Hula Valley in Israel, where a locally wet environment occurs within an otherwise desert setting (Jones 1940) and in the high mountains of tropical Costa Rica (Cohen *et al.* 1984).

Soil and water chemistry can also restrict the kinds of source plants and composition of a peat formed in an area. If, for example, the subpeat surface is a limestone, only certain types of plants may occur; whereas, if it is a granite, other plant species might be expected.

Thus, although climate, hydrology, and even soil and water chemistry can play important roles in restricting the types of plants contributing to a peat deposit, the variety of potential source plants is extensive. Furthermore, as is evident from the above discussion, the simple presence of a peat deposit (or coal seam) is not a good indicator of position of sea-level, altitude, climatic conditions, or even water chemistry. It is, therefore, imperative that the botanical composition of the peat or coal be determined before any significant conclusions are drawn regarding the depositional environment or ecological setting under which the peat or coal was formed (see also Collinson & Scott, *this volume*).

Mineralogical ingredients

All peats contain some mineral matter and ions that will become a portion of the 'ash' in a coal if the coal is burned (see also Casagrande, Spears, *this volume*). The solid inorganic constituents can be introduced initially into a peat deposit as: (1) detrital grains deposited by wind, water, or gravity or (2) authigenic materials produced in place by biological or chemical processes. Additional mineral matter or ions may be introduced at various stages after deposition and through the progressive stages of coalification by percolation of fluids from above into underlying peat layers (e.g. sulphur as described by Cohen *et al.* 1984).

The types of detrital mineral grains introduced into a deposit are controlled by the composition of the surrounding rocks, local weathering phenomena, and various transportation processes. Generally, nearness to stream channels, tidal channels or volcanoes will play the biggest role in controlling the amount of detrital input into a deposit.

Authigenic mineral matter occurs in all peat deposits but may become the dominant form in larger deposits, where peat is forming far from any detrital input (Fig. 1), or in the upper portions of domed deposits, where the deposit has risen above the influence of tides or stream flooding (Coleman *et al.* 1970). Doming and mineral leaching will be discussed in more detail in later sections. Authigenic mineral matter produced by plants or animals is of special significance in many peat deposits. Siliceous plant phytoliths from grasses, sedges, or palms (e.g. Cohen 1974b) have sometimes been found to be the dominant inorganic constituents in a peat deposit.

Processes of accumulation of peat ingredients

Normal sedimentation processes

Plant materials may accumulate by normal sedimentary processes, such as falling into or being washed into a deposit. In the former process, plant parts simply fall on to the surface of the peat deposit as they die, accumulating in a random pattern with respect to compass coordinates but tending to align themselves parallel to the surface (if flat or elongated).

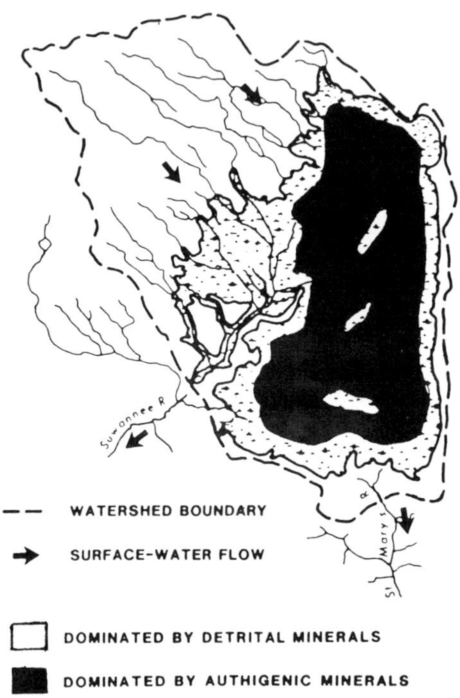

FIG. 1. Distribution of types of mineral matter in peats of the Okefenokee Swamp, Georgia, USA (after Andrejko 1986).

The latter process (i.e. accumulation by allochthonous processes) is much less common but does occasionally occur. A notable example of this type of deposit is the allochthonous ('coffee ground'-type) peat deposit described by Cohen (1970) from southern Florida. This deposit was formed as an organic 'beach sand' derived from streams carrying eroded peat particles from mangrove swamps into the Gulf of Mexico. In contrast with *in situ* deposits, current orientation of elongated particles is evident in this kind of deposit.

'Sedimentation' by plant growth

A form of organic sedimentation that has been greatly underestimated in the past as a means for increasing the size of peat deposits is 'sedimentation' by *growing* into a deposit (e.g. roots, rhizomes, or microorganisms). Petrographic studies of peats from southern Florida (Cohen 1968, Cohen & Spackman 1972, 1977), for example, revealed the dominance of roots composing many of these peats.

This form of sedimentation can not only increase the total volume of peat being accumulated but can also: (1) modify the texture of a peat; (2) increase the *in situ* density of the peat; (3) disrupt the bedding or microlamination of the peat; (4) cause compression of surrounding tissues by growth; (5) create new microenvironments within a deposit; (6) produce conduits for percolation of fluids or microorganisms from above (e.g. Raymond & Davies 1981); and (7) pull nutrients from lower layers up to the living plants, thus depleting lower layers of certain chemical constituents.

Furthermore, 'ingrown' ingredients do not go through a 'surface litter' stage, as do 'normally' sedimented plant fragments, so they may be well preserved even though surface decomposition processes may be too harsh to allow preservation of intact materials that accumulated by dropping on the surface. A good example of this process is illustrated by mangrove root peats from southern Florida (Fig. 2).

'Subterranean redeposition'

Another process that is peculiar to phytogenic organic sediments is the formation of floating mats (or peat 'batteries'). These batteries are separated from underlying peats by a water layer. Their morphologies and origins are described in

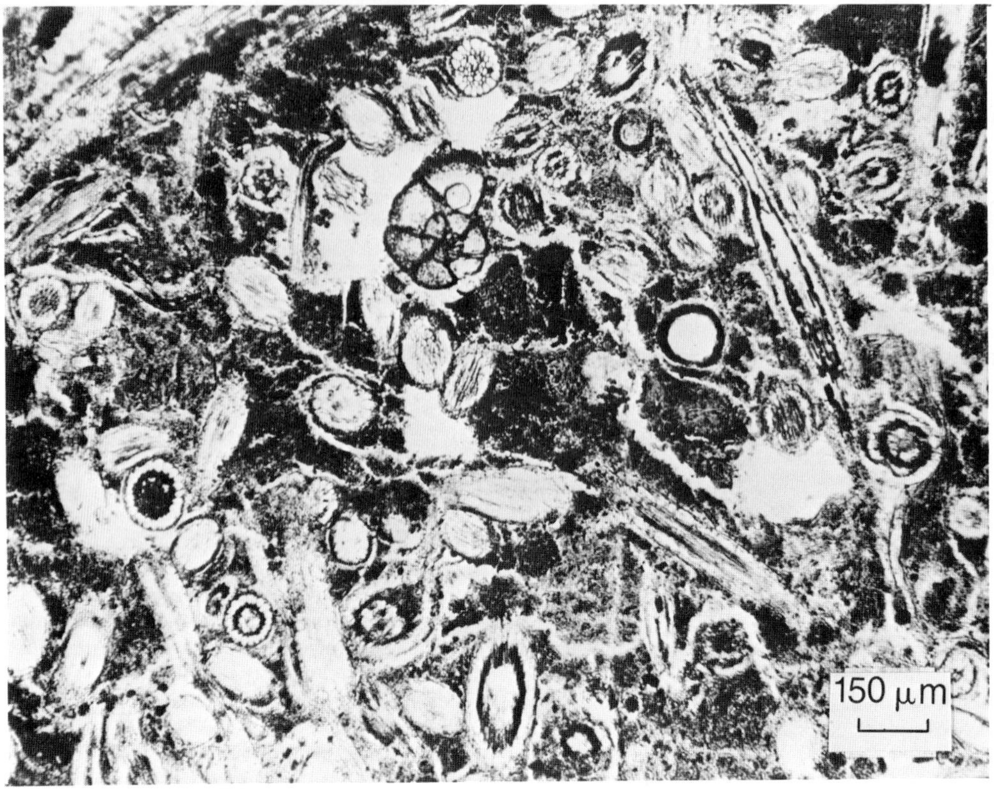

FIG. 2. Photomicrograph of an oriented microtome section of *Rhizophora* peat from southern Florida. Note that this peat is dominated by 'ingrown' sedimentary ingredients (roots).

detail in Cypert (1972) and Spackman *et al.* (1976).

Once these layers of peat separate from the underlying peat deposit or grow out into open water areas, organic components may filter down from them and through the water to form a new layer of fine-grained material between the mat and the underlying peat (i.e. a layer of redeposited material mixed with some recent algal and fecal material). This process may create some difficulties in interpreting the palaeoecological history of peat deposits and may also result in some anomalous radiocarbon dates. A physical result of this process may be the formation of distinct layering in certain peat deposits. This layering should be retained in the coal seams that might form from these deposits.

Processes of alteration of surface materials

Drying

If plant tissues become stranded on a dry surface (such as a tree stump) or fall to the surface and do not become buried, they may shrink because of moisture loss. For most dead plant tissues, this process is irreversible; i.e. rewetting will not cause the tissues to swell up again to their original shape. This shrivelling is most commonly observed for leaf tissues and is readily recognized when peats are observed microscopically using oriented microtome sections (Fig. 3). Occasionally a leaf will be observed that is shrivelled on one side and not shrivelled on the other. It is

FIG. 3. Photomicrographs of cross-sections of *Rhizophora mangle* leaves from southern Florida showing a modern leaf (a) and a shrivelled leaf of the same species in a microtome section of peat (b). The shrivelled leaf has undoubtedly gone through a surface litter stage in which it experienced drying and shrinkage.

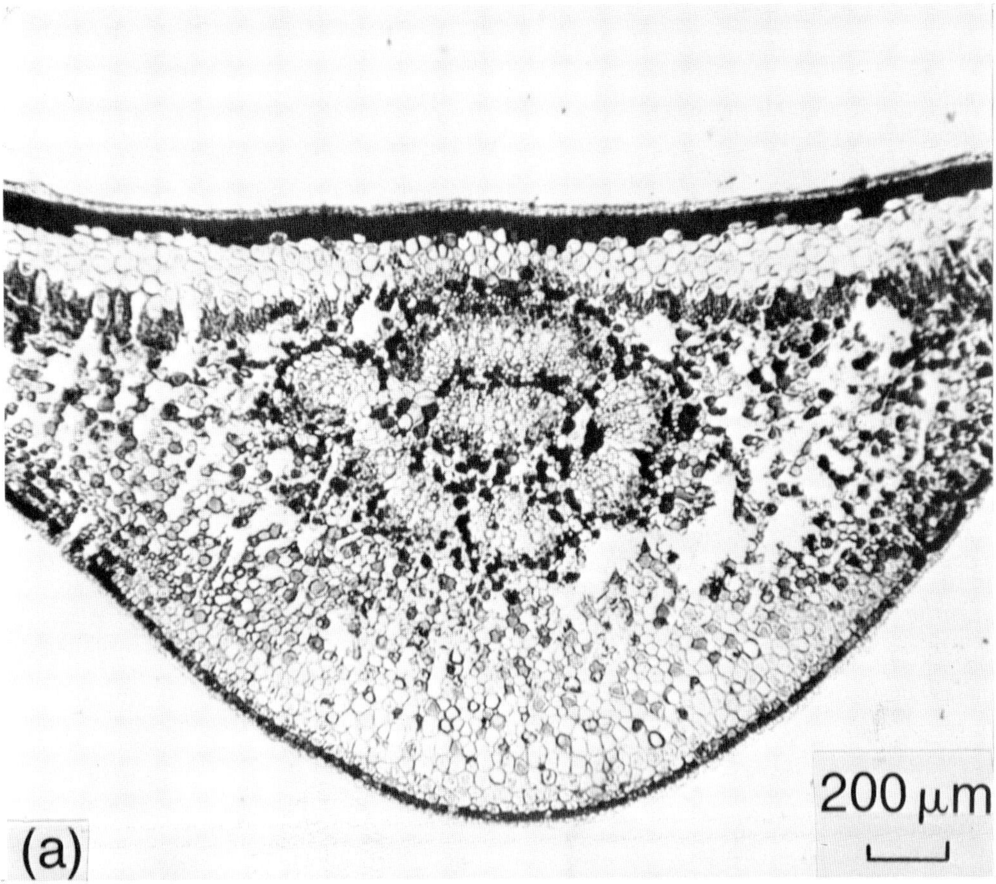
200 μm
(a)

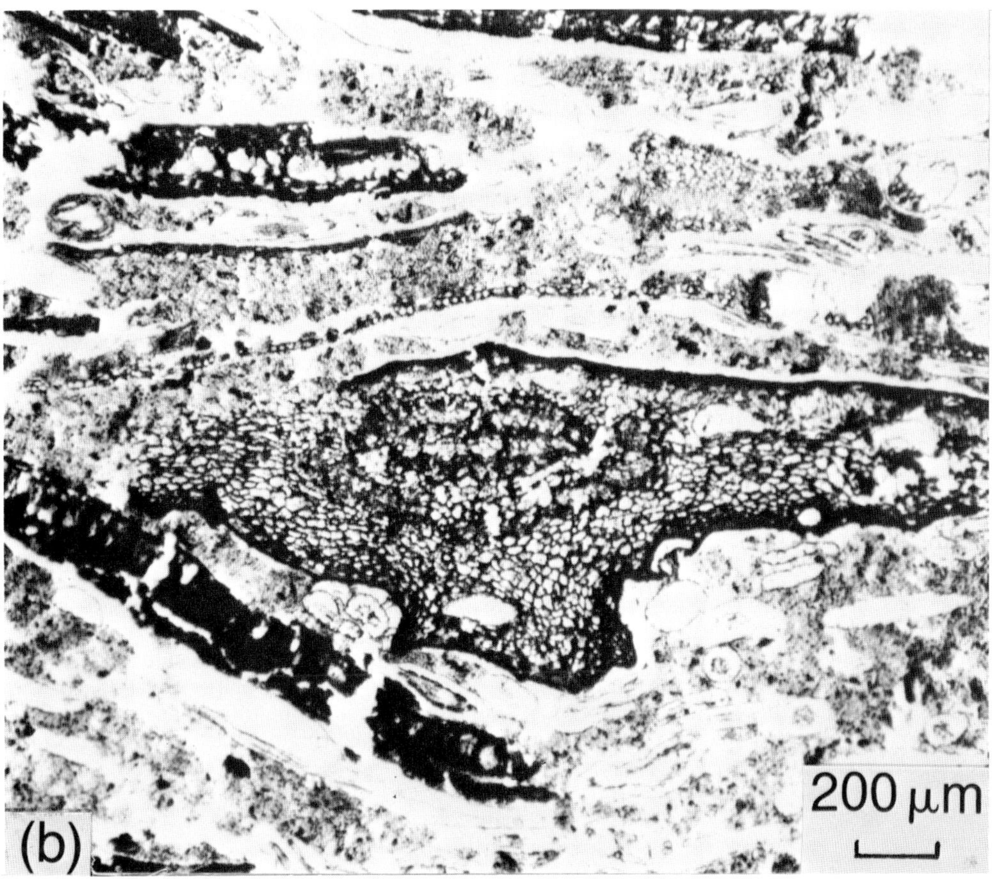
200 μm
(b)

evident that half of this leaf was exposed to the air and the other half was not.

Fires

Fires can affect living plant tissues, surface litter, or previously deposited peat. When living or recently dead tissues or surface litter are burned, varying amounts of charcoal (fusinite), semi-charcoal (semi-fusinite), and ash are produced depending upon the plant types being burned and the severity of the fire (Cohen 1974a, Cohen & Spackman 1977). Woody plants, for example, because of their density, will generally produce more charcoal than non-woody plants. However, whenever surface vegetation in a swamp is burned, regardless of the amount of charcoal produced, this charcoal always occurs in a matrix of unburned tissue fragments. On the other hand, when surface peat layers are burned, entire bands rich in fusinite may be produced. Burning of surface peat also causes a much more drastic change in the environment, sometimes even creating ponds or lakes when water levels return to normal (Cypert 1961), and always causing major vegetational changes that result in sharp breaks in a sequence and distinct banding (see also discussion of 'fire splays').

Fires may also cause changes in the chemistry of surface peat by releasing nutrients for growth of new plants, raising the pH, producing ephemeral minerals (Andrejko 1983), or concentrating unburnable ingredients such as sponge spicules or siliceous phytoliths (Fig. 4).

Ingestion or metabolism

Ingestion or metabolism of plant constituents can occur while the tissues are still attached to the living source plant, after tissues have fallen to the surface, or within the peat deposit. However, this process is most important in the surface portions of the deposit or before the ingredients are incorporated into the deposit.

Common metabolizers are animals, such as crabs, termites, worms or other arthropods. These organisms cause structural breakdown of tissues and cells as well as chemical breakdown during metabolism. They also produce fecal pellets (Fig. 5) and other waste materials.

Microorganisms (such as fungi and bacteria) can cause physical or chemical breakdown of cell components and production of new substances. Certain of these changes can be readily observed in microtome thin sections. These include: (1) darkening and increase in reflectance of cell walls around sites of fungal attack (Cohen & Spackman 1980), (2) disintegration of cell walls or cell fillings and (3) production of 'secondary' cell fillings that are darker and more resistant to decay, oxidation or chemical dissolution (Fig. 6) than are primary cell fillings (Cohen 1979b). In addition to the effects of the metabolism of microorganisms, the bodies of the microorganisms themselves can be incorporated into the deposit as spores, hyphae, or sclerotia, which are converted to the maceral sclerotinite in coals (see Scott, *this volume*, table 1.

Some petrographically observable differences between living tissues and 'peatified' tissues

Colour

When living tissues from most plants are viewed under the microscope, their cell walls are nearly transparent and require differential staining to be easily distinguished. Tissues collected from surface litter in a swamp are also generally very light in colour, except in the immediate vicinity of fungal attack, where they may be darkened. On the other hand, the cell walls of tissues collected from within the peat deposit are often 'naturally stained' to some degree, although the degree of staining varies from cell type to cell type, with the most lignin-rich cells often being the least stained. This 'staining' process (sometimes referred to as 'impregnation') is not well understood. It probably represents some kind of a chemical reaction between the lignocellulosic cell wall materials and the peat fluids.

It is interesting to note that the range of colours observed in all thin sections of peats is not much greater than that observed, for example, in thin sections of bituminous coals. Peat tissues can be found that are transparent, light-yellow, yellow, orange, reddish-brown, various browns (from light-brown to dark-brown), and various reds. The biggest difference between peats and coals is that peat tissues usually have more of the light-yellow, light-brown, red-brown, and brown colouration and less of the bright orange and red colouration. However, there are exceptions to this rule. For example, peat tissues from certain parts of North Carolina are nearly all bright red or orange and could easily be mistaken for lignites or bituminous coals on the basis of colour alone. Thus, increased temperature (e.g. by burial) is not necessarily a requirement for these colour changes to occur.

The cell walls of peat tissues are never all the same colour in any one view through the microscope but exhibit different colours depending on the following parameters: (1) the species

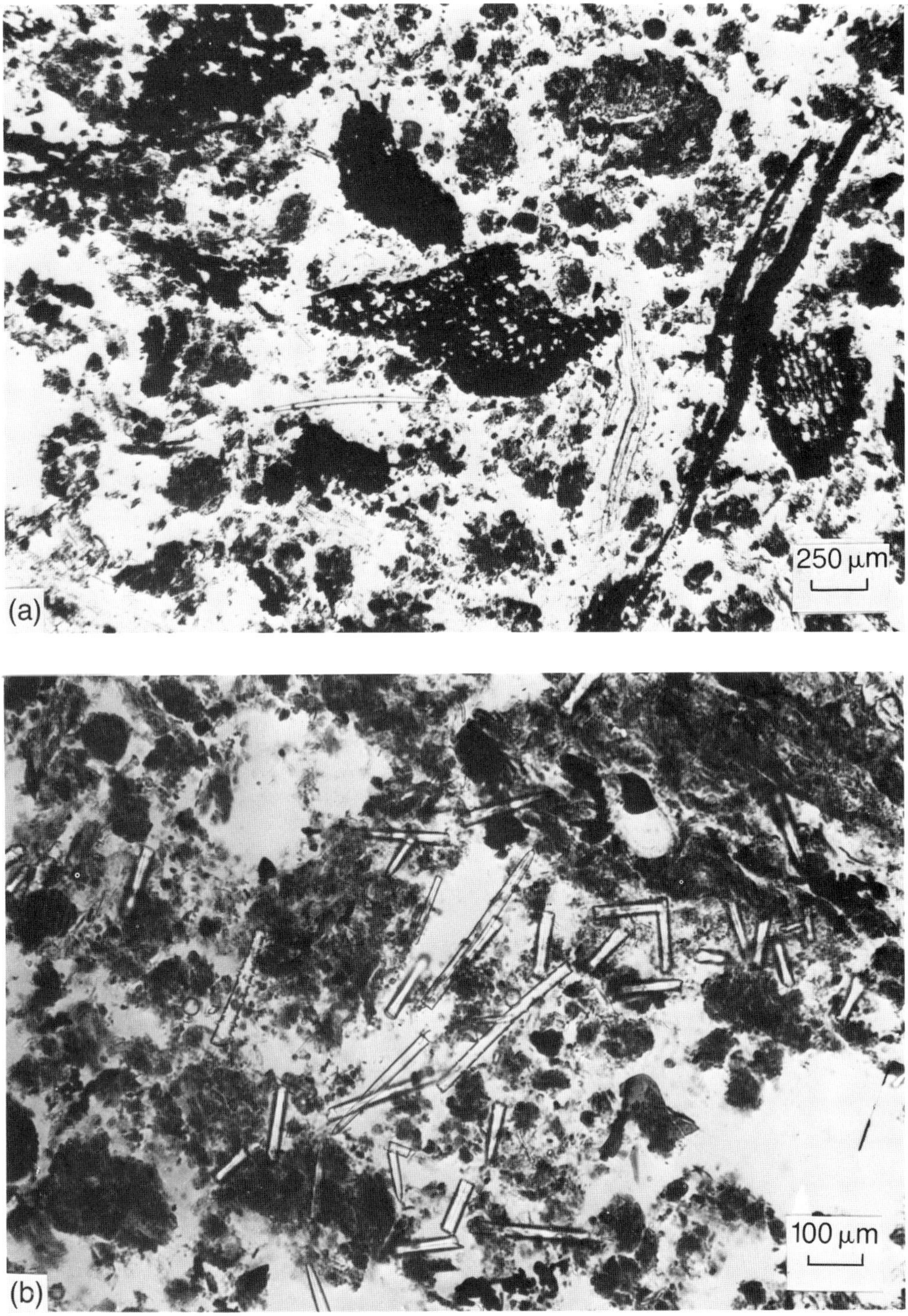

FIG. 4. Photomicrographs of microtome sections of peats from the Okefenokee Swamp that show evidence of fires. Often large pieces of charcoal are evident (a) and unburnable siliceous ingredients are sometimes concentrated (b) (after Spackman *et al.* 1976).

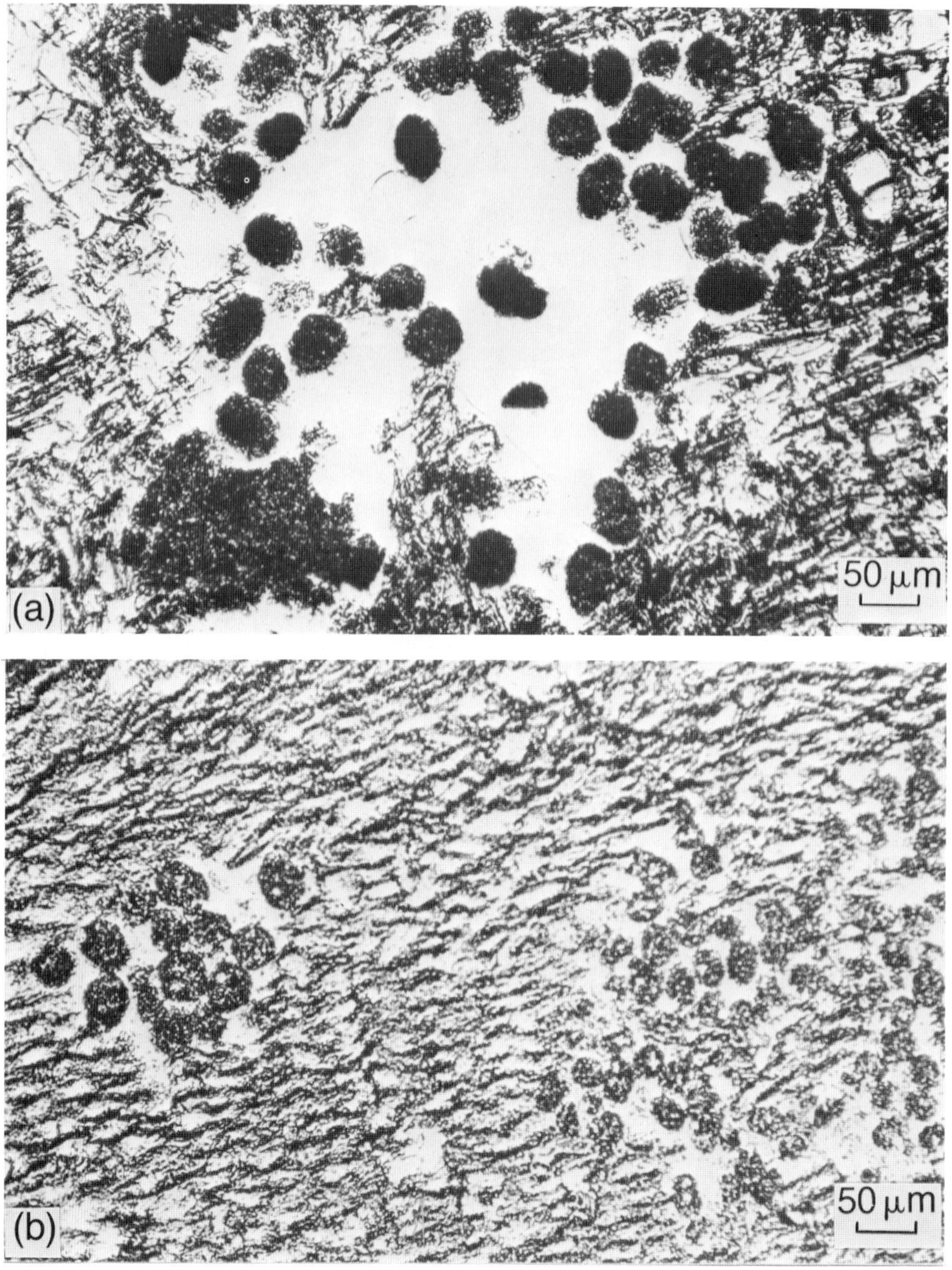

FIG. 5. Photomicrographs of fecal pellets. (a) Fecal pellets left behind after an arthropod has eaten its way through some peat tissue. (b) The same process exhibited in a coal ball peel of Pennsylvanian age from Illinois, USA.

of plant, (2) the organ or tissue type, (3) the cell type, (4) the stage of alteration or 'age' of the tissue, (5) their proximity to cell fillings or previous sites of cell fillings, (6) their proximity to microbiological activity, and (7) probably a host of unknown chemical processes affecting only small areas (microenvironments) and caused by accidents of association.

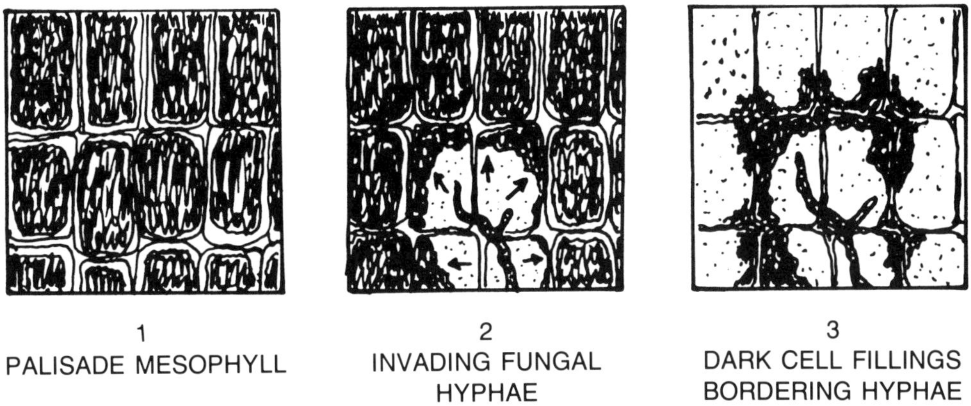

FIG. 6. Schematic diagram showing (1) the palisade mesophyll of a living *Persea* leaf, (2) the same leaf from surface litter collected from the Okefenokee Swamp showing attack by fungal hyphae and dissolution of fillings and (3) the same leaf at a later stage of alteration (in a peat) showing dissolution of all the cell fillings except for the 'secondary cell fillings' which were produced during the fungal activity.

These colour changes are a petrographic reflection of part of the process of formation of the vitrinitic coal macerals.

Opacity

Although a change from uncoloured to coloured cell walls darkens the walls of tissues viewed in thin section, this 'impregnation' process needs to be separated from the process of transformation from nearly transparent to nearly black (opaque) materials. This process is presumably the process whereby the inert coal macerals such as fusinite or macrinite are produced.

There is strong evidence that much, if not all, of this darkening process ('inertinization') occurs under more oxygenated conditions than those in which the vitrinitic macerals are formed. For example, the cell walls of certain portions of *Salix* leaves found in Everglades peats were dark-brown and granular whenever these leaves were shrivelled (due to drying); but when the leaves showed no signs of shrivelling, the same cell walls were always light yellow to orange (Cohen & Spackman 1980).

Tissues showing evidence of microbiological activity are often darkened around the area of attack. However, it is important to note that microbiological 'darkening' of tissues tends to destroy the tissues that are being attacked and to produce granular materials rather than leaving the cell walls intact. Microbiological darkening also seems to be restricted to a very narrow zone of activity within only a few micrometres of the site of microbiological attack. On the other hand, fires (as described previously) tend to darken tissues without necessarily destroying the physical structure of the cells, and this darkening effect covers a much wider zone.

Birefringence and fluorescence

Under cross-polarized light, cell walls of living plants exhibit birefringence. This birefringence is also exhibited by many tissues in peats. Since birefringence has been equated with cellulose content, it might be expected that birefringence would decrease with depth as a result of cellulose decomposition. However, unless the same type of tissue is measured at different depths, this relationship may be unclear. Figure 7, for example, shows percentages of birefringent organics from two cores from the Okefenokee Swamp in Georgia. Note that the peat sequences are reversed in the two cores and that the area percentages of birefringent organics correlate here with the peat types rather than with the depth. On the other hand, as pointed out by Stout (1985), if the same cell type from the same plant species is compared from the living plant to the peat to a lignite, the loss of birefringence is noticeable.

Another observation about birefringence is that it seems to be inversely proportional to the darkening of tissues, i.e. the darker the tissue the lower the birefringence. This is not surprising, as darker cells might be expected to have lost more of their cellulose content.

Many components of living plants are autofluorescent under blue light, and certain changes in this fluorescence are observable in peats. In living plants, most cell walls exhibit a green to yellow-green fluorescence, and tanniniferous cell fillings exhibit a more yellow or orange colour. Stout

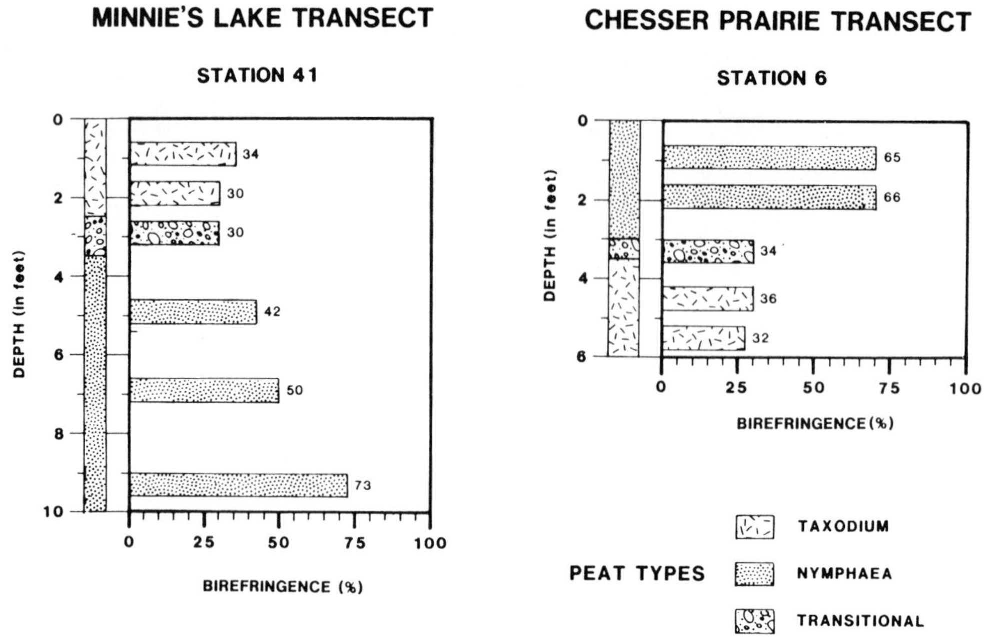

FIG. 7. Percentages of birefringent organics determined by area point-counts of microtome sections of peats from two stations in the Okefenokee Swamp. Note that the percentages of birefringent ingredients are related here to peat types rather than to depth. This diagram should not be misinterpreted as intensity of birefringence, which may show some other relationship (after Cohen & Andrejko 1984).

(1985) has shown that, in the same type of tissue, as coalification proceeds from living tissue to peatified tissue to lignitic tissue, fluorescence of xylem cell walls tends to change from green to yellow-green to yellow to orange.

As a sidelight of this study, Stout made the interesting observation that the more richly stained cell walls within the same tissue fragment exhibit a more orange fluorescence than do those that are less stained. This was also observed by Cohen & Andrejko (1984). This fluorescence colour is similar to that of many of the tanniniferous cell fillings of living plants and suggests to us that the more darkly stained walls have been impregnated with some substance that has changed their fluorescence properties. This substance could well be derived from cell fillings (see Cell-filling changes).

Reflectance

In a recent paper by Cohen *et al.* (1985b), mean random reflectances of 'huminitic' macerals were measured from 12 polished microtome sections of surface peats. These peats were collected from a wide variety of depositional and ecological settings, including grass/sedge environments in Florida, *Nymphaea* (deeper water aquatic) environments in Florida and Georgia, *Sphagnum* bogs in Minnesota and Maine, and various swamp-forest environments in Florida, Georgia, North Carolina, Minnesota, and New York.

At least two (and occasionally as many as four) distinct subtypes of huminitic macerals were evident in each peat sample. However, all huminitic macerals were grouped into two categories: (1) cell walls and cell wall fragments (humotelinites), and (2) cell fillings, secretions or amorphous masses (humocollinites).

The average reflectance of all huminitic macerals from all samples was 0.158. However, in any one peat sample, the average reflectance of the humocollinitic macerals was always significantly higher than that of the humotelinitic group.

It is interesting to note that in transmitted light, the humocollinitic macerals were always darker in colour than were the humotelinites. Also, within the humotelinitic group, the more darkly coloured cell walls had higher reflectances than did the clear or lighter-coloured walls. This was even evident in adjacent cell walls from the same tissue fragment.

Differences in plant source ingredients and depositional settings appeared to have greater effect upon reflectances of huminites than did differences in climate (e.g. huminites from tropi-

TABLE 1. *Mean random reflectances of huminitic macerals grouped according to source vegetation*

Vegetation type	Humocollinites	Humotelinites	Average all huminites
Swamp-forest	0.235	0.123	0.179
Grass/sedge marsh	0.264	0.122	0.193
Nymphaea marsh	0.210	0.099	0.154
Sphagnum bog	0.134	0.052	0.093

cal peats did not consistently have higher reflectances than those from cooler climates).

Table 1 shows the reflectances of huminitic macerals from peat samples grouped according to vegetational type. These differences in reflectances could result from differences in dryness of the depositional environment in which these peats formed, with peats from persistently wetter environments (such as *Nymphaea* marshes) having lower reflectances (due to less oxidation) than those from drier environments. In this regard, the highest average reflectance was from an Everglades sawgrass peat, which contained a significant quantity of associated inertinitic material. However, the second highest average reflectance was from a mangrove (*Rhizophora*) peat from Florida that contained no associated inertinite but in which most of the tissues had a reddish-brown colour.

Cell-filling changes

Secretions of various compositions occur within certain cavities in vascular plants. The most common type is the orange to reddish-brown 'tannin' fillings found within the lumina of cells. Petrographic changes in this type of filling are the focus of this discussion.

This kind of filling often takes the shape of the cell cavity that it fills in the living plant (i.e. rectangular, round, or elongated); but sometimes it only partially fills the lumin in which it formed, resulting in various irregular shapes.

Cell fillings in peats sometimes appear to have been very resistant to alteration. On occasion, rows of intact cell fillings are found in microtome sections of peat tissues even though the cell walls that surrounded them are almost completely decomposed. More commonly, cell fillings are observed to be pitted, partially dissolved, or completely missing from the cells that contained them when the plant was living.

The fact that some of these materials may be dissolved from plant tissues under certain circumstances should not be surprising, as tissues containing these substances have been used for tanning leather for many years.

As previously mentioned, fungal hyphae invading tissues containing cell fillings can cause dissolution of the fillings immediately surrounding the hyphae. This phenomenon was reported by Cohen (1979b) in a study of surface litter from the Okefenokee Swamp. Another observation of this study was that 'secondary cell-filling' material was often precipitated a short distance away from the invading hyphae. These fillings were slightly darker than the original fillings. In a later stage of alteration, surface litter of this same type was found to have all its normal cell fillings dissolved away with only the secondary cell fillings surrounding previously intruded fungal hyphae remaining. A similar phenomenon is illustrated in Fig. 8. In this case, the epidermal cells of living *Salix* leaves contain cell fillings. A leaf of this same species found in the peat no longer contained cell fillings in the epidermal cells. Instead, secondary cell fillings were present in a tissue area within the leaf that did not have this type of filling before. Thus, cell fillings may be dissolved away and reprecipitated in a more resistant form in other portions of the same tissue or perhaps even in other nearby tissues. One might speculate that this process might somehow be involved in the colour changes (staining or impregnation) described previously for peatified cell wall materials.

The formation of highly resistant, darkened, secondary cell fillings could account for what have been described as 'fusinized' cell fillings in coals and also for some macerals referred to as 'secretion sclerotinites'.

The precipitation of secondary cell fillings in tissue cavities undoubtedly plays an important (but rarely recognized) role, along with cell wall swelling and gelification, in the formation of certain types of vitrinitic macerals in coals.

Cell wall changes

In Cohen & Spackman (1980) the breakdown of multilayered cell walls of sclereids from water lily tissues is described in some detail. Two modes of alteration were discovered, one similar to that previously described by Barghoorn (1949) and

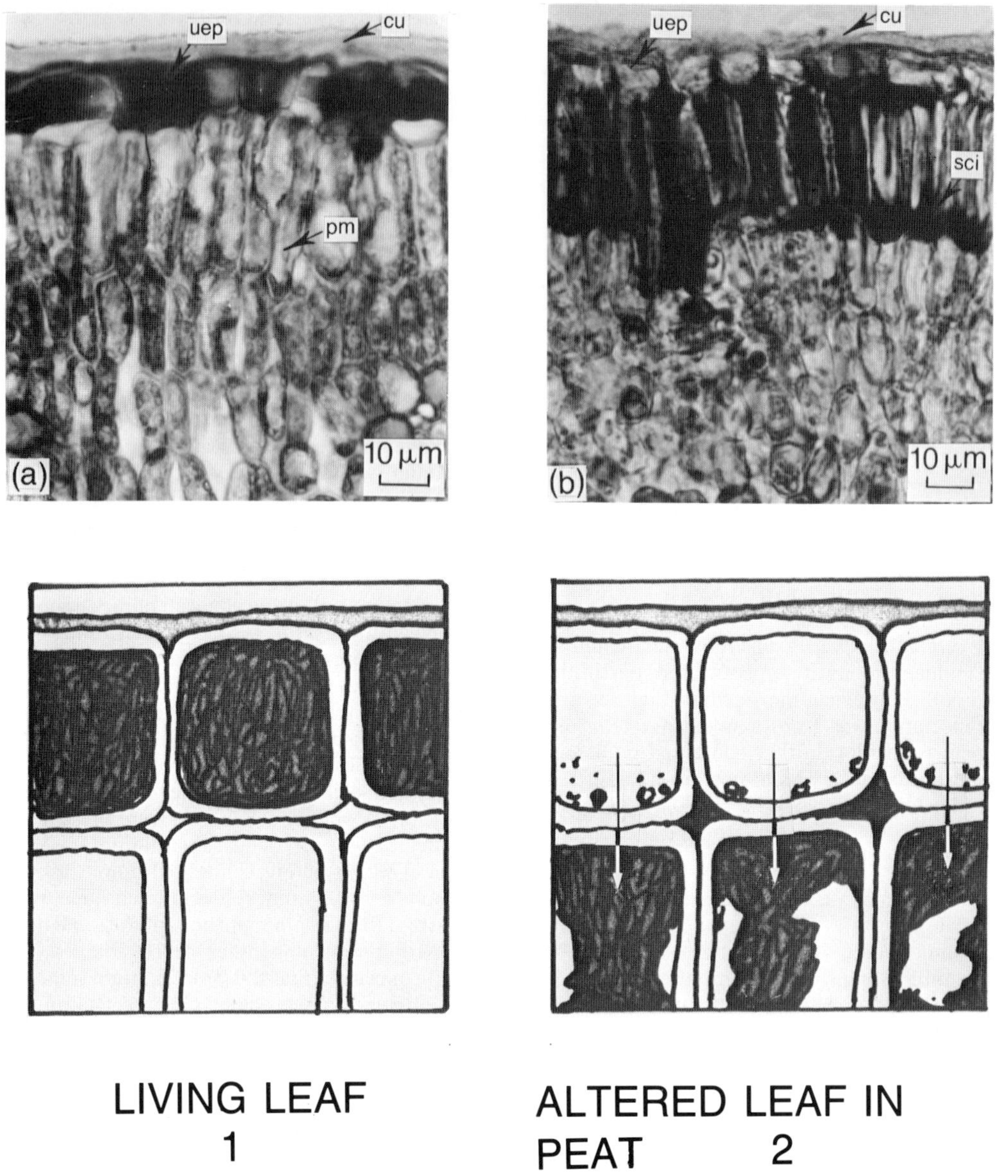

FIG. 8. Migration of cell filling materials and reprecipitation as 'secondary' cell fillings in previously empty cells. (a) and (b) are photomicrographs of *Salix* leaves and (c) and (d) are comparable schematic diagrams to illustrate the process. In step 1 cell fillings occur in the upper epidermis (uep) of the living leaf (pm = palisade mesophyll and cu = cuticle). In step 2 the cell fillings have disappeared from the upper epidermis and secondary cell fillings (sci) now occur in the palisade mesophyll (after Cohen & Spackman 1980).

the other being a new series of degradation steps. The discovery of this new sequence suggests that other sequences may be found as cell wall degradation processes are investigated for different types of cells from different plant species.

Spackman & Barghoorn (1966) describe a number of different substances filling cells of tissues from the Brandon Lignite in Vermont. Some of these substances are granular, and some are solid, and some are fixed to and indistinguishable from the cell walls. Cohen & Spackman (1980) describe some similar-looking materials in peats from the Everglades of Florida. Some possible stages in the formations of these materials are shown in Fig. 9. The first part of the process seems to be the formation of some unidentified material in the intercellular spaces. This is followed by a fusion of the cell walls with these infilling substances. The process may stop at this point (d), or the cell lumina may become filled with a light granular material that may also 'gelify' and fuse with the cell walls, producing a clear, solid mass of vitrinite-like light-orange material (e).

It is unclear where these infilling substances may have originated, perhaps partly from the cell walls themselves (as suggested by Stout 1985) or perhaps partly from the peat fluids. In any case, this process of 'gelification' is not widespread in peats, rather it seems to affect only small areas of certain tissues and is usually surrounded by tissue that has remained unaltered. Usually, these gelified masses: (1) are more orange in colour than the surrounding tissues (which are usually light-yellow to light-brown), (2) are more brittle than the surrounding tissues, (3) contain small cracks that look somewhat like desiccation cracks and (4) are found in peat types that have formed in deeper water aquatic habitats (such as *Nymphaea* peats).

Total and pyritic sulphur in peat deposits

The occurrence and distribution of pyrite and total sulphur in the peat deposits of southern Florida were investigated by Cohen *et al.* (1971, 1983, 1984, see also Casagrande, *this volume*). Pyrite was found to occur as individual euhedral crystals, framboids, and thin crusts or lenses or minute crystals. Pyrite tended to occur selectively in void spaces in or between peat tissues. There was also a strong correlation between the presence of microorganisms in these peats and the occurrence of pyrite.

With respect to peat types, pyrite was found to be highest in brackish peats, moderately high in marine peats, and lowest in strictly freshwater peats. Overlying marine or brackish peat was found to cause an enrichment of sulphur and/or pyrite in underlying freshwater peats. On the other hand, underlying high-sulphur peats seemed to have little to no effect on overlying low-sulphur peats.

Estuarine stream channels can play a part in introducing sulphur into a peat deposit. In fact they can act as conduits for brackish water, as sea-levels rise and fall. In the Snuggedy Swamp of S Carolina, for example, total sulphur was found to increase toward estuarine streams. Furthermore, overlying brackish, clay-rich splay deposits from these streams may cause an increase in pyrite and total sulphur beneath them, whereas freshwater splays may block the influence of overlying high-sulphur units on underlying low-sulphur peats.

Geomorphic features can play a part in restricting sulphur input into peat deposits. For example, freshwater-shedding/marine-blocking features, such as barrier-beach deposits, can reduce sulphur content of peats formed directly behind them. 'Doming' of a peat swamp in a shoreline area can result in a decrease in sulphur input towards the top centre of these deposits if the peat can dome high enough to restrict marine waters into the deposit (e.g. Cohen & Raymond, unpublished report; Raymond *et al. in press*).

Evidence of leaching of siliceous mineral matter from certain peat deposits

Petrographic analyses of peat cores from the Okefenokee Swamp (Cohen 1974b, Yaekel 1981) revealed the presence of abundant biogenic siliceous mineral matter in most areas of the swamp (primarily plant phytoliths and sponge spicules). However, petrographic and SEM observations of samples from some of the deepest sites (Cohen 1974b, Andrejko 1983, 1986) indicated that certain zones within the peat sequence appeared to be very low in these siliceous ingredients. Furthermore, the last petrographically recognizable spicules just above these 'spicule free zones' appeared to be highly pitted or corroded. SEM observations confirmed the pitted and fragmented nature of the spicules and phytoliths in these zones (Andrejko 1983). The mechanism for this destruction of siliceous ingredients is still uncertain, but Andrejko (1986) suggests a dissolution phenomenon resulting from the metabolic activities of microorganisms, such as bacteria, endolithic fungi, or diatoms. Whatever its cause, the loss of mineral matter from this peat sequence may be highly important since it suggests that mineral matter can be significantly reduced in certain peat deposits, increasing the potential for them to produce low-ash coals. A similar circumstance was recently revealed in a study of a tropical domed peat in Panama (Cohen & Raymond, unpublished report). In this case, although the petrographic analysis indicated the presence of peat types that should have had a significant content of authigenic silica, no authigenic silica was observed petrographically, and the ash content of these peats was very low (in

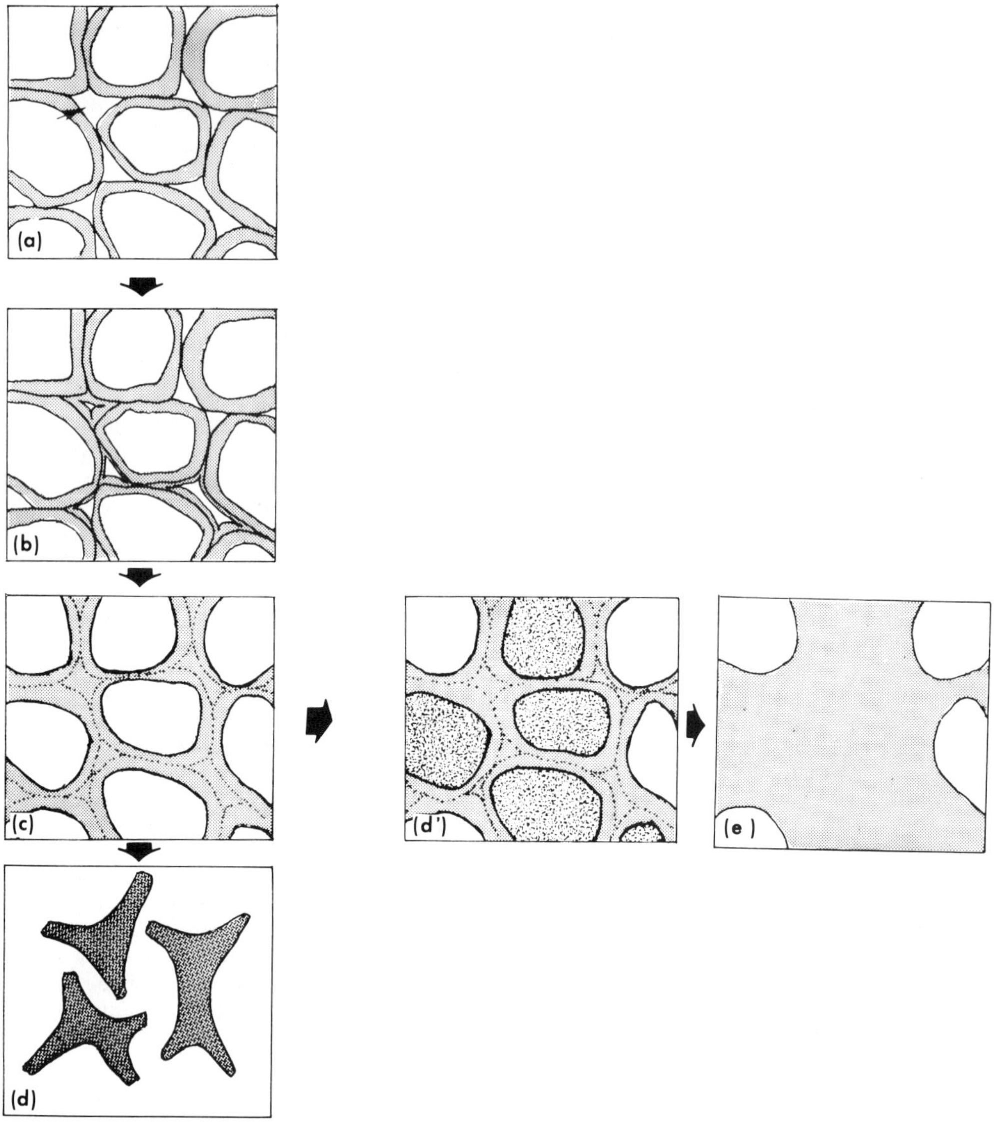

FIG. 9. Two alternative paths of alteration of the cortical cells of *Nymphaea* roots (modified from Cohen & Spackman 1980). (a) Schematic diagram of a group of cells from the cortex of a modern root of *Nymphaea* sp. Note the large intercellular spaces (arrow). (b) Schematic diagram of the same group of cells as illustrated in (a) but now showing the beginning of the alteration of the tissue in the peat. The intercellular spaces have become filled with a substance presumably derived, at least in part, from the peat fluids, since the cell walls do not appear to be altered. (c) Schematic diagram illustrating the next phase in the alteration of the tissue in the peat. The intercellular spaces have become completely filled and are no longer readily distinguishable from the cell walls. (d) Schematic diagram illustrating one possible course in the further transformation of the tissue. The tissue becomes orange to red and brittle and breaks into angular fragments. (d′) Schematic diagram illustrating an alternative path of alteration of the tissue. The lumina of the cells become filled with a light-yellow, granular material of unknown origin. (e) Schematic diagram illustrating the next stage in the transformation of the tissue shown in (d′). The granular material within the lumina of the cells has become clear and orange in colour. There is no longer any distinction between this material, the cell walls, or the intercellular material. This clear, orange mass may then become dark orange to red and brittle (as in stage d).

some samples less than 2% by dry weight). There is little doubt that the mineral matter was 'leached' from these low-ash zones, since the plants that produced these peats are similar to ones that we had previously ashed from other areas and had found to contain as much as 10% ash.

The further potential for leaching of inorganic material from peats has recently been demonstrated in laboratory experiments on Louisiana peats (Kosters & Bailey 1986). Ash contents were reduced by as much as one-third by simply leaching with distilled water, although the reduction in this case is attributable to a reduction in water-soluble salts rather than silica.

Formation and composition of 'fire splays'

Staub & Cohen (1979), in a study of the peat deposits of the Snuggedy Swamp of S Carolina, discovered that there were two types of splay deposits that interfinger with the peat. The first type, the crevasse splay, formed when flood waters broke through either a peat or a clay levee. This type of deposit is composed mainly of clay and silt. The second type, which had not been described before and which can occur only where there are peat deposits, was caused by fire (Fig. 10). Its origin appears to be as follows. During dry periods in the peat swamp, fires start either by spontaneous combustion or by lightning, which ignites the surface vegetation. As these fires burn across the swamp, they also burn down into the peat. This causes topographically low areas in the peat body. At some point the fire may approach a stream channel and may do one of two things. It can burn up to a clay levee and weaken it by burning the peat out behind and possibly below it, eventually causing failure and consequent flooding. Or, if a channel is actively eroding the edge of the peat swamp, the fire may burn to the edge of the channel, which will cause flooding. Eventually freshwater vegetation recolonizes the burned area, and peat accumulation continues.

At the base of all fire splays is a fusinite (charcoal) zone, which is black and granular. This fusinite commonly grades up into clayey peat, which in turn grades up into peaty clay, but one or the other of the clay units may be missing in this type of splay. Fire splays are generally larger than crevasse splays. They are very irregular in areal distribution due to their fire origin and show no preferred orientation with respect to the tidal channels. This type of splay should be present as a split or parting in a coal seam. A good example of a probable fire splay was described by Kravitz & Crelling (1981) in the Herrin (No. 6) coal in southern Illinois.

Depositional models

Deltaic environments

As pointed out by Cohen (1979a, 1984) and McCabe (1984), deltaic models for coal formation

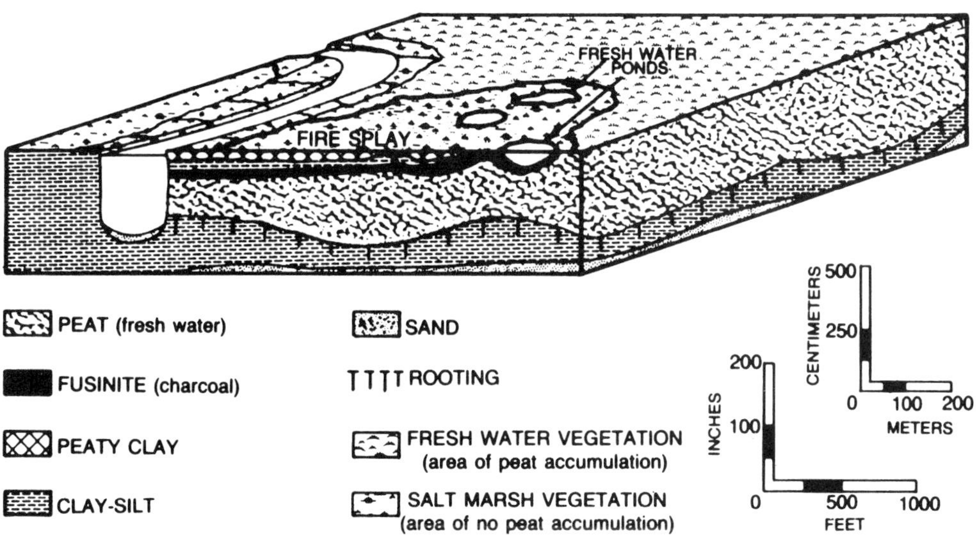

FIG. 10. Characteristics of a fire splay in the Snuggedy Swamp of South Carolina, USA (modified from Staub & Cohen 1979).

have been greatly over-applied by geologists in the past. In fact, the vast majority of economic coal deposits have been interpreted, at least at some time in the past, as being deltaic in origin. However, most modern deltas have been found to be poor places to find good peat deposits because of the dominance of inorganic sediments at these sites. Peat deposits from both the Mississippi and Niger deltas, for example, have been used in the past as modern coal analogues, but neither site has been found to contain very much peat (contrary to reports by many researchers, e.g. Frazier *et al.* 1978, Horne *et al.* 1978). Somewhat more peat was recently found in the Mississippi delta area by Kosters (1983), but its ash content and thickness do not suggest that these peats would form economically exploitable coal seams. However, some plausible hypotheses can be constructed, based on studies of modern peats, which could make deltaic models (and, for that matter, any other models for coal formation) more applicable to the ancient. The major factor seeming to preclude the use of modern deltaic environments for coal-forming models is the amount of mineral matter (or 'ash') that is found in modern analogues. Some possible ways to reduce mineral matter might be: (1) doming the deposit, (2) leaching of mineral matter from previously deposited peat or (3) depositing the peat on old surfaces where inorganic sedimentation processes are no longer active.

Doming of peat would seem to be especially significant since this process is widespread in modern peat areas, especially where rainfall is much higher than evaporation. Some excellent examples of domed (or raised) peat deposits can be found in Malaysia (as described by Anderson 1964, Coleman *et al.* 1970 and McCabe 1984, see also Moore, *this volume*). A cross-section of one of these deposits, modified from Coleman *et al.* (1970), is shown in Fig. 11. The raising of the bog above high tide level (and stream flood level) restricts the input of detrital mineral matter into the upper layers of the deposit causing them to be much lower in ash content. This process can allow low-ash peats to form in many different kinds of settings (including deltaic settings—see also discussion of interbarrier coastal plain models below). It should also be noted that domed peats are not simply restricted to tropical areas but are present in nearly all temperature zones. For example, domed deposits have been described in Panama (Cohen & Raymond, unpublished report), in N Carolina (Otte & Ingram 1980), and in Minnesota (Malterer 1980), to name only a few places.

The second hypothesis that could explain why mineral matter content is lower in coals than in many modern peats is the possibility that mineral matter leached from peat deposits after they formed. This hypothesis was discussed previously.

The third hypothesis involves formation of peat on top of deposits that once represented areas of active inorganic sedimentation but which are now only abandoned surfaces upon which the peat can form. In the Okefenokee Swamp, for instance, the marine (lagoonal) surface upon which the peat has formed is much older than the first peat (perhaps hundreds of thousands of years

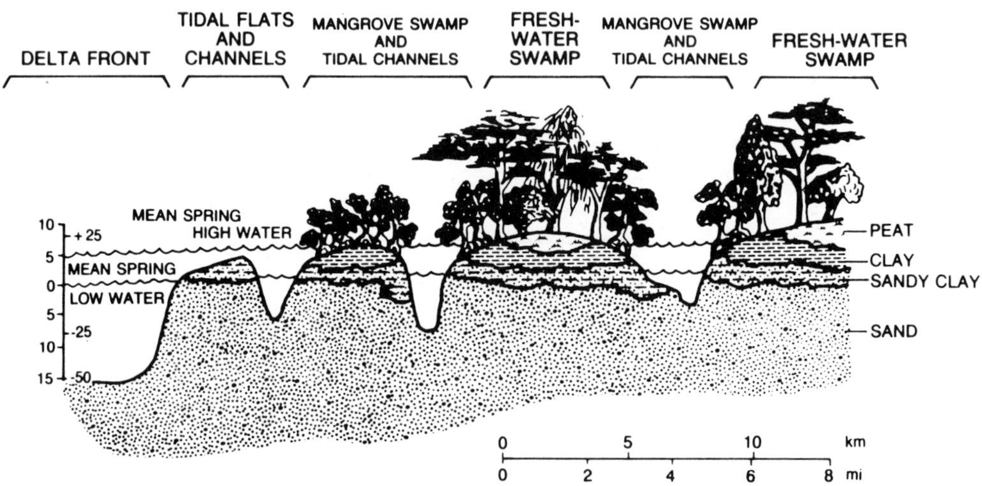

FIG. 11. Diagrammatic cross-section of Klang-Langat Delta, Malaysia, showing the development of raised peat deposits (modified from Coleman *et al.* 1970 by McCabe 1984).

BP instead of thousands of years BP). Despite this long hiatus, there are no obvious indicators of an unconformity between the peat and its 'seat earth'. It would, therefore, be difficult to recognize this time gap in the ancient. Peat deposits could thus have formed on long-abandoned delta lobes and could easily be misinterpreted in a coal sequence as having been associated with the active delta.

Interbarrier ('back-barrier') coastal plain environments

Numerous papers have been published in the last several years concerning the presence of quartz-rich arenites (orthoquartzites), interpreted as beach-barrier deposits, in coal-bearing strata (e.g. Hobday 1974, Horne *et al.* 1978, Masters 1966, Roehler 1976). However, the model which is usually shown depicts a peat-forming environment directly behind an active coastal beach-barrier deposit. Although many mineable coals are found in juxtaposition with barrier deposits (e.g. the Beckley Seam of Virginia), modern back-barrier settings (salt marsh or brackish lagoonal equivalents) have generally been found to be very poor places to produce anything more than thin lenses of high-sulphur, high-ash peat. (See also the discussion of doming.) On the other hand, the Okefenokee Swamp peats, for example, would (if buried) produce a high-quality, low-sulphur coal seam behind an orthoquartzitic beach-barrier sequence. However, as previously mentioned, the barrier and underlying lagoonal-salt marsh sediments are many thousands of years older than the peat. This disconformity is subtle enough that it would probably not be noticeable in the rock record. This observation suggests that this phenomenon might have occurred in the past and that all ancient 'back-barrier' coals may not have formed near an active shoreline. Furthermore, other than its large size, the Okefenokee Swamp setting is not really unusual in modern coastal plains. Stranded barrier-beach features are found on coastal plains throughout the world and are especially well developed in areas of small tidal influence (microtidal areas). Generally each barrier has a relatively flat, relatively impermeable (former) salt marsh or lagoonal surface behind it. It appears that when (and if) hydrological and climatic conditions become favourable for peat deposition, these surfaces can become ideal places for peat deposition to start. Peat deposition might also start at several of these sites at nearly the same time but not necessarily at the same elevation.

In light of these findings, two three-part models have been constructed to exhibit the characteristics of undomed and domed, back-barrier peat-forming environments on coastal plains. Figure 12, from Cohen (1984), gives some of the characteristics of peats (coals) that might form in these settings. Some of the parameters used in this model include: (1) cross-sectional geometry, (2) thickness, (3) continuity, (4) orientation, (5) basal contacts, (6) types of splits or partings, (7) types of splays, (8) composition of seat earths, (9) vertical and lateral variations in petrographic composition, and (10) vertical and lateral varia-

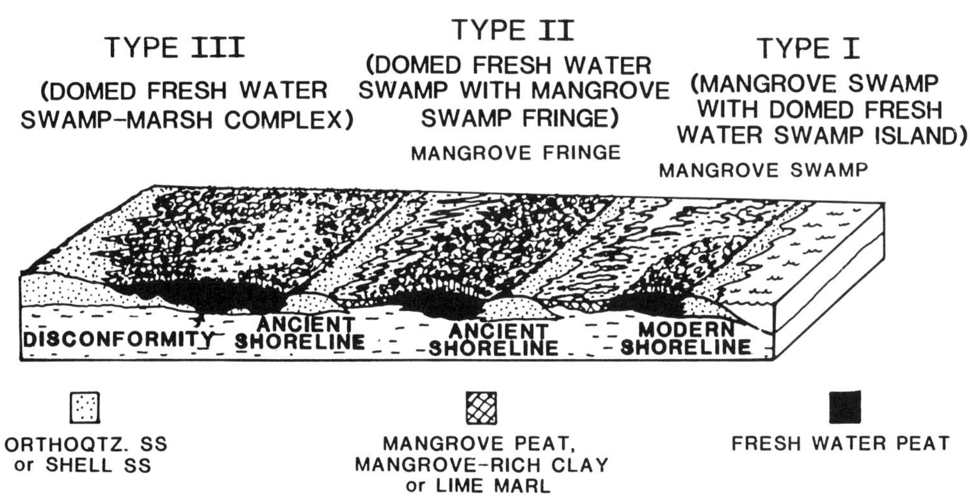

FIG. 12. Interbarrier ('back-barrier') coastal plain model for an area with raised ('domed') peat swamps. Type III has no marine influence, type II has some marine influence and type I has a strong marine influence and is the most similar to that illustrated by Fig. 11 except that it is behind a beach-barrier shoreline that helps protect it and provides additional freshwater runoff (after Cohen 1984).

tions in 'proximate' chemistries. The detailed descriptions of coals that might form in interbarrier settings are too many to describe here (see table 1 in Cohen 1984); however, many coal seam parameters can be predicted (and then tested) by construction of models of this type.

Conclusion

Although detailed studies of modern peat deposits have been relatively few compared with studies of other types of sedimentary deposits, such studies contain a great many interesting new discoveries about coal formation and composition.

Rather than attempting to present in this paper all of the new discoveries, we have given examples of new ideas or lines of research that we feel will prove to be particularly significant with regard to interpreting the characteristics of coal seams. In this effort, we have drawn heavily upon our own research or upon that of our students.

The significance of these studies and the need to continue them is clear when one considers the many new hypotheses about coal origins that have come solely from studies of modern peat deposits.

ACKNOWLEDGEMENTS: This work was funded in part by the Office of Basic Energy Sciences of the US Department of Energy and in part by National Science Foundation Grants EAR79–26382 and EAR82–01088.

References

ANDERSON, J. A. R. 1964. The structure and development of the peat swamps of Sarawak and Brunei. *Journal of Tropical Geography*, **18**, 7–16.

ANDREJKO, M. J. 1983. Environment and geochemical role of silica in peat-producing sawgrass communities in the Florida Everglades. *Unpublished MS thesis*, University of South Carolina, Columbia, South Carolina.

—— 1986. Siliceous mineral matter in Okefenokee peats. *Unpublished PhD thesis*, University of South Carolina, Columbia, South Carolina.

BARGHOORN, E. S. 1949. Degradation of plant remains in organic sediments. *Botany Museum Leaflets, Harvard University*, **14**, 1–20.

CAMERON, C. C. 1970a. Peat deposits of northeastern Pennsylvania. *U.S. Geological Survey Bulletin* 1317-A, 1–90.

—— 1970b. Peat deposits of southern New York. *U.S. Geological Survey Bulletin* 1317-B, 1–31.

COHEN, A. D. 1968. Petrology of some peats of southern Florida: with special reference to the origin of coal. *Unpublished PhD thesis*, Pennsylvania State University, State College, Pennsylvania.

—— 1970. An allochthonous peat deposit from southern Florida. *Geological Society of America Bulletin*, **81**, 2477–2482.

—— 1974a. Evidence of fires in the ancient Everglades and coastal swamps of southern Florida. *In: Environments of South Florida, Present and Past*, pp. 213–218. Miami Geological Society Memoir 2.

—— 1974b. Petrology and paleoecology of holocene peats from the Okefenokee swamp-marsh complex of Georgia. *Journal of Sedimentary Petrology*, **44**, 716–726.

—— 1979a. The Okefenokee Swamp: a low sulfur end-member of a depositional model for coastal plain coals. *Compte Rendu Neuvième Congrès International de Stratigraphie et du Géologie du Carbonifère*, **4**, 289–298.

—— 1979b. Resinoid precursors in Okefenokee and Everglades peats. *9th International Carboniferous Congress Abstracts*, p. 41.

—— 1984. The Okefenokee Swamp: a low-sulphur end-member of a shoreline-related depositional model for coastal plain coals. *In:* RAHMANI, R. A. & FLORES, R. M. (eds) *Sedimentology of Coal and Coal-bearing Sequences*, pp. 231–240. Special Publication of the International Association of Sedimentologists **7**.

—— & ANDREJKO, M. J. 1984. Premaceral contents of peats correlated with proximate and ultimate analyses. American Chemical Society Symposium Series 252. *Chemistry and Characterization of Coal Macerals*, pp. 21–32.

——, BARDIN, S. W. & RAYMOND, R. Jr 1983. Distribution of sulfur in Snuggedy Swamp peats. *Geological Society of America Abstracts with Program*, **15**:6, 546.

—— & RAYMOND, R. Jr, unpublished report. *Preliminary evaluation of the peat resources of Panama.*

——, ——, ARCHULETA, L. & MANN, D. 1985b. Preliminary study of the reflectance of huminitic macerals in recent surface peats. *Proceedings of the Society of Organic Petrologists*, pp. 30–31.

——, ——, MORA, S., ALVERADO, A. & MALAVASSI, L. 1985a. Economic characteristics of the peat deposits of Costa Rica: Preliminary Study. *In: Tropical Peat Resources—Prospects and Potential*, pp. 146–169. International Peat Society.

—— & SPACKMAN, W. 1972. Methods in peat petrology and their application to reconstruction of paleoenvironments. *Geological Society of America Bulletin*, **83**, 129–142.

—— & —— 1977. Phytogenic organic sediments and sedimentary environments in the Everglades–mangrove complex, part II. Origin, description, and classification of the peats of South Florida. *Paleontographica B*, **162**, 1–61.

—— & —— 1980. Phytogenic organic sediments and sedimentary environments in the Everglades-mangrove complex, part III. Decomposition of plant tissues and the origin of coal macerals. *Paleontographica B*, **172**, 125–149.

——, —— & Dolsen, C. P. 1971. Occurrence and distribution of sulfur in peat-forming environments of southern Florida. *Geological Society of America Abstracts with Program*, **3**:7, 528.

——, —— & —— 1984. Occurrence and distribution of sulfur in peat-forming environments of southern Florida. *International Journal of Coal Geology*, **4**, 73–96.

Coleman, J. M., Gagliano, S. M. & Smith, W. G. 1970. Sedimentation in a Malaysian high tide tropical delta. *In: Deltaic Sedimentation, Modern and Ancient*, pp. 185–197. Special Publication of the Society of Economic Paleontologists and Mineralogists **15**.

Cypert, E. 1961. Effects of fires in the Okefenokee Swamp in 1954 and 1955. *American Midland Naturalist*, **66**, 483–503.

—— 1972. Origin of tree houses in the Okefenokee prairies. *American Midland Naturalist*, **82**, 448–458.

Frazier, D. E., Osanik, A. & Elsik, W. C. 1978. Environments of peat accumulation—coastal Louisiana. *Proceedings of the Gulf Coast Lignite Conference*, pp. 5–20. Texas Bureau of Economic Geology Report Investigation 90.

Hobday, D. K. 1974. Beach and barrier island facies in the Upper Carboniferous of northern Alabama. *Special Paper of the Geological Society of America*, **148**, 209–223.

Horne, J. C., Ferm, J. C., Caruccio, F. T. & Baganz, B. P. 1978. Depositional models in coal exploration and mine planning. *Bulletin of the American Association of Petrologic Geologists*, **62**, 2379–2411.

Jones, B. F. 1940. Report of the Percy Sladen expedition to Lake Huleh. *Journal of Ecology*, **28**, 357–376.

Kosters, E. C. 1983. Louisiana peat resources. *Technical Report of the Louisiana Geological Survey* DOE/FE/05113.

—— & Bailey, A. 1986. A reassessment of Louisiana peat resources based on leaching experiments. *Louisiana Geological Survey, Coastal Geology Technical Report* No. 2.

Kravitz, C. M. & Crelling, J. C. 1981. Effects of overbank deposition on the quality and maceral composition of the Herrin (#6) coal of southern Illinois. *Geological Society of America Abstracts*, **12**:7, 465.

Malterer, T. J. 1979. Peat resource estimation program in Minnesota. *In: Peat as an Energy Alternative*, pp. 69–75. Institute of Gas Technology, Chicago.

—— 1980. Distribution of raised bogs in Minnesota. *Proceedings of the 6th International Peat Congress*, pp. 64–67. Duluth, Minnesota.

Masters, C. D. 1966. Sedimentology of the Mesaverde Group and of the upper part of the Mancos Formation, northwestern Colorado. *Unpublished PhD thesis*, Yale University.

McCabe, P. J. 1984. Depositional environments of coal and coal-bearing strata. *In:* Rahmani, R. A. & Flores, R. M. (eds) *Sedimentology of Coal and Coal-bearing Sequences*, pp. 13–42. Special Publication of the International Association of Sedimentologists 7. Blackwell Scientific Publications, Oxford.

Otte, L. J. & Ingram, R. L. 1980. *Peat Resources of North Carolina*. University of North Carolina Press, Chapel Hill, North Carolina.

Radforth, N. W. & Browner, C. D. (eds) 1977. *Muskeg and the Northern Environment in Canada*. University of Toronto Press.

Raymond, R. Jr, Cameron, C. C. & Cohen, A. D. in press. Relationship between peat geochemistry and depositional environments, Cranberry Island, Maine. *International Journal of Coal Geology*.

—— & Davies, T. 1981. Mangrove root intrusion: a means of enriching sulfur in underlying peats. *Scanning Electron Microscopy* **I**, 651–656.

Roehler, H. W. 1976. Lagoonal origin of the coals in the Almond Formation in the Rock Springs Uplift, Wyoming. *Symposium on Geology of Rocky Mountain Coals*, pp. 85–89. Colorado Geological Survey.

Spackman, W. & Barghoorn, E. S. 1966. Coalification of woody tissue as deduced from a petrographic study of the Brandon Lignite. *Coal Science: Advances in Chemistry*, **55**, 695–707. American Chemical Society.

——, Cohen, A. D., Given, P. H. & Casagrande, D. J. 1976. *Comparative study of the Okefenokee Swamp and the Everglades-Mangrove complex of southern Florida*. Coal Research Section, Pennsylvania State University, State College, Pennsylvania.

Staub, J. R. & Cohen, A. D. 1979. Snuggedy Swamp of South Carolina. A back-barrier estuarine coal-forming environment. *Journal of Sedimentary Petrology*, **49**, 133–143.

Stout, S. A. 1985. A microscopic investigation of the fate of secondary xylem during peatification and early stages of coalification. *Unpublished MS thesis*, Pennsylvania State University, State College, Pennsylvania.

Yaekel, J. D. 1981. Geochemistry and petrography of peats from the Okefenokee Swamp, Georgia and the Everglades and coastal regions of southern Florida. *Unpublished PhD thesis*. Pennsylvania State University, State College, Pennsylvania.

D. Cohen, Los Alamos National Laboratory, Los Alamos, New Mexico 87545, USA.

William Spackman, The Pennsylvania State University, University Park, Pennsylvania 16802, USA.

Robert Raymond Jr, Los Alamos National Laboratory, Los Alamos, New Mexico 87545, USA.

Recent advances in coalification studies and their application to geology

Marlies Teichmüller

ABSTRACT: Recent advances in coalification studies since 1970 are reported, concerning microscopic methods, causes of coalification, the coalification process as such and application to geology. The degree of coalification ('rank') is mostly determined on the basis of vitrinite reflectance measured on finely dispersed material in rocks. For rocks bare of vitrinite other microscopic methods are used, such as fluorescence properties of liptinites and graptolite reflectance under incident light or coloration of microfossils like palynomorphs and conodonts in transmitted light. The validity of the various methods is discussed.

Maximum rock temperature and its duration are the main causes of chemical coalification reactions. Heating rates during subsidence influence the attainment of rank equilibrium. Contact-metamorphic coals and natural cokes formed from bituminous coals are characterized by a relatively high reflectance and anisotropy. Strong frictional heat may cause local increase of rank adjacent to thrust planes. Graphitization is thermodynamically not possible in nature without pressure and shearing. Confining pressure retards the chemical coalification reactions. Tectonic stress, especially shearing, promotes the increase of maximum reflectance and of anisotropy. It commonly leads to bi-axial optical properties instead of the more typical uni-axial negative geometry of vitrinite in undisturbed coal seams. Bi-axial optical properties are common for vitrinite inclusions in clastic rocks.

On the basis of combined fluorescence microscopical and geochemical studies, 'bituminization' has been recognized as *the* characteristic part of coalification in the range between sub-bituminous and bituminous coal. It is characterized by the generation of petroleum-like substances which enable the coal to soften and to agglomerate during carbonization. Hydrogenation properties are favoured by the bituminization process. The different liptinite macerals behave differently during coalification and pass coalification 'jumps' at different rank stages. A considerable part of total inertinites is formed during early coalification (brown coal to sub-bituminous coal stage) as 'rank inertinite'. This new knowledge appears to become important in relation to hydrocarbon exploration ('diagenetic methane').

Some examples for the application of coalification studies to geology refer to vitrinite reflectance as an indicator of rock diagenesis and metamorphism, palaeogeographical and structural history, palaeogeothermics and hydrocarbon exploration. Large-scale areal coalification studies in foredeeps and backdeeps of ancient and recent orogens suggest estimations of crustal thicknesses and relations to plate tectonics.

Keywords: coalification, vitrinite reflectance, graptolite reflectance, conodont colour alteration, illite crystallinity, bituminization, rank inertinite, basin studies, palaeogeothermics, plate tectonics, foredeeps of Alps, Apennines, Rhenohercynicum (Ruhr Carboniferous), gas deposits of NW Germany

The application of coalification studies to geology began with White's (1935) 'carbon-ratio theory' according to which close relations exist between degree of coalification ('rank') and oil and gas occurrences. These relations are based on the fact that both coalification and maturation of oil and gas depend on the maximum rock temperature and its duration. Since 1950 rank determinations have been made not only on coal seams but also on finely dispersed coaly material, especially vitrinitic inclusions, in other rocks (Teichmüller & Teichmüller 1950). McCartney (1952) introduced the photomultiplier for reflectance measurements on vitrinite. This method became the most common one used for coalification studies on rocks other than coal seams and since about 1970 has been introduced by many oil companies for maturation studies in order to facilitate oil and gas exploration. The history and state of the art for microscopic maturation studies was described recently (Teichmüller 1986a).

The number of publications on geologically applied coalification studies have increased considerably since 1975, especially with regard to hydrocarbon exploration, but also in relation to estimations of the degree of rock diagenesis and metamorphism as well as to questions of basin palaeogeography, tectonic history and palaeogeothermics. In this paper only some results can be reported. For more information refer to other recent reviews in the *Textbook of Coal Petrology* (Stach *et al.* 1982, pp. 361–374 and 381–413) and those published by Teichmüller & Teichmüller (1981), Robert (1985), Murchison *et al.* (1985) and Bustin *et al.* (1985).

Usage of methods

For geological studies vitrinite reflectance is normally used as the rank parameter. In contrast to chemical parameters it has the great advantage that it can be measured not only on coal layers but also on clastic rocks which usually contain finely dispersed vitrinite. In calcareous rocks and in pre-Devonian rocks vitrinite is rare or absent. Therefore other microscopic methods are also applied. Table 1 shows the main procedures. Detailed descriptions of the methods of reflectance and fluorescence measurements are given in the *International Handbook of Coal Petrology* (1971, 1985), in the *Textbook of Coal Petrology* (Stach *et al.* 1982), by Ragot (1977), Robert (1985) and Bustin *et al.* (1985).

The determination of the Thermal Alteration Index (TAI) (Staplin 1969) or similar methods (Correia 1967, Burgess 1974) have long been applied by petroleum geologists studying the organic matter after the minerals have been removed. They also use the organic matter concentrates for palynological and geochemical studies. Relatively new methods are the determination of the Conodont Alteration Index (CAI), developed by Epstein *et al.* (1977), and reflectance measurements on graptolites introduced by Watson (1976). Fluorescence measurements on liptinites were developed for coalification studies by Jacob (1964) and Ottenjann *et al.* (1974, 1975).

Figure 1 shows a correlation of the different rank parameters in relation to rank of coal and zones of hydrocarbon generation and destruction according to Dow (1977).

Determinations in transmitted light (CAI and TAI methods) are rather subjective. Measurements of light absorption and translucency (Gutjahr 1966, Grayson 1975) were not accepted in practice because they are too time-consuming. The *fluorescence measurements* are also rather time-consuming and therefore used mostly only to confirm or as a substitute for vitrinite reflectance. As with the TAI method, they are applicable only at low rank stages up to about 1.5% R_m (random or mean vitrinite reflectance), and have their greatest validity for source rocks in the 'oil window' (0.5–1.3% R_m), where the intensity decreases drastically and fluorescence colour shifts to the red (Teichmüller & Ottenjann 1977).

In contrast to the fluorescence methods and to the TAI method, the conodont method (CAI) is applicable up to the greenschist facies of rock metamorphism, i.e. to more than 5% R_{max} (maximum vitrinite reflectance). Stages 6 to 8 are caused by loss of organic carbon, release of water of crystallization and, finally, by recrystallization of the inorganic skeleton consisting originally of carbonate-apatite. As conodonts are most abundant in marine carbonate rocks where vitrinite is very rare and palynomorphs are poorly preserved, the CAI method offers a good replacement. This is valid especially for pre-Devonian rocks in which terrestrial vitrinite is absent.

Although *vitrinite reflectance* remained the most widely used rank parameter for geological studies (Teichmüller 1982a) it became popular to challenge its validity (Jones & Edison 1979, Powell *et al.* 1982, Price & Barker 1985, Durand *et al.* 1985). Petroleum geochemists especially point to weaknesses of the method, recommending their own alternatives.

TABLE 1. *Microscopic rank parameters applied to geological studies*

I	Incident light methods (polished sections)
	Reflectance of vitrinite (R_m, R_{max}, R_{min})
	Reflectance of exsudatinites ('migrabitumen')
	Reflectance of coalified microfossils (graptolites, acritarchs, chitinozoa)
	Fluorescence of liptinites (mainly alginites, sporinites)
	Intensity ($I_{546\,nm}$)
	Spectral maximum (λ_{max})
	Red/green quotient $\left(Q=\frac{\text{intensity at 650 nm}}{\text{intensity at 500 nm}}\right)$
	Alteration (A_I, 546 nm) during irradiation (15–30 min)
	Fluorescence intensity ($I_{546\,nm}$) of acritarchs and dinoflagellate cysts
II	Transmitted light methods (strew slides of isolated organic matter)
	Colour and translucency of palynomorphs (Thermal Alteration Index = TAI)
	Colour and translucency of conodonts (Conodont Alteration Index = CAI)

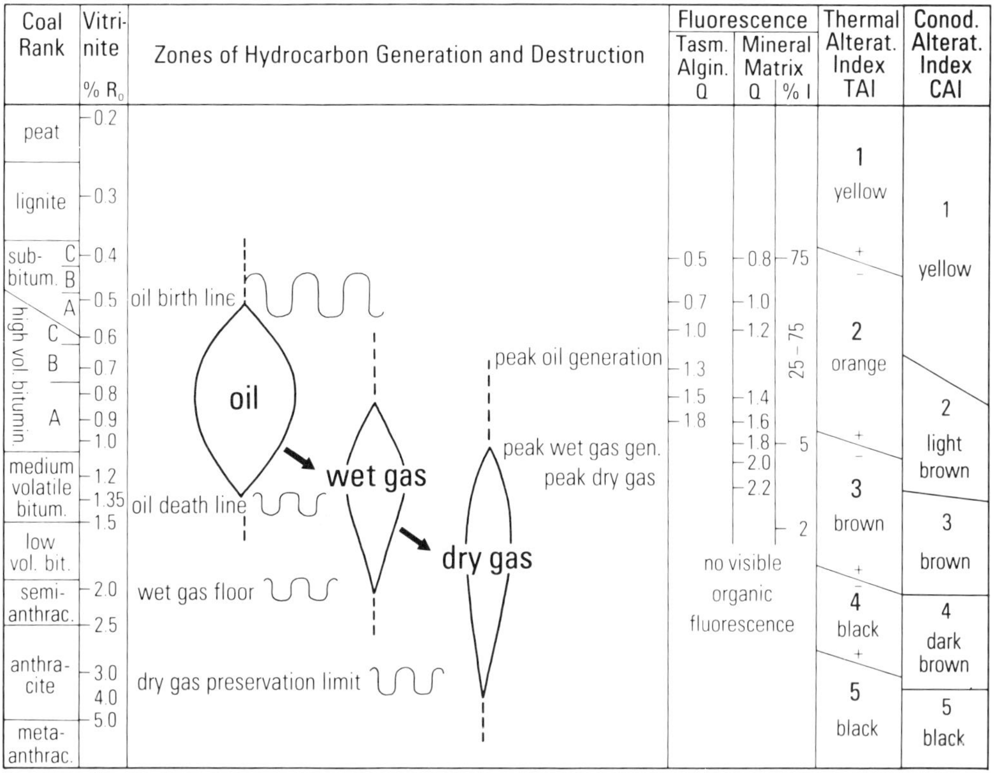

FIG. 1. Correlation of rank stages determined through microscopic rank parameters in relation to oil and gas maturity. After Dow (1977), modified and with supplements.

There is no question that both sets of methods should be used as is the practice in almost all petroleum companies. But it must be borne in mind that chemists usually analyse the bulk of organic matter and, therefore, may obtain the same average value from very different associations of constituents which can be recognized only under the microscope. Nevertheless, there exist some limitations of vitrinite reflectance which should be taken into account for coalification studies in geology.

(1) There is no question that measurements on organic matter concentrates (as usually applied by petroleum geologists) are much less reliable than measurements on the untreated rock. It is very difficult to distinguish between primary and recycled or oxidized vitrinites, exsudatinites or even certain inertinites (macrinite) in concentrates, whereas in polished sections of whole rocks primary vitrinite can be recognized on the basis of its elongated form, arranged parallel to the bedding (Fig. 2).

(2) The validity of vitrinite reflectance depends on the rank stage. At low rank stages the values scatter due to primary differences of the plant material and its degree of humification and gelification attained in the peat stage. Moreover, the rank gradient, i.e. the increase of vitrinite reflectance with depth, is relatively slow in low

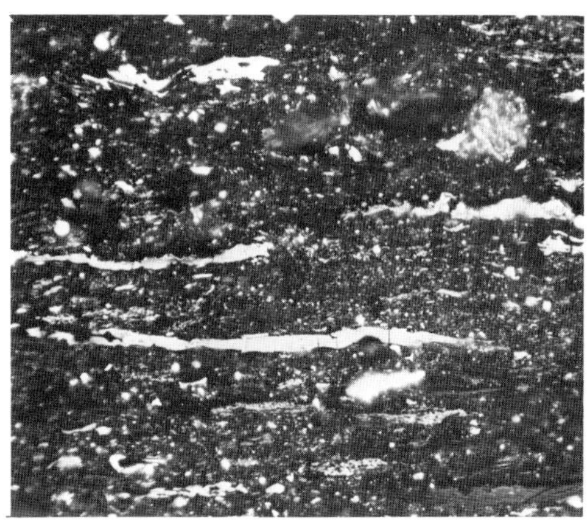

FIG. 2. Vitrinite (grey streaks) in claystone. Photomicrograph, polished section, oil immersion, 170 ×.

rank coals with less than about 1% R_m vitrinite reflectance, because aromaticity and ring condensation, which are both decisive for vitrinite reflectance, do not change much. Changes in these properties become significant in the stages of medium and low volatile bituminous coals and, especially during anthracitization, as is shown in Fig. 3, a standard plot of % R_m versus relative depth for W German boreholes. The diagram can be used to extrapolate R_m values and thus to predict rank of unknown levels from top to bottom or in the reverse direction.

Difficulties arise at high rank stages from the growing anisotropy, expressed as the difference between maximum and minimum reflectance ($R_{max} - R_{min}$), which are both measured under polarized light with rotation of the microscope stage, whereas the so-called random or mean reflectance (R_m) is measured in non-polarized light. The opinion that R_{max} can be measured in all positions of the vitrinite particles is out of date since it is known that, especially in clastic rocks, vitrinite usually has optical bi-axial properties instead of the normal uni-axial geometry (Stone & Cook 1979, Teichmüller & Teichmüller 1979). In clastic rocks, the position of the optical axes is caused not only by the overburden pressure but very often also by local diagonal directed, deformational pressure from surrounding minerals which are pressed into the vitrinite particle due to rock compression. Therefore, a considerable scatter of R_{max} values is normal for anchi-metamorphic and metamorphic rocks. In order to approach the real R_{max} value it is necessary to average not more than the 10–20% highest values measured (Stone & Cook 1979).

According to a rank profile of the 6000 m deep borehole Münsterland 1 (Teichmüller *et al.* 1979b), R_{max} values increase strongly in shear zones whereas R_{min} values decrease (Fig. 4). Moreover, R_{max} appears to change little after having reached values of 6–7%. For all these reasons R_{max} is not as good as has been claimed for rank measurements, especially in clastic rocks.

For real R_{min} values which, even at low rank stages, can be measured only in one direction (in coal seams normally perpendicular to the bedding) the 10–20% lowest values measured must be averaged. R_{min} becomes a good parameter for high rank meta-anthracites (>5% R_m) and for semi-graphites because, at these rank stages, it decreases instead of the former increase (Ragot 1977, Teichmüller *et al.* 1979b).

Up to about 4% R_{max}, the mean or random reflectance can be used as a good rank parameter (Fig. 5).

Anisotropy as such is a poor rank parameter. As will be described later, its value is influenced not only by the maximum temperature (as the main cause of coalification) but, to a high proportion also, by tectonic stress, especially shearing. Moreover, the primary composition of vitrinite, especially its hydrogen content, influences the degree of anisotropy attained. It is interesting that anisotropy appears to become constant or even to decrease in meta-anthracites between about 4 and 6% R_{max} as is shown in Fig. 6.

Figure 7 shows a folded vitrinite which reached the stage of semi-graphite. The degree and direction of anisotropy have been strongly influenced by the tangential tectonic pressure.

(3) The influence of hydrogen content on reflectance values of vitrinite has been discussed recently in many papers. It has long been known that perhydrous vitrinites, e.g. those occurring in the marine-covered Katharina seam of the Ruhr Basin, are distinguished by relatively low reflectance values. The same observation has been made for 'saprovitrinites' (Hamberger 1977) in sapropelic coals. Hutton & Cook (1980), Kalkreuth (1982) and Price & Barker (1985) pointed to relationships in coals and oil shales between the amount of alginite and vitrinite reflectance. Low reflecting vitrinites are common in oil source rocks of type I and II kerogen. In all these cases an anoxic or relatively anoxic milieu of deposition is characteristic and, obviously, has caused enrichment of hydrogen and—depending on it—the low reflectance of vitrinites. Another type of anomalously low reflecting vitrinites comprises the jets, drift woods in source rocks which have adsorbed and/or incorporated *secondary* bitumen from the surrounding organic matter. All these perhydrous vitrinites are distinguished by relatively strong fluorescence.

To avoid mistakes it is best to measure vitrinite reflectance *not* in the source rock itself but in normal clastic rocks which occur in closest proximity of the source rock. Moreover, fluorescence data can be used to check the reflectance values from source rocks. The fluorescence alteration (change of intensity during irradiation) is a good means of distinguishing between the immature, mature and overmature stages (Teichmüller & Ottenjann 1977, Leythaeuser *et al.* 1980). For rocks devoid of vitrinite, like many calcareous rocks and pre-Devonian rocks, either fluorescence of alginites or of other microfossils like dinoflagellatae cysts (Triassic–Recent) and acritarchs (pre-Cambrian–Recent) have been recommended as rank parameters (Hufnagel 1977, Kantsler & Cook 1979).

Vitrinite-like 'giant filaments' which were found in the Proterozoic (1300 Ma) of northern Australia by Peat *et al.* (1978) and which are

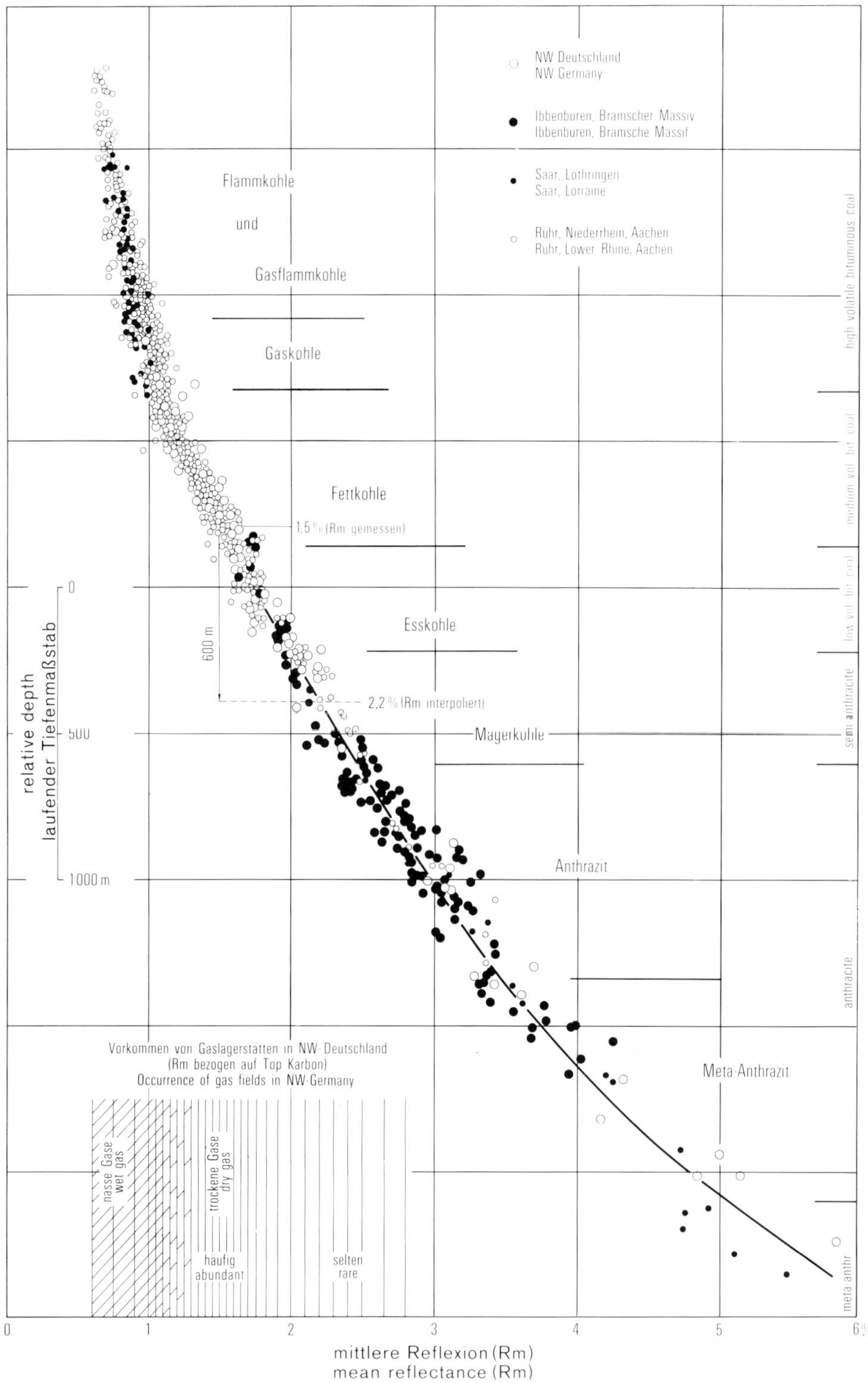

FIG. 3. Relation between vitrinite reflectance and relative depth in deep boreholes of western Germany. After Teichmüller *et al.* (1984). The occurrence of gas fields in relation to vitrinite reflectance is indicated at the bottom left. With permission from the Geologisches Landesamt Nordrhein-Westfalen.

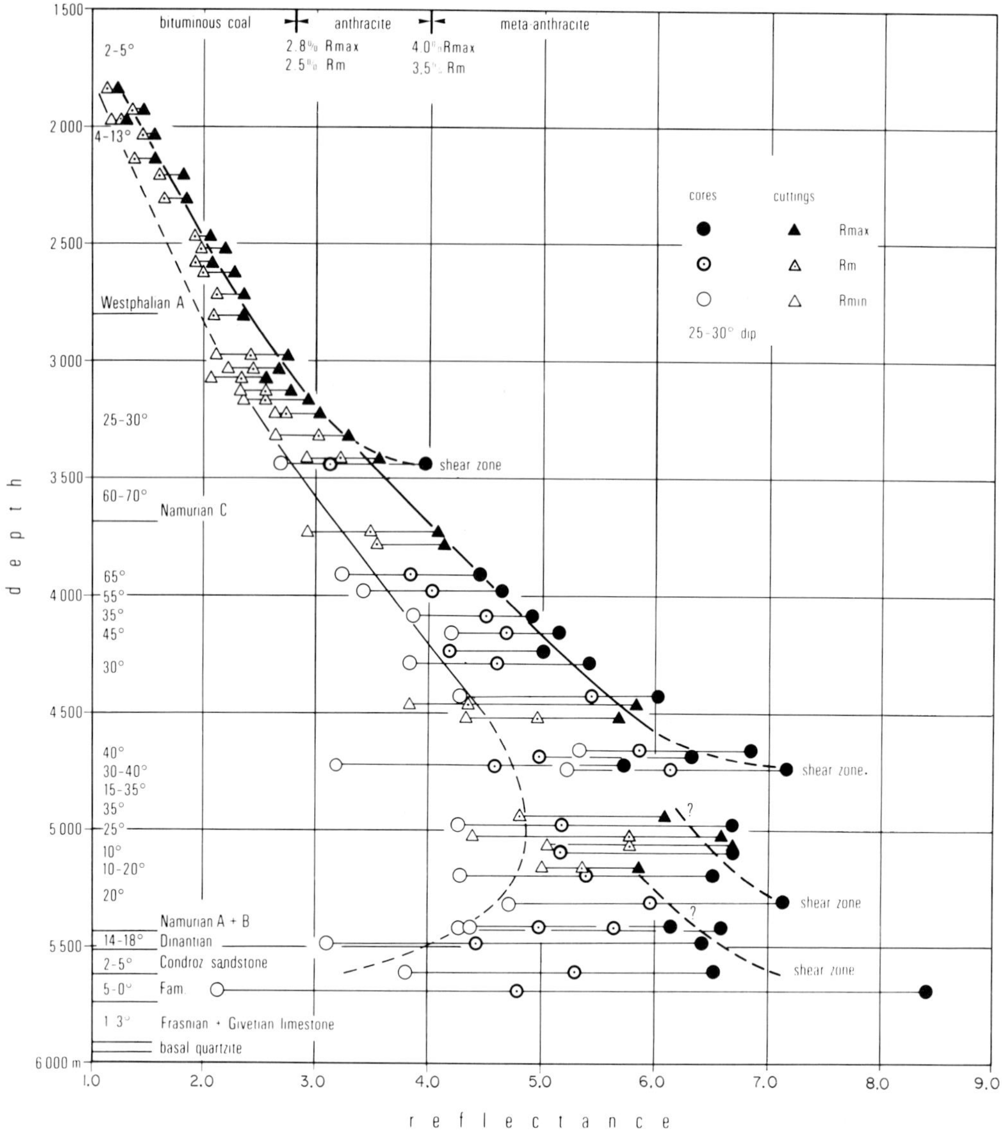

FIG. 4. Increase of mean (or random) (R_m), maximum (R_{max}) and minimum (R_{min}) vitrinite reflectance and of anisotropy ($R_{max} - R_{min}$) with depth in the Münsterland 1 borehole. After Teichmüller *et al.* (1979b). With permission from the Geologisches Landesamt Nordrhein-Westfalen.

assumed to be derived from blue-green algae, probably may also be used for comparative reflectance measurements.

At higher rank stages reflectance measurements on exsudatinites ('migrabitumen' of Jacob 1985) may sometimes be applied when vitrinite is lacking. But it must be borne in mind that exsudatinites display widely varying reflectance values at low rank stages, and that at higher rank stages the impsonites (exsudatinites with more than 1.5% R_m) normally give higher values than the corresponding vitrinites.

Recently *graptolite reflectance* has been used to replace vitrinite reflectance in Ordovician and Silurian rocks. This method was first used by Watson (1976) in Scotland. Watson found that graptolite reflectance increases towards dykes and plutons in the Southern Uplands. Later, independent of the unpublished work of Watson, this method was applied in Australia (Kurylowicz

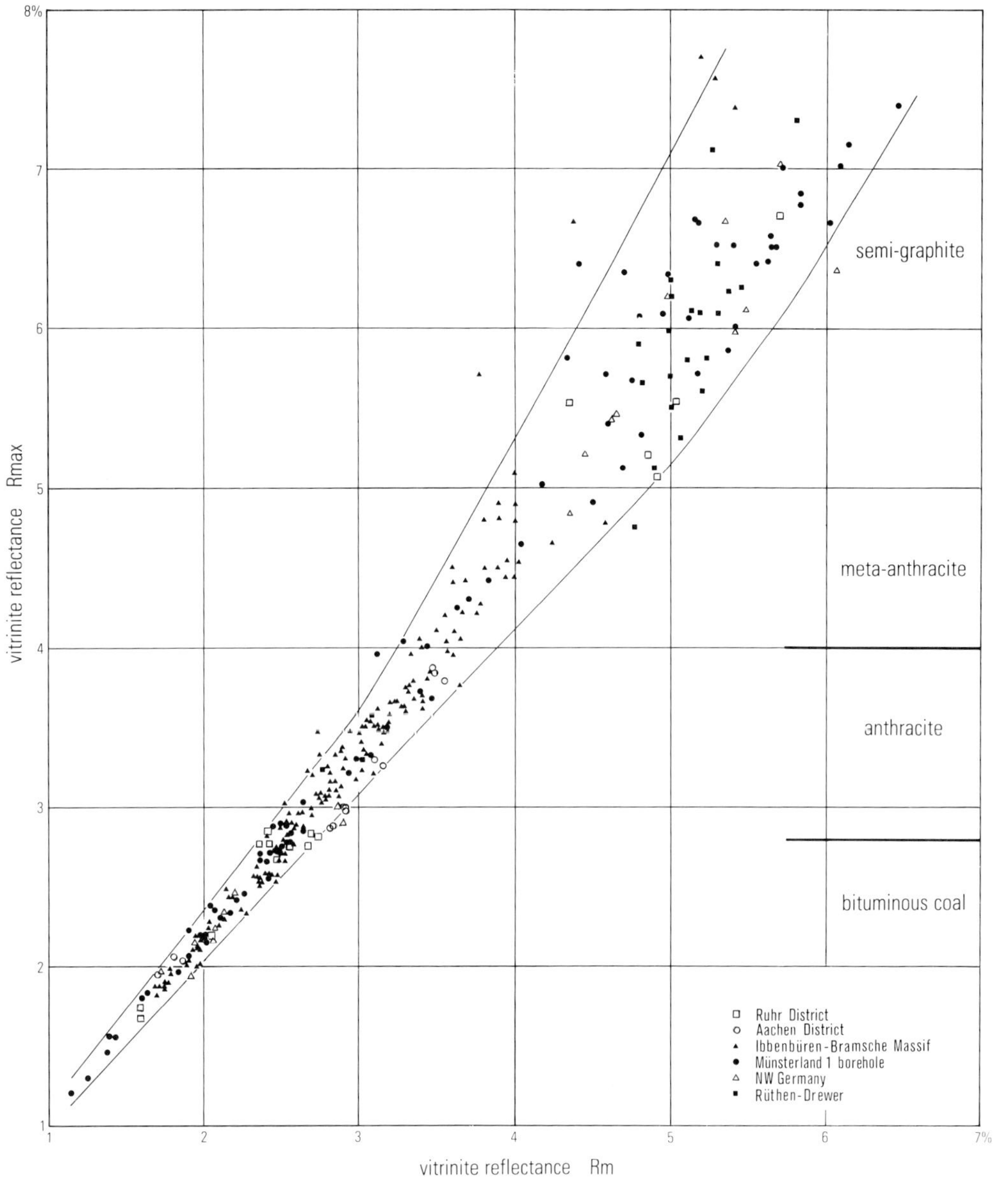

FIG. 5. Relation between maximum (R_{max}) and mean or random (R_m) vitrinite reflectance according to measurements on coals and coaly inclusions in clastic rocks from the Federal Republic of Germany. After Teichmüller *et al.* (1979b). With permission from the Geologisches Landesamt Nordrhein-Westfalen.

et al. 1976), in Germany (Teichmüller 1978, Teichmüller & Teichmüller 1982, Clausen & Teichmüller 1982), in Turkey (Goodarzi 1984), in Canada (Goodarzi & Norford 1985, Goodarzi *et al.* 1985) and in the British Isles (Kemp *et al.* 1985, Oliver 1987). Figure 8 shows high rank graptolite periderm (8.2% R_{max}) from the Silurian of Thuringia.

Reflectance studies on Ordovician and Silurian chitinozoans carried out recently by Goodarzi (1985b) indicate another fossil group with potential for use as a maturity and rank parameter.

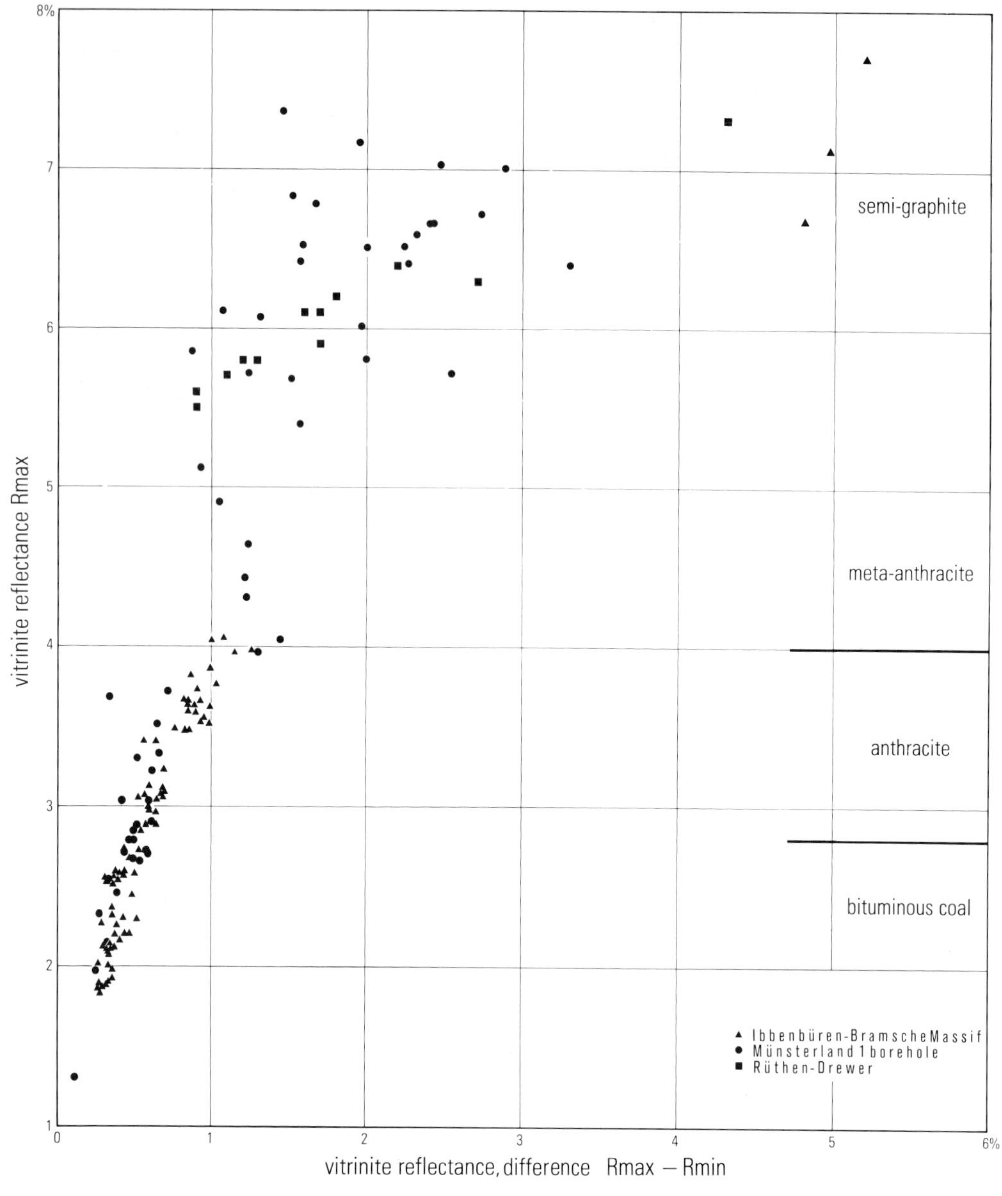

FIG. 6. Relation between maximum (R_{max}) reflectance of vitrinite and anisotropy (difference $R_{max} - R_{min}$) according to measurements on coals and coaly inclusions in clastic rocks from the Federal Republic of Germany. After Teichmüller *et al.* (1979b). With permission from the Geologisches Landesamt Nordrhein-Westfalen.

Causes of coalification

Temperature is the main factor for coalification reactions. Normally, it depends on the depth of subsidence, the geothermal gradient and the heat conductivity of underlying and accompanying rocks. The predominant role of temperature for the chemical reactions during coalification has been confirmed by many new investigations. Most convincing are results obtained from very young basins where the influence of heating time is well known. According to Barker (1979), in the Cerro Prieto geothermal field of Baja California (Mexico) young sediments of the Colorado delta

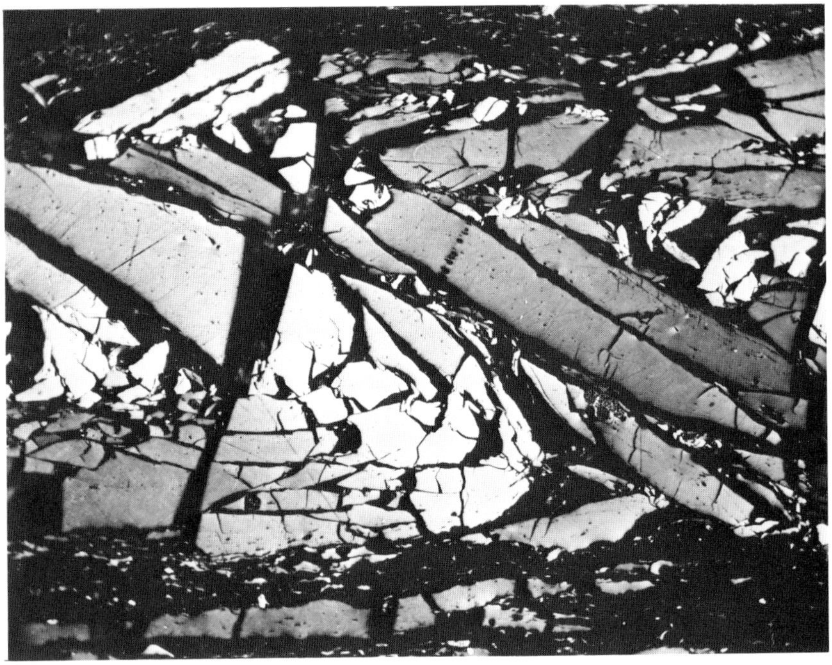

FIG. 7. Folded and sheared vitrinite layers in the stage of semi-graphite (6.8% R_{max}, 1.5% R_{min}), embedded in Lower Devonian phyllite of the Our Valley, Luxembourg. Photomicrograph, polished section, oil immersion, 265 ×, 1 polar. Note the strong, tectonically influenced anisotropy.

are in the process of being coalified. The area belongs to a rift system of high geothermal gradients (up to 160°C km^{-1}) caused by magmatic intrusions less than 6 Ma old. The rocks are filled with hot water and the vitrinite reflectance increases from 0.12% R_m (peat stage) at 240 m depth to 4.1% R_m (anthracite) at 1700 m depth where a temperature of 350°C was measured. The coalification gradient reaches 0.27% R_m/100 m. At 2000 m depth minerals of the greenschist facies (biotite, actinolite) indicate the onset of epi-metamorphism, although the random reflectance is only 4% R_m. Single values lie around 6% R_m, but the anisotropy is weak, probably due to lack of overburden pressure.

In Europe, the relations between temperature and vitrinite reflectance were studied in the rift system of the Upper Rhine Graben, where many boreholes were drilled and geothermal gradients vary considerably (Teichmüller 1979). Figure 9(a)

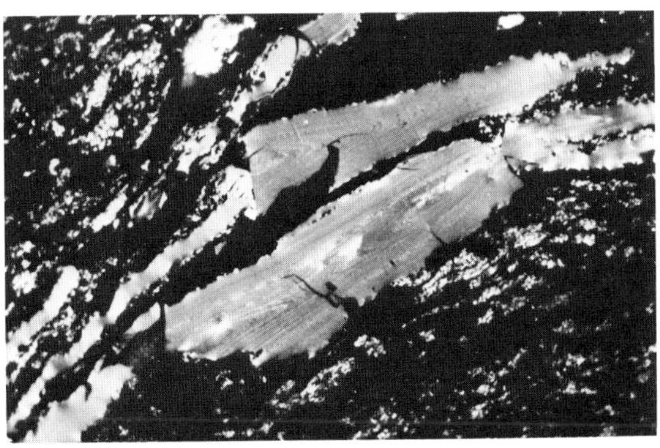

FIG. 8. Strongly coalified graptolite periderm (8% R_{max}) with strong anisotropy in Silurian shale from Thuringia. Photomicrograph, polished section, oil immersion, 1500 ×, 1 polar.

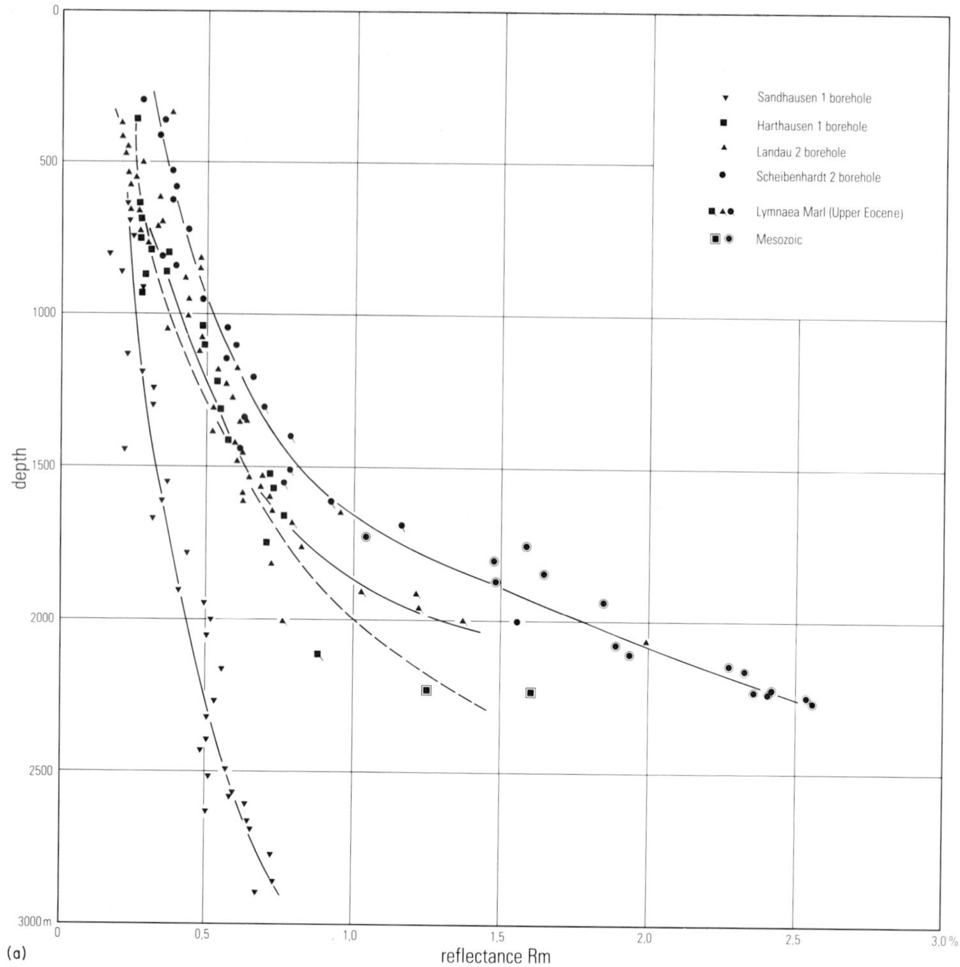

FIG. 9. (a) Increase of vitrinite reflectance with depth in four boreholes of the Upper Graben with different geothermal gradients (Sandhausen 1: 42°C km^{-1}, Harthausen 1: 67°C km^{-1}, Landau 2: 77°C km^{-1}, Scheibenhardt 2: 77°C km^{-1}).

shows that rank increase with depth depends on the geothermal gradients: the coalification gradient for the Sandhausen borehole with a geothermal gradient of 4°C/100 m is much lower than the coalification gradients for boreholes with geothermal gradients of about 7–8°C/100 m. Figure 9(b) shows the same boreholes, but with reflectance data plotted versus borehole temperatures instead of depths. The correlation is much closer than in Fig. 9(a).

The influence of heat conductivity on vitrinite reflectance, formerly described by Damberger (1968) for the Teufelspforte borehole of the Saar District, has been invoked again recently by Hacquebard (1985) who explains high reflectance values measured in areas of low gravity in New Brunswick, Canada, by a higher heat conductivity of igneous rocks at depth, which intruded before the coals were deposited.

The influence of *heating time* is still contested. There is no doubt that the influence of time is greater at higher temperatures. For temperatures less than 50°C the time factor can be neglected, as for the Mississippian lignites of the Moscow Basin.

The best examples of high temperatures are the contact metamorphic coals and natural cokes.

In the Carboniferous of the Lower Rhine District heating times of some days up to one week were calculated *inter alia* on the basis of electron spin resonance (ESR) for a natural coke (Fauth 1973). This coke surrounds a diabase dyke

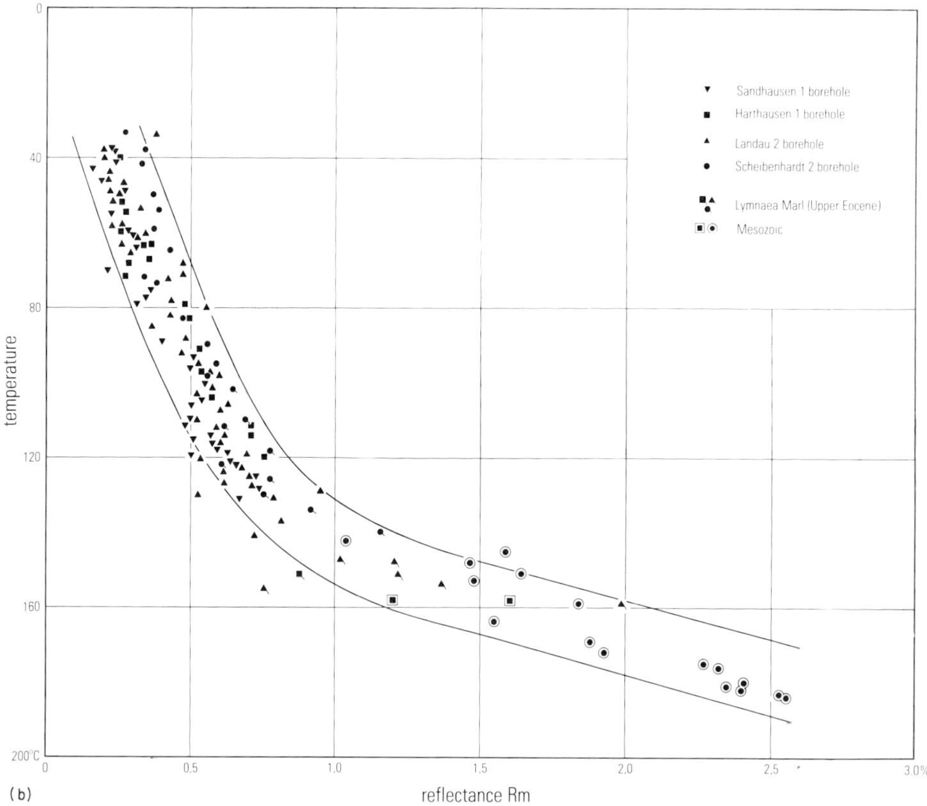

FIG. 9. (b) Increase of vitrinite reflectance with temperature in the same four boreholes as (a). Both figures after Teichmüller (1979). With permission from the Geologisches Landesamt Nordrhein-Westfalen.

of 1 m width, crossing a seam of coking coal with 1.4% R_m (Teichmüller 1973). A maximum reflectance of 12% R_{max} was reached near the contact. Figure 10(a–d) shows the influence of increasing coking temperature from the unaltered coal to the contact with the magma. All stages of coke formation, beginning with the plastic phase and ending with the phase of high temperature coke are preserved in a model manner, including the generation of first tiny semi-coke nuclei at the end of the plastic phase (Fig. 10a) within a temperature range of only 15°C and not higher than 300–315°C! The varying coke structure of different vitrinite types is of special interest: the relatively H-rich desmocollinite forms coarser graphoid 'crystals' than many telinites and semifusinites (Fig. 10d). Such pictures are unknown from technical cokes.

Figure 11 shows the temperature–time diagram for different distances from the contact. Interestingly, after 8 days, the temperature was about the same (*c*. 350°C) at the contact and 40 cm away from the contact, as a result of the low heat conductivity of coal. In comparison with 'normally' coalified coals of the same C-content, the natural coke is distinguished by much higher reflectance values (up to 12.2% R_{max}) and a stronger anisotropy (9.8% $R_{max} - R_{min}$). The presumed reason is the anticipation of graphitoid ordering promoted by a highly plastic phase under pressure.

A similar effect of short heating times at high temperatures can be observed in thermo-metamorphic anthracites and semi-anthracites such as those from Bohmte, the Piesberg and Erkelenz in W Germany which are all distinguished by high values of reflectance and anisotropy compared with the chemical rank parameters (Teichmüller 1973).

Bustin (1983) measured anomalously high reflectance values (3% R_m versus the normal 0.8% R_m) in very narrow coal films immediately adjacent to thrust planes or within the corresponding shear zones of the Canadian Rocky Moun-

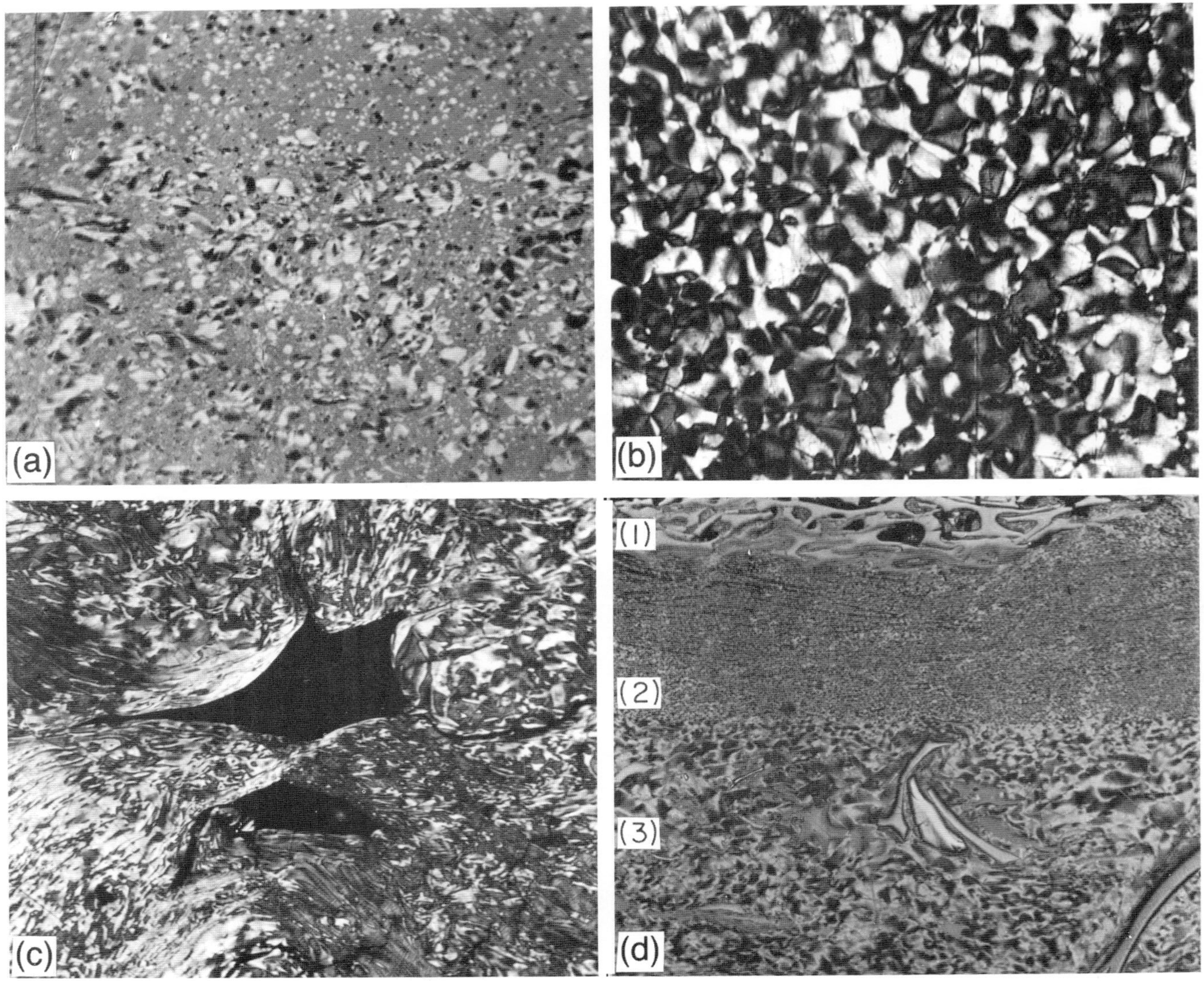

FIG. 10. Photomicrographs showing the transition from a Carboniferous coking coal (1.6% R_{max}) to natural coke at the contact of a diabase dyke, Friedrich Heinrich Mine, Lower Rhine District. Polished sections, oil immersion. (a) Appearance of first, strongly anisotropic coke nuclei (white and black) in vitrinite (grey.) 610 ×, 1 polar. (b) Highly anisotropic semi-coke with typical mosaic structure. 610 ×, crossed polars. (c) Highly anisotropic coke with fluidal structures. In the centre two pores (black).

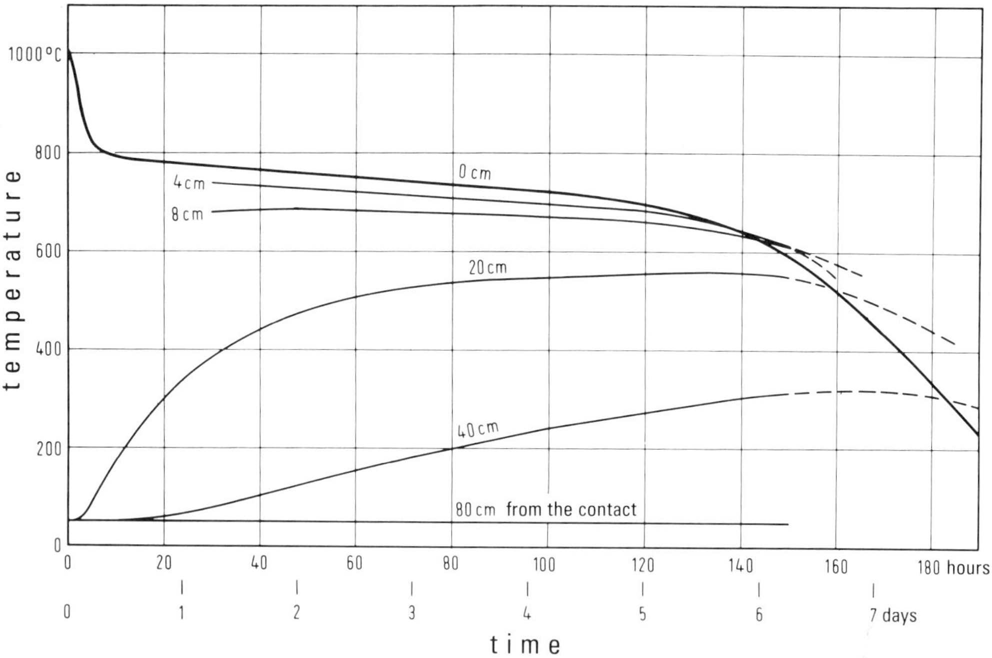

FIG. 11. Relations between temperature and time after intrusion of a magmatic dyke into a coking coal seam of the Lower Rhine district, western Germany. After Teichmüller (1973), based on Fauth (1973). The relations are different for different distances from the contact.

tains. He explains the local strong rise of reflectance as a result of local and short frictional heat generated during stick-slip faulting.

A well known time–temperature–reflectance diagram, based on former studies of Karweil (1956), Lopatin (1971) and others but with the concept of 'effective heating time' (Hood *et al.* 1975) has been published by Bostick *et al.* (1979) (Fig. 12). The effective heating time is the time during which the coal was within 15°C of its maximum burial temperature. This means that, usually, the *maximum* temperature plays the decisive role for the degree of coalification.

Yet, the time factor, implicit in the heating rate during subsidence, has to be considered too. According to Australian coal petrologists (Kantsler *et al.* 1978, Shibaoka *et al.* 1978) there is a lag of vitrinite reflectance behind temperature in the offshore area of the Gippsland Basin which has undergone extremely rapid sedimentation during the Middle and Upper Tertiary. During the Tertiary in the rapidly sinking Upper Rhine Graben a duration of heating of 2.3 Ma was insufficient for rank equilibrium to be attained (Espitalié 1979, Teichmüller 1979). Suggate (1982) suggests a duration of 1–10 Ma for the stabilization of organic matter reactions. Barker (1983) found that in liquid-dominated (hydrothermal) geothermal systems a reaction duration of 10 000 yr is sufficient to complete coalification reactions.

Two factors have to be considered: (1) in rapidly sinking basins the equilibrium temperature for a given depth and geothermal gradient may not yet have been reached, and (2) the equilibrium of coalification for a given temperature may not yet have been attained. According to Lopatin (1976a, b) subsidence rates need to be lower than 10 m Ma^{-1} if reflectance equilibrium is to be maintained.

As to the influence of *pressure* on coalification, the opinion that static pressure retards chemical coalification reactions has been confirmed again recently by carbonization experiments under various pressures (Horváth 1983, Goodarzi 1985a). In any case, the reflectance was higher at atmospheric pressure than at elevated pressures for the same heating temperatures and heating times. Epstein *et al.* (1977) obtained similar results when heating fresh conodonts under different pressures. They found that confined pressure retards colour alteration.

On the other hand, new results have confirmed that *tectonic stress* influences the R_{max} values and the anisotropy of vitrinites as, e.g. in the Münsterland 1 borehole (Teichmüller *et al.* 1979b)

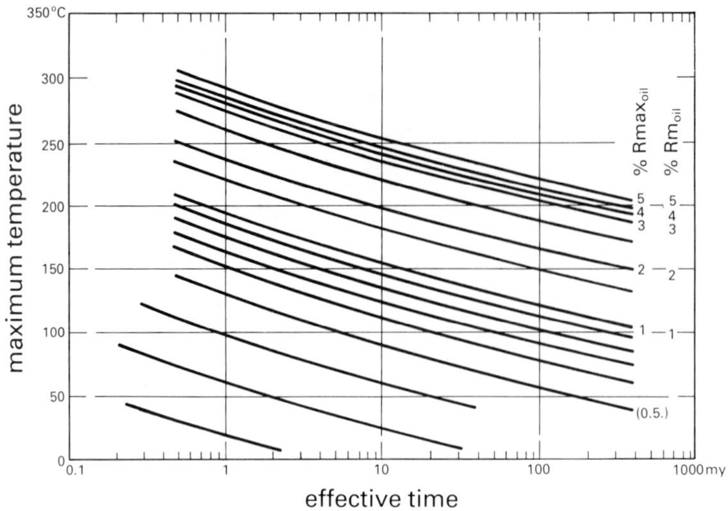

FIG. 12. Relations between vitrinite reflectance (R_m, R_{max}), maximum coalification temperature and effective coalification time. After Bostick *et al.* (1979), supplemented. With kind permission from N. H. Bostick.

(cf. Fig. 4). The optical character of vitrinite is also changed. In strongly folded coals the position of the indicating surfaces may even have been changed from the more typical uni-axial negative to a bi-axial positive geometry, as described by Levine & Davis (1984) from the Broad Top Coal Field in the Valley and Ridge Province of the Pennsylvanian Appalachians. There, the R_{max} axes are oriented parallel to the fold axes (Fig. 13). In the southern anthracite field of Pennsylvania, Hower & Davis (1981) found the maximum-intermediate reflectance plane to be oriented parallel to the axial plane of the synclinorium. Stone & Cook (1979) detected bi-axial optical properties which were related to the stress field of faults in the Sydney Coal Basin of Australia. Figure 14 shows R_{max} directions (in one vitrinite band!) of two coal samples taken near a fault.

Undoubtedly semi-graphitization and graphitization are promoted by tectonic stress. In experiments Oberlin & Terrière (1975) and Bonijoly Roussel (1980) obtained semi-graphite from anthracite at a temperature of 1000°C under a pressure of 5 kbar, in contrast to a temperature of 2200°C without pressure. Graphitization occurred at 1600°C instead of 3000°C. Bonijoly *et al.* (1982) state that, in nature, progressive graphitization is thermodynamically impossible, even at the highest rock temperatures and longest heating times. It is only together with shear that graphitization takes place in nature at temperatures of the order of 550–700°C according to Hamilton *et al.* (1970), at 450°C and 2–6 kbar according to Landis (1971), at 390°C and 6.3 kbar according to Diessel *et al.* (1978), and even at 300°C according to Taylor (1971). Actual graphite indicates the epi-zone of rock metamorphism according to Schüller (1961).

The coalification process

The most important new results obtained since 1970 are those concerning the coalification process as such. 'Bituminization' (Teichmüller 1974a, b) has been recognized as *the* characteristic part of coalification in the range between sub-bituminous and high volatile bituminous coal. Its detection has been favoured by collaboration between coal petrologists and organic geochemists, as well as by the rapid development of fluorescence microscopy, which rendered visibility of newly generated bitumen in polished sections of coal. Such material includes the maceral exsudatinite filling cracks and pores (Fig. 15a, b), strong fluorescing liquid expulsions (Fig. 16), smear films on vitrinites (Fig. 17) and a striking secondary fluorescence of vitrinite (Fig. 18), beginning at the sub-bituminous stage (about 0.5% R_m) and reaching its maximum at about 1.1% R_m at the boundary between high and medium volatile bituminous coals. Fluorescence measurements on liptinites and vitrinites revealed characteristic changes or 'jumps' (Fig. 19) which coincide with well known geochemical changes of source rocks in the 'oil window', like peaks of extracts, of aromatic hydrocarbons and

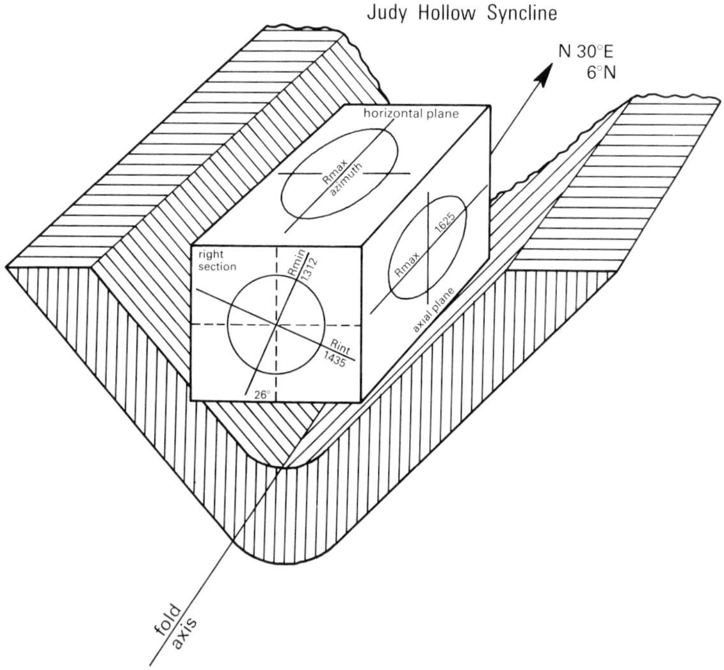

FIG. 13. Relation between the optical axes and indicating surfaces of vitrinite in coal samples and the direction of the axis of a syncline in the Pennsylvanian Appalachians. After Levine & Davis (1984). With kind permission from the authors and the Geological Society of America.

n-alkanes (Fig. 20), a reversal of the pristane/phytane ratio (Fig. 21), a gradient change of the CPI-decrease and other indications (Radke *et al.* 1980, Teichmüller 1982b). The bitumen in coal is obviously generated from liptinites as well as from vitrinites (especially perhydrous vitrinites) and is trapped in the ultra-fine pore structure of vitrinites. Thus, according to electron-microscopic studies of Rouzaud (1984), liquid-like droplets become visible in vitrinites in the rank range of bituminization. Their morphology and microstructure resemble those of 'resin *β*' (which is insoluble in toluene but soluble in anthracene oil) of certain coal tars. The size of the solid particles (surrounded by the liquid) increases with rank from 50 Å in high volatile bituminous coals to 1 μm in low volatile bituminous coals. After disappearance of the liquid particles the 'coals have over-stepped the semi-coke stage' (Rouzaud & Oberlin 1985) and the oriented 'molecules' grow to about 100 Å size.

Figure 22 shows the range of bituminization in relation to various rank parameters and to the amount of extractable hydrocarbons from coals. In the last column the maturity stages for oil and gas source rocks are indicated. According to Fig. 22, bituminization in coals coincides with the 'oil window'. The onset of bitumen generation in coal appears to cause the diagenetic gelification which is the most striking petrographic alteration during the whole coalification process. Later, newly generated bitumen obviously causes the coking properties of bituminous coals. It enables the coal to soften and agglomerate. These relationships, suggested since 1974, have been confirmed recently by comparative studies of vitrinite fluorescence and coking parameters (Creaney *et*

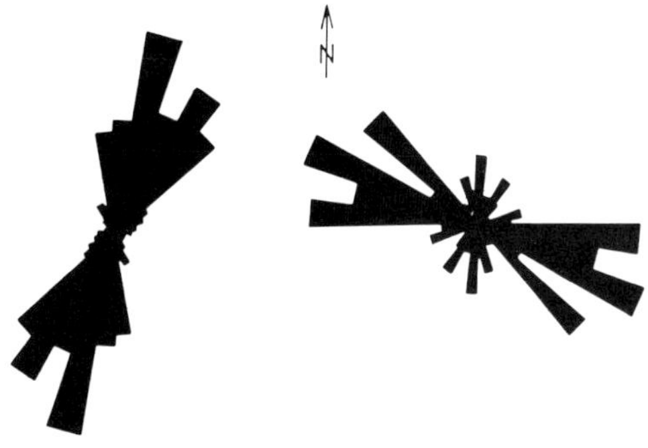

FIG. 14. Direction of maximum reflectance in one band of vitrinite from two coal samples, taken near a tectonic fault in the Sydney Basin, Australia. After Stone & Cook (1979). With kind permission from A. C. Cook.

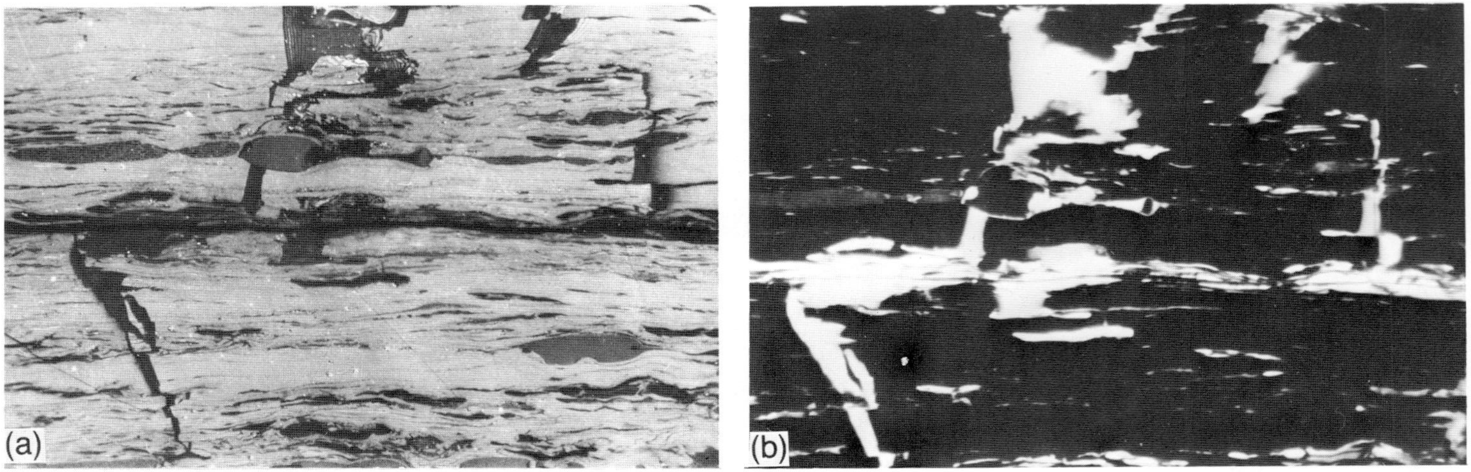

FIG. 15. Strongly fluorescing exsudatinite filling cracks in telinitic vitrinite with resinous cell inclusions. Oligocene sub-bituminous coal from Rumania. Polished sections, 430 ×. (a) Under normal white light, oil immersion. Exsudatinite black, resins dark grey. (b) Under blue light irradiation, dry lens. Exsudatinite white.

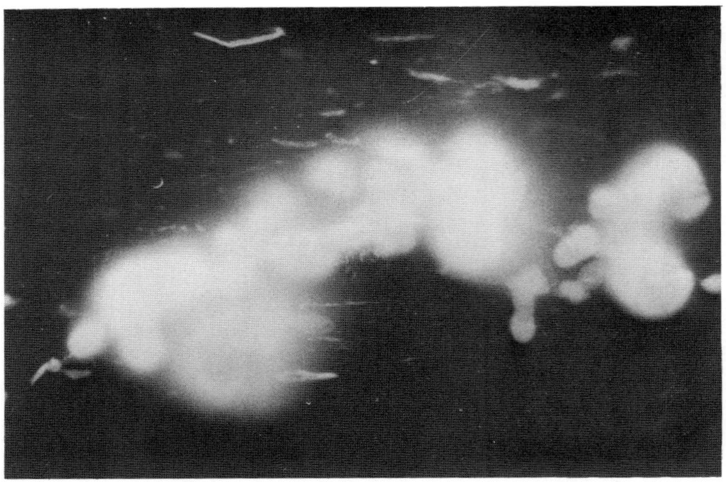

FIG. 16. Strongly fluorescing droplets exuding from a fissure in an Oligocene sub-bituminous coal from Upper Bavaria. Photomicrograph, polished section, oil

FIG. 17. Vitrite with smear film in a Carboniferous high volatile bituminous coal from Illinois. The left part of the picture was irradiated with blue light for 1 hr.

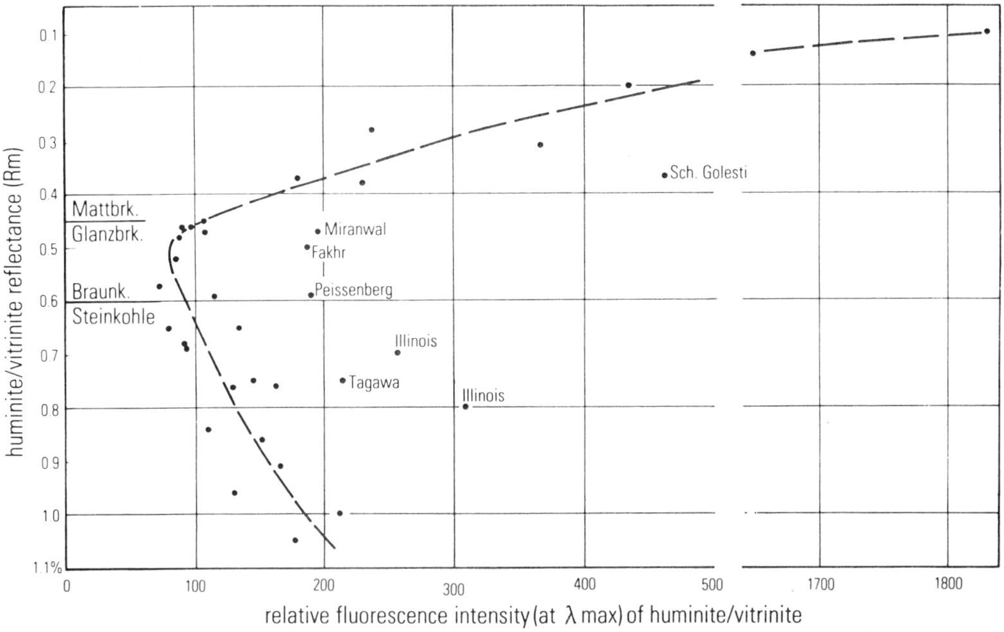

FIG. 18. Change of fluorescence intensity of fluorescing huminites/vitrinites with increasing rank (reflectance). After Teichmüller (1982b). With permission from the Geologisches Landesamt Nordrhein-Westfalen.

al. 1980, Ottenjann *et al.* 1981, 1982, Teichmüller 1982b, Lin *et al.* 1986). Moreover, coals lying in the bituminization range of coalification (0.5–1.3% R_m) are especially suitable for hydrogenation.

It is not yet well known among coal petrologists that certain coals are acknowledged as source rocks for oil (see also Murchison, Khorasani, *this volume*). These coals are distinguished by deposition in marine environments and by perhydrous vitrinites, e.g. the Tertiary coals of the Mahakam delta of Borneo (Oudin & Picard 1982) and the Jurassic coals of the North Sea.

No doubt that our new knowledge about the coalification process, especially about the bituminization range, profited considerably from the rapid, recent development of organic geochemistry and organic petrology of source rocks. Organic geochemists consider vitrinite as type-III kerogen, whereas most liptinite macerals occurring in coals can be grouped roughly under type-II organic matter. Figure 23 shows the maturation tracks for these types of organic matter in the van Krevelen diagram H/C:O/C.

Bitumen generation in coals obviously causes the 'liquid structure' which, based on X-ray studies, Hirsch described as early as 1954 for bituminous coals. It facilitates the ordering and arrangement of the aromatic lamellae parallel to the bedding plane as a consequence of overburden pressure. Therefore reflectance anisotropy becomes clearly noticeable in the coking coal stage. This ordering is relatively easy after the oxygen-rich groups have disappeared as H_2O and CO_2 during early coalification, as is demonstrated by two models of Oberlin *et al.* (1960) in Fig. 24: the uppermost model shows low rank coal with many cross-wise arranged O-rich groups which cause a high porosity and a rigid structure. The lower model is comparable with a high rank vitrinite (about 2% R_m) in which the aromatic lamellae could arrange themselves parallel to the bedding because the hindering O-rich groups have disappeared.

It should be mentioned here that this ordering is easier for macerals with a primary high H-content, such as liptinites, perhydrous vitrinites and also exsudatinites. Conversely one can conclude that highly anisotropic vitrinites (which may occur even in relatively low rank coals) were richer in hydrogen than 'normal' vitrinites with the same temperature history.

Quantitative fluorescence microscopy (cf. Table 1) permitted us to differentiate and to follow the coalification track of the various liptinite macerals (Ottenjann *et al.* 1974, 1975, Teichmüller 1982b), also in oil source rocks (Teichmüller & Ottenjann 1977). Suberinite and resinite alter strongly already in the sub-bituminous stage, releasing hydrocarbons (Powell *et al.* 1978,

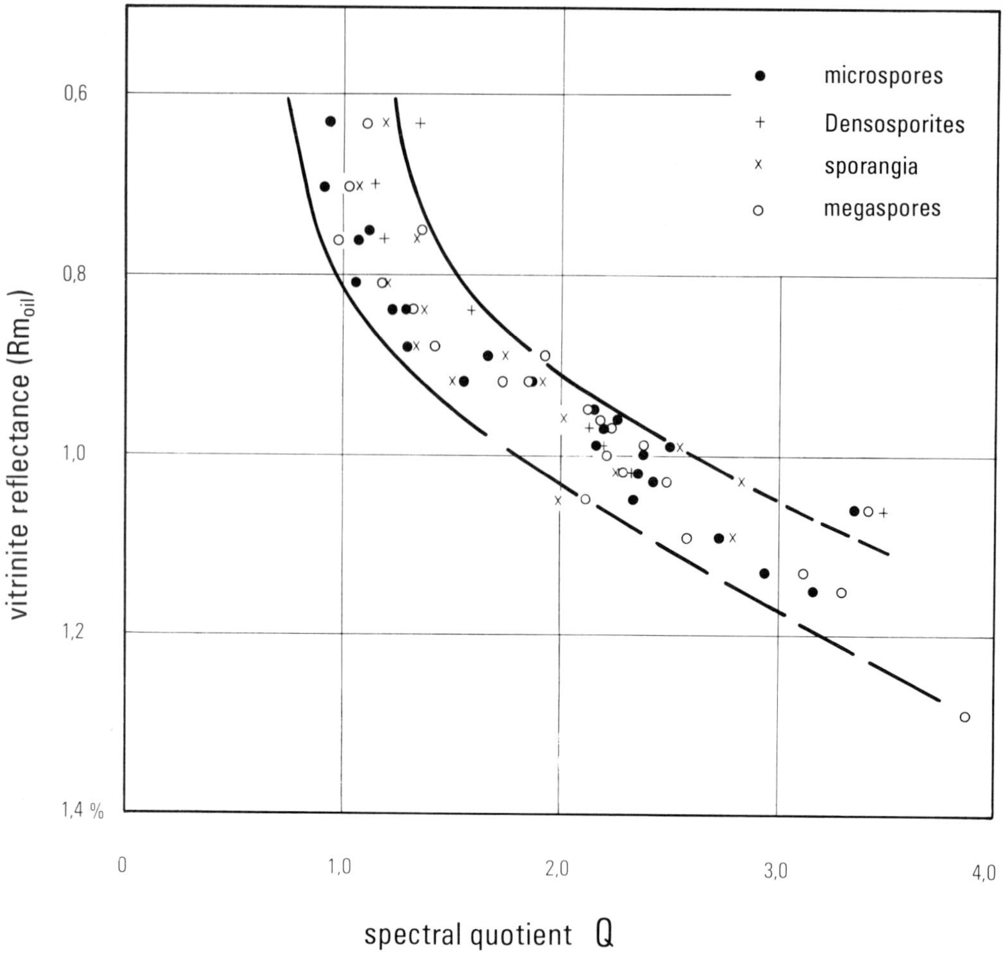

FIG. 19. Relation between fluorescence spectral quotient of sporinites and vitrinite reflectance in Carboniferous coals of the Ruhr Basin. After Teichmüller (1982b). With permission from the Geologisches Landesamt Nordrhein-Westfalen.

Teichmüller 1982b), whereas sporinite and cutinite pass additional 'coalification jumps' later (cf. Fig. 22).

A new aspect with regard to the coalification process appears likely to become important in the near future: the *coalification track of inertinites*. Up to now, most organic petrologists thought that macerals of the relatively C-rich and highly reflecting inertinite group do not change much during coalification. Only a few workers were thinking of 'inertinitization' during coalification. Today, at least part of the inertinite maceral micrinite is acknowledged as a coalification product left behind from formerly H-rich macerals, like bituminite, after oil generation. In the *Textbook of Coal Petrology*, Teichmüller (1975a, pp. 220–222) introduced the concept of 'rank fusinite', pointing, *inter alia*, to the scarceness of fusinites and other inertinites in brown coals and lignites. Recently, Smith & Cook (1980) studied low rank Australian coals under the same aspect and showed that the reflectance of many inertinites increase drastically in the range between brown coal and high volatile bituminous coal (0.2–0.9% R_{max} vitrinite) as is demonstrated in Fig. 25. These findings imply the generation and release of coalification gases or liquids in the same rank range. From this point of view a relationship between high inertinite content and the readiness for methane outbursts in a seam with 1.1% R_m vitrinite reflectance of the Bowen Basin, Australia, is of interest (Teichmüller 1975b, unpublished). Nandi & Montgomery (1975) and Diessel (1983) have shown that a

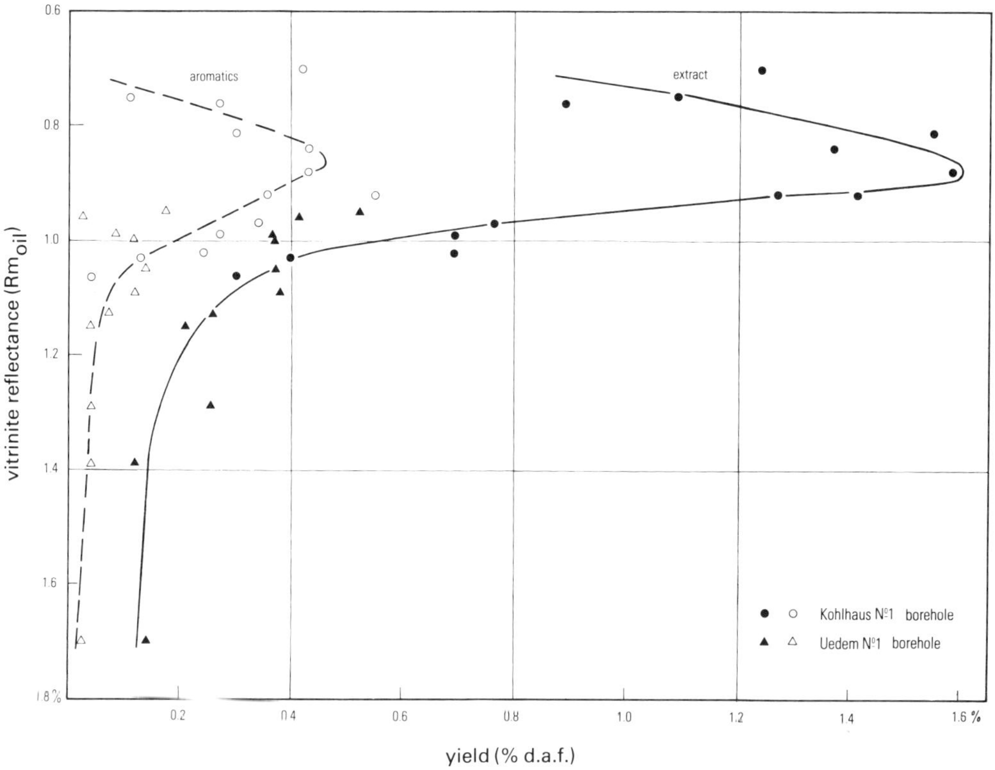

FIG. 20. Relationship between yield of extract (full circles and triangles) and of extractable aromatics (empty circles and triangles) and vitrinite reflectance in Ruhr coals. After Teichmüller (1982b). With permission from the Geologisches Landesamt Nordrhein-Westfalen.

considerable part of inertinites in Cretaceous coals of Canada and in Gondwana coals of Australia do not behave inert during carbonization. These inertinites even fluoresce under green light excitation (Diessel 1985, Brown *et al.* 1985). As a consequence of all these observations we must assume a 'coalification jump' for a great part of the inertinites, the jump lying in the range between brown coal and high volatile bituminous coal. Clearly, this jump is not valid for pyrofusinites which owe their high reflectance to primary charring, but it may well be responsible for the rich occurrences of 'diagenetic methane' as, e.g. in the Canadian Arctic and elsewhere. Recently, Smith *et al.* (1984) gave an example from the Cooper Basin, Australia.

Questions of *anthracitization and semi-graphitization* as well as of the *formation of real graphite* could be clarified, at least partly, by the usage of quite new methods, i.e. high resolution transmission electron microscopy in combination with

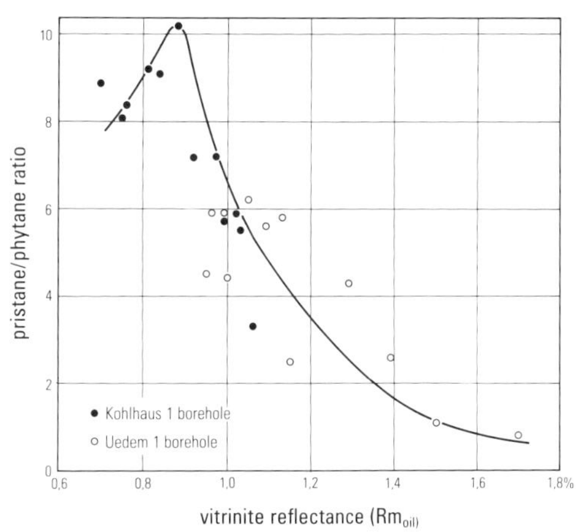

FIG. 21. Relationship between pristane/phytane ratio and vitrinite reflectance in Ruhr coals. After Teichmüller (1982b). With permission from the Geologisches Landesamt Nordhein-Westfalen.

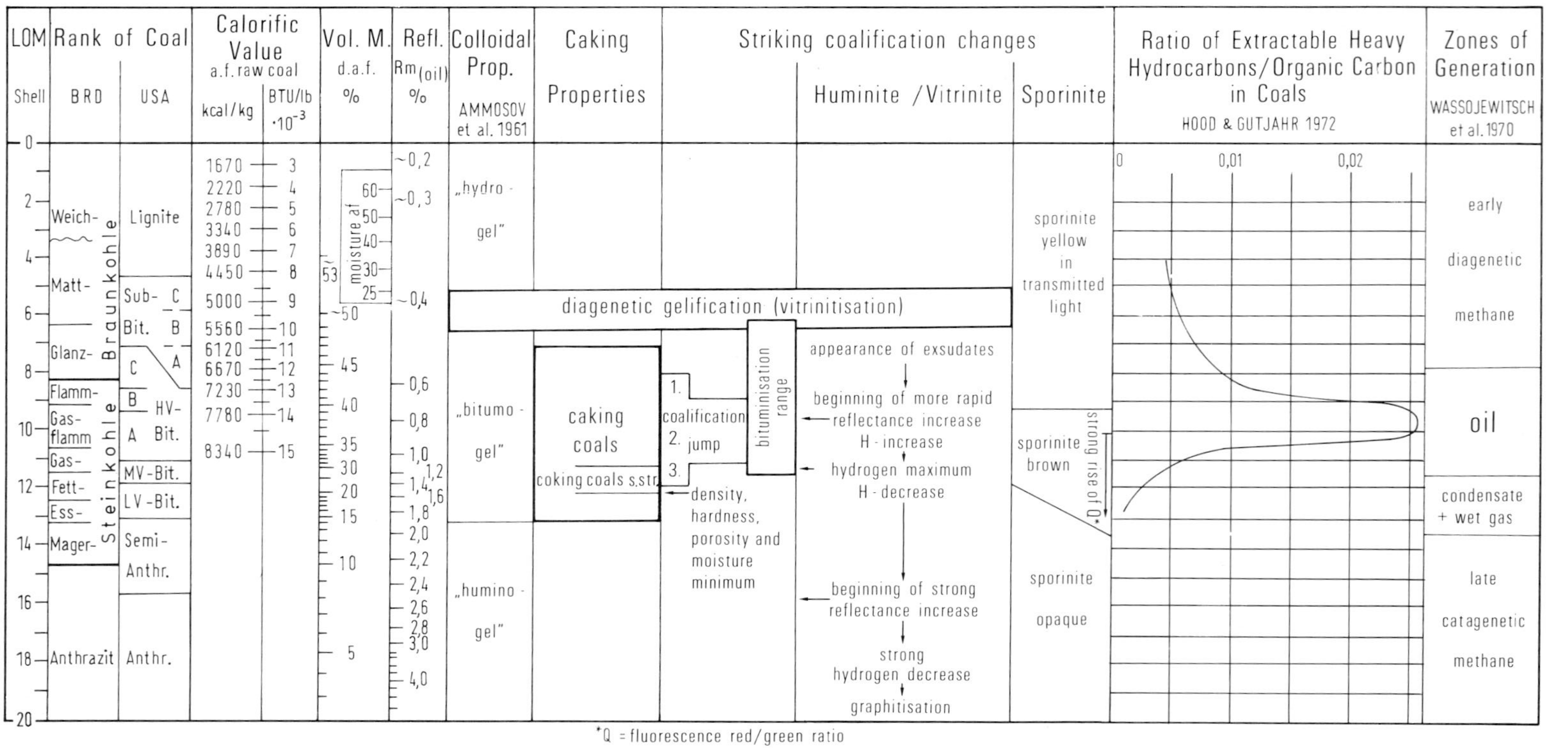

FIG. 22. The bituminization range of coalification in relation to diagenetic gelification, caking properties, coalification jumps of liptinites, yield of extracts, and oil generation in source rocks. After Teichmüller (1974a, b, supplemented 1982b). With permission from the Geologisches Landesamt Nordrhein-Westfalen.

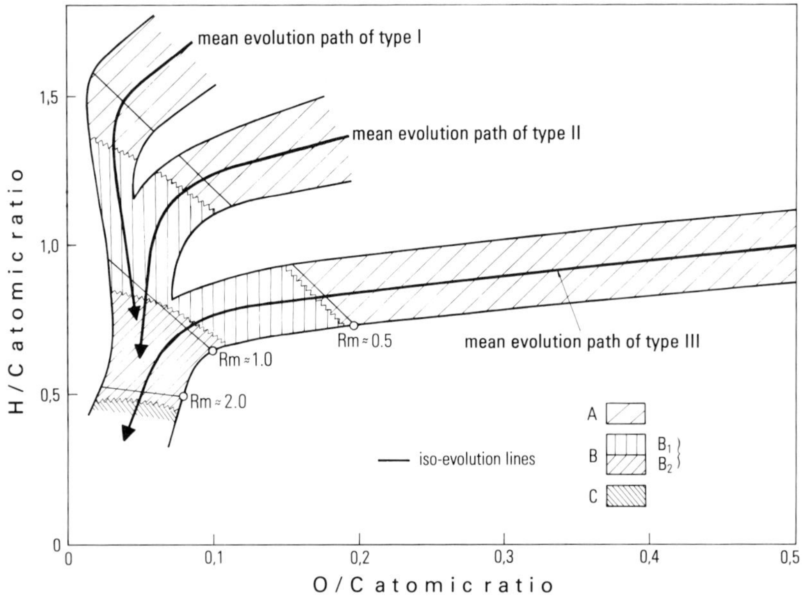

FIG. 23. Evolution paths of different types of organic matter in source rocks. After Durand (1980). A = immature, B = mature, C = overmature. Maximum of oil generation between B1 and B2. With kind permission from B. Durand.

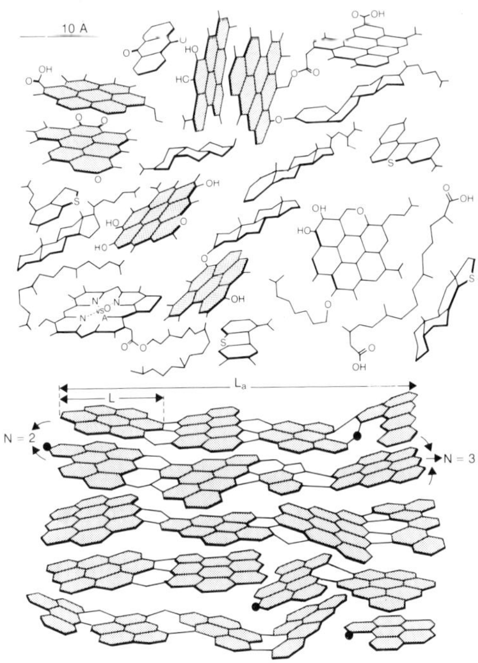

FIG. 24. Structural models of low rank coal (above) and of the turbostratic arrangement as found in high rank coals (below). After Oberlin *et al.* (1980). Stacks of aromatic layers are shaded. N = number of layers per stack, L = diameter of aromatic units, La = diameter of a wrinkled layer of aromatic units. With kind permission from A. Oberlin and B. Durand.

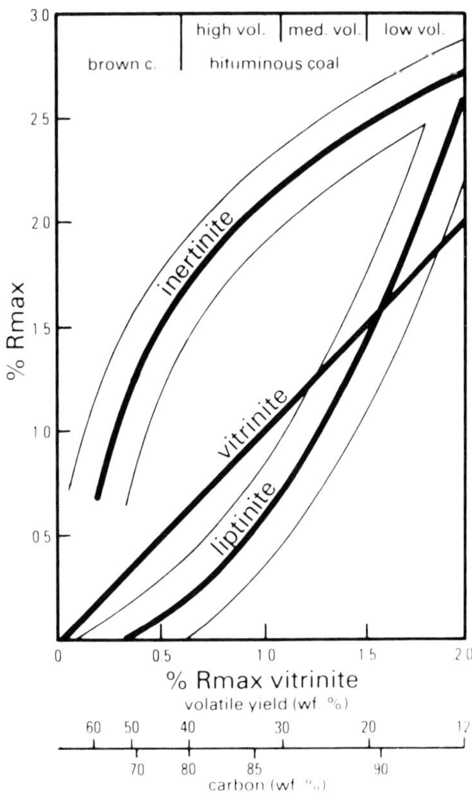

FIG. 25. Coalification tracks of the three maceral groups according to Smith & Cook (1980), modified. With kind permission from A. C. Cook.

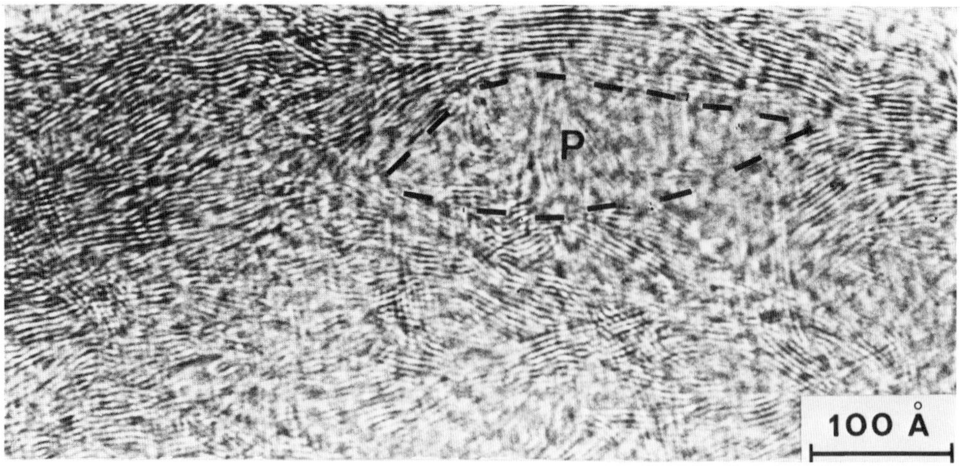

Fig. 26. High resolution electron microscope photomicrograph (bright field) of an anthracite, 1 500 000 ×. After Bonijoly (1980), by courtesy of Professor Oberlin, Orléans. Single aromatic layers (black) are arranged around walls of large pores (P).

X-ray and electron diffraction, as well as with optical diffraction using a laser beam. Such studies were carried out in the Institut Marcel Mathieu at Orléans by Oberlin and her co-workers. Bonijoly Roussel (1980) found that semi-graphitization and graphitization occur suddenly as 'jumps' and are thermodynamically impossible in nature without pressure and shearing stress. The bright field electron microscopic photomicrographs of anthracites and graphites are a sensation for coal petrologists because they picture the single aromatic lamellae, as may be seen from Fig. 26 of an anthracite with a magnification of 1 500 000 and a resolution of 3 Å! In this anthracite the pores are surrounded by aromatic layers which are still scrumpled and very small, compared with the straight, large lamellae of a graphite. The large size of the pores is typical, and explains the relatively high moisture content of anthracites.

Examples of the application of coalification studies to geology

Vitrinite reflectance as an indicator of rock diagenesis and metamorphism

Relationships between vitrinite reflectance and mineralogical changes have been studied in many regions and in sequences from deep boreholes. Marked alterations in sedimentary clastic rocks begin only when coaly inclusions have passed through stages equivalent to the entire range from peat to the beginning of the anthracite stage. Physical parameters of diagenesis, like porosity, change much earlier.

Figure 27 shows that in deep boreholes of the Saar District (W Germany) sandstone porosity decreases from 11 to 13% in the high volatile bituminous coal stage to less than 1% in the low volatile stage (1.5% R_m). The kaolinite zone extends down to the boundary between high and medium volatile bituminous coals (about 1.2% R_m). The zone of sericite blastesis corresponds to the low volatile bituminous and the anthracite stages, and chlorite blastesis occurs where the coal reaches the stages of high rank anthracites and meta-anthracites. The boundary between diagenesis and anchi-metamorphism (or 'very low grade metamorphism') is reached at about 3.5% R_m (4.0% R_{max}) according to Breitschmid (1982) for the Swiss Alps and according to Teichmüller *et al.* (1979b) for Westphalia. According to Buntebarth (1983) this boundary, reached at 3470 m in the Saar 1 borehole, corresponds to a palaeotemperature of 220°C, in contrast to the present temperature of only 95°C. Epidote, indicating the epi-zone of rock metamorphism, was found at 5037 m depth in the Upper Devonian. No reflectance measurements were possible in these rocks but very fine inclusions of highly anisotropic graphite or semi-graphite were observed under the microscope.

Many new papers deal with the correlation between vitrinite reflectance and illite crystallinity (Wolf 1975, Kübler *et al.* 1979, Teichmüller *et al.* 1979b, Teichmüller & Teichmüller 1979, 1980, Kisch 1980, Frey *et al.* 1980, Hesse & Ogunyomi 1980, Kemp *et al.* 1985, Oliver 1987). According

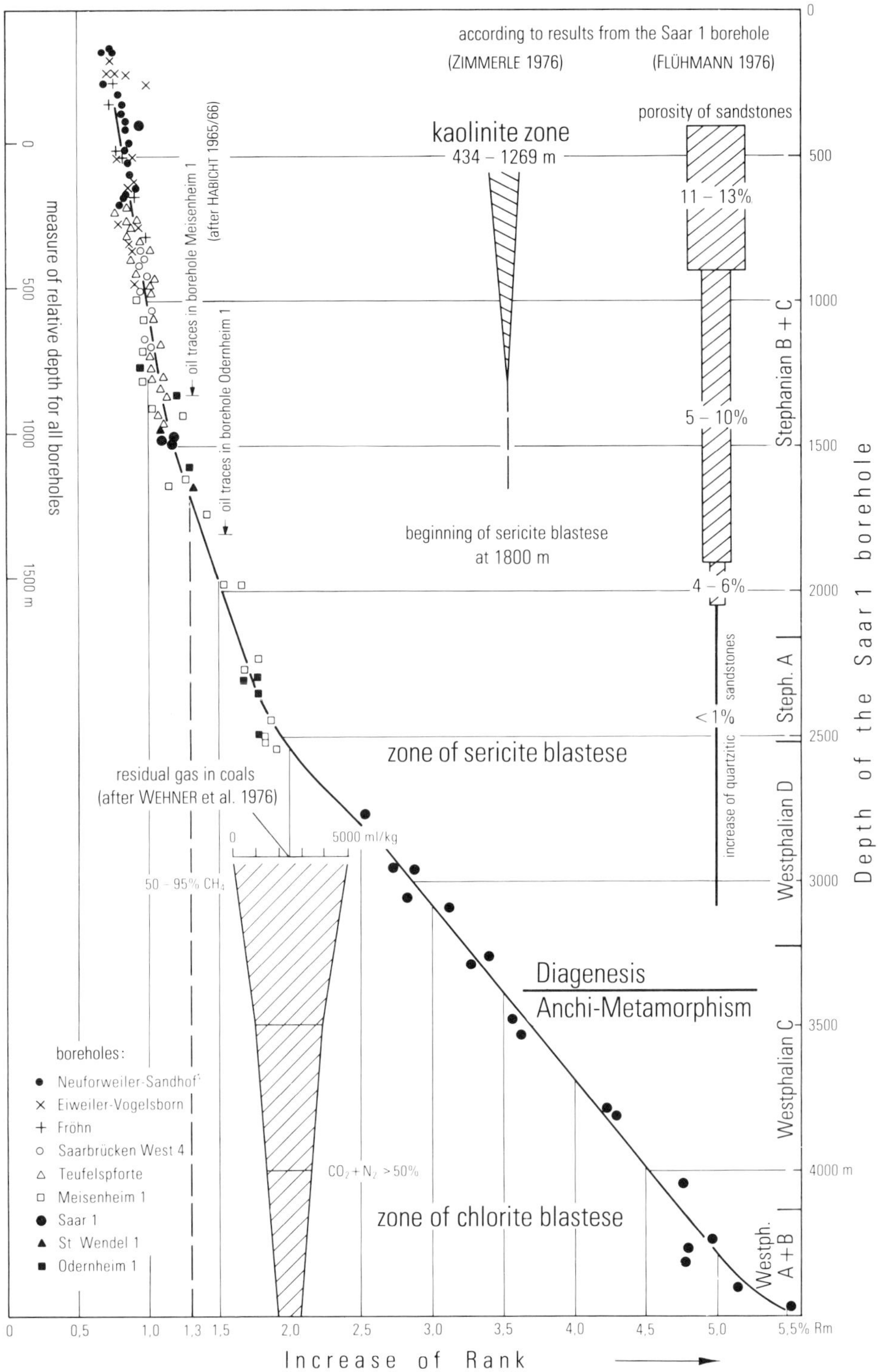

FIG. 27. Increase of vitrinite reflectance and of rock diagenesis and metamorphism with depth in boreholes of the Saar District (western Germany). After Teichmüller *et al.* (1983). With permission from the Deutsche Geologische Gesellschaft.

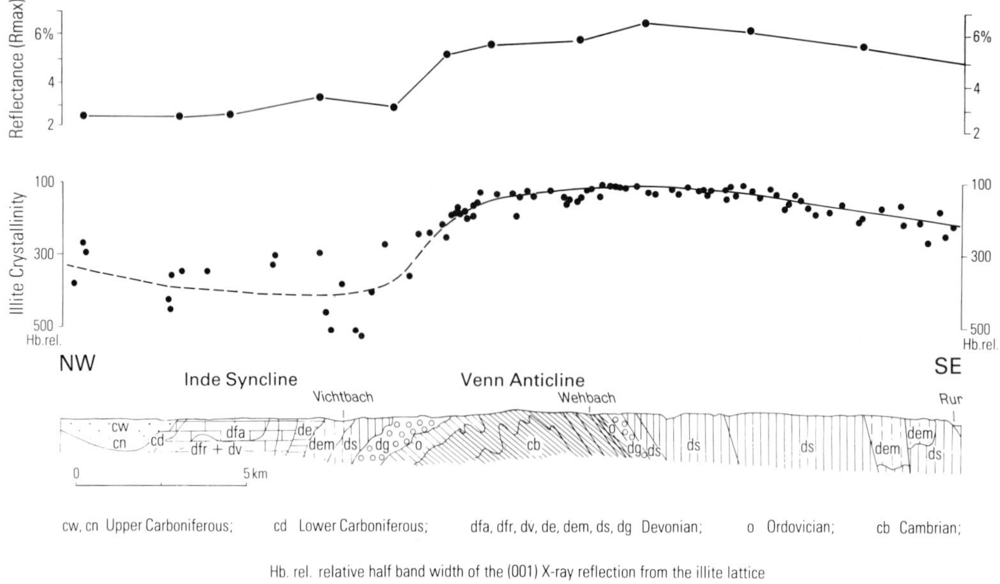

FIG. 28. Vitrinite reflectance and illite crystallinity in Palaeozoic rocks of the Venn (Stavelot) anticline. After Teichmüller & Teichmüller (1979). Geological section and illite crystallinity after Kasig & Späth (1975). With permission from the Geologisches Landesamt Nordrhein-Westfalen.

to Fig. 28 a good parallelism exists between the two parameters in a cross-section through the Venn Anticline. Figure 29 shows the correlation of illite-crystallinity versus % R_{max} for two cross-sections through the Helvetic Alps (Kien Valley of the Bernese Oberland and a section between Lake Lucerne and the Reuss Valley) after Frey *et al.* (1980). Although the values scatter, there is a fairly good correlation. Interestingly, R_{max} values lying above the correlation curves stem from localities with particular high shear stress. Fluid inclusion data from the same cross-sections show a relatively good correlation with IC and Rm values. The transition from the HHC zone to the CH_4 zone, indicated by a hatched line, corresponds with 2.5–3.0% R_m (i.e. late diagenesis

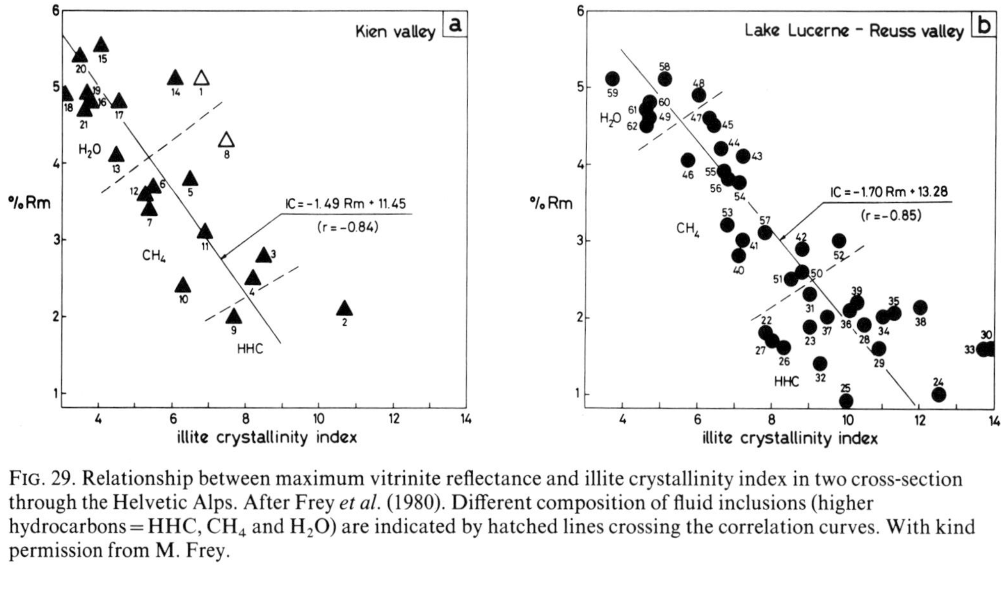

FIG. 29. Relationship between maximum vitrinite reflectance and illite crystallinity index in two cross-section through the Helvetic Alps. After Frey *et al.* (1980). Different composition of fluid inclusions (higher hydrocarbons = HHC, CH_4 and H_2O) are indicated by hatched lines crossing the correlation curves. With kind permission from M. Frey.

with *c.* 200°C and 1.2 kbar) whereas the transition from the CH_4 to the H_2O zone corresponds to 4.5–5.5% R_m (i.e. with the deep anchi-zone (270°C, 1.7 kbar)).

Comparative studies in Westphalia (Teichmüller *et al.* 1979b) resulted in the conclusion that in the zone of rock diagenesis (up to about 200°C and 4% R_{max}), vitrinite reflectance is a much better indicator than illite crystallinity whereas for the anchi-zone of rock metamorphism illite crystallinity is superior. In the epi-zone both parameters lose their significance.

A fundamental paper dealing with relations between rock metamorphism, coalification and graphitization on one side and rock temperature and pressure on the other was published by Diessel *et al.* (1978) on Tertiary high pressure schists of New Caledonia, Australia. According to these studies the first semi-graphite appears, still associated with particles at the stage of meta-anthracite, together with lawsonite as an index mineral, at about 250°C. True graphite is formed at the basis of the blueschist facies with garnet, later epidote as index minerals at estimated temperatures of 370°C. Figure 30 shows these relationships. The authors conclude that the high pressure (caused by nappe piling) accelerated graphitization and 'retarded coalification'.

According to Harris (1979) organic indices are generally used for assessing temperatures of <300°C, whereas mineral indices are used for higher temperatures.

Coalification studies in relation to palaeogeography, tectonics and geothermics

Since coal reacts very sensitively to temperature rise, coalification studies permit conclusions on former burial depth of sediments, their tectonic history and palaeogeothermics. According to Middleton (1982) 'the signature of the thermal

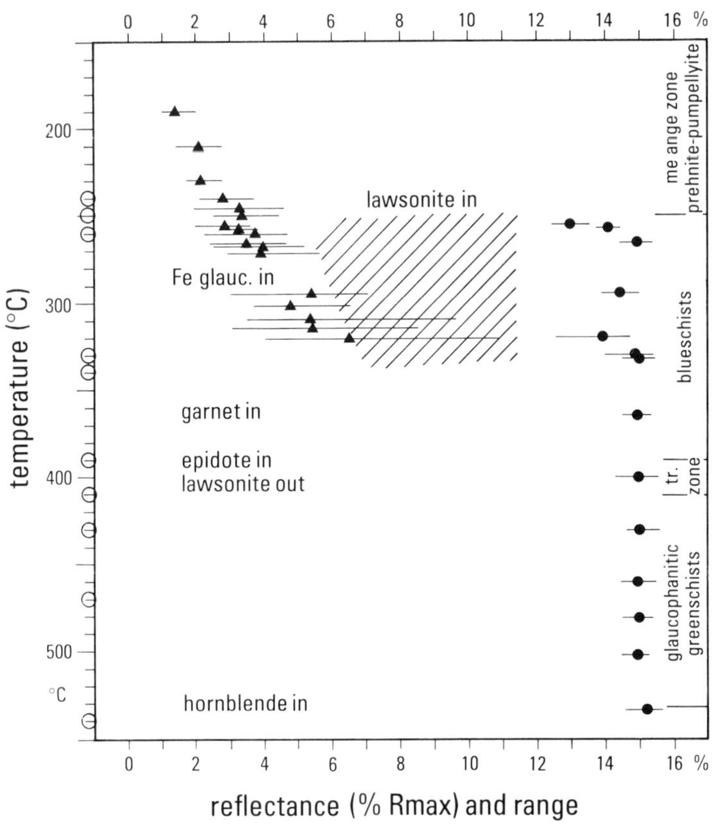

FIG. 30. Relationship between maximum reflectance of organic constituents, rock metamorphism (with index minerals) and maximum rock temperature in Tertiary high pressure schists of New Caledonia, Australia. After Diessel *et al.* (1978), modified. Triangles indicate coal stages, circles the stage of graphite. Transition zone shaded. Open circles at the ordinate indicate temperature values obtained through the oxygen isotope method. With kind permission from C. F. K. Diessel.

and tectonic history of a sedimentary basin is preserved by the reflectance of vitrinite . . .'

Coalification profiles from deep boreholes and coalification maps for certain stratigraphic horizons are used to reconstruct the ancient outline of basins and their tectonic and geothermal history, mostly on the basis of vitrinite reflectance and subsidence curces as shown in Figs 34 and 36. In many cases thermodynamics are also considered. Corresponding computer models were developed to facilitate the estimations of palaeogeothermics, thickness of eroded sediments and timing of tectonic events (Karweil 1956, Lopatin 1971, Bostick 1973, Tissot & Espitalié 1975, Burne & Kantsler 1977, Buntebarth 1978, 1979, Waples 1980, Welte & Yükler 1981, Falvey & Deighton 1982, Middleton 1982, Ritter 1985). In the following sections some examples for geologically applied coalification studies are reported.

Basin studies

In a detailed study of late Cretaceous and Tertiary coals of the Canadian Arctic Archipelago Bustin (1986) estimated from rank gradients maximum palaeothicknesses of sediments and thus could outline basins with well defined margins. Palaeogeothermal gradients varying between 15 and 55°C km^{-1} were deduced from rank gradients on the basis of numerical modelling.

Kalkreuth & McMechan (1984) found a westward decrease of coal rank in Upper Cretaceous–Tertiary deposits as a consequence of the timing of Laramide deformation across the area extending from the core of the Alberta Syncline to the western edge of the Foothills of the Canadian Rocky Mountains. In the west, early Laramide uplift prevented deep burial. The conditions seem comparable with those at the southern margin of the Ruhr Carboniferous (Teichmüller & Teichmüller 1971).

Oncken (1982) attempted to reconstruct the development of a geosyncline in the Variscan Rhenish Mountains evaluating coalification studies carried out by Wolf (1972) who had already drawn conclusions on the tectonics of the same region. Estimations of total thickness of the geosynclinal sediments and the heat flow history as well as of post-orogenic uplift (degree of erosion) were made.

In the British Isles, Oliver (1987) and Kemp *et al.* (1985) studied the metamorphism of the paratectonic Caledonides on the basis of graptolite reflectance, as well as illite crystallinity and index minerals. In Wales for example, graptolite reflectance in rocks of Arenig and Wenlock age increases from 0.5% R_{max} in the borderland to 8–9% R_{max} in the Welsh Basin due to increasing depths of former burial and palaeotemperatures. The same trend was found on the basis of conodont colour by Aldridge (1986) who estimated from CAI values geothermal gradients in the order of 60°C km^{-1} either in Caledonian or in Hercynian times. Figure 31 shows the change of graptolite reflectance from the Welsh Borderland to the Welsh Basin in relation to index minerals of rock metamorphism according to Oliver (1987).

Epstein *et al.* (1977), who developed the method to determine the conodont alteration index (CAI), depicted maps with CAI isograde lines for pre-Devonian rocks in the Appalachian Basin, showing the increase of rank from NW to SE across the basin, as a consequence of the increase of maximum burial depth and thus of maximum temperature. Other isograde maps on the basis of conodont colour alteration index (CAI) were published by Harris *et al.* (1980) for Ordovician to Triassic rocks in Nevada, Utah, Idaho and California as well as by Wardlaw & Harris (1984) for Arizona and by Patrick *et al.* (1985) for Alaska. Bergström (1980) used the same method to assess palaeotemperatures of Ordovician rocks in Scandinavia and the British Isles with the result that the Baltoscandian-platform rocks of southern Sweden were not heated above 90°C, whereas in the Oslo, Trondheim and Jämtland regions and in the British Isles (except for the Welsh Borderland and the Girvan area of Scotland) temperatures of more than 300°C have been assessed.

According to Aldridge (1984) colours of Silurian conodonts in the Oslo region display close correlations to the proximity of Permian intrusions.

Based on vitrinite reflectance, thicknesses of eroded sediments and palaeotemperatures were

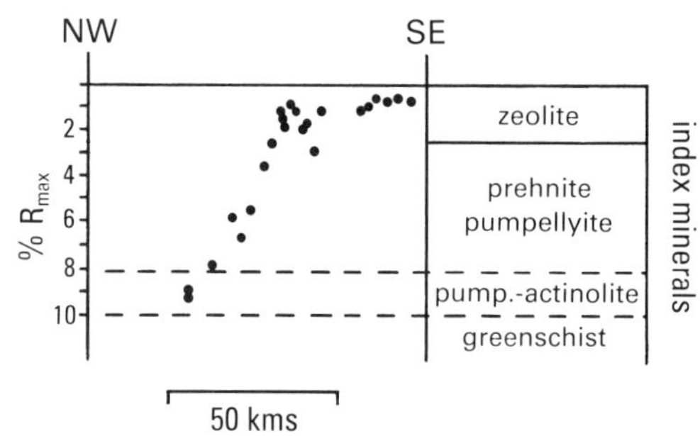

FIG. 31. Increase of graptolite reflectance (R_{max}) in a cross-section from the Welsh Borderland in the SE to the Welsh Basin in the NW, in relation to change of index minerals. After Oliver (1987), with kind permission from the author.

estimated for the Sydney coal basin, Australia by Middleton & Schmidt (1982). The geothermal results match results from palaeomagnetic studies (overprinting of magnetization of late Permian and Mesozoic rocks), according to which palaeotemperatures were the highest (45–63°C km^{-1} effective geothermal gradients) at the coast line. The authors explain these high temperatures with the onset of seafloor spreading of the Tasman Sea 76 Ma ago.

Tectonics

The timing of coalification (pre-, syn- or post-kinematic) in relation to tectonic events (folding, thrusting, faulting) can best be clarified using geological cross-sections with iso-reflectance lines (cf. Fig. 33). Such studies have been carried out, *inter alia*, by Hacquebard (1975), Bustin (1984) and Pearson & Grieve (1986) in the Canadian Rocky Mountains, by Hesse & Ogunyomi (1980) in the northern Appalachians and by Reutter *et al.* (1983) in the northern Apennines. 'Transported coalification' has been described from the Helvetic Alps where the Axen nappe with strongly coalified organic matter overlies much less coalified Helvetic Flysch (Frey *et al.* 1980). New results on relations between orientation of vitrinite anisotropy and the direction of tangential pressure (Stone & Cook 1979, Hower & Davis 1981, Levine & Davis 1984) have been discussed already (p. 140).

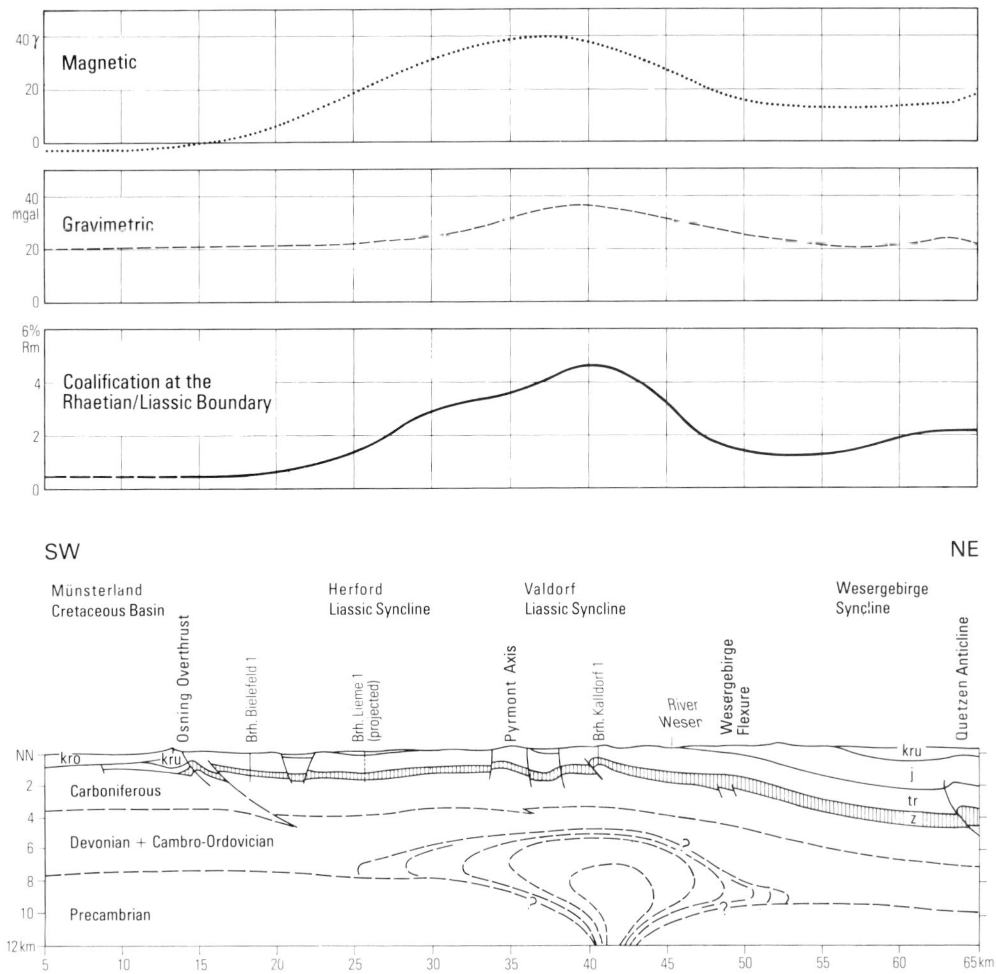

FIG. 32. Section across the Vlotho Massif on the basis of magnetics, gravimetry and vitrinite reflectance. After Deutloff *et al.* (1980). With kind permission from O. Deutloff.

Large-scale coalification studies in relation to geothermics and plate tectonics

Large-scale geothermal coalification studies ought to be combined with geophysical studies (gravimetry, magnetics, seismics, tellurics), if possible, as has been practised, e.g. in western Germany. Deep seated magmatic intrusions were detected there and their age, depth and size could be estimated, as for the Bramsche and Vlotho Massifs (Teichmüller & Teichmüller 1950, 1984, Bartenstein *et al.* 1971, Deutloff *et al.* 1980) which are distinguished by very high coalification peaks. Figure 32 demonstrates the close relationship between magnetic, gravimetric and reflectance data in a cross-section through the Vlotho Massif, with the presumed magmatic body at a depth of 6–8 km.

A comparison of the coalification patterns in foredeeps and backdeeps of ancient and recent orogens, combined with geophysical results, may permit conclusions to be drawn on crustal thicknesses and plate tectonics, as will be reported in more detail in the following.

The Hercynian foredeep of the Ruhr Basin and the Alpidic foredeep of the Alps differ by exhibiting very different coalification patterns. The coalification gradients of the Carboniferous molasse of the Ruhr are much higher than those of the molasse in the Alpine foredeep, although

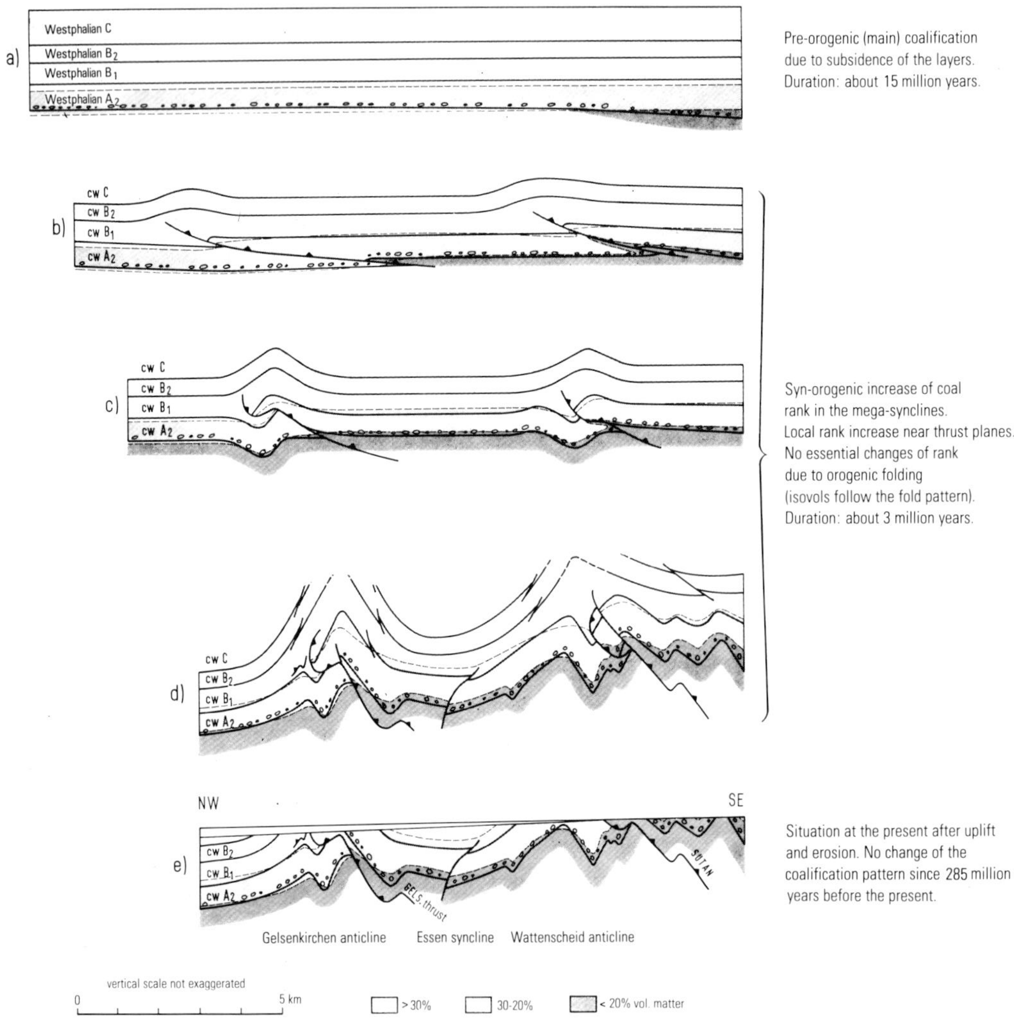

FIG. 33. Coalification history of the Ruhr Carboniferous according to Teichmüller & Teichmüller (1971). With permission from the Geologisches Landesamt Nordrhein-Westfalen.

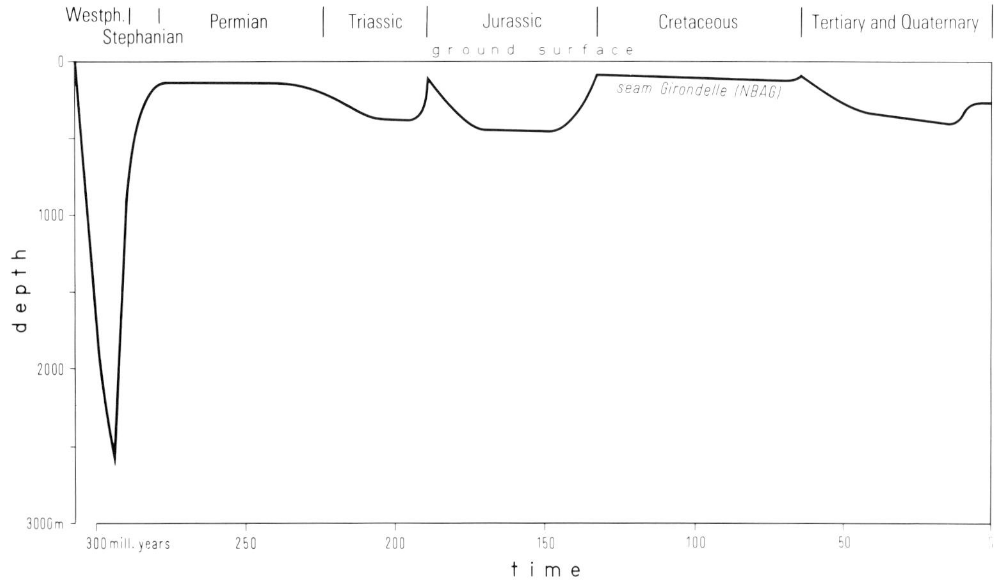

FIG. 34. Burial history of a seam (Girondalle) in the Westphalian A of the Lower Rhine Basin. After Teichmüller & Teichmüller (1986). With kind permission from G. Buntebarth and Springer-Verlag.

the present geothermal gradients are the same and the depths of burial as well as the duration of maximum heat exposure were similar.

We know from earlier studies (Patteisky *et al.* 1962) that coalification in the Ruhr district was mainly pre-orogenic, i.e. it took place before the Asturian folding and uplift. Figure 33 shows the development of coalification in relation to thrusting and folding. Zones of equal rank run parallel with the stratification. The pre-orogenic coalification is also evident from very low rank Permian (Kupferschiefer), overlying strongly coalified Westphalian. The Cretaceous cover was not thick enough to promote coalification, comparable to that of the Carboniferous.

Figure 34 shows a typical subsidence/uplift history for a coal seam in the Lower Rhine Basin: deep subsidence during the Westphalian and rapid uplift even at the Westphalian–Stephanian boundary. Such subsidence curves are based on the thickness of sediments deposited in the various geological time spans. Together with the coalification gradients (deduced from measurements of vitrinite reflectance in 53 deep boreholes) such subsidence curves allow reconstructions of the geothermal conditions during the Carboniferous in the Ruhr (Buntebarth *et al.* 1982). The coalification gradients vary between about 0.5% and 1.0% R_m km^{-1}, the higher gradients being found in the earlier Westphalian. Based on these fundamentals, Koppe (1980) and Buntebarth *et al.* (1982) computed geothermal gradients ranging on average between 60 and 70°C km^{-1} for the Westphalian (Fig. 35). Higher gradients, up to 90°C km^{-1} were computed north of Krefeld and near Westerholt where indications suggest additional heat input from intrusive bodies at great depths.

In the foredeep of the northern Alps coalification is very young. In the molasse, burial depth (2.5 km for the molasse base near Munich) and effective coalification time (folding and uplift 8–12 Ma before the present) are similar to figures deduced for the Ruhr molasse during the Carboniferous. According to Fig. 36 the burial time for the maximum depth, and thus for the maximum rock temperature, was even longer in the Alpidic molasse as may be seen from a comparison between the subsidence curves for the Anzing 3 borehole (Alps) and the Mathilde seam of the Ruhr Carboniferous. Nevertheless, the rank attained by the Mathild seam is much higher (1.25% R_m) than the rank reached by the Tertiary molasse (0.5% R_m).

Boreholes drilled by the oil industry through the folded and imbricated molasse of the foreland, as well as through the folded and imbricated molasse of the northern Alps (where the molasse is overlain by imbricated Helveticum, Flysch and Austroalpine nappes), were studied for an increase of vitrinite reflectance with depth. The coalification gradients proved to be surprisingly low. They are more than one order of magnitude

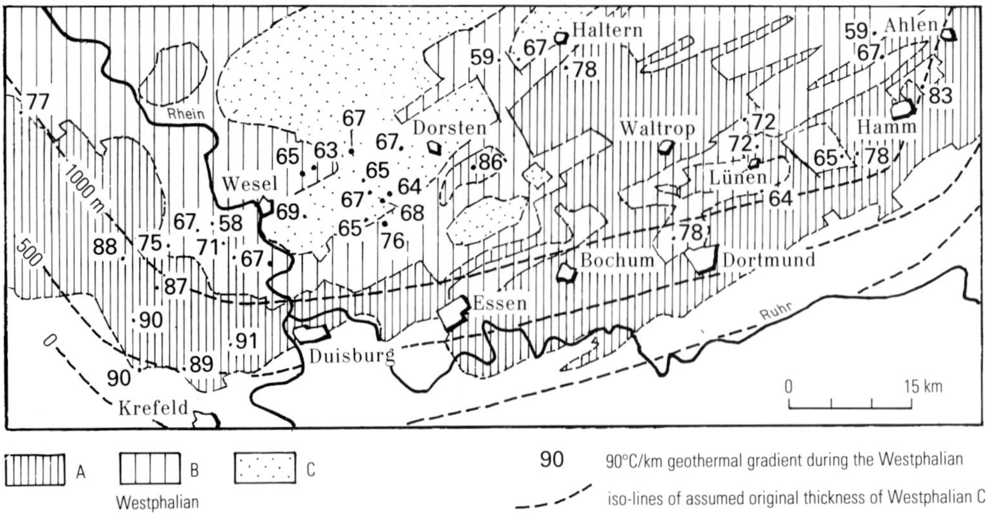

FIG. 35. Palaeogeothermal gradients (°C km^{-1}) for the Westphalian of the Ruhr and Lower Rhine Districts. Values computed from coalification gradients measured in 53 boreholes. After Buntebarth *et al.* (1982). With kind permission from G. Buntebarth.

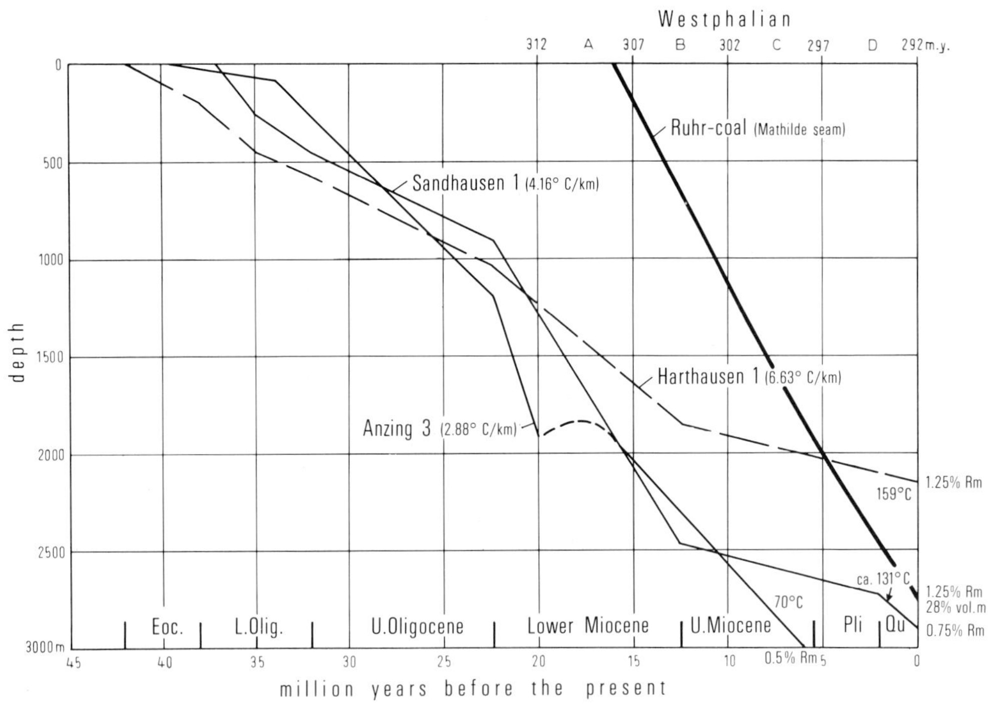

FIG. 36. Comparison of subsidence histories in the Alpine foredeep (Anzing 3 borehole) and the Hercynian foredeep of the Ruhr area (Mathilde seam). After R. Teichmüller (1973) and Teichmüller & Teichmüller (1975), modified. Time-scale for the Westphalian on the upper abscissa. Sandhausen 1 and Harthausen 1 are boreholes in the Upper Rhine Graben. With permission from Geologica Bavarica.

lower (0.03–0.09% R_m km^{-1}) than the coalification gradients of the Hercynian molasse (0.5–1.0% R_m km^{-1}), with the result that the stage of bituminous coal is very seldom reached in the sub-alpine molasse.

According to recent coalification studies of Jacob *et al.* (1982) the coalification gradients decrease from north to south, especially when the pre-tectonic, non-imbricated position of the nappes is considered (Fig. 37). In the south, within the folded and imbricated molasse, the coalification gradients are lower than would be expected when compared with the present geothermal gradients, which vary between 19 and 23°C km^{-1}. Therefore the authors assume a pre-orogenic coalification which took place 10–20 km further south where temperature gradients were lower in pre-thrusting times and are lower at present as well.

The coalification gradients and geothermal gradients are in accordance with the present crustal thickness which increases from north to south, due to subduction of the northern crust. Figure 38 shows the dipping of the Moho from 33 to 40 km between the foreland and the northern rim of the Alps.

The northern Apennines are a very young orogen with the foredeep of the Po Plain in the NE. Folding and nappe piling occurred in Miocene times. According to Reutter *et al.* (1983) the main coalification is younger than the emplacement of the nappes, but older than the last thrust movements and the final tensional tectonics which took place, together with magmatic intrusions, in Pleistocene times in the backdeep in the SW. Within the single nappes the rank increases from the foredeep of the Po Plain in the east to the backdeep at the Tyrrhenian coast in the west. Within the nappe piles the rank increases from the uppermost to the lowermost nappe. Thus, the main regional thermal event has to be considered as late synorogenic (coalification started not earlier than the latest Miocene). Therefore a close relationship exists between the present heat flow and the coalification pattern.

The iso-lines of present heat flow density (Fig. 39) show an increase from 30 to 110 mW m^{-2} from NE to SW where coal of the uppermost Miocene reached the stage of bituminous coal (0.74% R_m) at Ribolla, whereas vitrinite of the same age is still in the brown coal stage (0.22% R_m) in the foredeep of the Po Plain. There, a very low coalification gradient was determined in the Ponte del'Olio borehole (0.07% R_m km^{-1} (Fig. 40). The geothermal gradient is also very low (19°C km^{-1}), whereas on the other side of the orogen much higher gradients were measured, e.g. 0.19% R_m km^{-1} and 32°C km^{-1} in the Martina borehole south of Elba Island.

The section of Fig. 41 shows the subduction of the Adriatic crust and the intensive imbrication of nappes under the Po Plain, causing a thickness

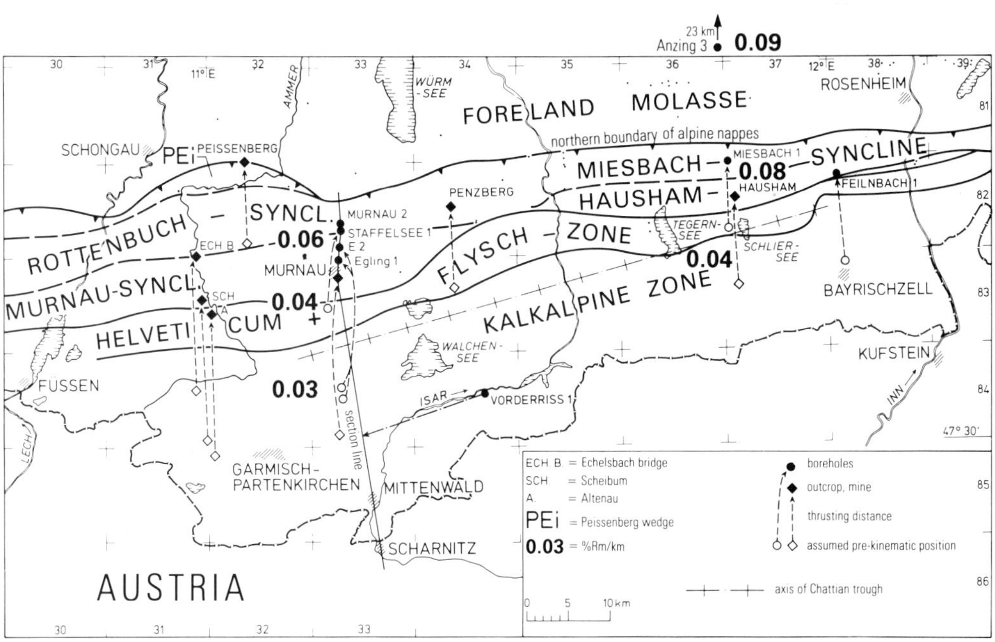

FIG. 37. Decrease of coalification gradients in the molasse of the Alpine foredeep from north to south, approaching the centre of the orogen. After Jacob *et al.* (1982), modified. With kind permission from M. Müller.

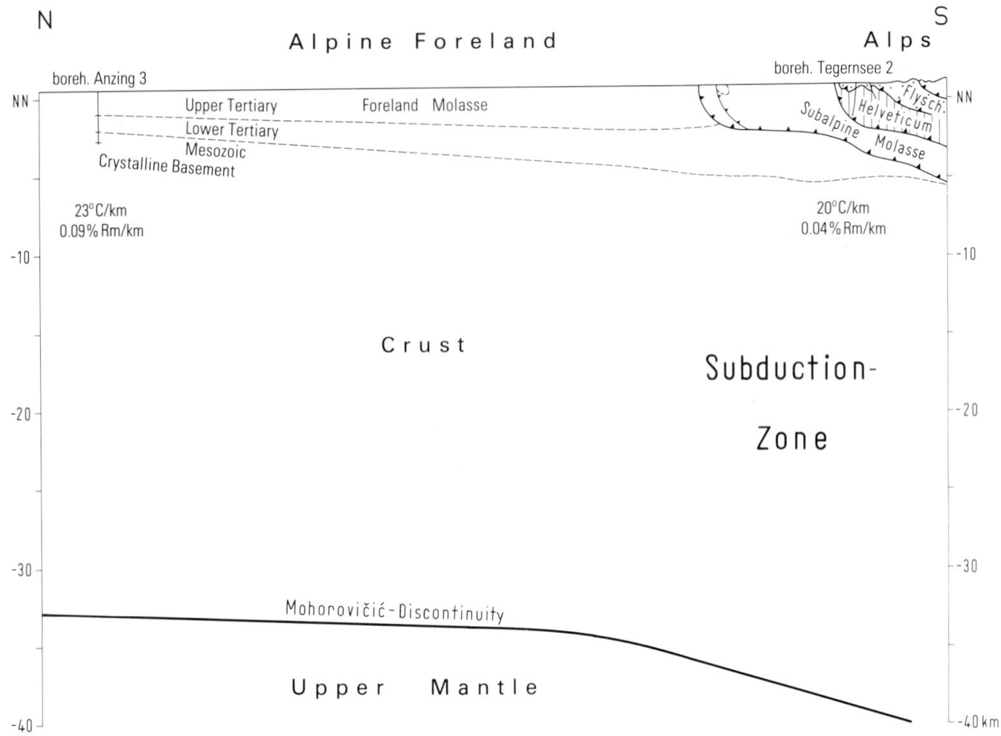

FIG. 38. Subduction zone at the northern margin of the Alps. After Teichmüller & Teichmüller (1975). Depth of the Moho according to Giese & Stein (1971). With permission from Geologica Bavarica.

of more than 5000–6000 m of Pliocene and Pleistocene sediments and a deep position for the Moho (38 km near Faenza). From there, the Moho ascends towards the west up to 20 km under Elba Island. The heat flow rises correspondingly from 40 mW m^{-2} in the Po Plain to 110 mW m^{-2} at the Tyrrhenian coast, with an anomalously high maximum of 320 mW m^{-2} near Larderello where geothermal energy is used as an energy source. Vitrinite reflectance in this section rises from 0.5% R_m in the NE to 5.6% R_m on Elba Island. In summary, the cross-section demonstrates the contrast between:

(1) subsidence, compression, thickening of the crust and weak coalification in the foredeep, and

(2) uplift, tension, plutonics and strong coalification in the backdeep of the northern Apennines.

For comparison, Fig. 42 shows a geophysical model of Giese (1978), demonstrating the different thicknesses of the crust in the foreland and in the backland of an orogen.

This leads to a diagram (Fig. 43) based on Čermák (1979) who originally pointed to the inverse relationship between recent heat flow density and crustal thickness for the Bohemian Massif, the Pannonian Basin and the Upper Rhine Graben. Koppe (1980) introduced values for the Westphalian of the Ruhr Basin. She estimated the heat flow densities from the computed temperature gradients and a mean thermal conductivity (of $K = 1.6$ W mK^{-1}) for the Ruhr Carboniferous. Her results correspond with a crustal thickness of 19–23 km during the Westphalian, in contrast to thickness of 29 km at the present time. Similar and much higher values of heat flow were obtained by Buntebarth (1983) for the Carboniferous of the Saar Basin and the Permian of the Nahe Basin respectively, both basins representing part of the sub-Hercynian backdeep. Together with the new data from the northern Alps and the northern Apennines the diagram of Fig. 43 suggests:

(1) a much thinner crust in the sub-Hercynian foredeep of the Ruhr Basin than in the Alpidic foredeeps of the Alps and the northern Apennines,

(2) higher heat flow densities and thinner crusts in the backdeeps of the Nahe Basin and the northern Apennines, compared with the corresponding foredeeps.

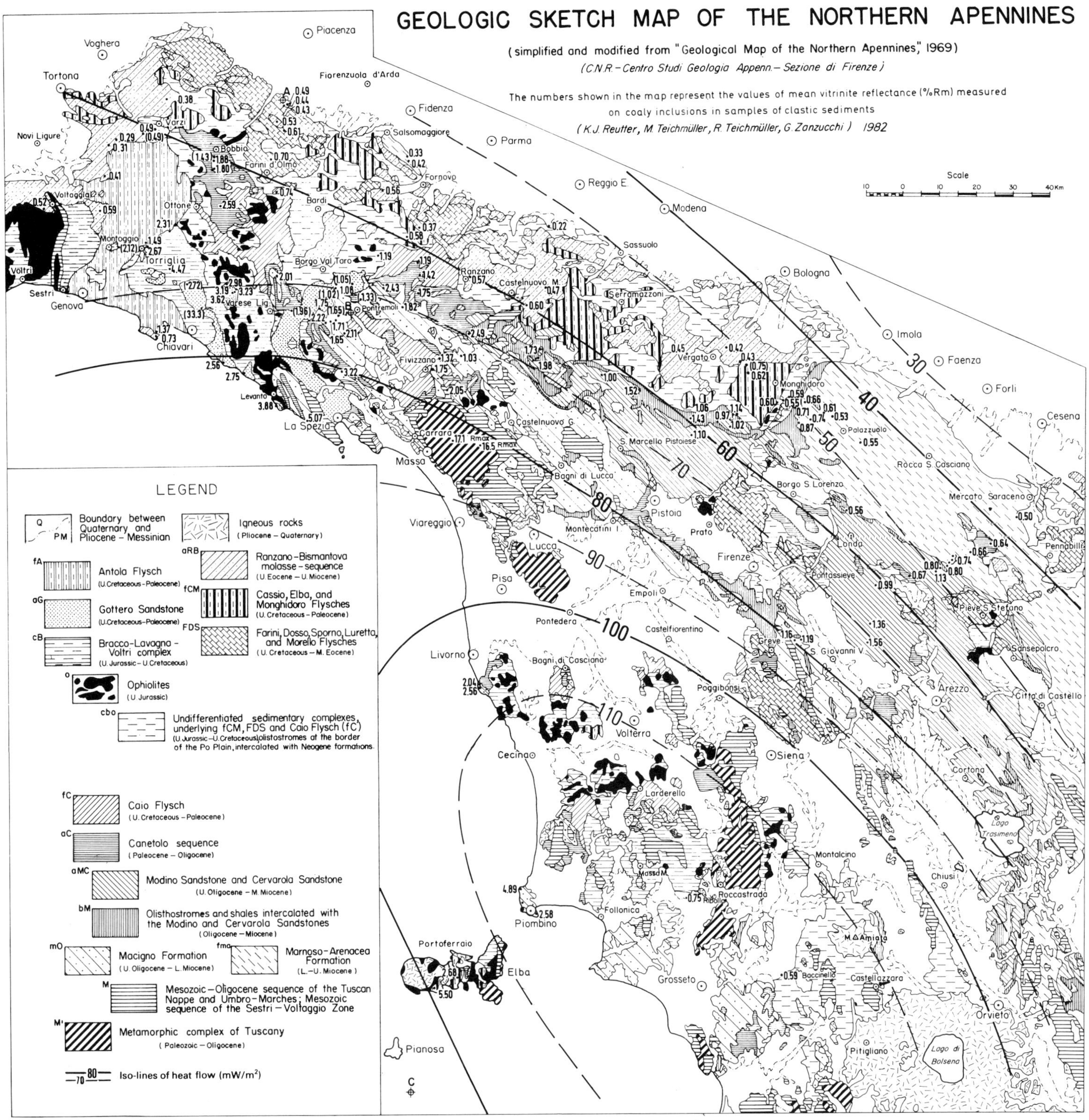

FIG. 39. Coalification map of the northern Apennines (after Reutter *et al.* (1983) with iso-lines of heat flow density (after Čermák & Hurtig 1979). With kind permission from K. J. Reutter and the Geologische Rundschau.

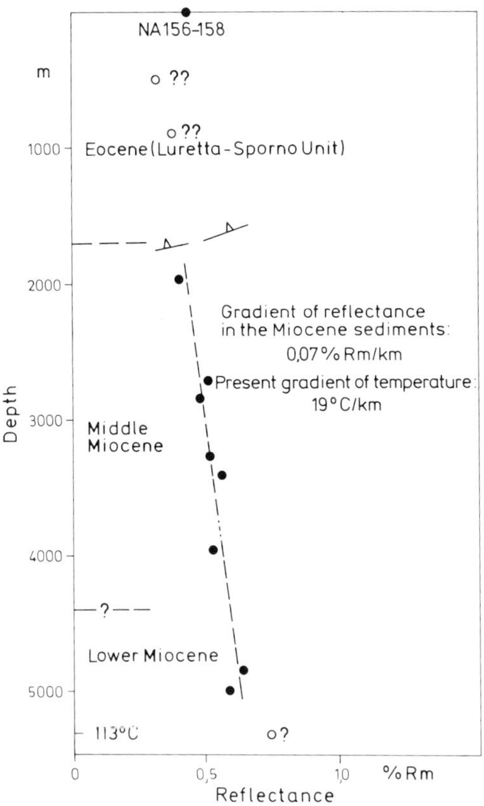

FIG. 40. Increase of vitrinite reflectance with depth in the Ponte del'Olio borehole. After Reutter *et al.* (1983). Note the low rank (0.6% R_m) at 5000 m depth. With kind permission from K. J. Reutter and the Geologische Rundschau.

Regarding the model of Giese (Fig. 42), the reason for the high heat flow in the Carboniferous foredeep at the Ruhr does not appear feasible, although Giese *et al.* (1983, p. 423) assume a crustal thickness of about 20 km for the sub-Hercynian foredeep in general in late Carboniferous times. From the standpoint of plate tectonics Ritter (1986) compares the Ruhr foredeep with 'foredeep-like structures at pacific-type continental margins, such as the Java and Sunda Seas'.

High heat flows in the European Hercynian in general have been suggested by many authors, *inter alia*, by Zwart (1967, 1976) who points to the shallow depth of rock metamorphic zones ('low pressure metamorphism') and to the unusually large number of granites and granodiorites. Zwart suggests geothermal gradients between 40 and 60°C km^{-1}, locally up to 100°C km^{-1}, as a consequence of 'activity in the mantle', possibly as mantle diapirs.

The relatively very low geothermal and coalification gradients in the foredeeps of the Alps and the northern Apennines clearly correspond with the thickening of the crust due to convergence and subduction of continental plates.

Coalification studies in relation to hydrocarbon exploration

The great practical significance of coalification studies lies in the application to hydrocarbon exploration. Therefore, since about 1975, most coalification studies have been carried out in relation to oil and gas occurrences, especially in:

Australia (by Kurylowicz *et al.* 1976, Burne & Kantsler 1977, Kantsler *et al.* 1978, Shibaoka *et al.* 1973, 1978, Kantsler & Cook 1979, McKirdy & Kantsler 1980, Cook & Kantsler 1980, Smith & Cook 1984, Kantsler *et al.* 1986),

Canada (by Hacquebard 1975, Kalkreuth & McMechan 1984),

France (by Alpern & Cheymol 1977, Robert 1980, 1985),

Germany (by Teichmüller & Teichmüller 1950, Bartenstein *et al.* 1971, Bartenstein & R. Teichmüller 1974, Teichmüller & Ottenjann 1977, Teichmüller *et al.* 1979a, 1984, Jacob *et al.* 1982, Teichmüller 1986b),

Japan (Asakawa & Fujita 1979),

USA (by Hood & Castano 1974, Epstein *et al.* 1977, Barrows *et al.* 1979, Bostick 1979, Bostick *et al.* 1979, Harris 1979) and

USSR (*inter alia*, by Ammosov & Sharkova 1975, Golytsin 1975, Lopatin 1976a, b).

Moreover, many organic geochemists have used coalification studies on the basis of vitrinite reflectance, besides chemical maturity parameters. Since it is impossible to quote all papers, the reader is referred to books and reviews, like Tissot & Welte (1978), Hunt (1979), Héroux *et al.* (1979), Teichmüller & Teichmüller (1981) and Robert (1985).

Only one recent example may be treated here briefly: the relations between coalification and gas deposits in northwestern Germany. A new coalification map of the top surface Carboniferous in NW Germany (Teichmüller *et al.* 1984, Teichmüller 1986b) shows the relationships between the present depth of the top Carboniferous, vitrinite reflectance referred to the top Carboniferous, and gas fields sourced by coalification gases from the Carboniferous, with reservoirs in the Upper Carboniferous, Permian (mainly Rotliegendes) and Lower Triassic (Bunter). The stratigraphic units subcropping at the top Carboniferous are indicated in various colours. The scale is 1:500 000.

The main results are demonstrated in the cross-

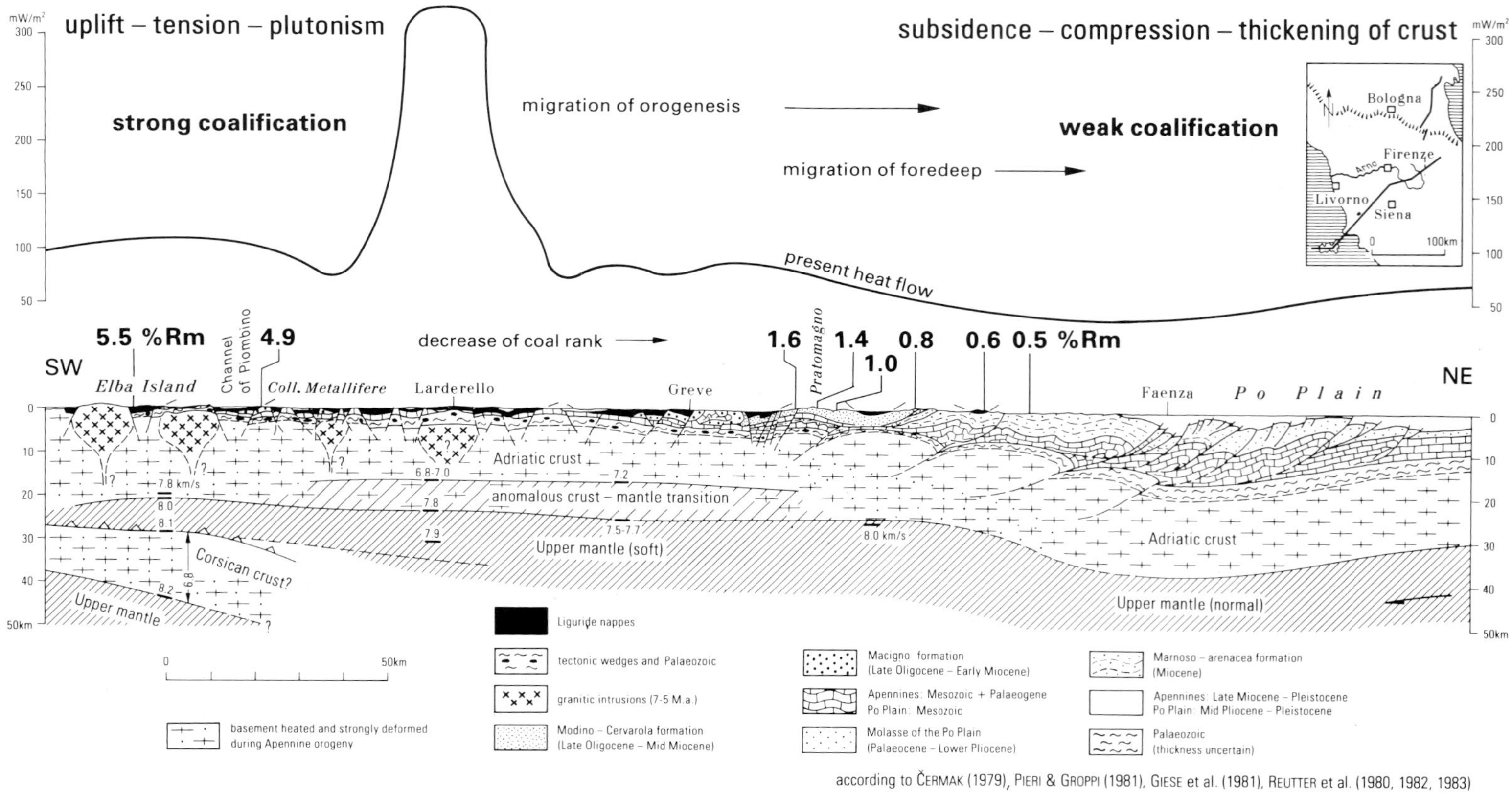

FIG. 41. Cross-section through the northern Apennines between the Po Plain (foredeep) and the Tyrrhenian Sea (backdeep) with coalification and heat flow data. After Teichmüller & Teichmüller (1986). With kind permission from G. Buntebarth and Springer-Verlag.

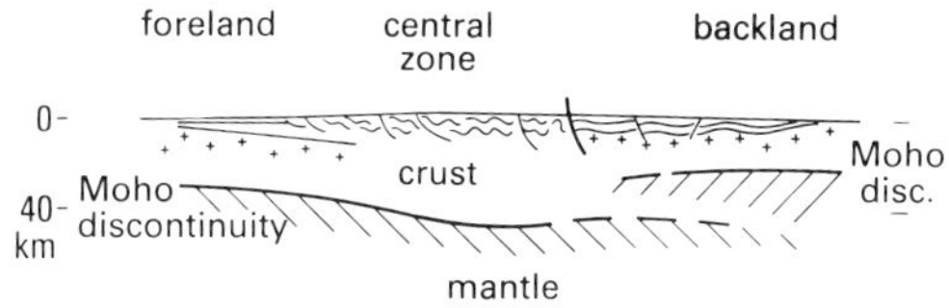

FIG. 42. Simplified model for the crustal structure of an alpine orogen. After Giese (1978). The crustal thickness increases from the foreland to the central zone and decreases rapidly from the central zone to the backland. With kind permission from P. Giese and the Deutsche Geologische Gesellschaft.

section of Fig. 44 which runs from the Hercynian Mountains in the south to the North Sea. The gradual slope of the top of the Carboniferous towards the north is evident. It is interrupted by a sharp uplift in the horst of Ibbenbüren where Westphalian C and D lie at the ground surface. Nevertheless, there we find the strongest coalification with meta-anthracites in the Westphalian D (Teichmüller & Teichmüller 1985). This high degree of coalification is caused by the Bramsche intrusion which led to additional heat input in Upper Cretaceous times (Bartenstein *et al.* 1971).

In the Ruhr District and the Münsterland Basin the present depth of the top Carboniferous is indecisive for the rank of coal. There, the reflectance values (0.7–1.1% R_m) rather reflect the age of the Westphalian, as a consequence of pre-orogenic coalification (see p. 155). In the Lower Saxony Basin, however, a post-Carboniferous coalification could take place due to deep subsidence and the resulting thick cover of Mesozoic and Cainozoic sediments amounting to 4000–5000 m. This re-coalification, along with the good seal by evaporites of the Zechstein and the Bunter Formations, explains the occurrence of many gas fields in the Südoldenburg High which, moreover, is characterized by a flat doming of the Carboniferous, having promoted gas accumulation. Near the coast of the North Sea the R_m values decrease again to 1.0–1.3 because, there, coalification obviously is not yet in equilibrium with the present depths and temperatures, due to a strong but too short period of subsidence in the late Quaternary.

In summary, the gas fields are restricted to areas where a late, post-Carboniferous re-coalification could take place, either due to a deep subsidence in Cainozoic or late Mesozoic times or due to an additional heat supply at the northern margin of the Bramsche Massif. Clearly no gas could accumulate in the roof of the intrusive body because porosity was lost as a consequence of strong heating. In terms of vitrinite reflectance, most gas fields lie between 1 and 2% R_m, some between 2 and 3% R_m (cf. Fig. 3). This is in accordance with the internationally acknowledged limits for gas deposits.

Suggestions for further work

Results of coalification studies in the last 15 years raised new problems whose clarification might be possible by the following studies:

(1) *Methods*

(a) More intense application of high resolution transmission electron microscopy, in combination with light microscopy and organic geochemistry, especially with regard to the coalification tracks of various vitrinite types (sub-macerals).

(b) Behaviour of vitrinite reflectance (R_{max}, R_{min}, R_m) with increasing temperature and pressure at very high rank stages ($>5\%$ R_{max}) with regard to stagnations, jumps, reversals.

(c) Correlation of graptolite reflectance with vitrinite reflectance using Devonian rocks (which may bear both constituents).

(d) Clarification of the chemical and physical causes of maceral fluorescence.

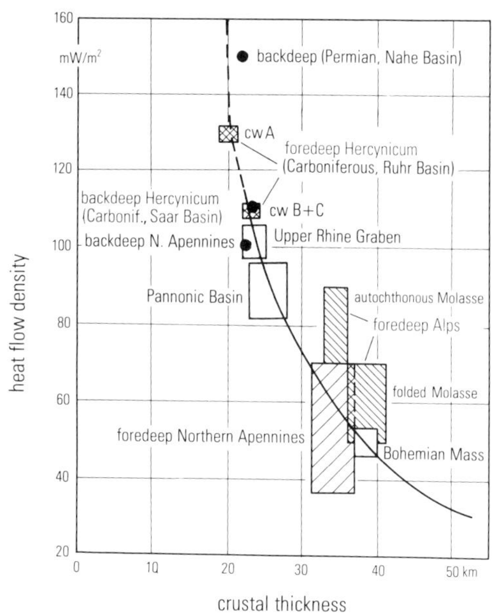

FIG. 43. Relationship between crustal thickness and heat flow density for different areas and geological times according to Čermák (1979), Koppe (1980), Buntebarth *et al.* (1982) and the present author. Foredeeps as shaded areas, backdeeps as black circles.

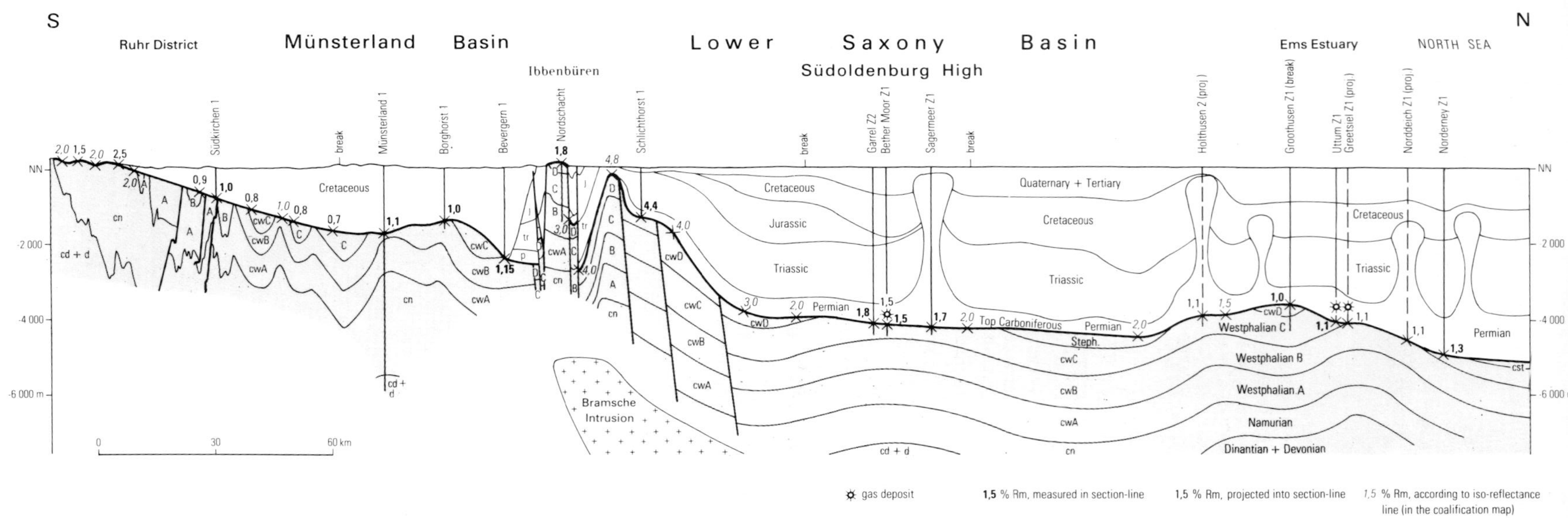

FIG. 44. Coalification at the top surface Carboniferous and gas deposits in a cross-section running from the Variscan Rhenish Mountains to the North Sea in western Germany. After Teichmüller *et al.* (1984) and Teichmüller (1986). With permission from the Geologisches Landesamt Nordrhein-Westfalen.

(2) *Causes of coalification*

Clarification of the role of heat conductivity of coal and other rocks for the increase of rank.

(3) *Coalification process*

(a) Clarification of relations between bituminization, 'liquid structure', and development of stronger anisotropy of vitrinites.

(b) Clarification of the process and the causes of diagenetic gelification ('vitrinitization') through, e.g. experiments with mixtures of huminites from brown coals and early generated bitumen from oil source rocks under geologically reasonable temperatures and pressures.

(c) Which coals are source rocks for oil? Oil-prone vitrinites besides liptinites? Migration of oil in coal?

(d) Systematic study of coalification tracks of single macerals and maceral types, especially of 'normal', sub-hydrous and per-hydrous vitrinites (see 1a, above).

(e) Studies concerning inertinitization during early coalification. Precursors of 'rank inertinites' (e.g., cellulose-rich and/or lipid-rich cell walls and/or humic detritus?). Relations to early generated, non biogenetic methane?

(4) *Application of coalification studies to geology*

Intensification of large-scale regional (coalification maps) and vertical (boreholes) coalification studies in combination with geophysical research in order to clarify problems of plate tectonics, depths of the crust–mantle boundary and magmatic intrusions.

References

ALDRIDGE, R. J. 1984. Thermal metamorphism of the Silurian strata of the Oslo region, assessed by conodont colour. *Geological Magazine*, **121**, 347–349.

—— 1986. Conodont palaeobiogeography and thermal maturation in the Caledonides. *Journal of the Geological Society of London*, **143**, 177–184.

ALPERN, B. & CHEYMOL, D. 1978. Réflectance et fluorescence des organoclastes du Toarcien du Bassin de Paris en fonction de la profondeur et de la température. *Revue de l'Institut Français du Pétrole*, **33**, 515–535.

AMMOSOV, I. I. & TAN SJU, I. 1961. The coalification process and the paragenetic behaviour of fossil fuels. *Izvestiya Academiya Nauk USSR* (in Russian).

AMMOSOV, I. I. & SHARKOVA, L. S. 1975. Paleotemperatures, lithification and oil and gas occurrences in Jurassic deposits in western Kasakhstan and central Asia. *In:* EREMIN, I. V. (ed.) *Paleotemperatures of Zones of Oil Formation*, pp. 60–69, Moscow, (Russian).

ASAKAWA, T. &. FUJITA, Y. 1979. Organic metamorphism and hydrocarbon generation in sedimentary basins of Japan. *CCOP Technical Publication, Series No. 6*, pp. 142–162.

BARKER, C. E. 1979. Vitrinite reflectance geothermetry in the Cerro Prieto geothermal system, Baja California, Mexico. *Master's Thesis*, Department of Earth Science, University of California.

—— 1983. Influence of time on metamorphism of sedimentary organic matter in liquid-dominated geothermal systems, western North America. *Geology*, **11**, 384–388.

BARROWS, M. H., CLUFF, R. M. & HARVEY, R. D. 1979. Petrology and maturation of dispersed organic matter in the New Albany shale group of the Illinois Basin. *In:* BARLOW, H. (ed.) *Proceedings of the Eastern Gas Shale Symposium*, pp. 85–114. Illinois Geological Survey, Preprint 1979 N.

BARTENSTEIN, H. & TEICHMÜLLER, R. 1974. Inkohlungsuntersuchungen, ein Schlüssel zur Prospektierung von paläozoischen Kohlenwasserstoff-Lagerstätten? *Fortschritte in der Geologie von Rheinland und Westfalen*, **24**, 129–160.

——, TEICHMÜLLER, M. & TEICHMÜLLER, R. 1971. Die Umwandlung der organischen Substanz im Dach des Bramscher Massivs. *Fortschritte in der Geologie von Rheinland und Westfalen*, **18**, 501–538.

BERGSTRÖM, S. M. 1980. Conodonts and palaeotemperature tools in Ordovician rocks of the Caledonides and adjacent areas in Scandinavia and the British Isles. *Förhandlingar vid Geologiska Föreningens i Stockholm*, **102**, 377–392.

BONIJOLY ROUSSEL, M. 1980. Evolution des substances carbonées naturelles. Formation de graphite. *Thèse*, Université d'Orléans.

——, OBERLIN, M. & OBERLIN, A. 1982. A possible mechanism for natural graphite formation. *International Journal of Coal Geology*, **1**, 283–312.

BOSTICK, N. H. 1973. Time as a factor in thermal metamorphism of phytoclasts (coal particles). *Compte rendu du septième Congrès international de Stratigraphie et de Géologie du Carbonifère*, Krefeld 1971, **2**, 183–193.

—— 1979. Microscopic measurements of the level of categenesis of solid organic matter in sedimentary rocks to aid exploration for petroleum and to determine former burial temperatures—a review. *Society of Economic Paleontologists and Mineralogists, Special Publication*, **26**, 17–43.

——, CASHMAN, S. M., MCCULLOW, T. H. & WADDELL, C. T. 1979. Gradients of vitrinite reflectance and present temperature in the Los Angeles and Ventura Basins, California. *In:* OLTZ, D. F. (ed.) *Low Temperature Metamorphism of Kerogen and Clay Minerals*, pp. 65–96. Soc. Econ. Paleont. Miner., Pacific Sect., Los Angeles.

BREITSCHMID, A. 1982. Diagenese und schwache Metamorphose in den sedimentären Abfolgen der

Zentralschweizer Alpen (Vierwaldstätter See, Urirotstock). *Eclogae geologicae Helveticae*, **75**, 331–380.

BROWN, K., DIESSEL, C. F. K., WOLF, M. & WOLF-FISCHER, E. 1985. The fluorescence properties of Carboniferous and Permian coking coals. *Proceedings of the International Conference on Coal Science*, Sydney, NSW, pp. 649–652. Pergamon Press, Oxford.

BUNTEBARTH, G. 1978. The degree of metamorphism of organic matter in sedimentary rocks as a palaeogeothermometer, applied to the Upper Rhine Graben. *Paleogeophysics*, **117**, 83–91.

—— 1979. Eine empirische Methode zur Berechnung von paläogeothermischen Gradienten aus dem Inkohlungsgrad organischer Einlagerungen in Sedimentgesteinen mit Anwendung auf den mittleren Oberrheingraben. *Fortschritte in der Geologie von Rheinland und Westfalen*, **27**, 97–108.

—— 1983. Zur Paläogeothermie im Permokarbon der Saar-Nahe-Senke. *Zeitschrift der deutschen geologischen Gesellschaft*, **134**, 211–223.

——, KOPPE, I. & TEICHMÜLLER, M. 1982. Palaeogeothermics in the Ruhr Basin. *In:* ČERMÁK, V. & HAENEL, R. (eds) *Geothermics and Geothermal Energy*, pp. 45–53. Schweizerbart, Stuttgart.

BURGESS, J. D. 1974. Microscopic examination of kerogen (dispersed organic matter) in petroleum exploration. *In:* DUTCHER, R. R., HACQUEBARD, P. A., SCHOPF, J. M. & SIMON, J. A. (eds) *Carbonaceous Materials as Indicators of Metamorphism*, pp. 19–30. Geological Society of America, Special Paper 2.

BURNE, R. V. & KANTSLER, A. J. 1977. Geothermal constraints on the hydrocarbon potential of the Canning Basin, western Australia. *BMR Journal of Australian Geology and Geophysics*, **2**, 271–288.

BUSTIN, R. M. 1983. Heating during thrust faulting in the Rocky Mountains: friction or fiction? *Tectonophysics*, **95**, 309–328.

—— 1984. Coalification levels and their significance in the Groundhog coal field, north-central British Columbia. *International Journal of Coal Geology*, **4**, 21–44.

—— 1986. Organic maturity of Late Cretaceous and Tertiary coal measures, Canadian Arctic Archipelago. *International Journal of Coal Geology*, **6**, 71–106.

——, CAMERON, A. R., GRIEVE, D. A. & KALKREUTH, W. D. 1985. Coal petrology: its principles, methods and applications. *Geological Association of Canada Short Course Notes 3*, 2nd revised edn.

ČERMÁK, V. 1979. Heat flow map of Europe. *In:* ČERMÁK, V. & RYBACH, L. (eds) *Terrestrial Heat Flow in Europe*, pp. 3–40. Springer-Verlag, Berlin.

—— & HURTIG, E. 1979. Heat flow map of Europe. *In:* ČERMÁK, V. & RYBACH, L. (eds) *Terrestrial Heat Flow in Europe*. Springer-Verlag, Berlin.

CLAUSEN, C. D. & TEICHMÜLLER, M. 1982. Die Bedeutung der Graptolithen-Fragmente im Paläozoikum von Soest-Erwitte für Stratigraphie und Inkohlung. *Fortschritte in der Geologie von Rheinland und Westfalen*, **30**, 145–167.

COOK, A. C. & KANTSLER, A. J. 1980. The maturation history of the epicontinental basins of western Australia. Economic Commission for Asia and the Far East, Technical Bulletin, **3**, 171–195. Committee for Co-ordination of Joint Prospecting for Mineral Resources in Asian Offshore Areas.

CORREIA, M. 1967. Relations possibles entre l'état de conversation des éléments figurés de la matière organique et l'existence de gisements d'hydrocarbures. *Revue de l'Institut Français du Pétrole*, **22**, 1285–1306.

CREANEY, S., PEARSON, D. E. & MARCONI, L. G. 1980. Anomalous coking properties of the Wolgan seam, NSW Australia. *Fuel*, **59**, 438–440.

DAMBERGER, H. 1968. Ein Nachweis der Abhängigkeit der Inkohlung von der Temperatur. *Brennstoff-Chemie*, **49**, 73–77.

DEUTLOFF, O., TEICHMÜLLER, M., TEICHMÜLLER, R. & WOLF, M. 1980. Inkohlungsuntersuchungen im Mesozoikum des Massivs von Vlotho (Niedersächsisches Tektogen). *Neues Jahrbuch für Geologie und Paläontologie, Monatshefte*, pp. 321–341.

DIESSEL, C. F. K. 1983. Carbonization reactions of inertinite macerals in Australian coals. *Fuel*, **62**, 883–892.

—— 1985. Fluoremetric analysis of inertinite. *Fuel*, **64**, 1542–1546.

——, BROTHERS, R. N. & BLACK, P. M. 1978. Coalification and graphitization in high-pressure schists in New Caledonia. *Contributions to Mineralogy and Petrology*, **68**, 63–78.

DOW, W. G. 1977. Kerogen studies and geological interpretations. *Journal of Geochemical Exploration*, **7**, 79–99.

DURAND, B. (ed.) (1980) *Kerogen. Insoluble Organic Matter from Sedimentary Rocks*. Technip, Paris.

DURAND, B., ALPERN, B., PITTION, J. L. & PRADLER, B. 1985. *Reflectance of vitrinite as a control of thermal history of sediments*. Institut Français du Pétrole, Ref. 33 729, Projet L31 07401, Géologie no. 27407.

EPSTEIN, E., EPSTEIN, J. B. & HARRIS, L. D. 1977. Conodont colour alteration—an index to organic metamorphism. *Geological Survey, Professional Paper* 995.

ESPITALIÉ, J. 1979. Charakterisierung der organischen Substanz und ihres Reifegrades in vier Bohrungen des mittleren Oberrheingrabens sowie Abschätzung der paläozoischen geothermischen Gradienten. *Fortschritte in der Geologie von Rheinland und Westfalen*, **27**, 87–96.

FALVEY, D. A. & DEIGHTON, I. 1982. Recent advances in burial and thermal geohistory analysis. *Australian Petroleum Exploration Association, Journal*, **22**, 65–81.

FAUTH, G. 1973. Elektronen-Spin-Resonanz-Untersuchungen zu den Verkokungstemperaturen der Naturkokse im Flöz Präsident/Helene des Ruhrkarbons von Kamp-Lintfort (Linker Niederrhein). *Geologische Mitteilungen*, **12**, 277–300.

FLÜHMANN, P. 1976. Auswertung der Bohrlochmessungen (Sandsteinhorizonte und Kohlenflöze). *Geologisches Jahrbuch A*, **27**, 21–28.

FREY, M., TEICHMÜLLER, M., TEICHMÜLLER, R., MULLIS, J., KÜNZI, B., BREITSCHMID, A., GRUNER, U.

& SCHWIZER, B. 1980. Very low grade metamorphism in external parts of the Central Alps: illite crystallinity, coal rank, and fluid inclusion data. *Eclogae geologicae Helveticae*, **73**, 173–203.

GIESE, P. 1978. Die Krustenstruktur des Varistikums und das Problem der Krustenverkürzung. *Zeitschrift der deutschen geologischen Gesellschaft*, **129**, 513–520.

—— & STEIN, A. 1971. Versuch einer einheitlichen Auswertung tiefenseismischer Messungen aus dem Bereich zwischen der Nordsee und den Alpen. *Zeitschrift für Geophysik*, **37**, 237–272.

——, WIGGER, P., MORELLI, C. & NICOLICH, R. 1981. *Seismische Studien zur Bestimmung der Krustenstruktur im Bereich der geothermischen Anomalie der Toskana*. Schlussbericht Forschungsprogramm EC Contr. Nos 486–78–1 EG D and 487–78–1 EG I, Freie Universität, Berlin.

——, JÖDICKE, H., PRODEHL, C. & WEBER, K. 1983. The crustal structure of the Hercynian mountain system—a model for crustal thickening by stacking. *In:* MARTIN, H. & EDER, F. W. (eds) *Intracontinental Fold Belts*, pp. 405–426. Springer-Verlag, Berlin.

GOLITSYN, M. W. 1975. On the metamorphic gas formation in the Karaganda Basin. *Bulletin Academy of Sciences USSR, Geological Series*, **9**, 123–127 (in Russian).

GOODARZI, F. 1984. Organic petrography of graptolite fragments from Turkey. *Marine and Petroleum Geology*, **1**, 202–210.

—— 1985a. Optical properties of vitrinite carbonized at different pressures. *Fuel*, **84**, 156–162.

—— 1985b. Reflected light microscopy of chitinozoan fragments. *Marine and Petroleum Geology*, **2**, 72–78.

—— & NORFORD, B. S. 1985. Graptolites as indicators of the temperature histories of rocks. *Journal of the Geological Society of London*, **142**, 1089–1099.

——, SNOWDON, L. R., GUNTHER, P. R. & JENKINS, W. A. M. 1985. Preliminary organic petrography of Palaeozoic rocks from the Grand Banks, Newfoundland. *Marine and Petroleum Geology*, **2**, 254–259.

GRAYSON, J. F. 1975. Relationship of palynomorph translucency to carbon and hydrocarbons in clastic rocks. *In:* ALPERN, B. (ed.) *Pétrographie organique et potentiel pétrolier*, pp. 261–273. Centre national de la Recherche scientifique, Paris.

GUTJAHR, CH. C. M. 1966. Carbonization measurements of pollen grains and spores and their application. *Leidsche geologische Mededeelingen*, **38**.

HACQUEBARD, P. A. 1975. Pre- and postdeformational coalification and its significance for oil and gas exploration. *In:* ALPERN, B. (ed.) *Pétrographie organique et potentiel pétrolier*, pp. 225–241. Centre national de la Recherche scientifique, Paris.

—— 1985. Geological and geothermal effects on coal rank variations in the Carboniferous basin of New Brunswick, Canada. *Compte rendu du dixième Congrès international de Stratigraphie et de Géologie du Carbonifère, Madrid*, **3**, 383–397.

HAMBERGER, K. 1977. '*Saprovitrinite*'. Proposal to the International Commission of Coal Petrology.

HAMILTON, L. H., RAMSDEN, A. R. & STEPHENS, J. F. 1970. Fossiliferous graphite from Undercliff, New South Wales. *Journal of the Geological Society of Australia*, **17**, 31–37.

HARRIS, A. G. 1979. Conodont color alteration, an organo-mineral metamorphic index, and its application to Appalachian basin geology. *In:* SCHOLLE, P. A. & SCHLUGER, P. R. (eds) *Aspects on Diagenesis*, pp. 3–16. Society of Economic Paleontologists and Mineralogists, Special Publication 26.

——, WARDLAW, B. R., RUST, C. C. & MERRILL, G. K. 1980. Maps for assessing thermal maturity (conodont colour alteration index maps) in Ordovician through Triassic rocks in Nevada and Utah and adjacent parts of Idaho and California. *United States Geological Survey, Miscellaneous Investigations*, Map I, 1249, scale 2 500 000.

HÉROUX, Y., CHAGNON, A. & BERTRAND, R. 1979. Compilation and correlation of major thermal maturation indicators. *American Association of Petroleum Geologists, Bulletin*, **63**, 2128–2144.

HESSE, R. & OGUNYOMI, O. 1980. Präorogene Versenkungstiefe und Orogenese als Diagenesefaktoren für altpaläozoische Kontinentalrandbildungen der nördlichen Appalachen in Quebec, Kanada. *Geologische Rundschau*, **69**, 546–566.

HIRSCH, P. B. 1954. X-ray scattering from coals. *Proceedings of the Royal Society of London A*, **225**, 143–169.

HOOD, A. & GUTJAHR, C. C. M. 1972. Organic metamorphism and the generation of petroleum. Paper read at the *annual Meeting of the Geological Society of America*, November, Minneapolis.

—— & CASTANO, J. R. 1974. Organic metamorphism: its relation to petroleum generation and application to studies of authigenic minerals. *United Nations Economic Commission for Asia and the Far East, Commission for Coordination of Joint Prospecting for Mineral Resources, Technical Bulletin*, **8**, 85–118.

——, GUTJAHR, C. C. M. & HEACOCK, R. L. 1975. Organic metamorphism and the generation of petroleum. *Bulletin of the American Association of Petroleum Geologists*, **59**, 986–996.

HORVÁTH, Z. A. 1983. Study on maturation process of huminitic organic matter by means of high-pressure experiments. *Acta geologica Hungarica*, **26**, 137–148.

HOWER, J. C. & DAVIS, A. 1981. Vitrinite reflectance anisotropy as a tectonic fabric element. *Geology*, **9**, 165–168.

HUFNAGEL, H. 1977. Das Fluoreszenzvermögen der Dinoflagellaten-Cysten-ein Inkohlungsparameter? *Geologisches Jahrbuch D*, **23**, 59–65.

HUNT, J. M. 1979. *Petroleum Geochemistry and Geology*. Freeman, San Francisco.

HUTTON, A. & COOK, A. 1980. Influence of alginite in the reflectance of vitrinite from Joadja, N.S.W., and some other coals and oil shales containing alginite. *Fuel*, **59**, 711–716.

INTERNATIONAL COMMISSION OF COAL PETROLOGY (ed.) 1971. *International Handbook of Coal Petrography*

1971/1985, 1st Supplement to 2nd edn. Centre national de la Recherche scientifique, Paris (reprinted 1985, The University of Newcastle upon Tyne).
JACOB, H. 1964. Neue Erkenntnisse auf dem Gebiet der Lumineszenzmikroskopie fossiler Brennstoffe. *Fortschritte in der Geologie von Rheinland und Westfalen*, **12**, 569–588.
—— 1985. Disperse solid bitumens as an indicator for migration and maturity in prospecting for oil and gas. *Erdöl und Kohle*, **38**, 365.
——, KUCKELKORN, K. & MÜLLER, M. 1982. Inkohlung und Tektonik im Bereich der gefalteten Molasse. *Erdöl und Kohle*, **35**, 510–518.
JONES, R. W. & EDISON, T. A. 1979. Microscopic observations of kerogen related to geochemical parameters with emphasis on thermal maturation. *In:* OLTZ, D. F. (ed) *Low Temperature Metamorphism of Kerogen and Clay Minerals*, pp. 1–12, Symposium in Geochemistry, Pacific Section, Society of Economic Paleontologists and Mineralogists, Los Angeles.
KALKREUTH, W. 1982. Rank and petrographic composition of selected Jurassic–Lower Cretaceous coals of British Columbia, Canada. *Bulletin of Canadian Petroleum Geologists*, **30**, 112–139.
—— & MCMECHAN, M. E. 1984. Regional pattern of thermal maturation as determined from coal-rank studies, Rocky Mountain foothills and front ranges north of Grande Cache, Alberta—implications for petroleum exploration. *Bulletin of Canadian Petroleum Geologists*, **32**, 249–271.
KANTSLER, A. J. & COOK, A. C. 1979. Maturation patterns in the Perth Basin. *Australian Petroleum Exploration Association Journal*, **19**, 94–107.
——, SMITH, G. C. & COOK, A. C. 1978. Lateral and vertical rank variations: implications for hydrocarbon exploration. *Australian Petroleum Exploration Association, Journal*, **18**, 143–156.
——, COOK, A. C. & ZWIGULIS, M. 1986. Organic maturation in the Eromanga Basin. *Geological Society of Australia, Special Publication*, **12**, 305–322.
KARWEIL, J. 1956. Die Metamorphose der Kohlen vom Standpunkt der physikalischen Chemie. *Zeitschrift der deutschen geologischen Gesellschaft*, **107**, 132–139.
KASIG, W. & SPÄTH, G. 1975. Neue Ergebnisse über die Geologie der Kern-und Mantelschichten des Hohen Venns auf Grund von Profilaufnahmen bei der Verlegung der Erdgasleitung Aachen-Rheinfelden. *Zeitschrift der deutschen geologischen Gesellschaft*, **126**, 1–14.
KEMP, A. E. S., OLIVER, G. H. J. & BALDWIN, J. R. 1985. Low-grade metamorphism and accretion tectonics: Southern Uplands terrain, Scotland. *Mineralogical Magazine*, **49**, 335–344.
KISCH, H. J. 1980. Illite crystallinity and coal rank associated with lowest-grade metamorphism of the Taveyanne greywacke in the Helvetic zone of the Swiss Alps. *Eclogae geologicae Helveticae*, **73**, 753–777.
KOPPE, I. 1980. *Paläogeothermische Untersuchungen im Ruhrgebiet und am Niederrhein anhand von Inkohlungsgraden organischer Substanzen.* Diplom-Arbeit, Technische Universität Clausthal.
KÜBLER, B., HÉROUX, Y., PITTION, J. L., CHAROLLAIS, J. & WEIDMANN, N. 1979. Sur le pouvoir réflecteur de la vitrinite dans quelques roches du Jura, de la molasse et das nappes préalpines, helvetiques et penninique (Suisse occidentale et Haut-Savoie). *Eclogae geologicae Helveticae*, **72**, 347–373.
KURYLOWICZ, L. E., OZIMIC, S., MCKIRDY, D. M., KANTSLER, A. J. & COOK, A. C. 1976. Reservoir and source rock potential of the Larapinta Group, Amadeus Basin, Central Australia. *Australian Petroleum Exploration Association, Journal*, **16**(2), 49–65.
LANDIS, C. A. 1971. Graphitization of dispersed carbonaceous material in metamorphic rocks. *Contributions to Mineralogy and Petrology*, **30**, 34–45.
LEVINE, J. R. & DAVIS, A. 1984. Optical anisotropy of coals as an indicator of tectonic deformation, Broad Top coal field, Pennsylvania. *Geological Society of America, Bulletin*, **95**, 100–108.
LEYTHAEUSER, D., HAGEMANN, H. W., HOLLERBACH, A. & SCHAEFER, R. G. 1980. Hydrocarbon generation in source beds as a function of type and maturation of their organic matter: a mass balance approach. *Proceedings of the 10th World Petroleum Congress*, **2**, 31–41.
LIN, R., DAVIS, A., BENSLEY, D. F. & DERBYSHIRE, F. J. 1986. Vitrinite secondary fluorescence, its chemistry and relation to the development of a mobile phase and thermoplasticity in coal. *International Journal of Coal Geology*, **6**, 215–228.
LOPATIN, N. V. 1971. Temperature and geological time as factors in coalification. *Izvestija Akademii Nauk SSRS, Serija geologica*, **3**, 95–106 (in Russian, translated into English by N. W. Bostick, 1972).
—— 1976a. The influence of temperature and geologic time on the catagenetic processes of coalification and petroleum and gas formation. *In: Study of Organic Matter in Recent and Old Sediments*, pp. 361–366. Nauka Press (in Russian).
—— 1976b. Historical-genetic analysis of oil formation by use of a model of regular continuous subsidence of the source rock. *Izvestiya Academiya Nauk SSSR, Seriya geologiya no. 8*, pp. 93–101 (in Russian).
MCCARTNEY, J. T. 1952. A study of the Seyler theory of coal reflectance. *Economic Geology*, **47**, 202–210.
MCKIRDY, D. M. & KANTSLER, A. J. 1980. Oil geochemistry and potential source rocks of the Officer Basin, South Australia. *Australian Petroleum Exploration Association, Journal*, **20**, 68–86.
MIDDLETON, M. F. 1982. Tectonic history from vitrinite reflectance. *Geophysical Journal of the Royal Astronomical Society*, **68**, 121–132.
—— & SCHMIDT, P. W. 1982. Paleogeothermetry of the Sydney Basin. *Journal of Geophysical Research*, **87**, no. B7, 5351–5359.
MURCHISON, D. G., COOK, A. C. & RAYMOND, A. C. 1985. Optical properties of organic matter in relation to thermal gradients and structural deformation. *Philosophical Transactions of the Royal Society of London A*, **315**, 157–186.

NANDI, B. N. & MONTGOMERY, D. S. 1975. N; 229 thermal behaviour of semi-fusinite in C; 230 coal from western Canada. *Fuel*, **54**, 193- 231

OBERLIN, A., BOULMIER, J. L. & VILLEY, M. 1980. Electron microscopic studies of kerogen microtexture. Selected criteria for determining the evolution path and evolution stage of kerogen. *In*: DURAND, B. (ed.) *Kerogen, Insoluble Organic Matter from Sedimentary Rocks*, pp. 191–241. Technip, Paris.

—— & TERRIÈRE, G. 1975. Graphitization studies of anthracites by high resolution electron microscopy. *Carbon*, **13**, 367–376.

OLIVER, G. J. H. 1987. Arenig to Wenlock regional metamorphism in the paratectonic Caledonides of the British Isles: a review. *In*: HARRIS, A. L. & FETTES, D. J. (eds) *The Caledonian–Appalachian Orogen*. Special Publication of the Geological Society of London. Blackwell Scientific Publications. Oxford (in press).

ONCKEN, O. 1982. Zur Rekonstruktion der Geosynklinalgeschichte mit Hilfe von Inkohlungskurven (am Beispiel Ebbeantiklinorium, Rheinisches Schiefergebirge). *Geologische Rundschau*, **71**, 579–601.

OTTENJANN, K., TEICHMÜLLER, M. & WOLF, M. 1974. Spektrale Fluoreszenz-Messungen an Sporiniten mit Auflicht-Anregung, eine mikroskopische Methode zur Bestimmung des Inkohlungsgrades gering inkohlter Kohlen. *Fortschritte in der Geologie von Rheinland und Westfalen*, **24**, 1–36.

——, —— & —— 1975. Spectral fluorescence measurements of sporinites in reflected light and their applicability for coalification studies. *In*: ALPERN, B. (ed.) *Pétrographie organique et potentiel pétrolier*, pp. 49–65. Centre national de la Recherche scientifique, Paris.

——, WOLF, M. & WOLFF-FISCHER, E. 1981. Beziehungen zwischen der Fluoreszenz von Vitriniten und den technologischen Eigenschaften von Kohlen. *Proceedings der internationalen kohlenwissenschaftlichen Tagung in Düsseldorf 1981*, pp. 86–91. Glückauf, Essen.

——, —— & —— 1982. Das Fluoreszenzverhalten der Vitrinite zur Kennzeichnung der Kokungseigenschaften von Steinkohlen. *Glückauf Forschungshefte*, **43**, 173–179.

OUDIN, J. L. & PICARD, P. F. 1982. Genesis of hydrocarbons in the Mahakam delta and the relationship between their distribution and the overpressured zones. Paper read at the *11th annual Indonesian Petroleum Association Conference*, June.

PATTEISKY, K., TEICHMÜLLER, M. & TEICHMÜLLER, R. 1962. Das Inkohlungsbild des Steinkohlengebirges an Rhein und Ruhr, dargestellt im Niveau von Flöz Sonnenschein. *Fortschritte in der Geologie von Rheinlend und Westfalen*, **3**, 687–700.

PATRICK, B. E., EVANS, B. W., DUMOULIN, J. A. & HARRIS, A. G. 1985. A comparison of carbonate mineral and conodont color alteration index thermometry, Seward Peninsula, Alaska. *Geological Society of America, Bulletin*, **17**(6), 399.

PEARSON, D. E. & GRIEVE, D. A. 1986. Rank variation, coalification pattern and coal quality in the Crowsnest coalfield, British Columbia. *Bulletin of the Canadian Institute of Mineralogy and Metallurgy*, in press.

PEAT, C. J., MUIR, M. D., PLUMB, K. A., MCKIRDY, D. M. & NORVICK, M. S. 1978. Proterozoic microfossils from the Roper Group, Northern Territory, Australia. *BMR Journal of Australian Geology and Geophysics*, **3**, 1–17.

PIERI, M. & GROPPI, G. 1981. Subsurface geological structure of the Po Plain, Italy. *Consiglio nationale delle ricerche, Progetto finalizzato Geodinamica—Sottoprogetto 'Modello strutturale'*, **414**. AGIP, Milano.

POWELL, T. G., FOSCOLOS, A. E., GUNTHER, P. R. & SNOWDON, L. R. 1978. Diagenesis of organic matter and fine clay minerals: a comparative study. *Geochimica et Cosmochimica Acta*, **42**, 1181–1197.

——, CREANEY, S. & SNOWDON, L. R. 1982. Limitations of use of organic petrographic techniques for identification of petroleum source rocks. *Bulletin of the American Association of Petroleum Geologists*, **66**, 430–435.

PRICE, L. C. & BARKER, C. E. 1985. Suppression of vitrinite reflectance in amorphous rich kerogen—a major unrecognized problem. *Journal of Petroleum Geology*, **8**, 59–84.

RADKE, M., SCHAEFER, R. G., LEYTHAEUSER, D. & TEICHMÜLLER, M. 1980. Composition of soluble organic matter in coals: relation to rank and liptinite fluorescence. *Geochimica et Cosmochimica Acta*, **44**, 1787–1800.

RAGOT, J. P. 1977. Contribution à l'étude de l'évolution des substances carbonées dans les formations géologiques. *Thèse*, l'Université Paul Sabatier de Toulouse.

REUTTER, K. J., GIESE, P. & CLOSS, H. 1980. Lithospheric split in the descending plate: observations from the northern Apennines. *Tectonophysics*, **64**, T1–T9.

——, TEICHMÜLLER, M., TEICHMÜLLER, R. & ZANZUCCHI, G. 1982. Le ricerche sulla carbonificazione dei frustoli vegetali nelle rocce clastiche, come contributo ai problemi di paleogeotermia e tettonica nell'Appennino settentrionale. *Memoire della Societá geologica Italiana*, **21**, 111–126.

——, ——, —— & —— 1983. The coalification pattern in the northern Apennines and its palaeogeothermic and tectonic significance. *Geologische Rundschau*, **72**, 861–893.

RITTER, U. 1985. The influence of time and temperature on vitrinite reflectance. *In*: SCHENCK, P. A., LEEUW, J. W. & LIJMBACH, G. W. M. (eds) *Advances in Organic Geochemistry 1983*, pp. 473–480. Pergamon Press, Oxford.

—— 1986. Heat flow during the Carboniferous and Mesozoic of the Northwest German Basin. *Geologische Rundschau*, **75**, 293–300.

ROBERT, P. 1980. The optical evolution of kerogen and geothermal histories applied to oil and gas exploration. *In*: DURAND, B. (ed.) *Kerogen. Insoluble Organic Matter from Sedimentary Rocks*, pp. 385–414. Technip, Paris.

—— 1985. Histoire géothermique et diagenèse organique. *Bulletin des Centres de Recherches Exploration-Production ELF-Aquitaine, Mémoire* 8 Pau.

ROUZAUD, J. N. 1984. Relations entre la microtexture et les propriétés des matériaux carbonés—Application à la caractérisation des charbons. *Thèse*, Université d'Orléans.

—— & OBERLIN, A. 1985. Raw and pyrolyzed coals: a preferential field for transmission electron microscopy. *Abstracts of the International Conference on Coal Science*, Sydney, NSW.

SCHÜLLER, A. 1961. Die Druck-Temperatur- und Energiefelder der Metamorphose. *Neues Jahrbuch für Mineralogie, Abhandlungen*, **96**, 250–290.

SHIBAOKA, M., BENNETT, A. J. R. & GOULD, K. W. 1973. Diagenesis of organic matter and occurrence of hydrocarbons in some Australian basins. *Australian Petroleum Exploration Association Journal*, **13**, 73–80.

——, SAXBY, J. & TAYLOR, G. H. 1978. Hydrocarbon generation in Gippsland Basin, Australia. *Bulletin of the American Association of Petroleum Geologists*, **62**, 1151–1158.

SMITH, G. C. & COOK, A. C. 1980. Coalification paths of exinite, vitrinite and inertite. *Fuel*, **59**, 641–647.

—— & —— 1984. Petroleum occurrence in the Gippsland Basin and its relationship to rank and organic matter type. *Australian Petroleum Exploration Association, Journal*, **24**, 196–216.

SMITH, J. W., RIGBY, D., GOULD, K. W. & SMYTH, M. 1984. *Exploration, characterization and assessment of fossil fuels*, pp. 5–8. CSIRO Division of Fossil Fuels, Institute for Energy and Earth Sciences, Report of Research in the years 1982–1983.

STACH, E., MACKOWSKY, M. TH., TEICHMÜLLER, M., TAYLOR, G. H., CHANDRA, D. & TEICHMÜLLER, R. 1982. *Textbook of Coal Petrology*, 3rd edn. Borntraeger, Stuttgart.

STAPLIN, F. L. 1969. Sedimentary organic matter, organic metamorphism, and oil and gas occurrences. *Bulletin of Canadian Petroleum Geology*, **17**, 47–66.

STONE, I. J. & COOK, A. C. 1979. The influence of some tectonic structures upon vitrinite reflectance. *Journal of Geology*, **87**, 497–508.

SUGGATE, R. P. 1982. Low-rank sequences and scales of organic metamorphism. *Journal of Petroleum Geology*, **4**, 377–392.

TAYLOR, G. H. 1971. Carbonaceous matter: a guide to the genesis and history of ores. *Special Issue of the Society of Mineralogists and Geologists of Japan*, **3**, 283–288.

TEICHMÜLLER, M. 1973. Zur Petrographie und Genese der Naturkokse im Flöz Präsident/Helene der Zeche Friedrich Heinrich bei Kamp-Lintfort (Linker Niederrhein). *Geologische Mitteilungen*, **12**, 219–254.

—— 1974a. Entstehung und Veränderung bituminöser Substanzen in Kohlen in Beziehung zur Entstehung und Umwandlung des Erdöls. *Fortschritte in der Geologie von Rheinland und Westfalen*, **24**, 65–112.

—— 1974b. Generation of petroleum-like substances in coal seams as seen under the microscope. *In:* TISSOT, B. & BIENNER, F. (eds) *Advances in Organic Geochemistry 1973*, pp. 321–348. Technip, Paris.

—— 1975a. Origin of the petrographic constituents of coal. *In:* STACH, E., MACKOWSKY, M. TH., TEICHMÜLLER, M., TAYLOR, G. H., CHANDRA, D. & TEICHMÜLLER, R. (eds) *Textbook of Coal Petrology*, 2nd edn, pp. 176–238. Borntraeger, Stuttgart.

—— 1975b. *Untersuchung einer extrem gasausbruchsgefährlichen Kohle aus dem Flöz Gemini der Grube Leichhardt bei Blackwater, Queensland, Australien.* Bericht IIf/4506/75-M. Tei./Sc. im Archiv des Geologischen Landesamtes Nordrhein-Westfalen, Krefeld vom 5.12.1975, unpublished.

—— 1978. Nachweis von Graptolithen-Periderm in geschieferten Gesteinen mit Hilfe kohlenpetrologischer Methoden. *Neues Jahrbuch für Geologie und Paläontologie, Monatshefte*, pp. 430–447.

—— 1979. Die Diagenese der kohligen Substanzen in den Gesteinen des Tertiärs und Mesozoikums des mittleren Oberrhein-Grabens. *Fortschritte in der Geologie von Rheinland und Westfalen*, **27**, 19–49.

—— 1982a. Rank determination on sedimentary rocks other than coal. *In:* STACH, E., MACKOWSKY, M. TH., TEICHMÜLLER, M., TAYLOR, G. H., CHANDRA, D. & TEICHMÜLLER, R. (eds) *Textbook of Coal Petrology*, pp. 361–374. Borntraeger, Stuttgart.

—— 1982b. *Fluoreszenzmikroskopische Änderungen von Liptiniten und Vitriniten mit zunehmendem Inkohlungscrad und ihre Beziehunger zu Bitumenbildung und Verkokungsverhalten.* Geologisches Landesamt Nordrhein-Westfalen, Krefeld (publ.), Krefeld. English translation by N. H. Bostick appeared as Special Publication no. 1 of the Society for Organic Petrology, Houston 1984.

—— 1986a. Organic petrology of source rocks, history and state of the art. *In:* LEYTHAUESER, D. & RULLKÖTTER, J. (eds) *Advances in Organic Geochemistry 1985*, **1**, 581–599. Pergamon Press, Oxford.

—— 1986b. Coalification and natural gas deposits in northwestern Germany. *In:* BROOKS, J., GOFF, J. C. & VAN HOORN, B. (eds) *Habitat of Palaeozoic Gas in NW Europe*, pp. 101–112. Geological Society Special Publication 23. Scottish Academic Press, Edinburgh.

—— & OTTENJANN, K. 1977. Art und Diagenese von Liptiniten und lipoiden Stoffen in einem Erdölmuttergestein auf Grund fluoreszenzmikroskopischer Untersuchungen. *Erdöl und Kohle*, **30**, 387–398.

—— & TEICHMÜLLER, R. 1950. Das Inkohlungsbild des niedersächsischen Wealdenbeckens. *Zeitschrift der deutschen geologischen Gesellschaft*, **100**, 498–517.

—— & —— 1971. Inkohlung im Rhein-Ruhr-Revier. *Fortschritte in der Geologie von Rheinland und Westfalen*, **19**, 47–56.

—— & —— 1975. Inkohlungsuntersuchungen in der Molasse des Alpenvorlandes. *Geologica Bavarica*, **73**, 123–142.

—— & —— 1979. Ein Inkohlungsprofil entlang der linksrheinischen Geotraverse von Schleiden nach Aachen und die Inkohlung in der Nord-Süd-Zone der Eifel. *Fortschritte in der Geologie von Rheinland und Westfalen*, **27**, 323–355.

—— & —— 1980. A coalification section along a geotraverse west of the Rhine between Schleiden and Aachen. *In:* CLOSS, H., v. GEHLEN, K., ILLIES, H., KUNTZ, E., NEUMANN, J. & SEIBOLD, E. (eds) *Mobile Earth*, pp. 204–206. International Geodynamics Project, Harald Boldt Verlag, Boppard.

—— & —— 1981. The significance of coalification studies to geology—a review. *Bulletin des Centres de Recherches Exploration-Production Elf-Aquitaine*, **5**, 491–534.

—— & —— 1982. Das Inkohlungsbild des Lippstädter Gewölbes. *Fortschritte in der Geologie von Rheinland und Westfalen*, **30**, 223–239.

—— & —— 1984. Verbreitung und Eigenschaften tiefliegender Steinkohlen in der Bundesrepublik Deutschland. *Glückauf Forschungshefte*, **45**(3), 140–153.

——, —— & BARTENSTEIN, H. 1979a. Inkohlung und Erdgas in Nordwestdeutschland. Eine Inkohlungskarte der Oberfläche des Oberkarbons. *Fortschritte in der Geologie von Rheinland und Westfalen*, **27**, 137–170.

——, —— & —— 1984. Inkohlung und Erdgas—eine neue Inkohlungskarte der Karbonoberfläche in Nordwestdeutschland. *Fortschritte in der Geologie von Rheinland und Westfalen*, **32**, 11–34.

——, —— & LORENZ, V. 1983. Inkohlung und Inkohlungsgradienten in der Saar-Nahe-Senke. *Zeitschrift der deutschen geologischen Gesellschaft*, **134**, 153–210.

——, —— & WEBER, KL. 1979b. Inkohlung und Illit-Kristallinität. Vergleichende Untersuchungen im Mesozoikum und Paläozoikum von Westfalen. *Fortschritte in der Geologie von Rheinland und Westfalen*, **27**, 201–276.

TEICHMÜLLER, R. 1973. Die paläogeographisch-fazielle und tektonische Entwicklung eines Kohlenbeckens am Beispiel des Ruhrkarbons. *Zeitschrift der deutschen geologischen Gesellschaft*, **124**, 149–165.

—— & TEICHMÜLLER, M. 1985. Inkohlungsgradienten in der Anthrazitfolge des Ibbenbürener Karbons. *Fortschritte in der Geologie von Rheinland und Westfalen*, **33**, 231–253.

—— & —— 1986. Relations between coalification and palaeogeothermics in Variscan and Alpidic foredeeps of western Europe. *In:* BUNTEBARTH, G. & STEGENA, L. (eds) *Palaeogeothermics*, pp. 53–78. Lecture Notes in Earth Sciences 5. Springer-Verlag, Berlin.

TISSOT, B. & ESPITALIÉ, J. 1975. L'évolution thermique de la matière organique des sédiments: application d'une simulation mathématique. *Revue de l'Institut Français du Pétrole*, **30**, 743–777.

—— & WELTE, D. H. 1978. *Petroleum Formation and Occurrence*. Springer-Verlag, Berlin.

WAPLES, D. W. 1980. Time and temperature in petroleum formation: application of Lopatin's method to petroleum exploration. *Bulletin of the American Association of Petroleum Geologists*, **64**, 916–929.

WARDLAW, B. R. & HARRIS, A. G. 1984. Conodont based thermal maturation of Paleozoic rocks in Arizona. *Bulletin of the American Association of Petroleum Geologists*, **68**(9), 1101–1106.

WASSOJEWITSCH, N. B., KORCHAGINA, JU. I., LOPATIN, N. V. & CHERNYSHEV, V. V. 1970. Principal phase of oil formation. *International Geological Revue*, **12**, 1276–1296.

WATSON, S. W. 1976. The sedimentary geochemistry of the Moffat shales: a carbonaceous sequence in the Southern Uplands of Scotland. *Unpublished PhD thesis*, University of St Andrews, Fife.

WEHNER, H., DAMBERGER, H., LEYTHAEUSER, D. & WELTE, D. H. 1976. Organisch-geochemische Untersuchungen an Kohlen, Gesteinen und Restgasen aus der Bohrung Saar 1. *Geologisches Jahrbuch A*, **27**, 455–488.

WELTE, D. H. & YÜKLER, M. A. 1981. Petroleum origin and accumulation in basin evolution—a quantitative model. *Bulletin of the American Association of Petroleum Geologists*, **65**, 1387–1396.

WHITE, D. 1935. Metamorphism of organic sediments and derived oils. *Bulletin of the American Association of Petroleum Geologists*, **18**, 589–617.

WOLF, M. 1972. Beziehungen zwischen Inkohlung und Geotektonik im nördlichen Rheinischen Schiefergebirge. *Neues Jahrbuch für Geologie und Paläontologie, Abhandlungen*, **141**(2), 222–257.

—— 1975. Über die Beziehungen zwischen Illit-Kristallinität und Inkohlung. *Neues Jahrbuch für Geologie und Paläontontologie, Monatshefte*, 437–447.

—— & WOLFF-FISCHER, E 1984. Alginit in Humuskohlen karbonischen Alters und sein Einfluss auf die optischen Eigenschaften des begleitenden Vitrinits. *Glückauf-Forschungshefte*, **45**, 243–246.

ZIMMERLE, W. 1976. Petrographische Beschreibung und Deutung der erbohrten Schichten (in der Bohrung Saar 1). *Geologisches Jahrbuch A*, **27**, 91–305.

ZWART, H. J. 1967. The duality of orogenic belts. *Geologie en Mijnbouw*, **46**, 283–309.

—— 1976. Regional metamorphism in the Variscan orogeny of Europe. Franz Kossmat Symposium. *Nova Acta Leopoldina, Neue Folge No. 224*, **45**, 361–367.

MARLIES TEICHMÜLLER, c/o Geologisches Landesamt Nordrhein-Westfalen, de Greiff Str. 195, 4150 Krefeld, Federal Republic of Germany.

Mineral matter in coals, with special reference to the Pennine Coalfields

D. A. Spears

ABSTRACT: The minerals present in coals are grouped as either detrital or diagenetic. The detrital minerals undergo diagenetic modifications however, and provide one of the sources for elements incorporated into diagenetic minerals. The detrital minerals are present either dispersed or concentrated in the coal seams. The overall composition of the detrital minerals is a function of grain size and thus of current velocity. Clay mineral composition varies with the maturity of the sediment, and in particular whether this is reworked from within the basin or is more representative of the upland source. Dirt partings (intraseam mudrocks) represent either an increase in detrital sediment input or a reduction in the rate of organic accumulation, or even loss of organic matter. Wind blown sediment in coals are best represented by altered volcanic ashes, tonsteins, providing a means of correlation and minerals for absolute age determinations. The alteration of ash to kaolinite both in the peat and in the associated sediments demonstrates which clay mineral is stable in the pore waters. Diagenetic kaolinite, probably formed from elements organically derived, is also recognized, particularly in low ash coals. The elements present in diagenetic silicates, sulphides and carbonates originated from within the peat and also from mudrocks in the sequence. The diagenetic sequence of minerals in the cleat is equated with that of other clastic sediments and is similarly interpreted as a burial depth sequence. The residual, diagenetic porefluids resulting from burial diagenesis are the source of economically important, high Cl concentrations in the coal.

Keywords: mineral matter, coal, England, clay minerals, pyrite, siderite diagenesis

A variety of minerals are to be found associated with coal seams. Silicates, sulphides and carbonates are the most common, but oxides, phosphates and other minerals also occur. The presence of these minerals is reflected in the ash yield by the coal. The minerals are therefore of economic importance not only because of coal quality, but also because of ash properties (see Raask 1985, see also Moore, *this volume*). In addition elements are present within connate water. Although not strictly minerals these elements do contribute to the coal ash and are here included within the subject of mineral matter in coal.

Minerals occur in coals in either dispersed or concentrated forms. The former is the inherent mineral matter and the latter, which is less intimately associated with the coal and is therefore more amenable to physical separation, is the adventitious mineral matter (Raask 1985). The time of formation of the minerals also varies. There are those minerals transported by various means into the swamp and incorporated into the peat. The composition and distribution of these detrital minerals provides information on depositional processes and environments. After deposition the minerals suffered diagenetic alteration over a long time period. Modification of material within the coal-forming swamp was one source for the elements in pore solutions which led to the precipitation of early diagenetic minerals. With increasing burial and advancing coalification, fractures (cleat) developed in the coal and minerals precipitated in the voids. The diagenetic minerals, including those which precipitated in the cleat, therefore provide a record of pore fluid evolution. The final stage of this evolution may be represented by the brine trapped in the coal. The distribution of detrital and diagenetic minerals in coal is illustrated on Fig. 1.

The mineral matter in coal is of increasing economic importance as advances are made in coal utilization and plant operating conditions are made more stringent. The mineral matter also provides depositional and diagenetic information which should be included in a comprehensive coal petrology survey.

In this paper, the coal minerals will be examined on a genetic basis, thus linking mineral composition and mode of occurrence, rather than deal with mineral groups as in Mackowsky's excellent treatment in Stach's textbook (1982). Examples will be drawn mainly from the East Pennine and Midlands coalfields which were once part of the Carboniferous Pennine Basin in the north of England (see Love *et al.* 1983).

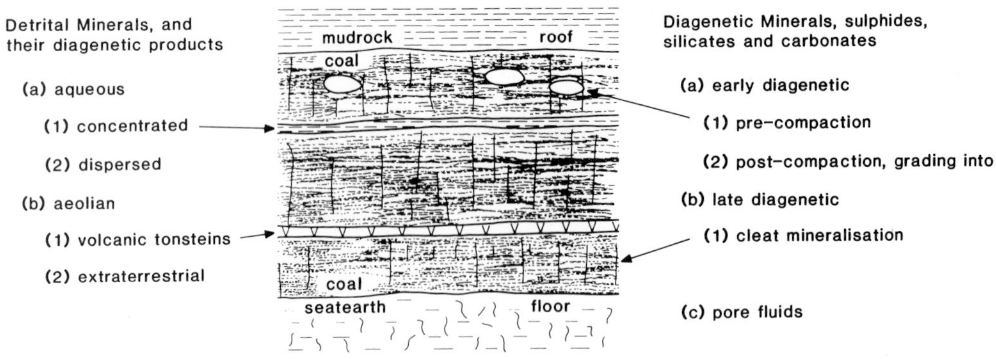

FIG. 1. The distribution of minerals in coals grouped according to depositional and diagenetic processes.

Detrital sediment

Aqueous transport

The detrital mineral matter is concentrated in the sediments adjacent to the coal seams and also in dirt partings within seams. The same minerals are present dispersed in the coal and a complete gradation exists from coals with a negligible detrital content through carbargilites to mudrocks with a minor organic content. Compositional information on those beds adjacent to coal seams, where the detrital minerals are concentrated, is of value in understanding the detrital minerals in coals which are more difficult to study because of their dispersed nature. In the work of Dixon *et al.* (1964) on the inorganic matter in coals from the E Midlands coalfields the average ash content of the seams mined was found to be 6.6%, whereas that of the marketed untreated smalls was 19.1%, indicating a substantial contribution of 'free dirt' (normal detrital sediment). These coals are mined and therefore include dirt partings and possibly some roof and floor materials in the 'run-of-mine' coal. The average ash content of British coals delivered to (CEGB) power stations in a typical year is 16.4% (Raask 1985) which again demonstrates that detrital mineral matter is a significant component of commercial coals.

The Carboniferous coal-forming swamps were predominantly low energy depositional environments and the detrital sediments associated with the coals were mainly fine grained. Mudrocks are thus generally associated with the coal seams, although siltstones and fine grained sandstones do occur, particularly in the roof-measures. The clay minerals are concentrated in the mudrocks and quartz is concentrated in the siltstones and sandstones, together with minor albite and K feldspar. The mudrock and siltstone boundary corresponds to a quartz content of 40% and to a grain size of approximately 0.01 mm (Spears 1970). In the Westphalian Coal Measures, the marine shales have quartz contents in the 10–20% range, whereas the range for the non-marine mudrocks is much greater. The non-marine mudrocks grade into siltstones (>40% quartz) but also very low quartz contents of less than 10% have been recorded for dirt partings, which are thought to represent very low energy depositional conditions (Spears & Kanaris-Sotiriou 1975). These low quartz mudrocks were mainly from dirt-partings within seams and were attributed to exceptionally low energy depositional conditions.

Compositional information on the detrital sediment was first obtained because of the former economic importance of fireclays. Fireclays were extensively analysed (e.g. Ennos & Scott 1964, Highley 1982) and were also the subject of pioneering X-ray powder diffraction studies of the clay minerals by Brindley & Robinson (1947). Important contributions to the knowledge of clay minerals in United Kingdom coalfields have also been made by Nicholls & Loring (1962) and Wilson (1965). More recent publications dealing with the clay mineralogy in the East Pennine Coalfields are Ashby & Pearson (1978) and Spears & Sezgin (1985). The latter study dealt with the mineralogy of stratotype section of the Subcrenatum Marine Band and associated sediments, including coal. The variation in the clay mineralogy through the stratotype section is thought to show the compositional range from mature clay sediment reworked from within the basin to less mature, clay sediment derived from the adjacent uplands (Fig. 2). Representative X-ray diffraction traces are shown on Fig. 3, to illustrate this compositional range from sample LN22 (mature) to sample LN9 (immature). Sample LN22 (seatearth) mainly consists of disordered kaolinite

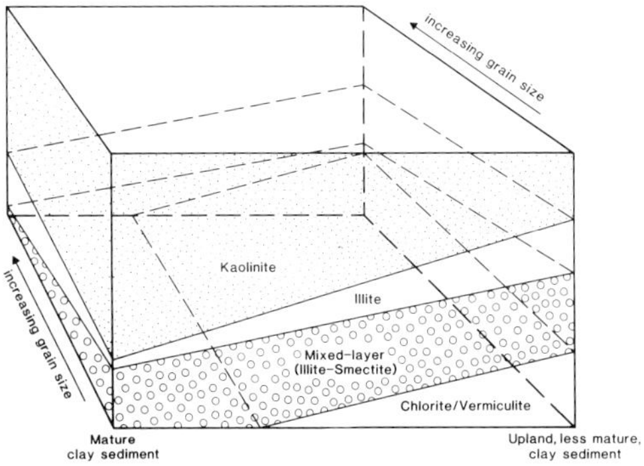

FIG. 2. To show the variation in clay mineralogy as a function of maturity and grain size. The variation in clay percentages from mature to upland, less mature clay sediment is based on the Subcrenatum stratotype at Langsett (Spears & Sezgin 1985). The changes with grain size, on the other hand, are qualitative.

(70%) with minor mixed-layer illite/smectite and only trace amounts of discrete illite. Sample LN9 (marine shale) contains less kaolinite (35%), more discrete illite (20%), mixed-layer illite/smectite (40%) and chlorite (5%) and a higher value of the ratio of illite to mixed-layer clay. These samples are from the Namurian–Westphalian boundary, whereas sample MM5 is a marine shale from the base of the Westphalian C. Sample MM5 (Fig. 3) contains more illite and less kaolinite and compared with sample LN9 demonstrates a variation with time in the composition of the upland derived clay sediment. Also shown diagramatically on Fig. 2 is the predicted variation in clay mineralogy as the grain size of the mudrocks increases towards siltstones based on the analysis of size fractions in Ashby & Pearson (1978), Spears & Amin (1981) and Caswell (1983). Figs 2 & 3 therefore illustrate the range in detrital clay mineralogy present in the sediments associated with coals attributed to, first, a balance between reworked mature sediment from within the basin and the less mature clay derived from the upland source and, second, grain size sorting of this material as a function of the energy of the depositional environment. This clay sediment is potentially available in the coal-forming swamps (a) for inclusion in the coal seems as either dispersed mineral matter or intraseam mudrocks (dirt partings), and (b) to form roof and floor measures.

Normal sedimentary processes, such as lateral migration of channels, will ensure reworking of mature clay sediment in the coal-forming swamp. The mode of occurrence of this detrital sediment will depend upon the balance between rates of detrital influx and organic accumulation. Dirt partings will result from sudden influxes of sediment-laden water into the swamp possibly related to the formation of crevasse splays in a fluviatile environment or widespread flooding. The latter control is one which would introduce a greater proportion of upland derived, less mature sediment into the swamp. On the other hand changes in the inorganic-organic proportions could be due to the rate of organic accumulation which is a balance between the rate of organic production and destruction (see Clymo, *this volume*). If the water table was below the surface of the peat the rate of organic oxidation would be greater and so too would be the potential loss due to fires. On the other hand an increase in the level of the water table, possibly linked with the rate of subsidence, could reduce the rate of organic accumulation thereby giving an increased concentration of detrital sediment in the final deposit. Detrital sediment, and particularly the clay minerals, whether dispersed through the coal seam or concentrated in intraseam mudrocks are therefore of interpretative value in unravelling the sedimentological history of the coals. Unfortunately this approach cannot be adopted at the present for the coals from the East Pennines and Midlands coalfields, because too little information is available on the composition and distribution of the detrital minerals.

Intraseam dirts in the E Pennines were found to be variable in composition (Dixon *et al.* 1970). Some were kaolinite rich and in others illite (mica) was more abundant than kaolinite (54% and 31% respectively). The latter illite:kaolinite proportion was comparable not only with the roof measures examined but also the floors. The important point to note is that roof and floor measures may have similar clay compositions and that many coals do not overlie well-developed seatearths with a high proportion of basin derived mature sediment. This is well illustrated by the Barnsley Coal, where in the Selby Coalfield the mudrocks immediately below the coal contain chlorite, the ratio of the mixed-layer clay to illite is low and kaolinite is less abundant than the sum of illite and mixed-layer clay (M. Dewison, personal communication). Such a sample would plot very much towards the right on Fig. 2. The mineralogy of six seams from the Cannock Coalfield was studied by Caswell (1983). The dirt partings were found to contain the maximum kaolinite and mixed-layer clay, the minimum illite and zero chlorite. The mineralogy is compatible with a fine-grain size and a significant contribution from reworked mature sediment. The highest illite values were recorded in the roof measures, whereas the floor measures were found to be more variable, particularly in the kaolinite

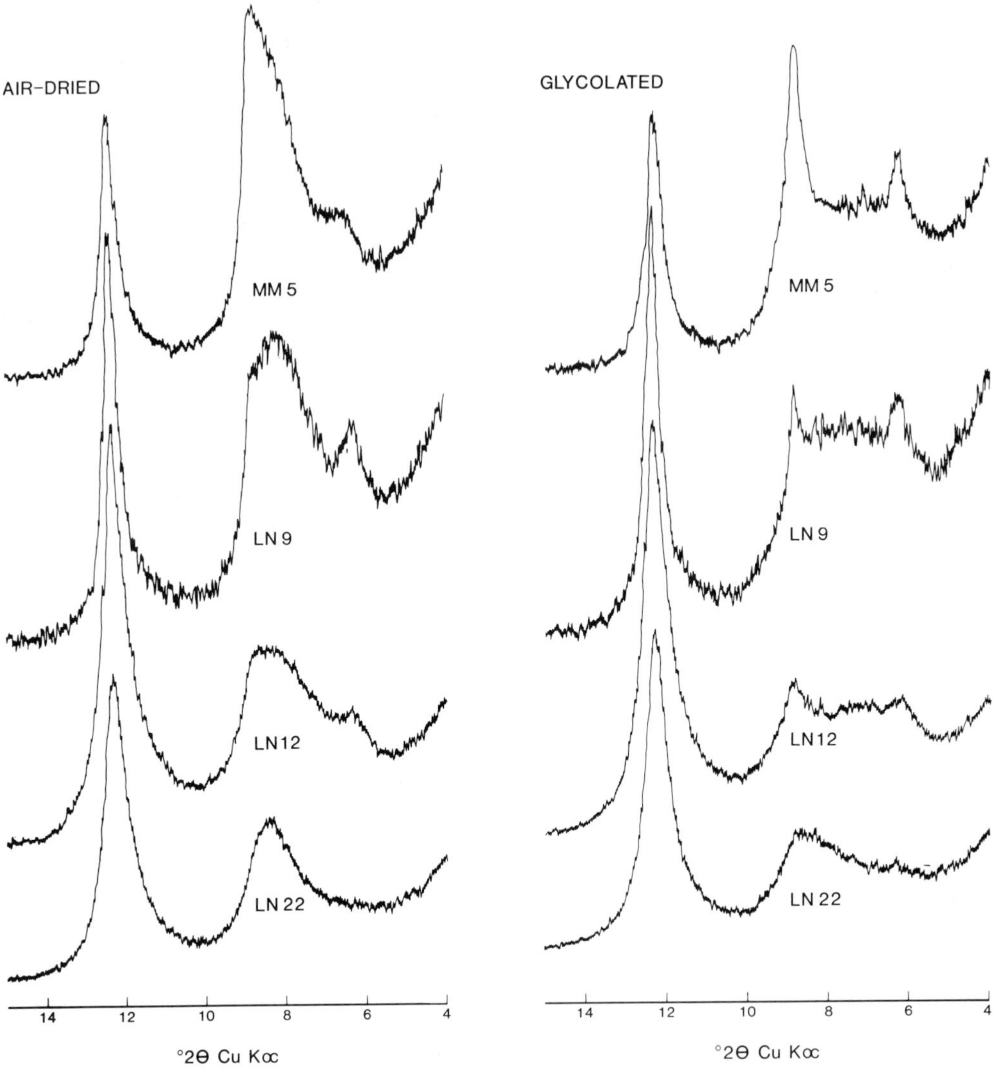

FIG. 3. X-ray diffraction traces of $<2\ \mu m$ clay fractions prepared as smear mounts to illustrate the clay mineral changes from mature clay sediment (sample LN22) to less mature, upland clay sediments (samples LN9 and MM5). The air-dried and glycolated traces are shown for each sample. The main features are from LN22 to MM5, increases in:

chlorite	(peak at $\sim 6.2°2\theta$)
discrete illite	(glycolated trace at $\sim 8.8°2\theta$)
illite/kaolinite ratio	(kaolinite at $\sim 12.4°2\theta$)

and the

illite/mixed-layer clay ratio (the mixed-layer clay at $\sim 8.2°2\theta$ air-dried, expands with glycolation).
Sample LN22, 12 and 9 are from the Subcrenatum stratotype and MM5 from the Aegiranum stratotype.

percentage with a range from ~ 20 to 45%. There is therefore a range in the mineralogy of the floor measures and a comparable range is to be expected in dirt partings and in the clays dispersed through the seam. In the Illinois Basin the immature clay suite is identified by Hughes *et al.* (*in press*) as the shale type suite and the mature basin derived sediment as the soil type suite. A third type is also recognized by these workers, the gley type suite consisting of mixed-layer illite/smectite and well-crystallized kaolinite, but this suite has not been positively identified in the Pennine Coalfields.

In mudrocks and carbargilites the minerals are

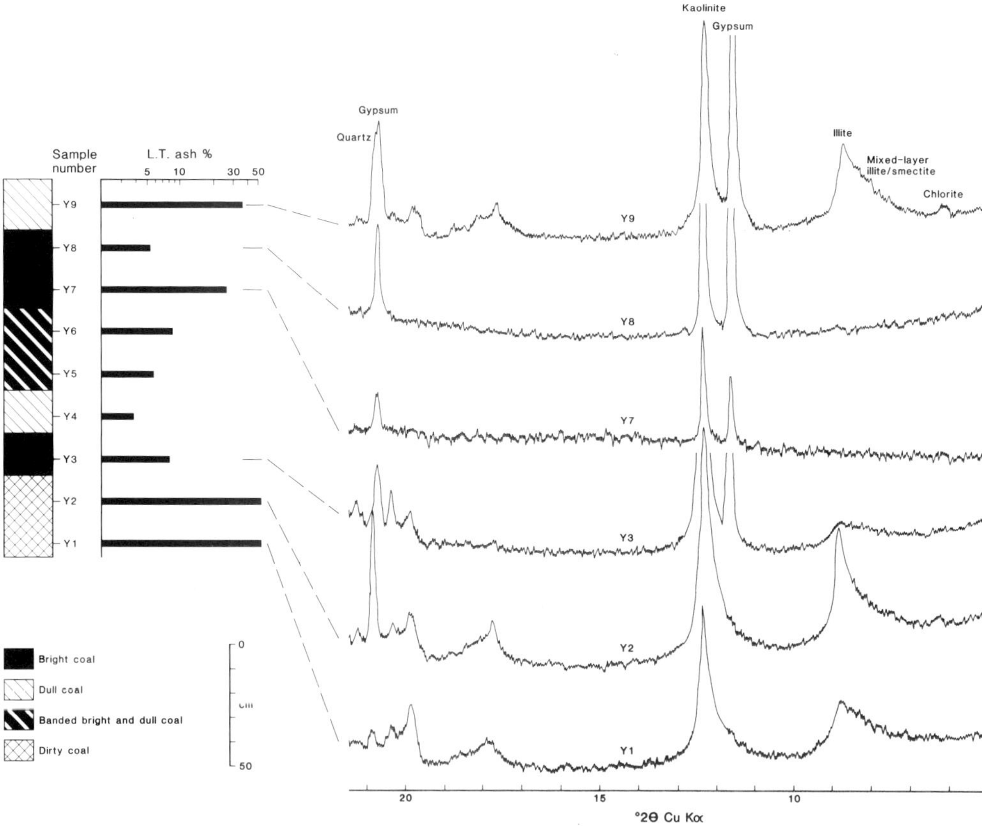

FIG. 4. Yard Seam, Lea Hall colliery, Cannock Coalfield, W Midlands; to show variation in low temperature (L.T.) ash% through seam and representative X-ray diffraction traces. The latter were run as air-dried smear mounts. Samples Y4, Y5 and Y6 are very similar to sample Y8 and the diffraction traces are not shown (Caswell 1983).

present in sufficient abundance that methods such as X-ray diffraction can be used without applying separation techniques. Separation techniques are required if the mineral abundance in the coals is low and the minerals are dispersed. This problem was overcome by Gluskoter (1965) using a low-temperature plasma asher. The organic matter is removed by oxidation and the minerals are concentrated with only minor modifications. Using this technique coals from the Cannock Coalfield were investigated (Caswell 1983) and work is currently in progress on coals from the East Pennine Coalfields. The clay minerals in the coal samples are identical with those in the associated mudrocks with the notable exception that the low-ash coals contain a higher proportion of well-crystallized kaolinite.

The variation in the clay mineralogy through the Yard Seam, Cannock Coalfield (Caswell 1983) is illustrated on Fig. 4. Those coal samples with an ash content (L.T. ash%) of >30% (samples Y1, Y2 and Y9) contain illite, mixed-layer illite/smectite and kaolinite as major constituents. Quartz is a major component in sample Y2 and is present in trace amounts in samples Y1 and Y9. Chlorite was also detected in samples Y2 and Y9. This is a typical detrital suite in the associated mudrocks and is similarly interpreted for these coal samples. In sample Y8, which contains ~5% L.T. ash, only kaolinite was detected from those minerals considered above. Samples Y4, Y5 and Y6, which also have low ash contents, have a very similar mineralogy to sample Y8. The X-ray diffraction trace for sample Y7 is again similar except the peak intensities are reduced and the background increased due to the high Fe (pyrite) content, which is responsible for the ash content of 26.6%. The kaolinite which dominates some of the low ash samples is thought to be diagenetic in origin based on the lack of such abundant, well-crystallized kaolinite in the associated detrital sediments together with direct evidence of diagenetic origin considered later when dealing with mode of occurrence and clay mineral stability. In sample Y3 (Fig. 4) which has a relatively low ash content of 8.4%, kaolinite is the major mineral but illite and mixed-layer clay are also present. These detrital minerals have been detected in coal samples from the Barnsley seam with ash contents as low as 2% (M. Dewison, personal communication). Prominent gypsum peaks are also observed in most of the L.T. ash diffraction traces (Fig. 4). The gypsum is produced during ashing and represents sulphate fixation from an organic S source (Miller *et al.* 1979). Other minerals recorded in the X-ray diffraction traces are pyrite in most samples, and calcite and ankerite in a few samples (Y5, Y6 and Y7). The origin of these non-detrital minerals is considered in the following sections.

Not considered in the above discussion is the possible alteration of clay minerals in the peat swamp. In modern peats the pH is low and organic acids are present, which favour clay alteration. Staub & Cohen (1978) reported that the clay immediately below the peat in the modern Snuggedy Swamp, Carolina, was kaolinite rich due to leaching from the peat, whereas lower in the sequence, and also where peat was absent, the clays were montmorillonite (smectite) rich. The latter clays were bedded and lacking in rootlets. The analogy with seatearths (underclays) seems clear and yet there are problems. Kaolinite-rich seatearths in the East Pennine Coalfields are relatively uncommon, more common is the shale-type suite of Hughes *et al.* (in press) present not only below the seam but in intimate contact with organic matter within the seam. Possibly the answer lies in the composition of the clay suite described by Staub & Cohen (1978). The kaolinite apparently formed from smectite and this mineral has not been recorded from the E Pennine Coalfields. However, this need not mean that smectite was never present. The diagenetic levels achieved by the Carboniferous sediments were sufficient to eliminate discrete smectite by illitization reactions. Although smectite is antipathetic in abundance to kaolinite in the clays of the Snuggedy Swamp illite apparently follows kaolinite (Staub & Cohen 1978, table 1). It is therefore conceivable that the rate of reaction of smectite in the swamp environment is greater than that of illite but, even if this were correct, it should lead in general to a higher kaolinite content of the Carboniferous mudrocks closely associated with coals, whereas such rocks are apparently the exception. Kaolinitization of normal sediments was one of the mechanisms considered by tonstein workers, but was rejected by Bouroz (1964) on the evidence of high illite concentrations in thin dirt partings. Only feldspar was believed to have altered to kaolinite.

In the paper by Renton & Cecil (1979) there are a number of interesting observations on the origin of minerals in coals. They propose that the clays and quartz are dominantly derived from inorganic materials originally contained within swamp plants. The possibility that the minerals described above are not detrital but diagenetic requires careful consideration. It is the contention of this paper that a clay mineral assemblage which can be recognized both in dirt-partings and dispersed through the coal is comparable with the detrital clay minerals in associated roof and floor measures. This clay mineral assemblage consists of kaolinite, illite, mixed-layer illite/smectite, chlorite and quartz and is not an equilibrium assemblage which could form in the peat swamp. Such an assemblage must be mainly derived from eroded soils in which recycling of existing minerals is an important process. That is not to deny, however, that plant derived elements are important in the formation of diagenetic minerals, and this will be examined in the following sections. The proposal by Renton & Cecil (1979) of mineral concentration by organic degradation is also important and is certainly a mechanism which must be considered for dirt-partings in addition to the more obvious catastrophic flooding events. In the Okefenokee Swamp, Davis *et al.* (1984) reported that in areas where the detrital inorganic input was minimal a positive relationship exists betwen ash content and the proportion of fine granular organic debris, the latter is a function of the degree of organic destruction. Biogenic silica is important in the Okefenokee Swamp although detrital quartz is also present. Davis *et al.* (1984) also provided additional information on the Waynesburg coal. The quartz present in the coal appears detrital in character and diagenetic evidence of silica precipitation was not observed, which led the authors to suggest that diagenetic silica could contribute to diagenetic clay minerals.

Wind transport

The coal-forming swamp and its environs appear at first sight to be unlikely areas in which to detect air-borne sediment. The rocks in the coal-bearing sequence are clastic dominated and were transported and deposited in water; the peat also accumulated with the water table at or above the surface of the peat. However, the rate of coal deposition is very slow, calculated rates are of the order of 4–100 yr mm^{-1} (McCabe 1984), and the energy of the depositional environment is low, which ensures low rates of supply and sedimentation of aqueous transported detrital sediment. Both these factors give rise to a high preservation potential for any air-borne sediment. In the coalfields of NW Europe stratigraphically important marker bands were noted in coal seams and in the adjacent sediments. These clay-rich mudrocks were characteristically hard with a conchoidal fracture and were thus termed tonsteins. The distinctive lithology, the sharp contacts with enclosing sediment, and the lateral continuity, possibly greater than that of the associated sedimentary facies, suggested a number of origins including alteration of volcanic ash layers (see reviews in Moore 1964, Williamson 1970). It is now generally agreed that many tonsteins do represent altered volcanic ashes.

In the Pennine coalfields a widespread 5-cm thick tonstein occurs above the Aegiranum Marine Band. The volcanic origin of this horizon was proved by: (a) the lateral passage into a K-bentonite consisting of mixed-layer illite/smectite; (b) a restricted heavy mineral suite consisting of zircon, apatite and opaques, which is characteristic of bentonites (Weaver 1963); and (c) aspects of the geochemistry which were typically volcanic rather than sedimentary (Spears 1970). One of the geochemical features of this horizon was the high concentrations of Th, and to a lesser extent U, which were responsible for significant γ-ray highs on the geophysical logs of boreholes. The geochemistry of this horizon was extended in Spears & Rice (1973) and an acid volcanic origin was established based on immobile trace elements. Additional tonsteins from the Pennine Coalfields and other European coalfields were studied in Spears & Kanaris-Sotiriou (1979). Two groups of tonsteins were recognized, the one found only in the UK in which the tonstein developed from a basic volcanic ash of local origin, and the other, found in all coalfields, of tonsteins formed from the alteration of acid volcanic ash. The dispersal of basaltic ash in the Carboniferous of Scotland, which included passage into tonstein, was restricted to approximately 50 km (Francis 1961). The Carboniferous isopach maps in Scotland and in eastern England have been used to augment dispersal data of recent Surtseyan activity (Francis 1985). Acid to intermediate magmas occur in Plinean eruptions, which characteristically are of great magnitude. The areas covered by these ash falls are comparable with some of the Upper Carboniferous tonsteins in NW Europe (Francis 1985).

Tonsteins are recorded from the majority of the world's coalfields (Burger 1985). In the Cretaceous and Tertiary coal beds of the Western Interior of North America laterally persistent tonsteins occur. The volcanic origin of these rocks was established (Ryer *et al.* 1980, Bohor & Triplehorn 1981) on several lines of evidence including the presence of sanidine and β-quartz form phenocrysts (α-quartz after β-quartz) characteristic of high temperature volcanic assemblages. The identification of the ash falls enabled Ryer *et al.* (1980) to construct a detailed facies history of the C coal bed in the Ferron Sandstone Member, utilizing the four ash partings within the coal as isochronous surfaces. The original impetus for tonstein studies, that of correlation, is thus continued but to a much more detailed level. There is also a continuing need for coal seam identification and tonsteins continue to be useful, as in the Donetz Basin (Zaritsky 1985). Tonsteins also have another stratigraphic use and that is in absolute age determinations. Lippolt & Hess (1985) examined a large number of Carboniferous tonsteins from Europe and found sanidine in the majority. Sanidines were separated from nine samples and $^{40}Ar/^{39}Ar$ ages determined resulting in a modification of the Carboniferous time-scale. The volcanic ash partings in Tertiary coals from Alaska are another example where original minerals have been used to obtain absolute age determinations (Triplehorn *et al.* 1984).

The presence of sanidine in tonsteins is important for absolute age determinations, but also its survival in the coal is convincing supporting evidence that alteration of detrital sediment was limited. Nevertheless the presence of diagenetic kaolinite in tonsteins shows that unstable material was originally present. This material was the volcanic glass and ash. Original textures may be preserved by the kaolinite and new diagenetic forms such as vermicular kaolinite may grow. The range of textures led to a descriptive classification by Schuller (1951), which is compared with Bouroz's genetic classification (1972) in Bouroz *et al.* (1983). Bouroz (1972) had the advantage of working on tonsteins from both paralic and limnic basins. In the latter alteration was less extreme and volcanic textures could be identified with certainty. In the Permian Coal Measures of New South Wales, the volcanic activity ranged from thin ash falls, now preserved as tonsteins, to interseam tuffs many metres in thickness (Diessel 1985). This range enables tonsteins to be compared with related, less altered tuffs, and furthers the understanding of diagenetic changes and the textural modifications. Grading was noted by Diessel (1985) from coarse crystal tuffs through vitric tuff to fine ashstone within a thickness of only a few centimetres. The abundance of biotite is noteworthy, as is their diagenetic destruction to produce the graupen texture documented by Schuller (1951). The ash partings in the Tertiary coals from Alaska (Triplehorn *et al.* 1984) also preserve stages of alteration. Recognition of these stages should lead to a more refined tonstein classification more closely related to pyroclastics.

The formation of tonsteins also provides evidence on the clay mineral stability in the coal-forming swamp. The unstable volcanic material provided an ideal ion source for the formation of diagenetic minerals. In coal-bearing strata kaolinite dominates and rarely is illite/smectite (smectite precursor) the most abundant mineral. A survey of European tonsteins (Spears & Kanaris-Sotiriou 1979) revealed only one example with a significant illite/smectite content. A rapid burial rate may have been responsible for the early establishment of a closed system producing the higher ionic activities in the pore fluids. These pore fluid conditions existed under marine conditions, as the presence of K-bentonites (mixed-layer illite/smectite) in the marine Edale Shales in the Pennine Basin demonstrates (Trewin & Holdsworth 1972). In non-marine, coal-bearing sequences, diagenetic kaolinite dominates and if illite/smectite is present it is minor. This assemblage is readily distinguished from detrital clay assemblages and particularly the less mature shale type. It is noteworthy that tonsteins in the coals do not differ compositionally to those in the associated sediments, suggesting uniformity in pore fluid composition with respect to clay stability fields.

In the earlier discussion it was noted that kaolinite is a normal component of the detrital sediment. Enrichment of kaolinite in dirt-partings is possible and confusion with volcanic tonsteins might well result if identification was based solely on kaolinite percentage. The hand specimen character of tonsteins is an expression of the more random orientation of the fine-grained diagenetic kaolinite, whereas in detrital sediments parallelism with the bedding is the norm. Tonstein is a useful description of rocks in the field and although strictly non-genetic the beds so named have proved to be volcanic in origin providing that identification was not solely based on the kaolinite percentage.

The abundance of volcanic ash partings in coals of all ages demonstrates a preservation potential for air-borne material. The possibility that this includes extraterrestrial material was raised by Finkleman & Stanton (1978) in a SEM based study on accessory minerals in coal and in the low-temperature plasma ash. In the ash 5 μm spheres of magnetite were identified as a meteor ablation product. Less exotic, but a very real possibility, is the presence of tropospheric dust containing a mixture of quartz and clay minerals, depending on the grain size.

Early diagenetic minerals

The minerals considered in this group are those which are pre-cleat formation and includes the post-compaction phase of formation. This corresponds to Mackowsky's (1982) first stage of coalification, the second stage includes the cleat mineralization and is here termed late diagenetic which grades into catagenesis.

Sulphides, mainly pyrite and marcasite, are important minerals in coals. A review of sulphur in coals is provided in this volume by Casagrande. In the Pennine Basin the petrography and isotope composition of the diagenetic pyrite in the Halifax Hard Bed Coal and associated sediments were described by Love *et al.* (1983). The sequence includes the *Gastroceras listeri* marine band and therefore coal is overlain by a marine roof, unlike the majority of coals in this coalfield. The pyrite is recognized as pre- and post-compaction. In an early pre-compaction stage the pyrite occurs as framboids and associated microcrystalline grains, and in a later stage the pyrite occurs as non-framboidal granular to massive aggregates and fibrous radiating microscopic pyrite infilling and replacing plant material and possibly subsidiary marcasite. The post-compaction pyrite is present in veins. The interpretation of the $\delta^{34}S$ values (Love *et al.* 1983) is that the earliest pyrite in the coal peat was linked with the composition of the overlying freshwater and the main phase of pyrite formation was associated with downwards diffusion from the overlying marine incursion.

The iron source for the pyrite was probably the detrital sediment in the sequence. Removal of iron from the pore fluids by sulphide precipitation was so effective that the carbonates which occur in that part of the sequence are iron-poor in contrast to siderites abundant elsewhere in the sequence (Love *et al.* 1983). The iron-poor carbonates occur in remarkable bullions and coal balls which are an unusual feature of the *G. listeri* horizon (Stopes & Watson 1908, Moore 1968). The carbonate concretions are pre-compactional and fossil material is well preserved enabling comparisons to be made with adjacent coal. In the Carboniferous of the northern hemisphere calcite and dolomite coal balls are mainly limited to seams with marine roofs (Teichmüller 1982). Scott & Rex (1985), in their comprehensive review of carbonate coal balls, noted the uniformity in mineralization across the Euramerican province. The predominant carbonate (usually calcite) occurs as fine, fibrous crystals. Ferroan calcite is a minor component in most coal balls and usually post-dates the calcite. In the dolomitic coal balls replacement of the fibrous calcite is believed to have taken place generally without loss of texture except in some Dutch coal balls. Coprecipitation of pyrite and carbonate is reported by Love *et al.* (1983), and it is interesting to consider if these reactions are directly linked. Microbial sulphate reduction generates reduced sulphur species, HCO_3^- and H^+, whereas reduction of Fe_2O_3 gives HCO_3^- and OH^- in addition to Fe^{2+}. If these two reactions are balanced a small excess of OH^- is a possibility (Curtis 1983a). However in modern marine sediments not all the reduced sulphur species are removed by precipitation (Berner & Westrich 1985) and therefore net H^+ is produced, thus inhibiting rather than favouring carbonate precipitation. Furthermore the iron reduction is not dependent on overlying marine waters and therefore in freshwater peats there should be OH^- generation from this source and yet carbonate precipitation is favoured in marine situations. Sulphide and carbonate precipitation must therefore be indirectly linked, the seawater independently providing a sulphate source and buffering capacity to give pH values >7 and thus promoting calcite precipitation. In marine influenced peats higher pH values are indeed recorded (Staub & Cohen 1979, Teichmüller & Teichmüller 1982 and references therein).

The observation above that pyrite is associated with iron-poor carbonates demonstrates how effective pyrite formation is in removing iron from solution. Siderite formation is only possible when the activity of reduced sulphur species is very low. This is achieved either in a freshwater environment where sulphate species are low initially or during diagenesis in marine sediments below the depth of sulphate diffusion. The precipitation of ferroan calcite in marine associated coal balls should, in general, post-date calcite precipitation and would be expected to take place in the peat below the limit of sulphate diffusion. The *G. listeri* sequence in addition to pyrite (Love *et al.* 1983) contains siderite in nodules and bands. The siderite formed with organic matter the principle carbon source (Curtis *et al.* 1972) and preserving clay fabrics at progressive stages of compaction (Oertel & Curtis 1972). The siderite therefore formed over a long time period, and the concentration into bands was attributed to preferential flow along siltier horizons (Curtis *et al.* 1975). The iron present in the siderite was probably associated initially with the clay minerals in the mudrocks. In the coals iron was available based on the pyrite enrichment but the only detrital source is unlikely to have been the clays within the coal on a mass basis and the associated mudrocks probably contributed. The majority of coals in the Pennine Basin are not overlain by marine strata. Although there is less pyrite present in these coals the amount remains significant in contrast to the non-marine mudrocks. In the Lea Hall Yard seam (Fig. 4), pyrite average 6.5% in those coal samples with low ash contents (samples Y4, Y5, Y6 and Y7) (data in Caswell 1983) whereas the average pyrite % of non-marine mudrocks is 0.8% (Spears & Amin 1981). The ratio of pyrite to detrital sediment (as a first approximation the sum of $SiO_2\% + Al_2O_3\%$ for the Lea Hall coal samples is 2.3:1). This contrasts with the marine shales (Spears & Amin 1981) where the ratio is 0.06:1 and thus the need to invoke an iron source in addition to that provided by detrital sediments within the coal. Concretionary pyrite also occurs in the coals but rarely in the non-marine mudrocks. It is difficult to attribute this to greater iron availability in the coal-forming peat than in the mudrocks when I have argued above for a detrital iron source. The common occurrence of siderite nodules and bands, not only in the *G. listeri* section, but through the Coal Measures as a whole, proves that iron was available in the mudrocks. Microbial activity and diagenetic conditions within the non-marine mudrocks were potentially suitable for pyrite formation. What was lacking was a suitable sulphate source. The concretionary pyrite in the coal had such a source and it is tempting to conclude, by elimination, that organically derived sulphur may have been involved. The time of pyrite formation in the coals may thus have overlapped that of siderite in the mudrocks. It is postulated that both minerals may have had the same cation and anion sources (detrital sediment and organic matter) but the increased organic content in the coal provided a sulphur source and thus siderite precipitation was retarded. The activities of reduced sulphur species are also a function of rate of production in the peat and rate of supply of ferrous iron, mainly from an external source. An increase in the relative rate of the latter would give siderite precipitation in the coal.

In the discussion on detrital input in coals reference was made to the organic matter as an element source for silicates. Biogenic silica is important in the Okefenokee Swamp (Davis *et al.* 1984) but organic matter also contains alumina which could be involved in mineral precipitation (Moore 1964, Renton & Cecil 1979). Diagenetic silicates occur in the coals of the Pennine Basin. In low ash coals with a reduced detrital input, well crystallized kaolinite is important as was noted earlier for samples from the Yard Seam

(Fig. 4). The X-ray characteristics of this material are comparable with kaolinite in tonsteins and unlike the kaolinite in seatearths. Under the petrographic microscope kaolinite and quartz are observed infilling cells and both occur in the cleat. Quartz-rich layers with unequivocal diagenetic textures are relatively rare; good examples are seen in the Clowne Seam from Whitwell Colliery and the Thick seam from Coventry Colliery. The relative importance of submicron-sized, diagenetic grains dispersed through the coal or associated with specific macerals is largely unknown for these coals. Low-ash coal samples from four seams in Staffordshire (Caswell 1983) are dominated by kaolinite with only minor quartz in most samples. This quartz does not occur with detrital clay minerals (illite, illite/smectite, chlorite and poorly crystalline kaolinite) and an authigenic origin is possible. In the Littleton Eight Feet seam more quartz was recorded and possibly this too includes diagenetic quartz. Generally, the quartz content is relatively minor compared with kaolinite in low-ash samples and loss of silica with respect to alumina from vegetation to coal would appear to have taken place; possibly at an early diagenetic stage as suggested by Davis *et al.* (1984).

Detrital and diagenetic quartz in Australian coals are reported to be isotopically similar (Botz *et al.* 1986) whereas cathodoluminescence petrography, based on studies of the Upper Freeport coal in Pennsylvania (Ruppert *et al.* 1985), would appear to be more promising in separating detrital and diagenetic quartz.

The presence of kaolinite infilling voids demonstrates mobility of dissolved Al and Si. The solubility of aluminium is pH dependent and in acid, peat swamp porewaters the solubility increases markedly as the pH falls, whereas the solubility of silica is independent of pH below 9. The Al in equilibrium with kaolinite, in a quartz saturated solution, decreases from approximately 300 ppm at pH 3 to 0.5 ppm at pH 4. Parameters defining the kaolinite stability field include low ratios of $[K^+]/[H^+]$, $[Ca^{2+}]/[H^+]$ and $[Na^+]/[H^+]$ which are therefore favoured by low pH values (data in Curtis 1983). Diffusion of ions or movement of pore fluids into more open pores and fractures in which there is even a small amelioration in acidity will favour kaolinite precipitation. An increase in pH might be achieved by the passage through the peat of externally derived solutions. In the earlier discussion it was proposed that the iron precipitated within the coal was externally derived and the production of those solutions involved hydroxyl generation thus reducing acidity. An evolutionary sequence can also be suggested in which the locus for precipitation moves from the pore spaces within the coal, as these are eliminated by compaction and infill, to the later fractures (cleat) cross-cutting the bedding. More difficult to explain are the quartz-rich layers in some coal seams. These might be attributed to an increase in a silica-rich source, such as volcanic ash, but although silica is lost from the ash to produce a tonstein, quartz layers and tonsteins are not associated. Furthermore the organic matter is a source of biogenic silica and apart from possible retention in kaolinite the silica was apparently mainly lost from the coal (by comparison with silica analyses in Moore 1964 and Renton & Cecil 1979). The formation of quartz at earth-surface conditions, which would be equivalent to early diagenesis in peat, is generally attributed to ageing of silica hydrogels (review *in* Wilding *et al.* 1977). Direct precipitation of quartz, which is less soluble, is kinetically constrained. One speculative possibility is that amorphous silica precipitated from pore solutions because concentrations increased not by increased silica input but by loss of water due to evaporation; climate or fires could conceivably be responsible.

Late diagenetic minerals

A distinctive feature of bituminous coals is the jointing perpendicular to the bedding. The frequency of this jointing (cleat) exceeds that in the associated mudrocks. The frequency is greatest in vitrain bands and is inversely related to lithotype thickness (Spears & Caswell 1986). The presence of fractures perpendicular to the bedding facilitates the migration of pore waters and leads to mineral precipitation within the cleat. The cleat is mineralized. In the Illinois coals textural relationships between cleat minerals enabled a paragenetic sequence to be established (Hatch *et al.* 1976). This sequence is sulphides, with pyrite overlapping sphalerite deposition, then silicates with quartz pre-dating kaolinite and finally calcite. A similar sequence was noted by Fellows (1979) and Caswell (1983) in the Pennine coalfields. The early vitrain cleat is dominated by sulphides, mainly pyrite and marcasite but including sphalerite and galena. The major cleat cutting several coal types is generally multimineralic. Kaolinite post-dates the sulphide but quartz is relatively rare and difficult to date with respect to the kaolinite. In the Pennine coals carbonates follow the silicates, as in the Illinois coals, but ankerite is present and pre-dates the calcite.

The cleat mineral sequence in interpreted (Spears & Caswell 1986) as precipitation from evolving porefluids during burial diagenesis, with ions originating from the coal and the associated sediments. This interpretation is based on comparisons with diagenetic sequences recorded in clastic sequences.

A review of diagenetic sequences in sandstones by Hurst & Irwin (1982) demonstrates that quartz is ubiquitous and generally pre-dates kaolinite except in freshwater fluvial sandstones; this silicate phase is commonly followed by carbonate cementation, mainly calcite. The cleat sequence differs in that early sulphide is important. This is attributed to reaction between organically derived reduced sulphur species and externally derived ferrous iron. This explanation invokes the same element sources as the pre-compaction concretionary pyrite and the same argument may be advanced as for the kaolinite to explain the shift in the locus of precipitation from coal to fractures. The simplified diagenetic carbonate sequence recorded by Matsumoto & Iijima (1981) from freshwater Japanese coal-bearing sequences is siderite, ankerite and calcite. The siderite developed at an early diagenetic stage and is equated (Spears & Caswell 1986) with the cleat sulphide stage. The coal is thought to have been a sulphur source hence precluding siderite precipitation as in the above discussion of concretionary pyrite. In the Japanese sequences ankerite forms in the depth range 500–1700 m and calcite below 2000 m ($\sim$85°C). Boles & Franks (1979) record, in the Wilcox sandstones, diagenetic calcite at 1768/1942 m ($\sim$100°C) and ankerite deeper ($>$2500 m, 120–130°C) but the latter is a replacement of the earlier calcite for which there is no evidence in the cleat textures. The HCO_3^- precipitated in the coal cleat probably resulted from organic fermentation and decarboxylation reactions in the coal. The cations, on the other hand, probably resulted from diagenetic modification of the detrital minerals including illitization of the smectite and albitization of the plagioclase (Boles & Franks 1979). These minerals are present within the coal but mass considerations and the presence of conductive fractures strongly suggest an origin in the associated clastics and the mudrocks in particular. A maximum temperature of about 110°C may be suggested for the cleat minerals which are associated with high volatile B bituminous coals.

The cleat minerals are thought to preserve a record of pore fluid evolution during basin formation. Minerals were precipitated over a depth/temperature/time interval and a gradation from early to late diagenetic events is envisaged. The period of cleat precipitation may well overlap post-compaction precipitation within the coal. Stress conditions producing jointing in sedimentary rocks are encountered on uplift (Price 1966). However, the extreme compaction factor from peat to coal, together with the heterogeneous nature of coal should enable comparable stress conditions to be achieved during burial.

Textural relationships and isotope geochemistry are combined in the work of Botz *et al.* (1986) on minerals in Australian coals. Fine-grained kaolinite disseminated throughout the coal and siderite nodules were precipitated early in diagenesis at calculated temperatures of 16°–32°C. In the Bowen Basin the cleat-filling kaolinite formed at a calculated temperature of 56°C which is equated by the authors to a burial depth of $\sim$900 m. The textural relationships of the cleat carbonates proved of greater interpretative value than the $\delta^{12}C$ and $\delta^{18}O$ values because of the range, nevertheless the authors did demonstrate progressive enrichment in ^{13}C as fractures developed in the coal starting at shallow depths. Botz *et al.* (1986) interpret textural relationships in a comparable diagenetic model to that proposed in this work with the additional strength of isotopic data.

Connate water

Pore fluids are involved in the precipitation of diagenetic minerals as the coal evolves. Pore fluids containing the residual solutes may therefore be trapped in the coal. Indirect evidence of coal pore fluid composition is obtained from groundwaters. Groundwaters extracted at depth from Lower Coal Measures sandstones in the E Midlands are Na-Cl brines (Downing & Howitt 1969). National Coal Board minewater analyses (Caswell *et al.* 1984a, table 2) also show Cl to be the major anion and Na the major cation. In high volatile bituminous coals in the UK, high Cl concentrations are encountered. These Cl rich coals ($>$0.4 wt% Cl) lead to serious corrosion problems in power station superheater tubes. In order to reduce Cl concentrations in coal preparation plants additional information on the location and nature of Cl in coals was required. In the work of Caswell *et al.* (1984a, b) six seams and adjacent sediments from the S Staffordshire Coalfield were investigated. The total Cl percentages for the coals were in the range 0.5–1.1%, whereas the mudrocks were $<$0.5%. These data are plotted on Fig. 5 against the moisture content. The regression equation shown is for the coals alone. Inclusion of the mudrocks would increase an already significant correlation. An increase in carbon content therefore results in an increase in the amount of water contained within the coal which is then reflected in an increased wt% Cl

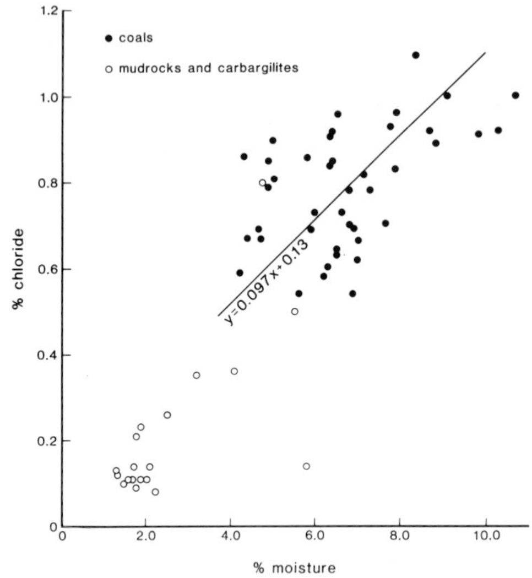

FIG. 5. Variation in wt % Cl and wt % moisture in four coal seams from the S Staffordshire coalfield (data from Caswell *et al.* 1984a). The reduced major axis regression equation for the coals alone is shown. The correlation is significant at the 99.9% level ($r = 0.51$). Inclusion of the mudrocks increases the correlation coefficient ($r = 0.85$) and the significance of the correlation.

for the whole rock. Variations in the Cl concentration of entrapped brines are not sufficiently large to obscure the relationship between water content and total Cl concentration. The moisture percentage shown on Fig. 5 is the 'as received content' and loss of moisture in the laboratory is a possibility. To eliminate this possibility samples were allowed to saturate in the laboratory and the moisture holding capacity was determined. This value is plotted against internal surface area (determined by CO_2 adsorption) on Fig. 6, where the expected relationship between water content and porosity is demonstrated. Additional direct evidence of pore fluid Cl in the coal was obtained by extracting water using a high-speed centrifuge and analysing the extract (Caswell *et al.* 1984a). Although Cl rich the solutions were less concentrated than calculations indicated, but the percentage water extracted was so low that the extract may not have been representative. Pore fluid concentrations in macropores and micropores could differ but interaction between fluid and coal on extraction cannot be eliminated.

The high Cl concentration in United Kingdom coals may be reduced by washing the coal. Fine grinding to an impracticable size is necessary however for the percentage Cl extracted to be $>50\%$ of the total Cl. Coals of a comparable rank also vary in their response to water extraction. The presence of carbonate minerals, for example increases the percentage Cl extraction, which is attributed to the anion exchange HCO_3^- for Cl^- (Daybell & Pringle 1958). The Cl may be present in micropores in solution originally but fine grinding would appear to create anion exchange sites (Caswell *et al.* 1984b).

Major element analyses of low-temperature coal ashes reflect the detrital and diagenetic mineral abundances. In samples with a negligible detrital content the diagenetic minerals achieve greater prominence. The distribution of sulphides and carbonates is less uniform in the coal than diagenetic silicates and in samples dominated by the latter, by non-occurrence of the other minerals, there is an important contribution from the contained brines. Under these conditions not only is Cl important in the ash but Na also. This contribution needs to be recognized in the calculation of mineral abundances from ash analyses. Comparing calculated abundances with those determined by other means, mainly X-ray diffraction, is one approach in determining the contribution of brines and even organic matter to the oxide analyses. Possibly non-crystalline sources other than brines are important. There is also potential in extracting and analysing pore fluids from coals subjected to different burial diagenesis within a basin and from coal microlithotypes in delineating pore fluid evolution. Detailed studies of coal will continue to make an important contribution to the understanding of sedimentary rock evolution in general and will

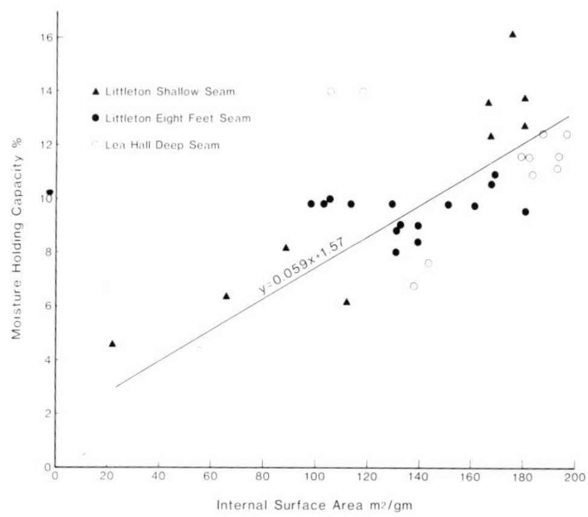

FIG. 6. To show the relationship between internal surface area (CO_2 adsorption) and moisture holding capacity (saturated water content) for three coal seam profiles, from the S Staffordshire coalfield. The reduced major axis regression equation is shown.

also be increasingly required in the future for more sophisticated coal utilization.

ACKNOWLEDGEMENTS: The author has benefited, over the years, from many fruitful discussions on sedimentary mineralogy with Professor C. D. Curtis. Past and present research students have helped to focus the author's attention on coal—for this and for the financial support of NERC, SERC, NCB and CEGB the author is extremely grateful. The views expressed in this paper are, however, the responsibility of the author and not necessarily those of either the NCB or the CEGB.

References

ASHBY, D. A. & PEARSON, M. J. 1978. Mineral distributions in sediments associated with the Alton Marine Band near Penistone, South Yorkshire. *In:* MORTLAND, M. M. & FARMER, V. C. (eds) *Proceedings of the International Clay Conference*, Oxford, pp. 311–321, Developments in Sedimentology 27. Elsevier, New York.

BERNER, R. A. & WESTRICH, J. T. 1985. Bioturbation and the early diagenesis of carbon and sulfur. *American Journal of Science*, **285**, 193–206.

BOHOR, B. F. & TRIPLEHORN, D. M. 1981. Volcanic origin of the flint clay parting in the Hazard No. 4 (Fire Clay) Coal Bed of the Breathitt Formation in eastern Kentucky. *Annual Geological Society of America, Coal Division Field Trip*, pp. 49–54. Kentucky Geological Survey.

BOLES, J. R. & FRANKS, S. G. 1979. Clay diagenesis in Wilcox sandstones of south west Texas: implications of smectite diagenesis on sandstone cementation. *Journal of Sedimentary Petrology*, **49**, 55–70.

BOTZ, R. W., HUNT, J. W. & SMITH, J. W. 1986. Isotope geochemistry of minerals in Australia bituminous coal. *Journal of Sedimentary Petrology*, **56**, 99–111.

BOUROZ, A. 1964. Les composants petrographiques principaux des schistes houillers et leur signification sedimentologique. *In: Compte Rendu 5me Congres International Stratigraphie Geologie Carbonifere*, Paris, 1963, **1**, 289–306.

—— 1972. Utilisation des marqueurs d'origine volcanique en stratigraphie. Exemples d'application dans les gisements houillers. *Memoire Bureau de Recherches Geologiques et Minieres*, **77**, 473–493.

——, SPEARS, D. A. & ARBEY, F. 1983. Essai de synthese des donnees acquises sur la genese et l'evolution des marqueurs petrographiques dans les bassins houillers. *Societe Geologique du Nord, Memoir* No 16.

BRINDLEY, G. W. & ROBINSON, K. 1947. X-ray study of some kaolinite fireclays. *Transactions of the British Ceramics Society*, **46**, 49–62.

BURGER, K 1985. Kohlentonsteine in Kohlenrevieren der Erde Erkenntnisstand 1983. *In: Compte Rendu 10 me Congress Internationale Stratigraphie Geologique Carbonifere*, Madrid, 1983, **1**, 155–174.

CASWELL, S. A. 1983. Geochemistry and mineralogy of coal and coal-bearing strata from the Cannock Coalfield with special reference to chlorine. *PhD thesis*, University of Sheffield.

——, HOLMES, I. F. & SPEARS, D. A. 1984a. Total chlorine in coal seam profiles from the south Staffordshire (Cannock) Coalfield. *Journal of the Institute of Fuel*, **63**, 782–787.

——, —— & —— 1984b. Water-soluble chlorine and associated major cratons from the coal and mudrocks of the Cannock and North Staffordshire coalfields. *Journal of the Institute of Fuel*, **63**, 774–781.

CURTIS, C. D. 1983a. Geochemistry of porosity enhancement and reduction in clastic sediments. *In:* BROOKS J. (ed.) *Petroleum Geochemistry and Exploration of Europe*, pp. 113–125. Special Publications of the Geological Society of London 12. Blackwell Scientific Publications, Oxford.

—— 1983b. The link between aluminium mobility and destruction of secondary porosity. *American Association of Petroleum Geologists Bulletin*, **67**, 380–384.

——, PEARSON, M. J. & SOMOGYI, V. A. 1975. Mineralogy, chemistry and origin of a concretionary siderite sheet (clay-ironstone band) in the Wetphalian of Yorkshire. *Mineralogical Magazine*, **40**, 385–393.

——, PETROWSKI, C. & OERTEL, G. 1972. Stable carbon isotope ratios within carbonate concretions: a clue to place and time of formation. *Nature, London*, **235**, 98–100.

DAVIS, A., RUSSELL, S. J., RIMMER, S. M. & YEAKEL, J. D. 1984. Some genetic implications of silica and aluminosilicates in peat and coal. *International Journal of Coal Geology*, **3**, 293–314.

DAYBELL, G. N. & PRINGLE, W. J. S. 1958. The mode of occurrence of chlorine in coal. *Journal of the Institute of Fuel*, **37**, 283–292.

DIESSEL, C. F. K. 1985. Tuffs and tonsteins in the Coal Measures of New South Wales. *In: Compte Rendu 10me Congrès Internationale Stratigraphie Géologique Carbonifére*, Madrid, 1983, **4**, 197–210.

DIXON, K., SKIPSEY, E. & WATTS, J. T. 1964. The distribution and composition of inorganic matter in British Coals: Part 1—Initial study of seams from the East Midlands Division of the National Coal Board. *Journal of the Institute of Fuel*, **37**, 485–493.

——, —— & —— 1970. The distribution and composition of inorganic matter in British coals. Part 2: Aluminosilicate minerals in the coal seams of the East Midlands coalfields. *Journal of the Institute of Fuel*, **43**, 124–128.

DOWNING, R. A. & HOWITT, F. 1969. Saline groundwaters in the Carboniferous rocks of the English East Midlands in relation to the geology. *Quarterly Journal of Engineering Geology*, **1**, 241–279.

ENNOS, F. R. & SCOTT, A. 1924. Refactory materials: Fireclays. Analyses and physical tests. *Memoir of the Geological Survey, Special Report, Mineral Resources of Great Britain* 28.

FELLOWS, P. M 1979. An investigation of the water

soluble elements in coal and coal bearing strata. *PhD thesis*, University of Sheffield.

FINKELMAN, R. B. & STANTON, R. W. 1978. Identification and significance of accessory minerals from bituminous coal. *Journal of the Institute of Fuel*, **57**, 763–768.

FRANCIS, E. H. 1961. Thin beds of graded tuff and tuffaceous siltstone in the Carboniferous of Fife. *Bulletin of the Geological Survey of Great Britain*, **17**, 191–215.

—— 1985. Recent ash fall: a guide to tonstein distribution. *In: Compte Rendu 10me Congrès International Stratigraphie Géologie Carbonifère*, Madrid, 1983, **4**, 189–195.

GLUSKOTER, H. J. 1965. Electronic low-temperature ashing of bituminous coal. *Journal of the Institute of Fuel*, **44**, 285–291.

HATCH, J. R., GLUSKOTER, H. J. & LINDAHL, P. C. 1976. Sphalerite in coals from the Illinois Basin. *Economic Geology*, **71**, 613–624.

HIGHLEY, D. E. 1982. *Mineral Dossier* No. 24—*Fireclay*. Mineral Resources Consultative Committee. HMSO, London.

HUGHES, R. E., DE MARIS, P. J., WHITE, W. A. & COWIN, D. K. in press. Origin of clay minerals in Pennsylvanian strata of the Illinois Basin. *Proceedings of the International Clay Conference*, Denver, 1985.

HURST, A. & IRWIN, H. 1982. Geological modelling of clay diagenesis in sandstones. *Clay Minerals*, **17**, 5–22.

LIPPOLT, H. J. & HESS, J. C. 1985. $^{40}Ar/^{39}Ar$ dating of sanidines from Upper Carboniferous tonsteins. *In: Compte Rendu 10me Congrès International Stratigraphie Géologie Carbonifère*, Madrid, 1983, **4**, 175–181.

LOVE, L. G., COLEMAN, M. L. & CURTIS, C. D. 1983. Diagenetic pyrite formation and sulphur isotope fractionation associated with a Westphalian marine incursion, northern England. *Transactions of the Royal Society of Edinburgh*, **74**, 165–182.

MACKOWSKY, M.-TH. 1982. Minerals and trace elements occurring in coal. *In:* STACH, E., MACKOWSKY, M.-TH., TEICHMULLER, M., TAYLOR, G. H., CHANDRA, D. & TEICHMULLER, R. (eds) *Textbook of Coal Petrology*, pp. 153–170. Gebruder Borntraeger, Berlin.

MCCABE, P. J. 1984. Depositional environments of coal and coal bearing strata. *In:* RAHMANI, R. A. & FLORES, R. M. (eds) *Sedimentology of Coal and Coal-bearing Sequences*, pp. 13–42. Special Publication of the International Association of Sedimentologists 7. Blackwell Scientific Publications, Oxford.

MATSUMOTO, R. & IIJIMA, A. 1981. Origin and diagenetic evolution of Ca-Mg-Fe carbonates in some coalfields of Japan. *Sedimentology*, **28**, 239–259.

MILLER, R. N., YARZAB, R. F. & GIVEN, P. H. 1979. Determination of mineral matter contents of coals by low temperature ashing. *Journal of the Institute of Fuel*, **58**, 4–10.

MOORE, L. R. 1964. Microbiology, mineralogy and genesis of a tonstein. *Proceedings of the Yorkshire Geological Society*, **34**, 235–292.

—— 1968. Some sediments closely associated with coal seams. *In:* MURCHISON, D. G. & WESTOLL, T. S. (eds) *Coal and Coal Bearing Strata*, pp. 105–123. Oliver & Boyd, Edinburgh.

NICHOLLS, G. D. & LORING, D. H. 1962. The geochemistry of some British Carboniferous sediments. *Geochemica et Cosmochimica Acta*, **26**, 181–222.

OERTEL, G. & CURTIS, C. D. 1972. Clay ironstone concretion preserving fabrics due to progressive compaction. *Bulletin of the Geological Society of America*, **83**, 2597–2606.

PRICE, N. J. 1966. *Fault and Joint Development in Brittle and Semi-brittle Rock*. Pergamon Press, Oxford.

RAASK, E. 1985. *Mineral Impurities in Coal Combustion*. Hemisphere Publishing Corporation, Washington.

RENTON, J. J. & CECIL, C. B. 1979. The origin of mineral matter in coal. *In:* DONALDSON, A., PRESLEY, M. W. & RENTON, J. (eds) *Carboniferous Coal Guidebook, B-37-1*, pp. 206–223. West Virginia Geological and Economic Survey.

RUPPERT, L. F., CECIL, C. B. & STANTON, R. W. 1985. Authigenic quartz in the Upper Freeport Coal bed, west-central Pennsylvania. *Journal of Sedimentary Petrology*, **55**, 334–339.

RYER, T. A., PHILLIPS, R. E., BOHOR, B. F. & POLLASTRO, R. M. 1980. Use of altered volcanic ash falls in stratigraphic studies of coal-bearing sequences: an example from the Upper Cretaceous Ferron Sandstone Member of the Mancos Shale in central Utah. *Bulletin of the Geological Society of America*, **91**, 579–586.

SCHULLER, A. 1951. Zur nomenklatur und genese der Tonsteine. *Neues Jahrbuch Mineralogie Monatsheften, 1961*, pp. 97–109.

SCOTT, A. C. & REX, G. 1985. The formation and significance of Carboniferous coal balls. *Philosophical Transactions of the Royal Society of London B*, **311**, 123–137.

SPEARS, D. A. 1970. A kaolinite mudstone (tonstein) in the British Coal Measures. *Journal of Sedimentary Petrology*, **40**, 386–394.

—— & AMIN, M. A. 1981. Geochemistry and mineralogy of marine and non-marine Namurian black shales from the Tansley Borehole, Derbyshire. *Sedimentology*, **28**, 407–417.

—— & CASWELL, S. A. 1986. Mineral matter in coals: cleat minerals and their origin in some coals from the English Midlands. *International Journal of Coal Geology*, **6**, 107–125.

—— & KANARIS-SOTIRIOU, R. 1975. Titanium in some Carboniferous sediments from Great Britain. *Geochimica et Cosmochimica Acta*, **40**, 345–351.

—— & —— 1979. A geochemical and mineralogical investigation of some British and other European tonsteins. *Sedimentology*, **26**, 407–425.

—— & RICE, C. M. 1973. An Upper Carboniferous tonstein of volcanic origin. *Sedimentology*, **20**, 281–294.

—— & SEZGIN, H. I. 1985. Mineralogy and geochemistry of the *G. subcrenatum* Marine Band and associated coal-bearing sediments, Langsett, South

Yorkshire. *Journal of Sedimentary Petrology*, **55**, 570–578.

STAUB, J. R. & COHEN, A. D. 1978. Kaolinite enrichment beneath coals; a modern analog, Snuggedy Swamp, South Carolina. *Journal of Sedimentary Petrology*, **48**, 203–210.

STOPES, M. C. & WATSON, D. M. S. 1908. On the present distribution and origin of the calcareous concretions known as 'Coal Balls'. *Philosophical Transactions of the Royal Society B*, **200**, 167–188.

TEICHMÜLLER, M. 1982. Origin of the petrographic constituents of coal. *In:* STACH, E., MACKOWSKY, M.-TH., TEICHMÜLLER, M., TAYLOR, G. H., CHANDRA, D. & TEICHMÜLLER, R. (eds) *Textbook of Coal Petrology*, pp. 219–294. Gebruches Borntraeger, Berlin.

—— & TEICHMÜLLER, R. 1982. The geological basis of coal formation. *In:* STACH, E., MACKOWSKY, M.-TH., TEICHMÜLLER, M., TAYLOR, G. H., CHANDRA, D. & TEICHMÜLLER, R. (eds) *Textbook of Coal Petrology*, pp. 5–86. Gebrucher Borntraeger, Berlin.

TREWIN, N. H. & HOLDSWORTH, B. K. 1972. Further K-bentonites from the Namurian of Staffordshire. *Proceedings of the Yorkshire Geological Society*, **39**, 73–91.

TRIPLEHORN, D. M., TURNER, D. L. & NAESAR, C. W. 1984. Radiometric age of the Chikaloon formation of south-central Alaska: location of the Paleocene–Eocene boundary. *Geological Society of America Bulletin*, **95**, 740–742.

WEAVER, C. E. 1963. Interpretative value of heavy minerals from bentonites. *Journal of Sedimentary Petrology*, **33**, 343–349.

WILDING, L. P., SMEEK, N. E. & DREES, L. R. 1977. Silica in soils: quartz, cristobalite, trydmite and opal. *In:* DIXON, J. B. & WEED, S. B. (eds) *Minerals in Soil Environments*, pp. 471–552. Soil Science Society of America, Madison.

WILLIAMSON, I. A. 1970. Tonsteins—their nature, origins and uses. *Mining Magazine*, **122**, 119–211.

WILSON, M. J. 1965. The origin and geological significance of the South Wales underclays. *Journal of Sedimentary Petrology*, **35**, 91–99.

ZARITSKY, P. V. 1985. A review of the study of tonsteins in the Donetz Basin. *In: Compte Rendu 10me Congrès International Stratigraphie Géologique Carbonifère*, Madrid, 1983, **1**, 235–241.

D. A. SPEARS, Department of Geology, Mappin Street, University of Sheffield, Sheffield S1 3JD, England.

Lycopod succession in coals: an example from the Low Barnsley Seam (Westphalian B), Yorkshire, England

K. M. Bartram

ABSTRACT: Lycopsid megaspores processed from petrologically defined samples of the Low Barnsley Seam, Westphalian B, Upper Carboniferous have vertical distributions arranged in repeated successions of phases. A total of six phases was distinguished producing a more detailed pattern of succession than has previously been identified using miospores. Two important phases were recognized for the first time, designated: Phase 0, characterized by the species *Cystosporites verrucosus* Dijkstra, *Lagenicula horrida* Zerndt, *Pseudovalvisisporites nigrozonalis* (Stach & Zerndt) Lachkar and *Cystosporites giganteus* (Zerndt) Schopf, interpreted as a colonizing assemblage, and Phase 2, characterized by the species *Triangulatisporites tertius* Potonié & Kremp and *Valvisisporites auritus* (Zerndt) Potonié & Kremp which is considered to represent an association of non-arborescent lycopods.

Phases are broadly related to petrographic type(s) but changes in phases and petrography are not necessarily coincident. Phasal changes of the kind described are believed to result from the formation of a raised bog.

Keywords: coal, megaspores, lycopods, environment, England, Carboniferous, Westphalian, succession

Interpretations of plant palaeoecology and floral changes through Upper Carboniferous coal seams of Euramerica have previously utilized two main approaches. The first relates miospore distribution to coal petrology (e.g. Smith 1957, 1962, 1968, Haquebard & Donaldson 1969, Navale 1964, Habib 1968), while the other study is based on the study of the permineralized vegetation of coal balls (Phillips & DiMichelle 1981). From a stratigraphic study of both coal ball macrofossils and palynological data, Phillips & Peppers (1984) recorded changing trends in the dominant coal swamp floras throughout the Upper Carboniferous of Euramerica. They reported that during the Westphalian B the major contributers to the formation of coals were lycopods. Palaeobotanical studies (e.g. Chaloner 1953a, b, 1954, Sen 1958, Abbot 1963, Balbach 1966, Leisman 1970, Leisman & Phillips 1979, Pigg & Rothwell 1983) have indicated that almost all of the coal lycopods were heterosporous, i.e. produced both large (> 200 μm) megaspores, and small (< 200 μm) microspores. However, the vertical distribution of megaspores through seams has received little attention, and the few existing studies (Pierart 1968, Scott 1978, Scott & King 1981) included little assessment of the relationship between megaspores and coal petrology. Megaspores being larger than microspores will have different dispersion patterns due to aerodynamic effects and/or transportation mechanisms. Scott & King (1981) suggested that megaspores did not undergo significant transport and should therefore give a more localized representation of parent plant distribution than miospores.

This paper describes the relationships of megaspores to coal petrology from the Low Barnsley Seam of the Yorkshire Coalfield, and attempts to recognize the floral development of the seam.

Materials and methods

A 288-cm length of core through the Low Barnsley Seam (Westphalian B) taken from St Aidans Opencast Mine, SW of Leeds (NGR SE 40 28) was made available for study by British Coal. At this locality the seam is composite comprising three leaves of coal (Bottom, Middle and Top) and a thin 'Special Category' coal (too thin to be economic) between the Bottom and Middle Leaves.

The coals between the clastic partings were removed from the core as units of variable thickness (2–26 cm) bound by fractures parallel to the bedding. Each unit was labelled and orientated prior to being coated with a 3-cm thickness of bonding plaster to prevent disintegration. The units were sectioned vertically; half was retained for palynological investigation whilst the other half was made into polished blocks for petrological study. Nineteen blocks were prepared using araldite resin (HY209) for embedding and alumina oxide powders for polishing.

The blocks were examined using oil immersion reflected light microscopy at a total magnification of × 240. A mechanical stage with a traversing range of 25 mm was employed to give a linear transect perpendicular to the bedding. The recommendations and descriptions of the ICCP (1963 and later supplements) were followed. The clastic partings between the coals were described prior to division into 4–6 cm intervals for palynological preparation.

Megaspores were macerated from the coals using fuming nitric acid followed by distilled water and 10% ammonia. Bromine pre-treatment (Corsin *et al.* 1967) was unnecessary. Twenty-gram subsamples of crushed coal (0.2–20 mm) were usually adequate except for material rich in pyrite, carbominerite or cutinite. One hundred megaspores from each sample were studied using a combination of scanning electron, reflected and transmitted light microscopy. The clastic sediments were digested using hydrofluoric acid. Miospores also macerated from each sample will be described in detail elsewhere.

The importance of recognizing the associations of spore assemblages with particular coal types is that the petrological composition of a coal is a product of several factors such as vegetation type, edaphic conditions, depositional environment and post-depositional coalification (Teichmüller & Teichmüller 1982). Environmental significance has been attributed to macerals and microlithotypes of coal by several authors (e.g. Karmasin 1952, Teichmüller & Thomson 1958, Teichmüller 1962, Haquebard *et al.* 1967, Styan & Bustin 1985). Identification of associations of megaspores to petrological 'types' may therefore allow the development of hypotheses concerning the environment of growth of the parent plants.

There have been two main techniques for recognizing the palynological composition of Carboniferous coal seams for palynological analyses. First, Smith (1957) employed a twofold method in his study of coals of the Yorkshire Coalfield. Initially a traverse of the coal blocks was employed to divide the seam into obvious, fairly homogeneous petrographic units. This gave intervals 1.5–7 cm thick. Coals from each of these subsections was then prepared for both palynology, and crushed for qualitative analyses of 500 positive identifications of microlithotypes.

Second, Haquebard & Donaldson (1969) also divided the seam into petrographic units. The top and bottom of each interval was delineated macroscopically by clastic partings and/or dull layers, producing a coarse (5–15 cm) subdivision of the seam. The petrology of each interval was then assessed using the aggregate thickness of each individual microlithotype as observed in a linear traverse perpendicular to the bedding. Such coarse division of the seam, based primarily on the macroscopic seam appearance, involves a high risk of combining both more than one palynological assemblage and petrographic sequence within one sample. Unpublished work by the author on interval sampling suggests that a maximum interval of 5 cm should be employed when sampling seams on the basis of their macroscopic appearance.

The technique used in this study was based upon the qualitative method of Smith (1957). Using the mechanical traverse of the stage perpendicular to the bedding, the thickness (in millimetres) of each microlithotype or association of macerals was recorded for each block. The microlithotype thicknesses ranged from 1 to 25 mm (a minimum thickness of 5 mm was required following subdivision of the core to yield sufficient megaspores). The thicknesses of the microlithotypes within each block were then plotted sequentially. The blocks were then divided into lithographic units representing a single dominant element (such as clarite) or of associations of macerals or microlithotypes (such as trimacerite in vitrite). The term clarite is used for a sporinite-vitrite association and cuticle clarite for a cutinite-vitrite association. The 12 lithographic units identified were 0.5–4.5 cm thick. The retained section of the core was divided for palynological preparations on an equal basis to the lithographic units.

Petrology of the Low Barnsley Seam

In order to examine the transition between clastic sedimentation and coal accumulation the petrological study of the Bottom Leaf included *c.*10 cm of the underlying carbonaceous shales. Observations on this part of the seam indicated a gradual change between the two environments. At the base of the core vitrite was absent, however, working up the core the clay mineral content gradually decreases (to $<2\%$) with increasing vitrite producing a series of carbominerite lithographic units. The petrology of the remainder of the Bottom Leaf was typical of a bright coal, being formed predominantly of vitrites and clarites. Towards the top of the leaf inertinite increases, either as trimacerite or as thin bands of semifusinite within clarites and vitrites. Framboidal pyrite was commonly observed. The termination of the leaf by Carbonaceous shales was abrupt, contrasting with the gradual reductions at the base of the leaf. The petrological log is shown in Fig. 1.

The Middle Leaf and Special Category coals

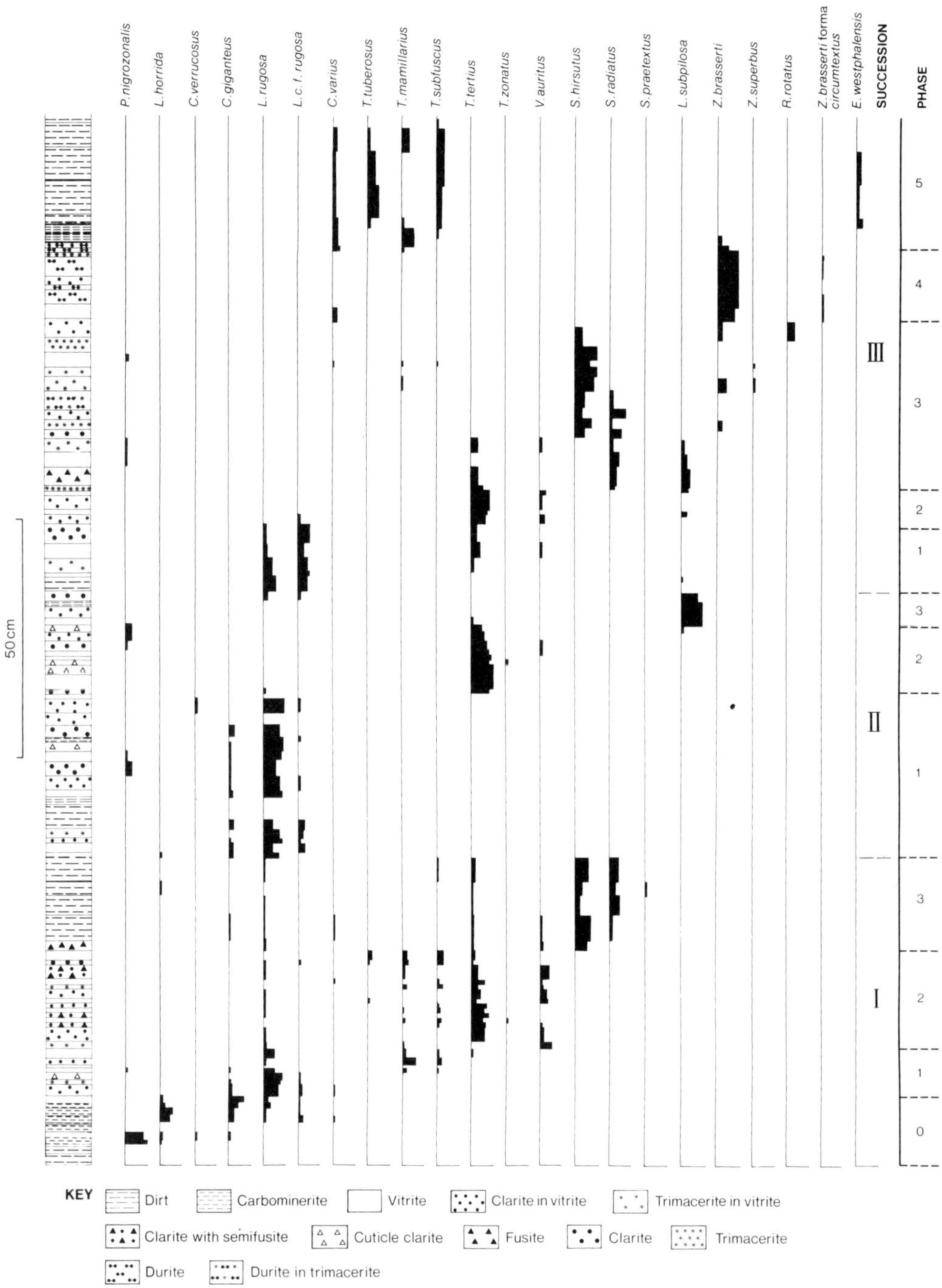

FIG. 1. The coal petrology, megaspore distribution and megaspore frequency of the Low Barnsley Seam profile.

were similarly bright, formed of vitrites and clarites. The Middle Leaf showed no increase in inertinite macerals towards the top of the leaf. Pyrite was observed in comparable quantities to that in the Bottom Leaf. The clay mineral content was negligible with the exception of a thin carbominerite near the base of the leaf. This may represent minor flooding of the swamp surface.

The Top Leaf was more petrologically diverse. Although the base of the leaf was composed primarily of clarites and vitrite, this is followed by a gradual increase in inertinite producing lithographic units with associated inertinite, trimacerites and finally durites. In the Top Leaf pyrite was rare and confined to the base. Clastic input at the top of the leaf was less gradational than the decrease in clastics at the base of the seam.

Distribution of megaspores

Megaspore counts from the maceration of 106 coal and sediment samples were plotted as a profile (Fig. 1) to show the vertical distribution and frequency of each megaspore species present at >5% of the sample count. Nine samples (predominantly clastics or carbominerites) that failed to yield 100 megaspores have not been plotted. Twenty megaspore species were identified. Their systematics will be described elsewhere. Figures 2 and 3 illustrate some of the megaspore species encountered in this study.

The profile shows that the distribution of megaspores within the Low Barnsley Seam is non-random. Successive changes in groups of spores may be observed. With reference to miospores, Smith (1957) defined dominant (or a group of dominant species), together with certain less abundant species which characterized part of a seam, as Phases. This concept may equally be applied to megaspores. Six megaspore Phases (0–5) were recognized from the Low Barnsley Seam Profile (Table 1). The order of succession of the phases is repetitive. Three sequences of change or 'Successions' are present. A Succession is hereby defined as a sequence of phases prior to a major reversal to a previously encountered phase. Successions I, II and III broadly correspond to the Bottom, Middle and Top Leaves respectively. Table 1 summarizes the species which represent each phase and the phases present in each succession.

Phase 0 was poorly represented and was only found in the carbominerites at the base of the seam, which generally yield insufficient megaspores. *Pseudovalvisisporites nigrozonalis* (Stach & Zerndt) Lachkar (Fig. 2a) was not restricted to Phase 0; it was also associated with developments of fusite within the seam. Phase 1 occurs at the base of each leaf and is related to the end of clastic input and the presence of vitrites and clarites. *Lagenicula rugosa* (Loose) Arnold (Fig. 2b) is used here for forms with laevigate exines. *L.*cf. *rugosa* is used for forms showing a diversity of ornament comparable to specimens observed in cones of *Flemingites olryi* (Zeiller) Brack-Hanes & Thomas. Phase 2 succeeds Phase 1 and is characterized by *Triangulatisporites tertius* (Potonié & Kremp) (Fig. 2e) and *Valvisisporites auritus* (Zerndt) Potonié & Kremp (Fig. 2d). However in Succession I, *Tuberculatisporites mamillarius* (Bartlett) Potonié & Kremp (Fig. 3f), *Tuberculatisporites subfuscus* (Wicher) Potonié & Kremp and *Tuberculatisporites tuberosus* (Ibrahim) Ibrahim are associated with this phase. This represents the only appreciable appearance of these spores within the coal. Petrologically, Phase 2 of Succession I is associated with coals containing more semifusite than encountered with Phase 2 of the other successions.

Phase 3 has a less clearly defined megaspore composition. It may be represented by one or more of the following dominant spores; *Setosporites hirsutus* (Loose) Ibrahim (Fig. 3b), *Spencerisporites radiatus* (Ibrahim) Felix & Parks and *Lagenicula subpilosa* (Ibrahim) Ibrahim (Fig. 3a). Phase 3 in Successions I and II is related to the input of clastic sediments above the Bottom and Middle Leaves, suggesting that the parent plants may have favoured a habitat associated with clastic input. The occurrence of Phase 3 in Succession III shows an association with trimacerite-rich coals with no clastic input. Phase 4 is restricted to Succession III where it immediately succeeds Phase 3. This phase is virtually monospecific, dominated by the spore *Zonalesporites brasserti* (Stach & Zerndt) Potonié & Kremp (Fig. 3c). It is associated with the presence of durites and trimacerites. Phase 5 occurs only in the sediments above the Low Barnsley Seam and may therefore represent the flora of a non-coal swamp environment.

Relationship of megaspore phases to coal petrology

There are no strong positive correlations between individual species and lithology. Durites in Succession III contain not only *Z. brasserti* of Phase 4 but *S. hirsutus* of Phase 3. Patterns of negative correlation are more obvious, for example neither *L. rugosa* or *T. tertius* were associated with durites. The dominant spores have ranges coinciding with several lithographic units; *T. tertius* in Succession II ranges over a series of vitrites, cuticle-clarites, clarites and fusite. The petrology of isolated samples of coal is therefore not necessarily helpful in the determination of the megaspore content. Smith (1957) also noted that dominant miospores of British seams similarly had large vertical ranges. Changes in megaspore frequencies related to petrology were

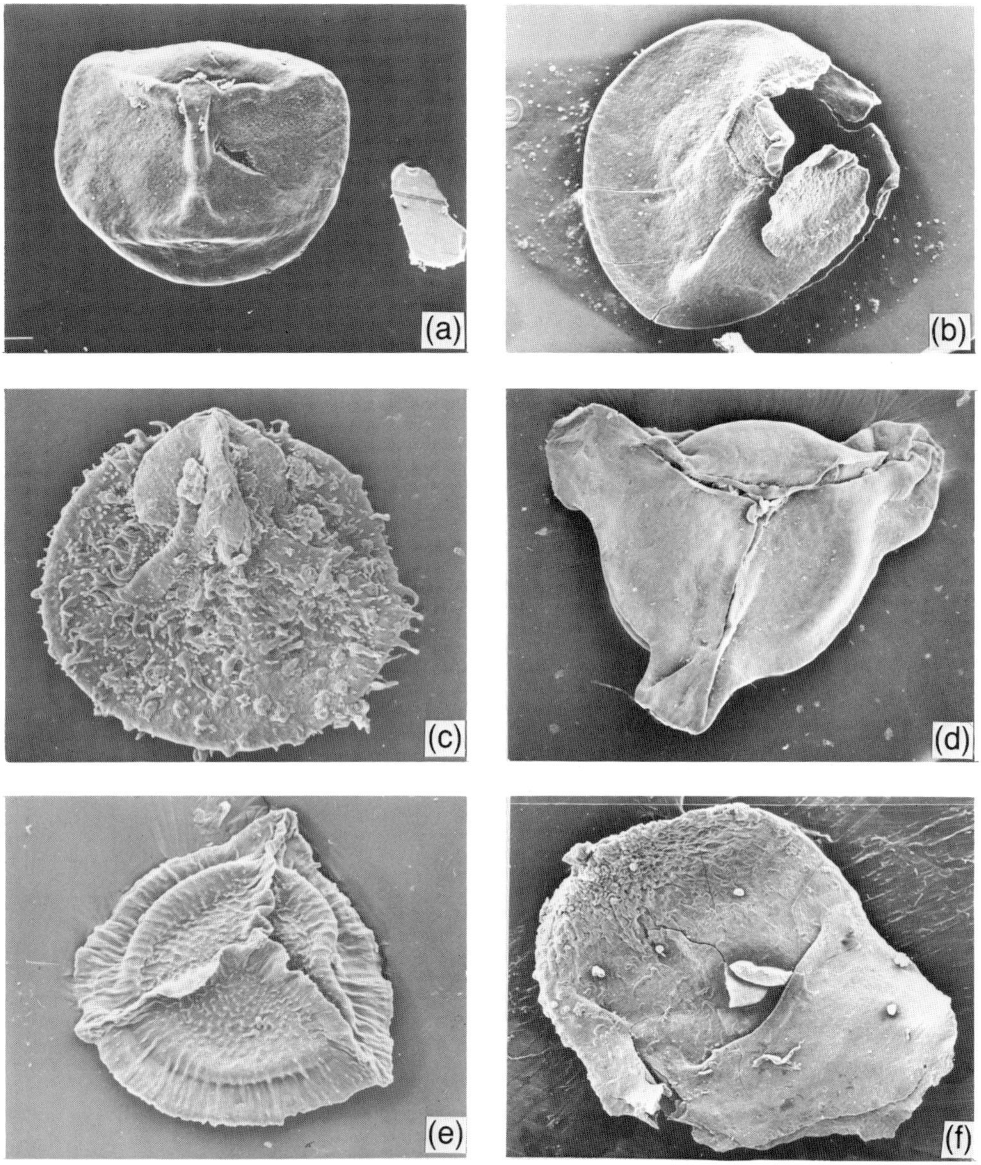

FIG. 2. SEM photomicrographs of selected megaspores from the Low Barnsley Seam: (a) *Pseudovalvisisporites nigrozonalis* (Stach & Zerndt) Lachkar × 50, (b) *Lagenicula rugosa* (Loose) Arnold × 50, (c) *Lagenicula horrida* Zerndt × 50, (d) *Valvisisporites auritus* (Zerndt) Potonie & Kremp × 50, (e) *Triangulatisporites tertius* Potonie & Kremp × 50, (f) *Cystosporites verrucosus* Dijkstra × 12.

most pronounced where there was a rapid alternation between inertinite-rich and intertinite-poor lithologies, for example the abundance of *S. hirsutus* and *S. radiatus* in Succession III vary as the proportions of clarite (*S. radiatus*—rich) and trimacerite (*S. hirsutus*—rich) fluctuate.

Despite the division of the seam into lithographic units, it was not possible, petrographically to identify indications of events which may have led to the introduction or the termination of a phase within the profile. Changes in phasal succession are not promoted by the occurence of fusinite, as may be formed by fire, flooding resulting in an input of clastic material, or by any other distinct petrological markers. Even when the boundaries between phases can be clearly

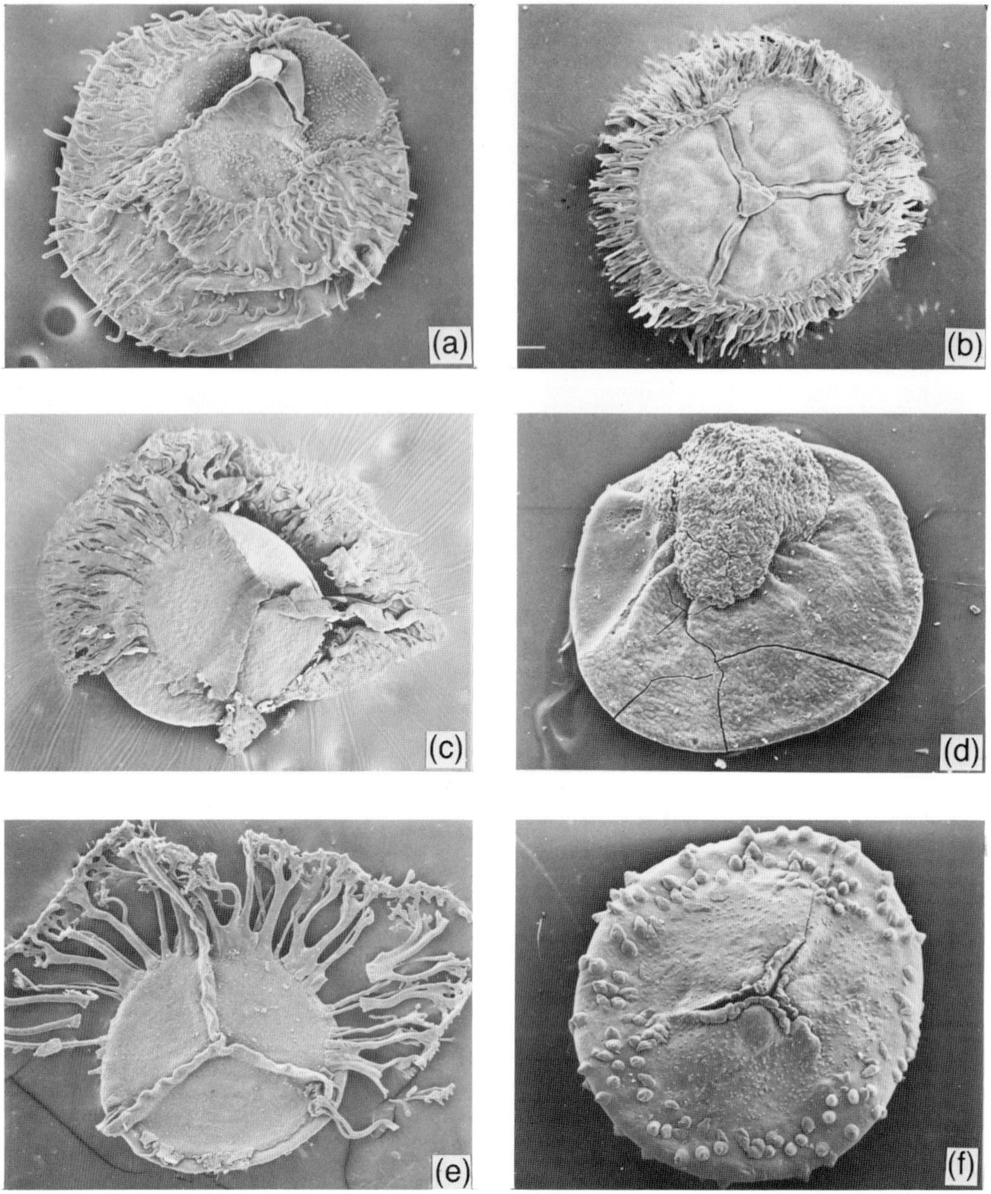

FIG. 3. SEM photographs of selected megaspores from the Low Barnsley Seam: (a) *Lagenicula subpilosa* (Ibrahim) Potonie & Kremp × 50, (b) *Setosisporites hirsutus* (Loose) Ibrahim × 50, (c) *Zonalesporites brasserti* (Stach & Zerndt) Potonie & Kremp × 25, (d) *Cystosporites varius* (Wicher) Dijkstra × 50, (e) *Zonalesporites rotatus* (Bartlett) Spinner × 50, (f) *Tuberculatisporites mamillarius* (Bartlett) Potonie & Kremp × 25.

defined (as in Succession II), they are related to common lithologies, in this case vitrite; however, the occurrence of vitrite within the seam is generally not indicative of a phase change. The difficulty of utilizing petrological discontinuities for the recognition of floral changes serves to emphasize the importance of closely-spaced sampling in order to identify megaspore phases.

The similar pattern of phases within each succession suggests the existence of a floral progression along a gradient of changing environment. If the idealized, uninterrupted sequence of phases through the Low Barnsley Seam with respect to time is ordered 0–5 then the overall progression of petrological variation may be discussed (Fig. 4).

TABLE 1. *The principal and associated megaspores of each phase (numbered 0–5) of the Low Barnsley Seam, and the successions in which they occur*

	Phase 0	Phase 1	Phase 2	Phase 3	Phase 4	Phase 5
Principal species	*Pseudovalvisisporites nigrozonalis* (Stach & Zerndt) Lachkar	*Lagenicula rugosa* (Loose) Arnold	*Triangulatisporites tertius* Potonie & Kremp	*Spencerisporites radiatus* (Ibrahim) Felix & Parks	*Zonalesporites brasserti* (Stach & Zerndt) Potonie & Kremp	*Tuberculatisporites subfuscus* (Wicher) Potonie & Kremp
	Lagenicula horrida Zerndt	cf. *Lagenicula rugosa*	*Valvisisporites auritus* (Zerndt) Potonie & Kremp	*Setosisporites hirsutus* (Loose) Ibrahim		*Tuberculatisporites mamillarius* (Bartlett) Potonie & Kremp
	Cystosporites verrucosus Dijkstra	*Cystosporites giganteus* (Zerndt)		*Lagenicula subpilosa* (Ibrahim) Potonie & Kremp		*Tuberculatisporites tuberosus* (Ibrahim) Ibrahim *Cystosporites varius* (Wicher) Dijkstra *Expansisporites westphalensis* (Bharadawaj) Loboziak
Associated species	*Cystosporites gianteus*		*Triangulatisporites zonatus* (Ibrahim)	*Setosisporites praetextus* (Zerndt) Potonie & Kremp		
			T. tuberosus *T. mamillarius* *T. subfuscus*	*Zonalesporites superbus* (Bartlett) Karczewska		
Succession	I	I, II, III	I, II, III	I, II, III	III	III

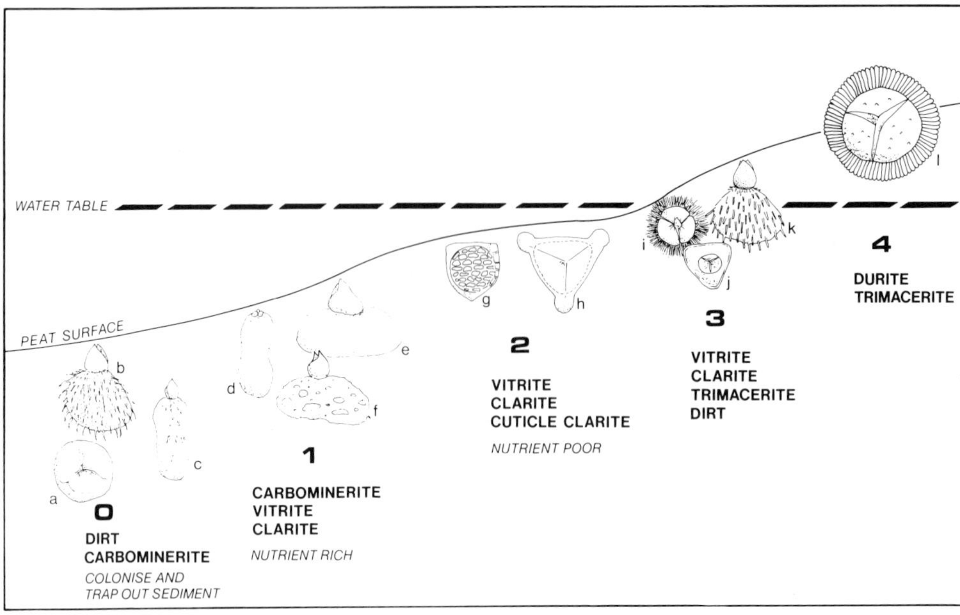

FIG. 4. A schematic diagram of an idealized uninterrupted sequence of megaspore assemblages through a seam (megaspores not to scale). a) *P. nigrozonalis*, b) *L. horrida*, c) *C. verrucosus*, d) *C. giganteus*, e) *L. rugosa*, f) *L.* cf. *rugosa*, g) *T. tertius*, h) *V. auritus*, i) *S. hirsutus*, j) *S. radiatus*, k) *L. subpilosa*, l) *Z. brasserti*.

Qualitative observations of the clastic content of the base of the seam showed that the clay mineral content gradually decreased to the gain of vitrite, indicating a gradual rather than abrupt transition between the clastic and coal environments. The spores of Phase 0 correspond only to this portion of the seam and, with the exception of *P. nigrozonalis*, are feebly represented elsewhere. This suggests that the parent plants probably favoured the conditions prevailing at the onset of peat formation, i.e. the presence of a raised water table and clastic input. Styan & Bustin (1985) from a study of the Fraser River Delta recognized sedges as a colonizing vegetation. As they became established the root systems trapped incoming sediment until a clastic-free environment was attained and succession of peat vegetation began.

Phases 1 and 2 are associated with lithologies in which vitrites, cuticle-clarites and clarites predominate. Phase 1 may also be associated with carbominerites at the base of the leaves, and semifusinite is also associated with Phase 2 of Succession II. Anaerobic conditions are a prerequisite for the formation of vitrite (Fischer 1952, Welte 1952). This is generally caused by the presence of a high water table isolating the accumulated organic matter from the atmosphere and bringing it into contact with oxygen-poor waters. The presence of pyrite in the lithographic units associated with these phases is further indicative of this sort of environment. The same conditions also apply to the formation of clarites and cuticle-clarite, as these lithologies merely represent the presence of liptinitic substances within vitrite. The association of the spores of Phases 1 and 2 to these lithologies implies that they were produced by plants growing on or close to waterlogged peats.

The presence of semifusinite in the lithologies associated with Phase 2 Development II may be related to the presence of *Tuberculatisporites* sp.. Smith (1962) postulated that semifusinite enrichment was symptomatic of an increase in the volume of water entering the swamp, causing concentration of inertinite via flotation. The increase in the through flow of water also aided the introduction of allochthonous elements. The only other significant development of *Tuberculatisporites* sp. is Phase 5 which is associated with a non-coal swamp environment. It is therefore possible that their association with semifusinite in Phase 2, in Succession II is indicative of an allochthonous element within the phase.

Phase 4 is only found in Succession III, which contains the only duritic-rich lithologies. Several types of durite may be recognized using the inertinite maceral composition. The durites of the Low Barnsley Seam profile were rich in the maceral macrinite. This type of durite corresponds to the crassi-durite of Smith (1962).

The origins of macrinite are not clearly understood; most workers (Timofeyev 1955, Stach 1952, Smith 1962, Littke 1986) regard it as an autochthonous deposit formed from peats that had undergone partial aerobic oxidation and from which there was some degree of drainage. Such conditions are typically found at the centre of raised ombrotrophic bogs (Moore, *this volume*). Other authors have used its association with boghead algae to postulate a subaquatic origin (Schopf 1952) or suggested formation from organic mud (Teichmüller 1962). A number of features of the durites of the Low Barnsley are suggestive of raised aerobic conditions. The absence of clay mineral content is characteristic of bogs formed above the flood level (McCabe 1984), syngenetic pyrite is absent and the presence of macrinite may indicate widespread oxidation.

The spores of Phase 3 are associated with either the introduction of clastic material at the top of the Bottom and Middle Leaves, or with trimacerite and inertinite-rich coals. Autochthonous trimacerites are formed in environments intermediate to (or alternating between) those which produce vitrites or durites. The spores of Phase 3 therefore represent lycopods which grew on a substrate at or near the water table in partially emergent conditions. This stage of peat development would appear to be particularly susceptible to flooding and clastic input.

A summary of the developmental history of the seam in terms of lycopod megaspore succession can therefore be presented (Fig. 4). At the onset of peat forming conditions the first lycopods to colonize the area were those represented by the spores of Phase 0. As these became established, clastic input was reduced and they were succeeded by spores of Phase 1. The absence of any petrological indicators to mark a change in succession from Phase 1 to Phase 2 suggests that the change in environment was subtle, possibly a change in nutrient supply or other edaphic control not indicated petrologically. Raising the peat to the level of the water table resulting in periodic oxidation prompted the succession of Phase 3 lycopods. Phase 4 lycopods exploited the conditions arising from a raised bog surface.

The environmental gradient as interpreted by coal petrology is the changing relationship of the peat to the position of the water table as in the development of a raised bog. Recent studies of tropical peats from SE Asia (Anderson 1964, Anderson & Müller 1974, Haseldonckx 1977) support the concept of a raised bog as a model for Carboniferous coal formation. The bogs are a product of peat formation derived from tropical rain forest vegetation. Profiles of both the vegetation (Anderson 1964) and the palynology (Anderson & Müller 1974, Haseldonckx 1977) of the peats have demonstrated that a catenary sequence of vegetation is present from the perimeter to the centre of each raised bog, accompanied by a decrease in species diversity. It is suggested that the megaspore developments represent a caternary sequence of coal swamp communities. The monospecific character of the Phase 4 megaspore assemblage would be consistent with a community of low species diversity occurring at the centre of the coal-forming swamp.

Discussion

Comparison of these results with ecological or environmental significance attributed to lycopods by other authors is difficult as previous detailed studies have been based on miospores or coal-ball macrofossils. Direct correlation between these and the present method of investigation is hampered by problems of establishing megaspore affinities to microspores or parent plants and comparison with data derived from coals of varying stratigraphic ages.

Smith (1962, 1968) recognized four phases of miospores. Three phases corresponded to changes in the edaphic conditions of the swamp as it progressed from submerged to emergent, to emerged (Lycospore, Transition and Densospore phases respectively). The fourth phase, the Incursion Phase, represented flooding and associated transport into the swamp of allochthonous material. If the microspore affinities are added to the megaspores (where known) a comparison of the megaspore phases with the expected miospore phases of Smith can be made (Table 2). Smith (1962), and Marshall & Smith (1965) failed to recognize a distinctive colonization phase. In the Low Barnsley Seam the colonization phase (Phase 0) is associated only with carbargillites and the dirts beneath the seam, emphasizing the importance of the palynology of the sediments in interpreting floral developments of the seam.

The microspores which have affinity to the megaspores of Phase 2 were not observed as 'dominant' components by Smith (1962). *Cirritriradiates saturni* (Ibrahim) Schopf, Wilson & Bentall was only associated with the Incursion Phase, in association with species of the Transition Phase. The miospore analysis of the same samples of Phase 2 spores also highlighted a discrepancy in the data yielded by mega- or microspores. In these samples *Cirritriradiates* does not exceed 7% of the total lycopod microspore flora whereas the megaspore *Triangulatisporites* may represent over 90%. *Endosporites* (Wilson & Coe) is encountered at 3% or less. The recognition of Phase 2 is considered important as the parent plants *Chaloneria* Pigg & Rothwell and *Selaginella* are non-arborescent. Miospore studies would not recognize the importance of this phase.

The species composition, vegetative structure, growth and reproductive biology of the plants and stratigraphic variation in species composition of coal ball vegetations have been used to reconstruct habitats and environments of lycopod growth (Phillips 1979, Phillips & DiMichele 1981, Phillips & Peppers 1984, DiMichele & Phillips 1985 and Phillips *et al.* 1985). Difficulties exist in making direct correlations between megaspores and their parent plants. For example, *C. giganteus* has been described *in situ* from *Lepidocarpon major* Hemingway, *Lepidocarpon waltoni* Chaloner and *Achlamydocarpon takhtajanii* Snigirevskaya, and as a result been given affinity to both *Lepidophloios* and *Lepidodendron*. Several similarities in environmental interpretations exist. The affinity of lycopod genera to megaspores is listed, where known, in Table 2.

Lepidophloios and *Lepidodendron* are interpreted from coal ball analyses as having wet, standing-water habitats. Phillips (1979) suggested that the sporangial units of both genera were adapted to aquatic dispersal. Phases 0 and 1 which included the megaspore *C. giganteus* were also associated with a high water table. The distribution of *L. rugosa* (Phase 1) supports the observation that *Paralycopodites brevifolium* (Williamson) DiMichele from coal balls occurs in habitats transitional from clastic to coal environments, or within coal measure sequences following disruption. *Chaloneria* Pigg & Rothwell was postulated by DiMichele *et al.* (1979) to grow in an open shallow water wetland marsh habitat. The association of *V. auritus* with *T. tertius* in Phase 2 supports the hypothesis of a predominantly non-arborescent lycopod vegetation. *Sig-*

TABLE 2. *The microspores and parent lycopods believed to have an affinity to the megaspores of each phase. L = Lycopore, T = Transition, D = Densospore and I = Incursion Phases of Smith (1962)*

Megaspore	Megaspore phase	Microspore	Miospore phase (Smith 1962)	Lycopod
Cystosporites varius	5	*Cappasporites* (*Granasporites irregularis* Smith pers. comm. 1986)	L	*Diaphorodendron* (DiMichele 1985)
Tuberculatisporites sp.	5	*Crassispora*	I	*Sigillaria* (Chaloner 1953a)
Zonalesporites brasserti	4	*Densosporites sphaerotrianularis*	D	*Sporangiostrobus* (Leisman 1970)
Lagenicula subpilosa	3	*Lycospora*		—
Setosisporites hirsutus	3	*Densosporites loricatus*	T	*Porostrobus* (Chaloner 1958)
Spencerisporites radiatus	3	—	T	*Spencerites* (Chaloner 1951)
Triangulatisporites sp.	2	*Cirritriradiates*	I/T	*Sellaginellites* (Hoskins & Abbott 1956)
Valvisisporites auritus	2	*Endosporites globiformis*	L/T	*Chaloneria* (Pigg & Rothwell 1983)
Lagenicula rugosa	1	*Lycospora*		*Paralycopodites* (DiMichele 1980)
Cystosporites giganteus	0/1	*Lycospora granulata* (*sensu* Kosanke)	L	*Lepidodendron* (Chaloner 1952)
Cystosporites verrucosus	0	—		—
Lagenicula horrida	0	cf. *Lycospora brevijuga*	L	*Lepidodendron sensu lato* (Chaloner 1953a)
Pseudovalvisisporites nigrozonalis	0	—		—

illaria, by its association with fusain-rich horizons and clastic input, was placed in a dry and/or clastic habitat. Phillips (1979) argued that dispersal of the megaspores with sporangial wall in attachment reduced dessication and was an adaption to drier habitats. The restriction of *Tuberculatisporites* to the clastic sediments above the seams is further supportive of a clastic environment for *Sigillaria*. *Sporangiostrobus* has no apparent adaptations for a standing water environment. The three layered sporangial wall and broad internal water absorbent cortex described by Remy & Remy (1975), were considered by Butterworth (1977) to be adaptations for ombrotrophic conditions. The megaspore, *Z. brasserti*, which characterized Phase 4 was restricted in occurrence to the presence of durites and trimacerites which are also interpreted as representing ombrotrophic conditions.

The postulated 'environmental' control on the distribution of megaspores would, therefore, appear to be consistent with data derived from the parent plants. One of the most notable observations on coal ball assemblages, however, is that the parent vegetations of Phases 2, 3 and 4 are generally rare components. Phillips & DiMichele (1981), from a study of vertical profiles of the Herrin No. 6 Coal Member, recorded that *Lepidodendron*, *Lepidophlois* and *Paralycopodites* accounted for over 90% of the lycopod peat volume. These lycopods are represented only in Phases 0 and 1 by *C. giganteus* and *L. rugosa*. This suggests that coal balls formed in waterlogged anaerobic conditions and the plants within them represent only the communities adapted to such environments. Coal ball vegetation should therefore be regarded as a special case for preservation rather than be used to represent the entire coal flora (see Scott & Rex 1985).

Pierart (1968), from a study of megaspores and miospores of the Beeringen 70 seam of Belgium, recognized two main associations of spores which he related to relative wetness. *Cystosporites*, *Triangulatisporites*, *Lagenicula* and *Tuberculatisporites* were associated with 'wet' environments and *Zonalesporites* and *Setosisporites* to less wet conditions; no attempt was made to place the spores into phases. Scott (1978) studied the distribution of megaspores through the Top Haigh Moor Seam (Westphalian B) from 1-cm thick samples of coal taken at 10-cm intervals. No details of the coal lithologies or petrology were given. Compared to the Low Barnsley Seam the profile obtained was species-poor, but the following sequence of megaspore dominance was observed; *Laevigatisporites glabratus* (Zerndt), *T. mamillarius*, *Triangulatisporites triangualatus* (Zerndt), *V. auritus*, *S. hirsutus* and finally *Z. brasserti*. *L. rugosa* was poorly developed, and as in the Beeringen 70 seam the genus encountered at the base of the seam was *Laevigatisporites*. The sequence of the succession of the other spores however follows the pattern observed from the Low Barnsley Seam with the exception of the dominance of *T. mamillarius* near the base of the seam. Scott & King (1981) used coarse interval sampling to plot the distribution of megaspores through the Nether Lount Seam of Leicestershire. They noted a division of the seam; *Lagenicula* and *Cystosporites* at the base, *Setosisporites* in the middle and then *Zonalesporites* associated with durain at the top of the seam. With the exception of *Triagulatisporites*, which was seldom encountered, Scott & King's observations broadly follow the phasal succession of the Low Barnsley Seam. A comparison of the Low Barnsley Seam with other profiles demonstrates that there are similarities in the pattern of floristic development. The significance of differences which do occur, e.g. *Laevigatisporites* at the base of the seam replacing *L. rugosa*, or the absence of particular phasal elements is not yet evident. Further studies on other seams and a series of profiles from an individual seam may demonstrate the consistency of floristic development within seams and establish the use of megaspore profiles from seam correlation.

Conclusions

The use of petrological discontinuities as guides for the subdivision of a seam for palynological samples are unsatisfactory; a given petrology cannot be used to predict the palynology accurately nor can phase changes be related to any given petrological 'markers'. This emphasizes the need for close incremental sampling of a seam in order to recognize valid spore phases. However, phase changes can often be related to changing petrological trends.

Megaspore assemblages are a useful additional tool in the interpretation of lycopod distribution and palaeoecology. This study of the Low Barnsley Seam illustrated a phasic sequence of megaspores from the base to the top of the seam. Three such sequences or successions were recognized. As suggested in the literature, the major control on phasal change is inferred to be one of changing water table as the peat developed into a raised bog. This results in changing edaphic conditions such as nutrient supply, pH and oxygen availability and, ultimately, vegetation changes.

More phases were identified by the examination of megaspores than have previously been recognized using miospores. This suggests that megaspores may offer a basis for detecting a more subtle pattern of change, possibly due to more localized distribution. Of the six phases recognized, two are particularly noteworthy; Phase 0 probably represents a colonizing community in the initial stages of peat development, being associated with carbominerites at the base of the seam. Phase 2 represents an important association of non-arborescent lycopods that has not been recognized on the basis of miospore studies. Finally megaspores can provide information on lycopod associations in environments not suited for coal ball formation.

Acknowledgements: I wish to thank British Coal for the donation of the Low Barnsley Seam Core, Professor D. M. Murchison and Dr J. M. Jones of the Organic Geochemistry Unit, University of Newcastle Upon Tyne for assistance and facilities for the petrological examination of the coals; Mr N. Holloway for the preparation of the coal blocks and Mr C. Hildrew and Mrs C. Flood for the preparation of the diagrams. I would like to thank Drs A. C. Scott and A. Buckley for commenting on the manuscript and NERC for the receipt of a research studentship.

References

Abbott, M. L. 1963. Lycopod fructifications from the Upper Freeport No. 7 in Southeast Ohio. *Palaeontographica B*, **112**, 93–118.

Anderson, J. A. R. 1964. The structure and development of the peat swamps of Sarawak and Brunei. *Journal of Tropical Geography*, **18**, 7–16.

—— & Müller, J. 1975. Palynological study of a Holocene peat bog and a Miocene coal deposit from N.W. Borneo. *Review of Palaeobotany and Palynology*, **19**, 291–351.

Balbach, M. K. 1966. Palaeozoic Lycopsid fructifications II. *Lepidostrobus takhtajani* North America and Great Britain. *American Journal of Botany*, **53**, 275–283.

Butterworth, M. A. 1977. *Palynology and Coal*. The 139th Annual Meeting of the British Association for the Advancement of Science, University of Aston.

Chaloner, W. G. 1951. On *Spencerisporites* gen. nov and *S. karczewskii* (Zerndt) the isolated spores of *Spencerites insignis* Scott. *Annals and Magazine of Natural History*, **4**, 861–873.

—— 1952. On *Lepidocarpon waltoni*. sp.N., from the Lower Carboniferous of Scotland. *Annals and Magazine of Natural History*, **5**, 572–582.

—— 1953a. On the megaspores of four species of *Lepidostrobus*. *Annals of Botany*, **17**, 263–273.

—— 1953b. On the megaspores of *Sigillaria*. *Annals and Magazine of Natural History*, **5**, 572–582.

—— 1954. Notes on the spores of two British Carboniferous Lycopods. *Annals and Magazine of Natural History*, **7**, 81–91.

—— 1958. *Polysporia mirabilis* Newberry, a fossil lycopod cone. *Palaeontology*, **32**, 199–209.

Corsin, P. M., Coquel, R. & Loboziak, S. 1967. Sur la composition palynologique verticale et horizontale d'un cycle sedimentaire du terrain houiller. *Palaeogeography, Palaeoclimatology, Palaeoecology*, **3**, 167–184.

DiMichele, W. A. 1980. *Paralycopodites* Morey and Morey, from the Carboniferous of Euramerica—a reassessment of generic affinities and evolution of *Lepidodendron brevifolium* Williamson. *American Journal of Botany*, **67**, 1466–1476.

—— 1985. *Diaphorodendron*, gen. nov., a Segregate from *Lepidodendron* (Pennsylvanian Age). *Systematic Botany*, **10**, 453–458.

——, Mahaffy, J. F. & Phillips, T. L. 1979. Lycopods of Pennsylvanian age coals: *Polysporia*. *Canadian Journal of Botany*, **57**, 1740–1753.

—— & Phillips, T. L. 1985. Lycopod reproduction and Paleoecology in a coal-swamp environment of late Middle Pennsylvanian Age (Herrin Coal, Illinois, U.S.A). *Review of Palaeobotany and Palynology*, **44**, 1–26.

Fischer, G. 1952. Studien uber den biologischen Abbau des Lignins durch Mikro-organisem. *Diss*. Techn Hochsch, Stuttgart.

Habib, D. 1968. Spore and pollen palaeoecology of the Redstone Seam (Upper Pennsylvanian) of West Virginia. *Micropalaeontology*, **14**, 199–220.

Haquebard, P. A., Birmingham, T. F. & Donaldson, J. R. 1967. Petrography of Canadian coals in relation to environment of deposition. *Proceedings of the Symposium of Science and Technology of Coal*, Ottawa, pp. 84–97.

—— & Donaldson, J. R. 1969. Carboniferous coal deposition associated with flood plain and limnic environments in Nova Scotia. *In:* Dapples, E. C. & Hopkins, N. E. (eds) *Environments of Coal Deposition*, pp. 143–191. Special Paper of the Geological Society of America 114.

Haseldonckx, P. 1977. The palynology of a Holocene marginal peat swamp in Jahore, Malaysia. *Review of Palaeobotany and Palynology*, **24**, 227–238.

Hoskins, J. H. & Abbott, M. L. 1956. *Sellaginellites crassicinctus*, a new species from the Desmoinsian series of Kansas. *American Journal of Botany*, **43**, 36–46.

International Committee for Coal Petrology 1963. *International Handbook of Coal Petrography*, 2nd edn. Centre nationale de la Recherche scientifique, Paris.

Karmasin, K. von 1952. Deutung des Facieswechel in der Frozen Erda und Aginir auf Grund mikropetrographischer Schitprobenuntersuchungen. *Bergbau Archiv*, **13**, 74–100.

Leisman, G. A. 1970. A petrified *Sporangiostrobus* and its spores from the middle Pennsylvanian of Kansas. *Palaeontographica B*, **129**, 166–177.

—— & PHILLIPS, T. L. 1979. Megasporangiate and microsporangiate cones of *Achlamydocarpon varius* from the mid-Pennsylvanian. *Palaeontographica B*, **168**, 100–128.

LITTKE, R. 1986. Petrological studies on Carboniferous coal-seams from West Germany. *Abstract of the International Symposium on Coal and Coal-bearing Strata*, p. 31. Royal Holloway and Bedford New College, London.

MCCABE, P. J. 1984. Depositional environments of coal-bearing strata. *In:* RAHMANI, R. A. & FLORES, R. M. (eds) *Sedimentology of Coal and Coal-bearing Sequences*, pp. 13–42. Special Publication of the Association of Sedimentologists 7. Blackwell Scientific Publications, Oxford.

MARSHALL, A. E. & SMITH, A. H. V. 1965. Assemblages of miospores from some Upper Carboniferous coals and their associated sediments in the Yorkshire coalfield. *Palaeontology*, **7**, 656–673.

NAVALE, G. K. B. 1964. Miospore patterns in Jura coals. *Palaeobotanist*, **13**, 1–17.

PHILLIPS, T. L. 1979. Reproduction of heterosporous arborescent lycopods in the Mississippian-Pennsylvanian of Euramerica. *Review of Palaeobotany and Palynology*, **27**, 239–289.

—— & DIMICHELE, W. A. 1981. Palaeoecology of Mid Pennsylvanian age coal swamps in Southern Illinois Herrin coal member at Sahara Mine No. 6. *In:* NICHOLAS, K. J. (ed.) *Palaeobotany, Palaeoecology and Evolution*, Vol. 1. Praegen, New York.

—— & PEPPERS, R. A. 1984. Changing patterns of Pennsylvanian coal swamp vegetation and implications of climatic change on coal occurrence. *International Journal of Coal Geology*, **3**, 205–255.

PHILLIPS, T. L., PEPPERS, R. A. & DIMICHELE, W. A. 1985. Stratigraphic and interregional changes in Pennsylvanian coal swamp vegetation: environmental inferences. *International Journal of Coal Geology*, **5**, 43–109.

PIERART, P. 1968. Les associations des microspores et des megaspores dans une couche (couche 70 Beeringen) du Westphalian A superieur de la Campine (Belgique). *Review of Palaeobotany and Palynology*, **7**, 275–283.

PIGG, K. B. & ROTHWELL, G. W. 1983. *Chaloneria* gen. nov., Heterosporous lycophytes from the Pennsylvanian of North America. *Botanical Gazette*, **144**, 132–147.

REMY, W. & REMY, R. *Sporangiostrobus puertollanensis* n. sp. und *Puertollania sporangiostrobifera* n. gen., n. sp. aus dem Stefan von Puertollano, Spanien. *Argumenta Palaeobotanica*, **4**, 13–29.

SCHOPF, J. M. 1952. Was decay important in the origin of coal? *Journal of Sedimentary Petrology*, **22**, 61–69.

SCOTT, A. C. 1978. Sedimentological and ecological control on Westphalian B plant assemblages from West Yorkshire. *Proceedings of the Yorkshire Geological Society*, **41**, 461–508.

—— & KING, G. R. 1981. Megaspores and coal facies; an example from the Westphalian A of Leicestershire, England. *Review of Palaeobotany and Palynology*, **34**, 107–113.

—— & REX, G. M. 1985. The formation and significance of Carboniferous coal balls. *Philosophical Transactions of the Royal Society of London B*, **311**, 107–113.

SEN, J. 1958. Notes on the spores of four Carboniferous lycopods. *Micropaleontology*, **4**, 159–164.

SMITH, A. H. V. 1957. The sequence of miospore assemblages associated with the occurrence of crassidurite in coal seams of Yorkshire. *Geological Magazine*, **94**, 345–363.

—— 1962. The palaeoecology of Carboniferous peats based on the miospores and petrography of Bituminous Coals. *Proceedings of the Yorkshire Geological Society*, **33**, 423–474.

—— 1968. Coal and seam profiles. *In:* MURCHISON, D. G. & WESTOLL, T. S. (eds) *Coal and Coal-bearing Strata*, pp. 31–40. Oliver & Boyd, London.

STACH, E. 1952. Die Vitrit-Durit Mischungen in der pertrographischen Kohlenanlyse. *Brennstof-Chemie*, **33**, 361–370.

STYAN, W. B. & BUSTIN, R. M. 1983. Petrography of some Frazer River Delta peat deposits; coal maceral precursors in temperate–climate peats. *International Journal of Coal Geology*, **2**, 321–370.

TEICHMÜLLER, M. 1962. Die Genese der Kohle. *Compte rendu 4th Congress of the stratigraphy and geology of the Carboniferous*, Heerlen, 1958, **3**, 699–722.

—— & TEICHMÜLLER, R. 1982. The geological basis of coal formation. *In:* STACH, E., MACKOWSKY, M.-TH., TEICHMÜLLER, M., TAYLOR, G. H., CHANDRA, D. & TEICHMÜLLER, R. (eds) *Stach's Handbook of Coal Petrology*, pp. 5–82. Gebruder Borntraeger, Berlin.

—— & THOMSON, P. W. 1958. Vergleichende mikroskopische und chemische Untersuchungen der wichigsten Fazies Typen im Hauptfloz der niederrheinischen Braunkohle. *Fortschritte in der Geologie von Rheinland und Westfalen*, **2**, 573–598.

TIMOFEYEV, P. P. 1955. Conditions of formation of genetic types of coal and their connection with cycles of sedimentational environment in the Dombass. *Dolkady Academiya Nauk SSSR*, **102**, 809–812.

WELTE, E. 1952. Uber die Enstehung von Huminsauren und Wege ihrer Reindarstellung. *Zeitschrift fur Planzenernahung und Dungung und Bodenkunde*, **56**, 105–139.

K. M. BARTRAM, Department of Geology, Royal Holloway and Bedford New College, University of London, Egham Hill, Egham, Surrey TW20 0EX, England.

Genesis of the Warwickshire Thick Coal: a group of long-residence histosols

I. M. Fulton

ABSTRACT: The Warwickshire Thick Coal is of Westphalian B age, and it developed at the southern margin of the Pennine Basin in central England. It usually consists of three to five leaves of coal which can total up to 8.5 m in thickness. Palynological analysis of the leaves has revealed the presence of four miospore types which occur in similar successions in each leaf. These coals are interpreted as long-residence histosols. Comparable successions have been recognized by other authors and a variety of mechanisms advanced for their formation. The model proposed to account for their origin is change in vegetation due to variations in water table (hydroseral growth), which occur during the development of ombrotrophic bogs similar to those found in the Far East. Many inter-related factors are believed to affect water table variation including climate, topography and raising of the bog surface (the latter controlled by both differential rate of subsidence and accumulation of plant debris). It is probable that the model proposed may be applicable to other coal seams, and this has implications for coal exploration. Estimation of regional rates of subsidence may help in the location of thick long-residence histosols, and examination of miospore cycles within leaves of coal may be useful in predicting the proximity of seam splits.

Keywords: coal, miospore, ombrotrophic bog, palaeoecology, exploration

Previous work on the Warwickshire Thick Coal (Thick Coal) has been concerned mainly with its tectonic structure, and correlation both within and outside the Warwickshire Coalfield (Howell 1859, Browne 1907, Shotton 1928, Barrow *et al.* 1919, Eastwood *et al.* 1923, Paget 1936, 1937, Mitchell *et al.* 1942, Smith & Butterworth 1967, Cope & Jones 1970). Correlation was generally achieved using lithostratigraphic means. However, Paget (1936, 1937), Butterworth (1956) and Smith & Butterworth (1967) used miospores for correlation of the seams comprising the Warwickshire Thick Coal. Detailed information on the vertical distribution of miospores within individual seams was published by Smith (1957, 1962, 1964) who worked on coal seams in the Yorkshire Coalfield, and by Corsin *et al.* (1967) in France and Habib (1966) in America. Recently much research has been carried out into the sedimentary environments in which modern peat, a histosol, has developed both in the USA, e.g. Coleman (1966) and Cohen (1974), and in the Far East, e.g. Anderson (1964, 1973), Anderson & Muller (1975) and Coleman *et al.* (1970).

Considerable exploration for coal by surface boreholes has been carried out by British Coal in the southern part of the Warwickshire Coalfield since the paper written by Cope & Jones (1970), so that the southern limit of the Thick Coal can now be defined. Palynological investigation of the Thick Coal was undertaken in one of these new boreholes (Longmeadow Wood, SP 2774/1614) so that a palaeoecological interpretation could be made. This has enabled comparison with previous interpretations of coal seams in Yorkshire and the USA. The palaeoecological interpretation has also been integrated with sedimentological data from the new boreholes, in order to understand both the depositional setting and mechanisms controlling the distribution of the leaves of the Thick Coal.

The Warwickshire Coalfield

Geographical and structural setting

The Warwickshire Coalfield, within which the Thick Coal occurs, is located almost in the centre of England between Birmingham, Nuneaton and Banbury, and is surrounded by other coalfields (Fig. 1). It has the shape of an elongate triangle the apex of which lies to the north and contains sediments of Westphalian A and B age. In the extreme NW the Coalfield is bounded by faulting with a 'Church Stretton' NE–SW trend, whilst in the west it is limited by the N–S trending Western Boundary Fault. The Coalfield is bounded in the NE by faulting with a 'Charnoid' NNW–SSE trend, and to the SE it is limited by both outcrop and incrop beneath sediments of Triassic age. In the SE both Westphalian A and B sediments attenuate to zero, against the Barbant Barrier but in the SW thin Westphalian A and B sediments continue into the Oxfordshire Coalfield.

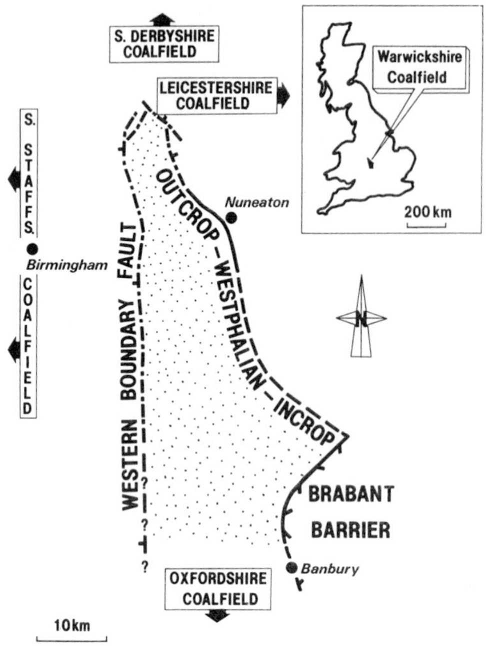

FIG. 1. Location and boundaries of the Warwickshire Coalfield, UK.

Palaeogeography and palaeoclimate

The position of the Warwickshire Coalfield on the southern edge of the Pennine Basin was recognized by Trueman (1947) and Wills (1948). Recently Guion (1987) produced a palaeogeographical map of the upper Westphalian A non-marine strata, which showed an embayment in the Wales–Barbant Barrier corresponding to the Warwickshire Coalfield. More recently NCB exploration data were used by Fulton & Williams (1986) to show that the southern edge of the Pennine Basin lay much further south during late Westphalian A and Westphalian B times (Fig. 2), and there may have been a link through the Wales–Brabant Barrier between the Pennine and southern England Basins. The edge of the Brabant Barrier in the SE of the Warwickshire Coalfield corresponds to the zero isopachyte for Westphalian A and B strata plotted by Fulton & Williams (1986).

Palaeomagnetic data have allowed the position of modern continents to be plotted through time (Smith *et al.* 1973, Ziegler *et al.* 1981). During Westphalian B times Britain lay on the eastern margin of a combined North American–N European continental plate. Recent palaeomagnetic evidence derived from Westphalian C rocks of the Warwickshire Coalfield has shown that the Pennine Basin lay within a few degrees of the equator (Turner *et al.* 1985). Reconstruction of ancient continents has permitted the modelling of atmospheric circulation, so that predictions of palaeoclimates can be made (Parrish 1982, Rowley *et al.* 1985). These have shown that the area of the Pennine Basin was likely to have been humid during Westphalian times with the possibility of heavy rainfall on the eastern edge of the continental plate which it occupied. Study of the tissues of Carboniferous arborescent lycopods indicates that some underwent sustained, rapid, uninterrupted growth (DiMichele & Phillips 1985), suggesting a climate with minimal seasonal change. From the above evidence the climate during Westphalian B times within the Pennine Basin is considered to have been tropical–pure equatorial (*sensu* Tricart 1972). This climate consisted of four alternating wet and dry seasons, in which the two dry seasons received enough precipitation to sustain the plant life which grew in the preceding wet seasons.

Environments of deposition

Westphalian B sediments in the Warwickshire Coalfield reflect deposition on an upper delta plain. Although no single, modern, geographical location displays all the aspects of the upper delta plain setting of the Pennine Basin, some of the characteristics are to be found in the modern

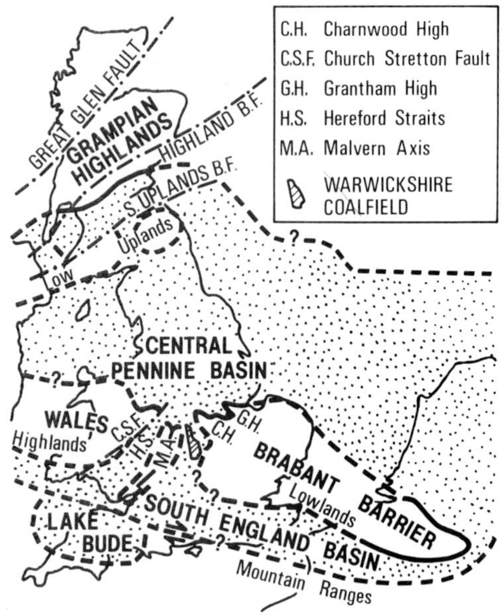

FIG. 2. Upper Westphalian A palaeography, for strata immediately preceding the Vanderbeckei Marine Band (after Guion 1987).

deltaic deposits of the SE USA (Coleman 1966, Frazier & Ozanik 1969, Cohen 1974) whilst others occur in the coastal plain deposits of the Far East (Anderson 1964, Coleman *et al.* 1970, Anderson & Muller 1975). The various environments within which deposition took place are similar to those described previously by authors working on Westphalian B sediments of the Pennine Basin (Elliott 1968, Scott 1978, 1979, Guion 1978, Haszeldine 1981, Fielding 1982, Williams 1986) and can be divided into four broad categories in order of volumetric importance (Table 1).

(1) *Lacustrine*

Sediments formed in this environment can be recognized by a suite of lithofacies which coarsen upwards from dark, massive or parallel laminated claystones often containing non-marine bivalves, to fine grained sandstones. They represent the products of a combination of many depositional processes by which the lakes were filled, among which are overbank flood, crevasse splay, and lacustrine distributary mouth bar deposits (Scott 1979, Fielding 1984a, Haszeldine 1984, Guion 1984).

(2) *Mire*

These deposits, often termed palaeosols, are characterized by features which demonstrate that the surface was either terrestrial or capable of supporting vegetation which may have been telematic or terrestrial. Although sometimes difficult to apply in this work, the FAO/UNESCO (1974) system is used in the classification of palaeosols. The Working Group (1971) on palaeosols believed that modern soil classification systems can be adapted to describe palaeosols. Palaeosols are divided into two types dependant on organic content (Table 1).

Those palaeosols with low organic contents, i.e. siliciclastic palaeosols (commonly called seatearths) can be recognized by a number of features including *in situ* roots, development of sideritic concretions associated with roots and pedogenic profiles. They have long been recognized by other workers in Westphalian sediments of the Pennine Basin (Elliott 1968, Guion 1978, Fielding 1982, Besly 1983). Many of these soils may be described as alluvial, *sensu lato* (Young 1976), and they have been classified under various headings mainly according to the processes required to form them. Those recognized in the Westphalian B sediments of the Warwickshire Coalfield have been interpreted as fluvisols, gleysols and podzols (FAO/UNESCO 1974).

In situ coal can be described as a floriclastic palaeosol, which is distinguished from siliciclastic palaeosols by its higher organic content (ash content less than 40% by weight on an air-dried basis, NCB 1972). Modern organic-rich soils have been classified as histosols (FAO/UNESCO 1974) on the basis of organic content and thickness. Unfortunately the definition of coal adopted by the NCB (1972) does not correspond to that of a histosol (FAO/UNESCO 1974) or to other classification systems of organic-rich soils and their sub-divisions into peat and muck (see review in Kearns & Davison 1983, see also Moore, *this volume*). However, in this work, the pedogenic term histosol will be used where appropriate to describe coal. When describing leaves of coal with less than 10% ash by weight, on an air-dried basis (a-db) the term low-ash histosol will be used (Table 1). Low-ash histosols can be divided further into long and short-residence types depending on their developmental history. Those whose evolution has included the development of dull coals dominated by *Densosporites* spp. are termed long-residence histosols, whilst those without this stage are short-

TABLE 1. *Depositional environments and their subdivisions*

1A: environments of deposition	*1B: subdivision of mire deposits*		
(1) Lacustrine	Floriclastic palaeosol (coal)		Siliciclastic palaeosol (dirt)
(2) Mire			>40% ash
(3) Channel	Low-ash histosol <10% ash	High-ash histosol 10–40% ash	
(4) Marine	Long-residence	Short-residence	

residence histosols. However, short-residence histosols are more usually of the high ash type. Both long- and short-residence histosols can be found in the Warwickshire Coalfield (Smith & Butterworth 1967).

(3) *Channel*

These range in size from small straight crevasse channels a few metres deep and less than 50 m wide to major sinuous channel belts (*sensu* Fielding 1986) of which the largest delineated is up to 1.2 km wide. They have erosive bases which may scour the underlying sediments for several metres and are often filled with fine grained sandstone. They have been recognized in sediments of Westphalian age within the Pennine Basin by Elliott (1965), Guion (1978), Scott (1978), Haszeldine (1983), Fielding (1986) and Williams (1986).

(4) *Marine*

Three clearly defined marine bands containing a marine/brackish phase (*sensu* Calver 1968) occur in sediments of Westphalian A and B age in the Warwickshire Coalfield. These are the Subcrenatum Listeri and Vanderbeckei Marine Bands, the former two being limited to the extreme north of the Coalfield, whilst the latter, with one exception (Fulton & Williams, *in preparation*) is confined to the northern third of the Coalfield.

The Warwickshire Thick Coal

Stratigraphy and geographical extent

The Thick Coal is of Westphalian B age, formed by a combination of many leaves of coal, although unlike the implication in Smith (1968), only rarely are they joined to form a single leaf. (Leaf is the term used to describe a single bed of coal separated from others by siliciclastic sediment. Seams are those single leaves or combinations of leaves and siliciclastic sediments, the names of which have mostly been inherited during mining.) For the purpose of this work those leaves forming the Thick Coal must be within 0.3 m of each other, or else they are considered separate leaves, lying therefore outside the area of the Thick Coal. Normally the Thick Coal is composed of three to five leaves totalling up to 8.5 m, separated by siliciclastic palaeosols (e.g. Longmeadow Wood borehole, four leaves; Fig. 3). The seams which form the Thick Coal were named following their mining in the north of the Warwickshire Coalfield and the majority are listed by Mitchell *et al.* (1942). Cope & Jones (1970) believed that two further seams should be included in the Thick Coal. First, they indicated that Mitchell *et al.* (1942, fig. 3) had omitted the High Main seam from their definition of the Thick Coal (although this seam had been included in the Slate seam at Exhall and Newdigate Collieries). Second, they included the Smithy seam, but in this paper it is excluded from the Thick Coal, because the area over which it is within 0.3 m of the lowest leaf of the Thick Coal is small and irregular, and its chemical properties are significantly different from the other leaves of the Thick Coal. The leaves of the Thick Coal which are all long-residence histosols usually have average sulphur contents of less than 4% by weight and average ash contents of less than 9% by weight (a-db) with most of the subsections not exceeding 5% by weight (a-db). These ash values would be even lower if the contribution made by oxidized pyrite to them were removed. The lowest seam (the High Main) of the Thick Coal is the uppermost seam of the Modiolaris Chronozone in Warwickshire and the other seams (Bottom Nine Feet to Thin Rider) occur in the Lower Similis–Pulchra Chronozone. Towards the centre of the Pennine Basin, in Yorkshire, about 100 m of siliciclastic sediment separates the Haigh Moor and Kents Thin seams which may temporally be the lateral equivalents of the lower and upper seams of Thick Coal.

The area covered by the Thick Coal is just over 100 km^2, and its boundaries are defined where siliciclastic palaeosols between the leaves increase in thickness beyond 0.3 m (Fig. 3). Thus an increase in thickness between Leaves 2 and 3 forms the northern and southern boundaries, that between Leaves 1 and 2 forms the western boundary, and the eastern boundary is formed by a combination of increase in thickness between Leaves 1 and 2, and Leaves 4 and 5. It must be noted that the split leaves of the Thick Coal extend for many square kilometres beyond the boundaries of the Thick Coal. To the SE the leaves of the Thick Coal become thin and are separated by increasing thicknesses of brown siliciclastic palaeosols, whilst in the north they also become thin but are separated by a much greater thickness of sediments deposited in a variety of environments.

Sedimentology and palynology at Longmeadow Wood borehole

At the Longmeadow Wood borehole the Thick Coal comprises four leaves numbered 1 to 4 from the base (Fig. 4) which vary in thickness from

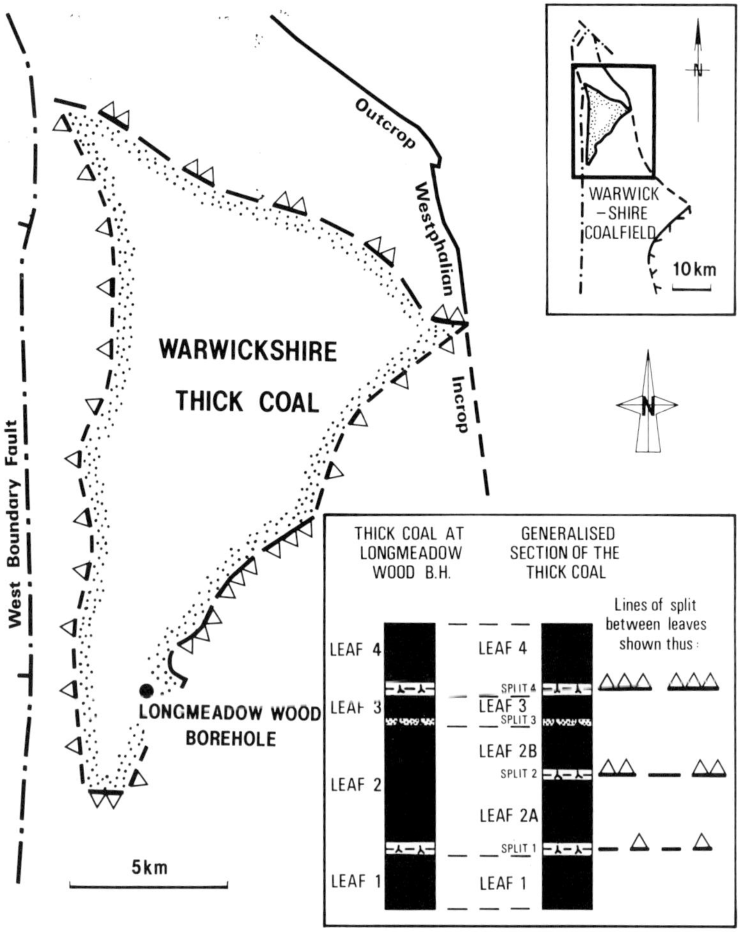

FIG. 3. Geographical extent of the Warwickshire Thick Coal. The lines of split between three of its leaves are used to define its extent.

0.34 to 2.09 m. Below Leaf 1 and between the other leaves a variety of siliciclastic palaeosols exist. These are formed of claystone except the palaeosol between Leaves 2 and 3 which is composed of fine grained sandstone. Above Leaf 1 is a dark grey claystone containing non-marine bivalves. The leaves of the Thick Coal and their associated siliciclastic sediments were sampled for their palynological content. Nine coal lithofacies and the occurrence of pyrite were used to divide the leaves into 104 subsections, with a minimum thickness of 5 mm and a maximum of 120 mm. Each lithofacies is characterized by a different combination of the lithotypes of Stopes (1919) and Cady (1942) with respect to their thickness, percentage and distribution. Siliciclastic subsections are formed by each interleaf bed (except between Leaves 1 and 2 which is divided into three subsections) and beds above and below the Thick Coal.

All subsections were analysed for ash and sulphur content. Using the methods outlined in Smith & Butterworth (1967) the siliciclastic palaeosols and dark grey claystone were macerated for their miospore content using hydrofluoric acid and fuming nitric acid, and a total of 83 coal subsections were macerated using fuming nitric acid. It was hoped that the choice of subsections would cover the range of coal lithofacies present and reflect the variability in chemical content and miospore population.

Systematics

Although nearly 150 miospore species were recognized, only those used in the present work

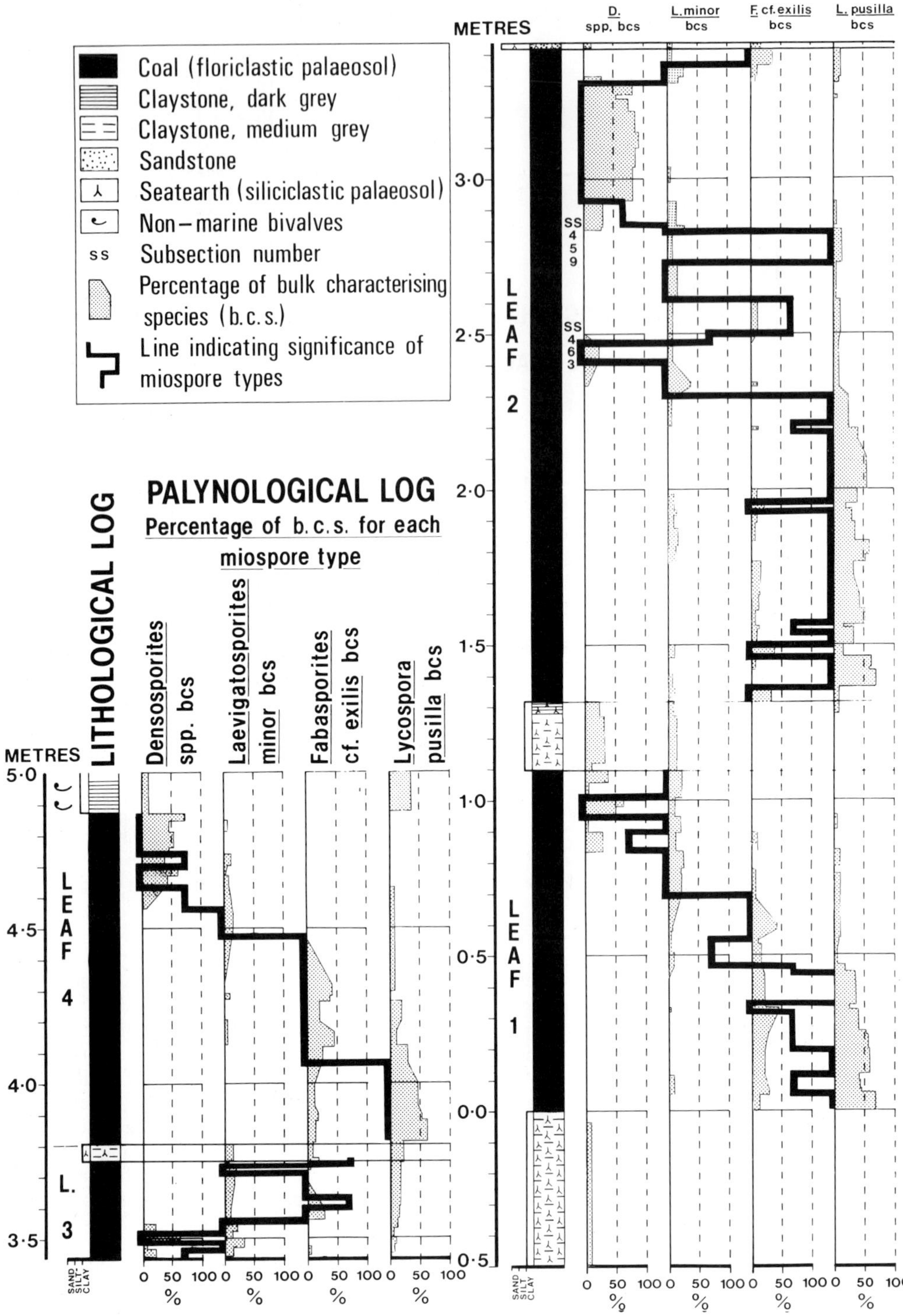

FIG. 4. Distribution of miospore types in the Warwickshire Thick Coal and associated siliciclastic sediments at Longmeadow Wood borehole. Leaves 1 and 4 display a normal miospore succession with a trend of significant miospore types from right to left. Leaf 2 contains two miospore cycles, the first from the base to subsection 459, and the second above to the top of the leaf. Leaf 3 displays a partial miospore cycle.

are treated systematically, principally to facilitate comparison with those recognized by other workers.

Genus *Densosporites* (Berry) Butterworth *et al.* (1964)

Densosporites anulatus (Loose) Smith & Butterworth (1967)

Densosporites cf. *anulatus* (Fig. 5a–d)

Remarks. Differs from *D. anulatus* in that some specimens possess rare grana, coni or verrucae on the distal surface of the cingulum, whilst others may have a thickening of the exoexine in the central area. The cingulum width has a larger range (20–46%) in terms of the mean spore diameter.

Densosporites gracilis Smith & Butterworth 1967 (Fig. 5e–h)

Densosporites sphaerotriangularis Kosanke 1950 (Fig. 5i–l)

Densosporites triangularis Kosanke 1950

Densosporites cf. *triangularis* (Fig. 5m–p)

Remarks. Differs from *D. triangularis* in its smaller size (35–59 μm), rounded triangular shape, and often specimens either possess a laevigate central area or rare minute coni on the distal surface.

Genus *Lycospora* (Schopf, Wilson & Bentall) Somers 1972

Lycospora pusilla (Ibrahim) Schopf, Wilson & Bentall 1944 (Fig. 5q–s)

Remarks. In this present work both *L. pusilla* and *L. pellucida* (Wicher) Schopf, Wilson & Bentall 1944 are recognized (distinguished on the basis of cingulum width) in preference to the revision of the species into two tendencies (Somers 1972). Using the provisional synonymy of Somers (1972) it is apparent that *L. pusilla* may be synonymous with *L. granulata* Kosanke 1950, a species used by several American authors (Ravn 1979, Ravn & Fitzgerald 1982) and *L. subjuga* Bharadwaj 1957, a species used by Peppers (1970).

Laevigatosporites minor Loose 1934 (Fig. 5t, u)

Fabasporites exilis Clendening 1967

Fabasporites cf. *exilis* (Fig. 5v–z)

Remarks. Differs from *F. exilis* in its smaller size (11–22 μm) and sometimes specimens possess a granulate ornament.

Comparison. F. pallidus Sullivan 1964 is often bean-shaped with only occasional crescentic folds.

Analysis of palynological data

The objective of this analysis was to define coal subsections of the leaves of the Thick Coal at Longmeadow Wood borehole in terms of their miospore content, so that a palaeoecological interpretation could be made. This type of analysis was attempted by Smith (1957) who studied five Westphalian B seams in the Yorkshire Coalfield in the centre of the Pennine Basin. He concluded that three phases could be recognized corresponding to high numbers of two species of the genus *Lycospora*, and one species from each of the genera *Laevigatosporites* and *Densosporites*, and that these phases occurred sequentially. Other authors have carried out similar analyses in France (Corsin *et al.* 1967), Canada (Hacquebard & Donaldson 1969) and Russia (Koval'chuk & Uziyuk 1973) but only recorded miospore genera. Use of miospore genera is likely to make the palaeoecological interpretation more difficult and it increases the possibility of grouping miospore species whose parent plants may have existed in different habitats. An advance was made by Smith (1962) working in the Yorkshire Coalfield, in the recognition of four phases, each characterized by an association of miospore species, although often one species was dominant. A similar analysis using miospore assemblages was carried out in the USA by Habib (1966).

The use of miospore associations is the best means of defining coal subsections by palynological methods, because it helps to ensure that the assemblage is autochthonous. However, it was decided to determine whether definition could be achieved using only a few miospore species. Twenty-one species occur abundantly (*sensu* Smith & Buttterworth 1967), i.e. greater than 10% in the leaves of Thick Coal, but all of the coal subsections can be defined using only seven of these species, viz. *Densosporites* cf. *anulatus* (Loose) Smith & Butterworth 1967; *D. gracilis* Smith & Butterworth 1967; *D. sphaerotriangularis* Kosanke 1950; *D.* cf. *triangularis* Kosanke 1950; *Lycospora pusilla* (Ibrahim) Schopf, Wilson & Bentall 1944; *Laevigatosporites minor* Loose 1934; and *Fabasporites* cf. *exilis* Clendening 1967. Because they make up the bulk of the miospore population they are called bulk characterizing species (bcs). These seven species can be arranged to give four miospore types. Three types are formed from individual bcs viz. *L. pusilla* bcs, *F.* cf. *exilis* bcs and *L. minor* bcs. When the distribution of the four species of *Densosporites* are examined it is found that they often occur together. For this reason they are grouped to give the fourth miospore type *Densosporites* spp. bcs.

Using the total miospore data gathered from the Thick Coal, relative percentages of the four miospore types were plotted against subsections in which they occurred (Fig. 4). Where subsections have not been counted, values have been interpolated from adjacent counted subsections. From Fig. 4 it can be seen that each coal subsection and many siliciclastic subsections

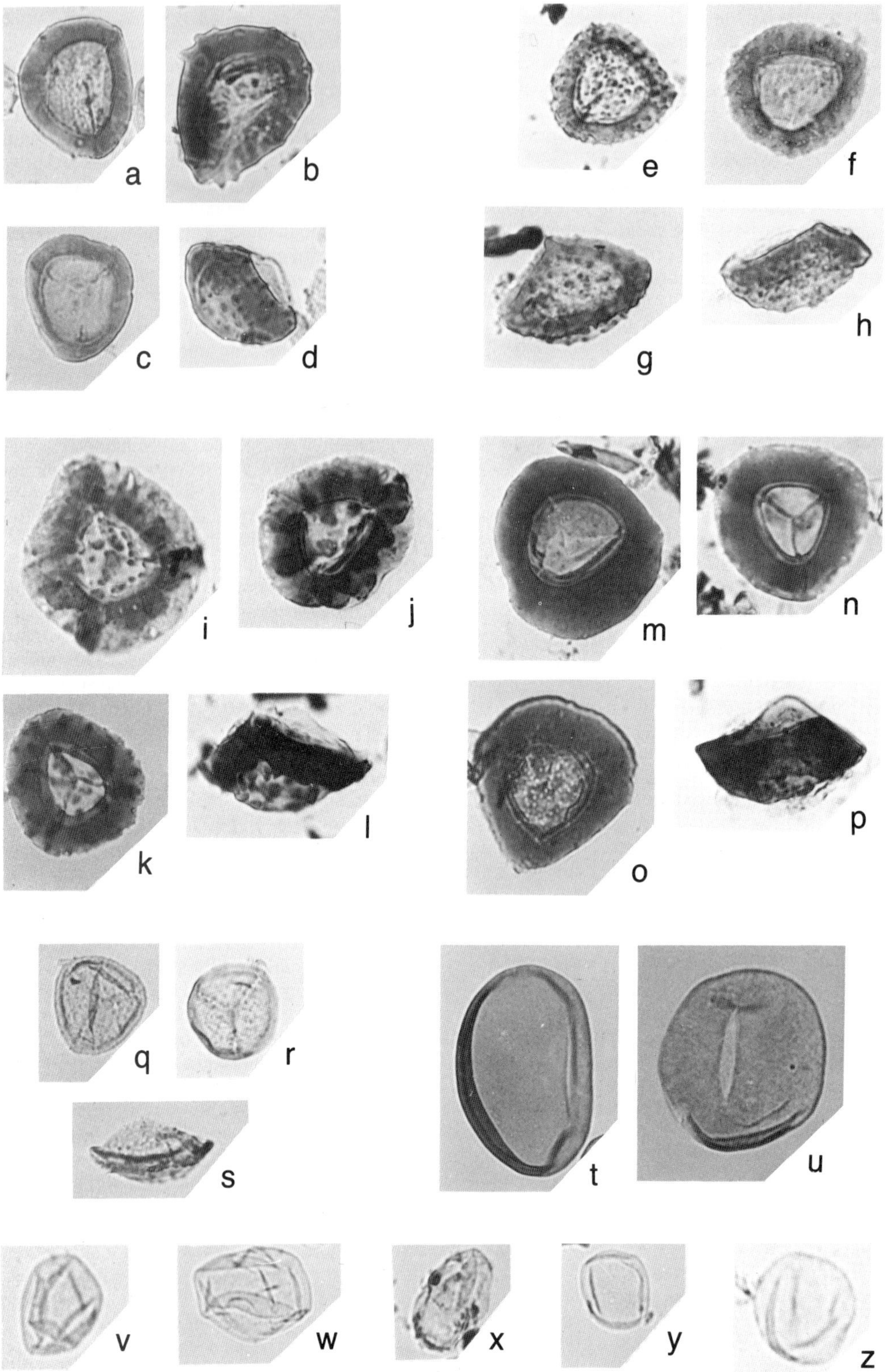
a
b
e
f
c
d
g
h
i
j
m
n
k
l
o
p
q
r
s
t
u
v
w
x
y
z

have at least one of the miospore types exceeding 10%. It is therefore necessary to devise a test of significance which can be applied to each subsection to determine which miospore type defines it. Because it is likely that different plants produced different numbers of miospores, and their dispersal pattern and rate of destruction before and after incorporation within the coal was also different, it is probable that the significance of each miospore type is governed by a different relative percentage. This relative percentage is calculated by averaging those subsections containing in excess of 10% (lower values are excluded because some may be allochthonous) for each miospore type, and using this value as a test of significance viz. 47% for *Densosporites* spp. bcs; 18% for *L. minor* bcs; 22% for *F.* cf. *exilis* bcs, and 32% for *L. pusilla* bcs. In Fig. 4 a thick line has been drawn beside each miospore type which is higher than the average value and therefore considered significant. Where two miospore types are significant the line has been drawn between them, and where a subsection contains no significant miospore type the line has been drawn nearest or between the most significant miospore types.

Examination of Leaves 1 and 4 (Fig. 4) reveals a trend from right to left, commencing with subsections defined by *L. pusilla* bcs (the initial miospore type) and passing upward into those defined by *F.* cf. *exilis* bcs, *L. minor* bcs and *Densosporites* spp. bcs (the climax miospore type). These leaves display what can be described as a 'normal miospore (type) succession'; although first at the base of Leaf 1 there is oscillation between subsections defined by *L. pusilla* bcs and *F.* cf. *exilis* bcs, and second at the top of the same leaf there is a return to subsections defined by *L. minor* bcs. Leaves of coal in which all four miospore types are present, usually in a normal miospore succession are termed 'long-residence' histosols. This normal succession compares well (apart from *F.* cf. *exilis* bcs) with the sequential relationship obtained by Smith (1957) of Lycospore to Transition to Densospore phases, and is similar to those observed by Corsin *et al.* (1967) in Westphalian C coals and Habib (1966) in Westphalian D age coals. Smith (1962) recognized a fourth phase, termed Incursion, which was characterized by the association of *Crassispora kosankei* (Potonie & Kremp) Smith & Butterworth 1967 and *Punctatosporites minutus* Ibrahim 1933. Although both of these species occurred abundantly in many subsections within the leaves of the Thick Coal, no association between them has been observed by the author. It is possible that the 'indeterminate thin exines' in the Lycospore phase (Smith 1962) may correspond to the species *F.* cf. *exilis*, although either this species may not have existed in such abundance, or may have been excluded during the preparation of the seams investigated by Smith (1962).

In Leaf 2 there appear to be two 'miospore (type) cycles', the first existing between the base of the leaf and subsection 459, and the second from subsection 459 to the top of the leaf. A cycle is formed when the sequence in a normal miospore succession is mirrored in coal subsections lying directly above it, in the order a, b, c, d, c, b, a. The first cycle commences at the base of the leaf with subsections defined by alternations of *F.* cf. *exilis* bcs and *L. pusilla* bcs, followed in the normal miospore succession by subsections defined by *L. minor* bcs and *Densosporites* spp. bcs to subsection 463. In the subsections immediately above this there are oscillations between various miospore types culminating in *L. pusilla* bcs in subsection 459. From this subsection a second cycle is initiated passing up through *L. minor* bcs to subsections dominated by *Densosporites* spp. bcs towards the top of the leaf. Above this to the top of the leaf a return through *L. minor* bcs to *F.* cf. *exilis* bcs is visible. This cyclic sequence is similar (apart from the recognition of *F.* cf. *exilis* bcs) to that observed by Smith (1962) within two Westphalian B seams from the Yorkshire Coalfield, i.e. The Swallow Wood and Hazel seams.

Leaf 3 is unusual in that it commences with subsections defined by *Densosporites* spp. bcs and *L. minor* bcs but returns through an oscillating sequence to subsections defined by *F.* cf. *exilis* bcs and *L. pusilla* bcs at the top of the leaf.

The siliciclastic palaeosols between the leaves contain moderate percentages of the miospore types which occur in the coal subsections immediately below or above. Between Leaves 1 and 2, where the siliciclastic palaeosol is divided into 3 subsections, a sequence is observable which reflects the bcs in both the under- and overlying leaves. Above Leaf 1 there is no relationship between the bcs in the underlying coal subsections and the overlying siliciclastic subsection. Subsections below Leaf 1 contain only low percentages of the bcs. It is unfortunate that the subsection immediately below the Thick Coal does not contain miospores.

FIG. 5. Bulk characterizing miospore species. All figures × 500, and of proximal surface unless otherwise stated. (a–d) *Densosporites* cf. *anulatus*, (d) equatorial view; (e–h) *D. gracilis*, (h) equatorial view; (i–e) *D. sphaerotriangularis*, (e) equatorial view; (m–p) *D.* cf. *triangularis*, (p) equatorial view; (q–s) *Lycospora pusilla*, (s) equatorial view; (t, u) *Laevigatosporites minor*; (v–z) *Fabasporites* cf. *exilis*.

Palaeoecological and sedimentological interpretation of the Thick Coal

Many of the palaeoecological interpretations of Westphalian coal seams were carried out during the 1960s (Butterworth 1964, 1966, Habib 1966, Habib & Groth 1967, Hacquebard & Donaldson 1969, Smith 1962, 1968). Since that time our understanding of the palaeoecology of modern histosols (Anderson & Muller 1975, Cohen & Spackman 1972, Haseldonckx 1977) and the environments in which they develop (Barber 1981, Cohen 1974, 1984, Coleman *et al.* 1970, McCabe 1984 and Moore & Bellamy 1974) has been considerably extended. With this documentation as a database it is now possible to attempt a more complete palaeoecological analysis of the miospore successions seen in the leaves of the Thick Coal, and to understand the determinants of the spatial and temporal distribution of those coal leaves.

Factors likely to affect miospore successions

The factors believed to influence the creation of miospore successions observed in the four leaves of the Thick Coal at Longmeadow Wood borehole are summarized in the form of a flow diagram (Fig. 6). They can be divided into climatic and edaphic factors, which are discussed with reference to sedimentation in the Pennine Basin, and particularly the formation of the Thick Coal.

(1) *Climatic factors*

Climatic control was proposed by Smith (1962) and endorsed by Chaloner & Muir (1968) to account for part of the sequence of palynological changes detected in Westphalian coal seams. Smith (1962) attributed the shift from his Transition to Densospore phase to an increase in both precipitation and atmospheric humidity. However, if histosol accumulation was more or less continuous at different localities throughout the Pennine Basin (Elliott 1985) and these exhibit similar miospore successions (Smith 1957, 1962, Bartram, *this volume*) it would be impossible for a change of climate to be the only factor controlling these successions within a leaf.

Climatic change is also believed to have occurred slowly during the Westphalian (Phillips & Peppers 1984), and not with the frequency necessary to allow the full development of sequences in a number of seams (Smith 1964) formed over relatively short intervals of geological time. Smith (1962) also suggested that the absence of the Densospore phase in Westphalian C coals in Britain, in contrast to its occurrence in seams of Westphalian A and B age, might be due to inadequate precipitation during this Stage of the Westphalian. This interpretation was supported by Butterworth (1966) who studied the world-wide distribution of densospores. Unfortunately, the evidence presented by Phillips & Peppers (1984) for a drier climate over the N American–N European continental plate during the Westphalian B and C when compared

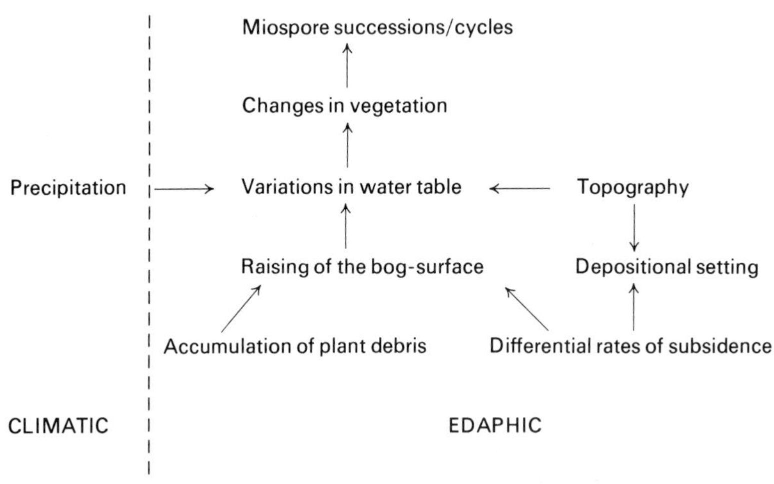

FIG. 6. Factors likely to affect the creation of miospore successions/cycles.

with the Westphalian A and D does not fit the interpretations of Smith (1962) and Butterworth (1966).

Histosols can develop in a variety of climates providing accumulation of plant debris exceeds its degradation. Decay of plant debris proceeds more rapidly under aerobic conditions but these are prevented by waterlogging, so that one of the critical factors in preservation is that part of the hydrological cycle which existed within the histosol. This has been summarized in the form of an equation by Bellamy (1972):

$$\text{Inflow} + \text{Precipitation} = \text{Outflow} + \text{Evapo(transpiration)} + \text{Retention}.$$

From this it may be seen that water may be supplied in two ways, either entirely by precipitation or by a combination of precipitation and groundwater. In this respect climate is partly responsible for the type of histosol which can develop, for only where precipitation is great enough to retain the water table near to the surface of a mire can an ombrotrophic (Du Rietz 1954) bog evolve from a rheotrophic (Moore & Bellamy 1974, p. 72) swamp (see Moore, *this volume*). Both of these types of mire would have been able to develop in the Warwickshire Coalfield because, as shown previously, the climate envisaged for Westphalian B times was tropical–pure equatorial.

(2) *Edaphic factors*

The interpretation of miospore successions within leaves of coal makes the fundamental assumption that the miospores forming them are autochthonous and represent a local population of parent plants. In both European (Tauber 1965) and tropical forests (Anderson & Muller 1975) it has been found that the majority of miospores have only travelled a few hundred metres from the plants which produced them. Changes in flora growing on Westphalian histosols are likely to have taken place because the plants were limited by their ability to cope with changing edaphic conditions.

Edaphic conditions in tropical histosols in the Far East have been studied by Anderson (1964, 1983). Initially he distinguished two types of mire (Anderson 1964): freshwater (rheotrophic) swamps which are subjected to regular flooding, have ash contents greater than 25%, pH values higher than 4.0 and an approximately level surface; and peat (ombrotrophic) bogs which are not subject to flooding, have ash contents less than 25%, pH values lower than 4.0 and a markedly convex surface. It is considered that the deposits of both types of mire can be recognized in Westphalian sediments of the Pennine Basin. Siliciclastic palaeosols and high-ash histosols are believed to have formed in rheotrophic swamps, whereas low-ash histosols formed mainly in ombrotrophic bogs (see Table 1 for properties). Anderson (1983) published the results of a survey of tropical peats of Western Malesia which showed that edaphic factors subject to the greatest variation in ombrotrophic bogs are water table position and the percentage of certain minerals. Variations in the water table during the dry season are minimal at the mire periphery (9–11 cm) and at a maximum in the centre (up to 19 cm), whilst during the wet season the water table may rise above the level of the mire surface at its periphery (Anderson 1964). Results from the central 19 km of a 25 km transect from the bog edge to centre, showed that phosphorous contents were reduced by over half (510 to 203 ppm) whilst potassium contents were reduced by over one-third (360 to 214 ppm). There was little variation in pH (3.3 to 3.6). It is believed that these edaphic variations are sufficient to produce the sequence of six plant communities recognized by Anderson & Muller (1975) in raised mires in NW Borneo, although the predominant control is probably that of water table variation. It should be noted that ash content along the 19 km transect (Anderson 1983) varied from 3.6% by weight towards the edge to 1.2% at the centre.

Rheotrophic swamps do not appear to create a sequence of numerous plant communities but instead produce an alternation of plant communities due to oscillations of wet and dry conditions. Cohen (1974) recognized two types of histosols in the Okefenokee Swamp on the basis of their macrofloral remains. Marsh plants produced histosols that were diagnostic of wet conditions, whilst trees and shrubs produced histosols diagnostic of drier conditions. A similar distribution of vegetation was observed by Spackman *et al.* (1969) in the freshwater environment of the Everglades in southern Florida, but in the coastal environment of the Everglades a separate type of vegetation recognized by Spackman *et al.* (1969) was shown to have formed under marine influence. Previously (Spackman *et al.* 1966) had shown the coastal vegetation to be transgressing on to the freshwater vegetation.

The transgression model above was used by Habib *et al.* (1966) and Habib & Groth (1967) to explain miospore successions in a Westphalian D age coal of Pennsylvania. They believed the change from a *Lycospore* assemblage zone at the base of leaves through intermediate zones to a *Densosporites* assemblage zone at the top could be explained by the increasing influence of salinity on the vegetation. This was because where this full succession occurred, beds of brackish/marine origin rested directly on top of the coal, whereas at other localities without the *Densosporites* assemblage zone the coal was overlain by beds of freshwater origin. The use of this model to explain similar miospore successions seen in the leaves of the Thick Coal can be rejected on the basis that these histosols were accumulating in a depositional setting distant from marine influence. An indicator of marine mudstones overlying coal is often a high level of sulphur within the coal (Williams & Keith 1963). Sulphur values for the leaves of the Thick Coal are low, e.g. at Longmeadow Wood borehole Leaf 1 is 2.5% (by weight), Leaf 2 is 3.4%, Leaf 3 is 3.5%, Leaf 4 is 2.8%, when compared with those values obtained from leaves lying below marine bands in the Warwickshire Coalfield (in excess of 9% by weight).

Smith (1962) initially believed climate to be partially responsible for the Densospore phase, but that changes in water level either as a result of isostacy, eustacy, or accumulation of peat were paramount in the formation of the other phases. Later Smith & Butterworth (1967, p. 86) and Smith (1968) postulated that a retreating waterfront would occur as peat built up from what can be interpreted as a rheotrophic swamp covered by a roughly constant depth of water, to an ombrotrophic bog with a water table below the peat surface. These changes in edaphic conditions were considered sufficient to account for the succession Lycospore to Transition to Densospore phases. An advancing waterfront caused by subsidence increasing above the rate of peat production would reverse the succession. Smith & Butterworth (1967) and Smith (1968) believed the Incursion phase to be produced by faster flowing floodwaters carrying siliciclastic sediment in suspension across the surface of the peat.

The miospore successions illustrated by Smith (1957, 1962, 1964) took place in leaves of coal which are low in ash content (generally less than 10% by weight) and this also true of the leaves of the Thick Coal. At Longmeadow Wood borehole the ash content of Leaf 1 is 6% by weight (a-db), of Leaf 2 is 7%, of Leaf 3 is 4% and of Leaf 4 is 3%. It is believed that in upper delta plain depositional settings, low-ash histosols are more likely to develop where the surface of the bog is raised above groundwater flood level, although it is possible that rheotrophic swamps could also produce low-ash histosols. Staub & Cohen (1979) suggested that clay could be flocculated by acid swamp waters, dumping it at the edge of the swamp, although the ash contents of the histosols which they studied were still in excess of 12% (Renton *et al.* 1979). Alternatively if deposition of siliciclastic sediment was not taking place near the swamp a low-ash histosol could be produced. Also, at some stage the luxuriance of the vegetation may have aided the exclusion of siliciclastic particles from the histosol by decreasing the flow rate of water passing across its surface. The effect of diagenetic processes outlined by Cecil *et al.* (1979) and Renton & Cecil (1979) may also have contributed to the production of low-ash histosols.

However, if the rheotrophic swamp was at first continually covered by water as proposed by Smith & Butterworth (1967), then in an active depositional setting within the Pennine Basin, it is believed that siliciclastic sediment would become incorporated in the mire to produce either a siliciclastic palaeosol or a high-ash histosol. The ash values for samples obtained by Spackman *et al.* (1969) in a rheotrophic swamp range from 10 to 56%. It appears therefore that the palynological successions observed by Smith (1957, 1962, 1964, 1968) and Smith & Butterworth (1967) took place within ombrotrophic bogs.

Interpretation of the Thick Coal at Longmeadow Wood borehole

The model proposed to account for miospore successions in the leaves of the Thick Coal at Longmeadow Wood borehole is illustrated by examining this sequence (including the siliciclastic sediments) from the base upwards. Prior to the formation of the Thick Coal at this locality, as the rate of subsidence on the delta plain reduced from its higher rate, it resulted in the near total filling of all waterlain areas by a mixture of channel- and lacustrine-derived sediment. This led to the creation of an environment in which a rheotrophic swamp developed. Sediment input was low and discontinuous resulting in the formation of a siliciclastic palaeosol. In contrast with those miospores found within leaves of coal it is believed that many of the miospores within siliciclastic sediment were allochthonous. One of the main reasons for this is because the water moving across the sediment surface would have been quite capable of transporting miospores. A discussion of the distribution of miospores within siliciclastic sediments has been given by Marshall & Smith (1965). The low values of all miospore types in the subsections below the Thick Coal are attributed to the great distance between this and the nearest ombrotrophic bog which was producing them.

When flooding from groundwaters carrying siliciclastic sediment ceased at this locality, the remaining area of shallow water was colonized by vegetation which produced *L. pusilla* bcs observed in subsections at the base of Leaf 1. This type of vegetation would be able to survive a partially exposed water table during the wet season (cf. Anderson 1964) although during the dry seasons the water table probably lay just below the surface of the bog. Initially, some of those mechanisms outlined previously to prevent siliciclastic sedimentation reaching the bog surface may have operated. The vegetation which produced *L. pusilla* bcs may be regarded as equivalent to one of the principal species recognized in phasic community I of Anderson & Muller (1975). An increase in variation of the water table may have resulted in the growth of vegetation producing *F.* cf. *exilis* bcs and where several adjoining subsections are defined by *F.* cf. *exilis* bcs it is presumed that the continued accumulation of the histosol was responsible for the change in hydrological conditions, as in the middle of Leaf 1. However, because of the oscillations between subsections defined by *L. pusilla* bcs and *F.* cf. *exilis* bcs towards the base of Leaf 1, the change in these conditions may only have been small and in this case may be accounted for by prolonged dry seasons.

As the surface of the bog became more raised it is postulated that the changes in hydrological conditions outlined by Anderson (1964) took place, creating a habitat for the growth of vegetation producing *L. minor* bcs following that of *F.* cf. *exilis* in the upper half of Leaf 1. These two miospore types are the intermediate stages of the miospore succession observed in Leaf 1, and may be comparable to several of the phasic communities (II–V) recognized by Anderson & Muller (1975). It is postulated that the creation of a flat-topped bog plain with the greatest variation in water table during the year enabled the climax vegetation to grow. This included those plants producing *Densosporites* spp. characterizing the miospore type, defining subsections towards the top of Leaf 1. It is possible that this vegetation could correspond to phasic community VI (Anderson & Muller 1975) which is found extensively in the centre of inland bogs in Sarawak.

The miospore succession observed in Leaf 1 is very similar to that in Leaf 4, except that the underlying siliciclastic palaeosol contains moderate percentages of three miospore types probably derived from adjacent histosols. In both of these leaves the succession is the simplest type formed during the progressive creation of an ombrotrophic bog. It is postulated that, when dealing with similar changes in vegetation, the most rapid creation of a fully raised bog (recognized in Westphalian coals of the Pennine Basin by the presence of *Densosporites* spp. bcs) will occur when the greatest differential exists between the lower rate of subsidence in the area of the bog and the higher rate outside it. Both Leaves 1 and 4 have similar thicknesses of coal defined by other miospore types (84 and 75 cm respectively) before *Densosporites* spp. bcs is reached. In Leaf 2 there is an extended sequence of subsections defined by *L. pusilla* bcs and *F.* cf. *exilis* forming 116 cm of coal before *Densosporites* spp. bcs is reached, implying a lower differential rate of subsidence for this leaf. However, those seams examined by Smith (1957, 1962) towards the centre of the Pennine Basin, have subsections containing abundant *Densosporites sphaerotriangularis* (one of the species characterizing *Densosporites* spp. bcs) at a much earlier stage in their growth, between 46 and 68 cm from their base. This may imply a greater differential rate of subsidence for the areas covered by these seams.

It is believed that reversal of the previously mentioned subsidence rates eventually resulted in the bog surface becoming lowered until groundwater carrying siliciclastic sediment flooded across it. The most rapid increase in rate of subsidence observed in the leaves of the Thick Coal is evident in Leaf 4, where the bog was terminated whilst covered in the climax vegetation producing *Densosporites* spp. bcs. Overlying this leaf the siliciclastic subsection has been interpreted as lacustrine in origin, with the large percentage of *L. pusilla* bcs derived from a distant bog in the appropriate developmental state. In contrast to this the miospore succession from *Densosporites* spp. bcs to *F.* cf. *exilis* bcs at the top of Leaf 2 may be a response to a slower rate of subsidence. The surface of the bog could have become depressed, causing water table variation to return to that suitable first for the vegetation which produced *L. minor* bcs and then *F.* cf. *exilis* bcs. A similar situation exists at the top of Leaf 3 where the uppermost coal subsection is defined jointly by *F.* cf. *exilis* bcs and *L. pusilla* bcs. At the top of Leaf 1 the rate of subsidence may have been slow at first (*Densosporites* spp. bcs to *L. minor* bcs) but then have increased more rapidly, although the overlying siliciclastic sediments contain moderate percentages of the miospore types found in the uppermost subsection of Leaf 2, testifying to the close proximity of vegetation producing these miospores.

The presence of two miospore cycles in Leaf 2 can be attributed to two episodes of reduced subsidence, separated by one of increased subsidence in this area. South of Longmeadow Wood borehole the increased rate of subsidence has resulted in a split in Leaf 2 comprising a siliciclastic palaeosol, which 3 km away has increased in thickness to 0.3 m (Fig. 3).

The effect of drainage in determining the type of vegetation growing on the bog may be confirmed by an interpretation of the base of Leaf 3. At its base the unusual occurrence of a subsection defined jointly by *Densosporites* spp. bcs and *L. minor* bcs, and the presence of these miospore types in the overlying subsections, may be related to the underlying siliciclastic palaeosol. This fine grained sandstone is penetrated by *in situ* roots and spread over a wide area. Although it appears that a slow increase in the rate of subsidence was taking place during the accumulation of the preceding leaf, it is postulated that this subsidence ceased prior to the deposition of the fine grained sandstone. At this time the bog surface could have been low enough to allow small channels, unconfined by steep levees, to move across it carrying a tractional bed load of fine grained sand. During high stage events sand would be transported from the channel in sheet floods forming a layer on the bog surface. When flow ceased and the surface became elevated the sand would form a well drained surface on which vegetation could grow. This surface was obviously suitable for the parent plants which produced both *Densosporites* spp. bcs and *L. minor* bcs.

Discussion and conclusions

The Warwickshire Thick Coal comprises a number of Westphalian B long-residence histosols which accumulated at the southern perimeter of the Pennine Basin. Subsections of the leaves of the Thick Coal can be defined by four miospore types using only seven bulk characterizing miospore species (bcs). Each type is characterized by one or more bcs, dispersed from parent plants which were dominant in vegetation, and collectively produced an abundance of miospores.

Three of the four types correspond closely to phases recognized by Smith (1962), i.e. *L. pusilla* bcs = Lycospore phase, *L. minor* bcs = Transition phase and *Densosporites* spp. bcs = Densospore phase. An additional type *F.* cf. *exilis* bcs is identified. Each of the leaves of the Thick Coal contains subsections defined by the climax miospore type (*Densosporites* spp. bcs) and is termed a long-residence histosol, to distinguish them from other short-residence histosols without this type which occur throughout the Pennine Basin. Under normal circumstances long-residence histosols occur in thicker coal leaves which indicates that an increased length of time was necessary for their formation when compared with that for short-residence histosols. The four miospore types occur in successions in the leaves of the Thick Coal which are comparable with those found in leaves of Westphalian coal at the centre of the Pennine Basin and in Westphalian analogues in France and America. It is therefore possible that the model proposed for the creation of miospore successions in the leaves of the Thick Coal may be applicable to leaves found in other localities.

The miospore successions and cycles observed in the leaves of the Thick Coal took place as a result of seral growth, corresponding to the way in which phasic plant communities develop in Sarawak (Anderson & Muller 1975). Growth may be termed hydroseral as it resulted mainly from variations in water table during the creation of an ombrotrophic bog, similar to those discussed by Anderson (1964, 1983) in the Far East. The growth of ombrotrophic bogs was permitted by the tropical climate in the central and southern Pennine Basin during Westphalian B times. Changes in climate were believed by Smith (1962) and Butterworth (1966) to be responsible for the stratigraphic distribution of *Densosporites*.

Examination of miospore successions within leaves of coal may disclose the existence of more than one miospore cycle within one leaf. This may indicate the proximity of a siliciclastic parting within that leaf, the presence of which is important in the evaluation of new coal reserves for prediction of coal quality.

The miospore successions within leaves of coal reveal differing rates of subsidence. When accumulation of plant remains kept pace with subsidence the same edaphic conditions and therefore the same vegetation persisted leading to increased thicknesses of coal of the same miospore type. A succession of miospore types was created due to change in edaphic conditions outlined above, when plant accumulation exceeded the rate of subsidence. The most rapid change in edaphic conditions, resulting in the fastest change from initial to climax miospore type, was brought about by a large differential between the lower rate of subsidence in the area of the bog and the higher rate outside it. Later when accumulation of plant remains was lower than the rate of subsidence it was possible to complete a miospore cycle, with the climax miospore type reverting to the initial type due to a reversal of edaphic conditions. However, if the rate of subsidence in the area of the bog was too great plant growth was terminated by drowning before reversion to the initial miospore type could take place.

Different rates of subsidence during the formation of Westphalian B sediments in the Pennine Basin, together with compaction (Fielding 1984b) and topography, controlled the distribution of the depositional environments. The rates of subsidence varied both spatially and temporally, although overall the rates of subsidence near the Pennine Basin depocentre between Yorkshire and Lancashire were much greater than at the periphery in, e.g. Warwickshire. At times of greater subsidence, deposition took place in lacustrine, channel and marine environments, whilst mires formed when reduced rates of subsidence occurred. For these reasons at the periphery of the Basin the interval between leaves of coal should be lower, and because reduced subsidence is likely to occur over longer periods of time the leaves should be thicker. The connection between reduced rates of subsidence near to stable landmasses and the formation of thick coal seams was made by Wills (1948, p. 47), although the Wales–Brabant landmass is now believed to be further south than he originally envisaged. This connection has considerable implications for depositional models constructed to aid in coal exploration, and is at variance with Fielding (1984b) who believed that thicker coal leaves might occur in the centre of the Pennine Basin. However, it is important to study rates of subsidence within individual coalfields because they can vary greatly (Fulton & Williams 1987) and result in dramatic change both in interseam thickness and thickness of coal leaves. The thinning and splitting of the leaves of the Thick Coal to the north, resulting in increased intervals between them is attributed to increased rates of subsidence. To the south, the thinning of the leaves with a resultant marginal increase in the interval between them is believed to be the product of a slight topographic elevation leading towards the Brabant Barrier. This is demonstrated by the presence of palaeosols indicative of better drained conditions than the normal grey siliciclastic palaeosols.

The greatest number of long-residence histosols is likely to occur in areas undergoing sustained low rates of subsidence, because of the increased probability of edaphic conditions changing from those suitable for vegetation producing the initial miospore type to the climax type. This helps to explain the increase in the number of *Densosporites sphaerotriangularis* within the leaves of the Thick Coal as they unite (Butterworth 1964). If the seams studied by Smith (1962) are typical of long-residence histosols in the central Pennine Basin, the reduced coal thicknesses necessary for the progression from initial to climax miospore type can be attributed to a greater differential rate of subsidence in that area.

A full analysis of the Warwickshire Thick Coal can only be made if all of the factors relating to its formation are taken into account. Whilst much is known about its thickness and chemical properties, the palynological investigation has only been undertaken in detail at one locality. However, it is believed that conclusions drawn from this investigation of four leaves of coal can be applied to other leaves developed in the Pennine Basin. It is hoped in the future to publish information from different localities relating miospore assemblages to the parent plants from which they were derived and the coal lithologies in which they were found.

ACKNOWLEDGEMENTS: This research is part of the author's current doctoral research undertaken at the University of Aston under the auspices of Professor D. D. Hawkes and supervised by Drs M. A. Butterworth and P. Turner. I would like to acknowledge the use of facilities at Aston University and at British Coal's Central Area Laboratory, Nuneaton, and the support of many colleagues within British Coal. I would like to thank Margaret Gregory for typing the manuscript and Susan Bourne for drafting the figures. Thanks also to my supervisors, and Harold Smith, Paul Guion, Huw Williams, John Rippon, Philip Jones, Andrew Scott and one anonymous referee for critically reviewing several versions of the manuscript. Publication has been approved by Mr L. Harris, Director of Central Area, British Coal: the views expressed in this paper are those of the author and not necessarily those of British Coal.

References

ANDERSON, J. A. R. 1964. The structure and development of the peat swamps of Sarawak and Brunei. *Journal of Tropical Geography*, **18**, 7–16.

—— 1973. An introduction to the ecology of the peat swamps of Sarawak and Brunei. *Proceedings of the International Peat Society Symposium on Classification of Peat and Peat Lands*, pp. 44–51.

—— 1983. Tropical peat swamps of Western Malesia. *In:* GORE, A. J. P. (ed.) *Ecosystems of the World. Vol. 4B, Mires: Swamp, Bog, Fen and Moor. Regional Studies*, pp. 181–199. Elsevier, Amsterdam.

—— & MULLER J. 1975. Palynological study of a Holocene Peat and a Miocene coal deposit from NW Borneo. *Review of Palaeobotany and Palynology*, **19**, 291–351.

BARBER, K. E. 1981. *Peat Stratigraphy and Climatic Change*. Balkema, Rotterdam.

BARROW, G., GIBSON, W., CANTRILL, T. C., DIXON, E. E. L. & CUNNINGTON, C. H. 1919. *The Geology of the Country around Lichfield*. Memoir of the Geological Survey of Great Britain.

BELLAMY, D. J. 1972. Templates of peat formation. *Proceedings of the International Peat Conference*, Helsinki, **4**, 7–18.

BESLY, B. M. 1983. The sedimentology and stratigraphy

of red beds in the Westphalian A to C of Central England. *Unpublished PhD thesis*, University of Keele.

BHARADWAJ, D. A. 1957. The palynological investigations of the Saar coals. *Palaeontographica B*, **101**, 73–125.

BROWNE, J. T. 1907. The Thick Coal of Warwickshire. *Transactions of the Institute of Mining Engineers*, **33**, 502–529.

BUTTERWORTH, M. A. 1956. The distribution of microspores in the coalfields lying to the west of the Pennines. *Unpublished PhD thesis*, University of Edinburgh.

—— 1964. The distribution of *Densosporites sphaerotriangularis* in the Westphalian B of the W Pennine coalfields of England. *Fortschritte in der Geologie von Rheinland und Westphalen*, **12**, 317–330.

—— 1966. The distribution of Densopores. *The Palaeobotanist*, **15**, 16–28.

——, JANSONIUS, J., SMITH, A. H. V. & STAPLIN, F. L. 1964. *Densosporites* (Berry) Potonie and Kremp, and related genera. C.I.M.P. Working Group No. 2, *Compte Rendu Cinquième Congrès Internationale de Stratigraphie et de Géologie du Carbonifière*, Paris 1963, **3**, 1049–1057.

CADY, G. H. 1942. Modern concepts of the physical constitution of coal. *Journal of Geology*, **50**, 337–356.

CALVER, M. A. 1968. Distribution of Westphalian marine faunas in Northern England and adjoining areas. *Proceedings of the Yorkshire Geological Society*, **37**, 1–72.

CECIL, C. G., STANTON, R. W., DULONG, F. R. & RENTON, J. J. 1979. Some geologic factors controlling mineral matter in coal. *In:* DONALDSON, A. C., PRESLEY, M. M. & RENTON, J. J. (eds) *Carboniferous Short Course and Guidebook*, pp. 43–56. American Association of Petroleum Geologists, Field Seminar, West Virginia Geological Economic Survey Bulletin, B-37-3.

CHALONER, W. G. & MUIR, A. 1968. Spores and floras. *In:* MURCHISON, D. G. & WESTOLL, T. S. (eds) *Coal and Coal-bearing Strata*, pp. 127–146. Oliver & Boyd, Edinburgh.

CLENDENING, J. A. 1967. Three new species of *Fabasporites* Sullivan 1964 from the Appalachian Basin. *Proceedings of the West Virginia Academy of Science*, **39**, 315–319.

COHEN, A. D. 1974. Petrography and paleoecology of Holocene peats from the Okefenokee swamp–marsh complex of Georgia. *Journal of Sedimentary Petrology*, **44**, 716–726.

—— 1984. The Okefenokee swamp: a low sulphur end member of a shoreline related depositional model for coastal plain coals. *In:* RAHMANI, R. A. & FLORES, R. M. (eds) *Sedimentology of Coal and Coal-bearing Sequences*, pp. 231–240. Special Publication of the International Association of Sedimentologists 7. Blackwell Scientific Publications, Oxford.

—— & SPACKMAN, W. 1972. Methods in peat petrology and their application to reconstruction of palaeoenvironments. *Bulletin of the Geological Society of America*, **83**, 129–142.

COLEMAN, J. M. 1966. Ecological changes in a massive fresh-water clay sequence. *Transactions Gulf Coast Association of Geological Societies*, **16**, 159–174.

——, GAGLIANO, S. M. & SMITH, W. G. 1970. Sedimentation in a high tide tropical delta. *Society of Economic Paleontologists and Mineralogists Special Paper*, **15**, 185–197.

COPE, K. G. & JONES, A. R. L. 1970. The Warwickshire Thick Coal and its mining environment. *Compte Rendu Sixième Congres Internationale de Stratigraphie et de Géologie Carbonifière*, Sheffield, 1967, **II**, 585–598.

CORSIN, P. M., COQUEL, R. & LOBOZIAK, S. 1967. Sur la composition palynologique verticale et horizontale d'un cycle sedimentaire de terraine houiller. *Palaeogeography, Palaeoclimatology, Palaeoecology*, **3**, 167–284.

DIMICHELE, W. A. & PHILLIPS, J. L. 1985. Arborescent lycopod reproduction and paleoecology in a coal-swamp environment of late Middle Pennsylvanian age (Herrin Coal, Illinois). *Review of Palaeobotany and Palynology*, **44**, 1–26.

DU RIETZ, E. 1954. Die Mineral Bodenwasserziegergrenze als Grundlage einer Naturlichen Zweigleiderung der Nord und Mitteleuropaischen Moore. *Vegetatio*, **5/6**, 571–585.

EASTWOOD, T., GIBSON, W., CANTRILL, T. C. & WHITEHEAD, T. H. 1923. *The Geology of the Country around Coventry*. Memoir of the Geological Survey of Great Britain.

ELLIOTT, R. E. 1965. Swilleys in the Coal Measures of Nottinghamshire interpreted as palaeo-river courses. *Mercian Geologist*, **1**, 133–142.

—— 1968. Facies, sedimentation successions and cyclothems in productive coal measures in the East Midlands of Great Britain. *Mercian Geologist*, **2**, 351–372.

—— 1985. Quantification of peat to coal compaction stages, based especially on phenomena in the East Midlands Coalfield, England. *Proceedings of the Yorkshire Geological Society*, **45**, 163–172.

FAO/UNESCO 1974. *Soil Map of the World, 1:5,000,000. Vol. 1 Legend CH.3. The Soil Units.* UNESCO, Paris.

FIELDING, C. R. 1982. Sedimentology and stratigraphy of the Durham Coal Measures, and comparisons with other British coalfields. *Unpublished PhD thesis*, University of Durham.

—— 1984a. Upper delta plain lacustrine and fluviolacustrine facies from the Westphalian of the Durham Coalfield, NE England. *Sedimentology*, **31**, 547–567.

—— 1984b. A coal depositional model for the Durham Coal Measures of NE England. *Journal of the Geological Society, London*, **141**, 919–931.

—— 1986. Fluvial channel and overbank deposits from the Westphalian of the Durham Coalfield, NE England. *Sedimentology*, **33**, 119–140.

FRAZIER, D E. & OSANIK, A. 1969. Recent peat deposits—Louisiana Coastal Plain. *In:* DAPPLES, E. C. & HOPKINS, M. E. (eds) *Environments of Coal Deposition*, pp. 63–85. Geological Society of America, Special Paper 114.

FULTON, I. M. & WILLIAMS, H. 1986. Controls on

Silesian sedimentation in the southern part of the Pennine Basin. *BSRG Conference. Controls of Upper Carboniferous Sedimentation, North-West Europe*, Keele, April 12–15, *Abstract*, pp. 38, 39.

—— & —— 1987. Palaeogeographical change and controls on Namurian, and Westphalian A and B sedimentation at the southern margin of the Pennine Basin. *In:* BESLEY, B. M. & KELLING, G. (eds) *Sedimentation in a Synorogenic Basin Complex—the Upper Carboniferous of North West Europe*.

GUION, P. D. 1978. Sedimentation of interseam strata and some relationships with coal seams in the East Midlands Coalfield. *Unpublished PhD thesis*, (C.N.A.A.) City of London Polytechnic.

—— 1984. Crevasse splay deposits and roof-rock quality in the Threequarters Seam (Carboniferous) in the East Midlands Coalfield, UK. *In:* RAHMANI, R. A. & FLORES, R. M. (eds) *Sedimentology of Coal and Coal-bearing Sequences*, pp. 291–308. Special Publication of the International Association of Sedimentologists 7. Blackwell Scientific Publications, Oxford.

—— 1987. *Palaeogeographical Atlas*. Geological Society of London.

HABIB, D. 1966. Distribution of spore and pollen assemblages in the Lower Kittanning Coal of Western Pennsylvania. *Palaeontology*, **9**, 629–666.

—— & GROTH, P. K. H. 1967. Paleoecology of migrating Carboniferous peat environments. *Palaeogeography, Palaeoclimatology, Palaeoecology*, **3**, 185–195.

——, RIEGEL, W. & SPACKMAN, W. 1966. Relationship of spore and pollen assemblages in the lower Kittanning coal to overlying faunal facies. *Journal of Paleontology*, **40**, 756–759.

HACQUEBARD, P. A. & DONALDSON, J. B. 1969. Carboniferous coal deposition associated with flood-plain and limnic environments in Nova Scotia. *In:* DAPPLES, E. C. & HOPKINS, M. E. (eds) *Environments of Coal Deposition*, pp. 143–191. Geological Society of America Special Paper 114.

HASELDONCKX, P. 1977. The palynology of a Holocene peat swamp environment in Johore, Malaysia. *Review of Palaeobotany and Palynology*, **24**, 227–238.

HASZELDINE, R. S. 1981. Westphalian B Coalfield sedimentology in NE England and its regional setting. *Unpublished PhD thesis*, University of Strathclyde.

—— 1983. Fluvial bars reconstructed from a deep, straight channel, Upper Carboniferous Coalfield of North-East England. *Journal of Sedimentary Petrology*, **53**, 1233–1247.

—— 1984. Muddy deltas in freshwater lakes, and tectonism in the Upper Carboniferous Coalfield of NE England. *Sedimentology*, **31**, 811–822.

HOWELL, H. H. 1859. *The Geology of the Warwickshire Coalfield*. Memoir of the Geological Survey of Great Britain.

IBRAHIM, A. C. 1933. *Sporenformen des Aegirhorizonts des Ruhr—Reviers*. Triltsch, Wurzeburg.

KEARNS, F. L. & DAVISON, A. T. 1983. Field classification system of organic-rich sediments. *In:* RAYMOND, R. & ANDREJKO, N. J. (eds) *Mineral Matter in Peat Workshop*, Los Alamos NM 2609083, Los Alamos National Laboratory, October, pp. 147–158 NTIS: LA-9907-OBES.

KOSANKE, R. M. 1950. Pennsylvanian spores of Illinois and their use in correlation. *Bulletin of the Illinois Geological Survey*, **74**, 1–128.

KOVAL'CHUK, G. M. & UZIYUK, V. I. 1973. Genetic relationship between petrographic and palynologic composition of coals, exemplified by coal seam 1_3 in the Krasndarmeyskiy region of the Donets Basin. *Akadamii Nauk, USSR Doklady*, **212**, 79–82.

LOOSE, F. 1934. Sporenformen aus dem Floz Bismark des Ruhrgebietes. *Arbeit Institut Palaeobotanik, Berlin*, **4**, 127–164.

MARSHALL, A. E. & SMITH, A. H. V. 1965. Assemblages of miospores from some Upper Carboniferous coals and their associated sediments in the Yorkshire Coalfield. *Palaeontology*, **7**, 656–673.

MCCABE, P. J. 1984. Depositional environments of coal and coal bearing strata. *In:* RAHMANI, R. A. & FLORES, R. M. (eds) *Sedimentology of Coal and Coal-bearing Sequences*, pp. 13–42. Special Publication of the International Association of Sedimentologists 7. Blackwell Scientific Publications, Oxford.

MITCHELL, G. H., STUBBLEFIELD, C. J. & CROOKALL, R. 1942. *The Geology of the Warwickshire Coalfield*, pp. 1–40. Geological Survey of Great Britain. Wartime Pamphlet No. 25.

MOORE, P. D. & BELLAMY, D. J. 1974. *Peatlands*. Elek Science, London.

NCB 1972. *Procedure for the assessment of reserves* PI.1972/4, pp. 1–8.

PAGET, R. F. 1936. The correlation of coal seams by microspore analysis: the Seams of Warwickshire. *Transactions of the Institute of Mining Engineers, London*, **92**, 59–88.

—— 1937. Correlation of coal seams by microspore analysis: the northern part of the Warwickshire field and some collieries in South Derbyshire. *Colliery Guardian*, **154**, 823–826.

PARRISH, J. T. 1982. Upwelling and petroleum source beds with reference to the Paleozoic. *Bulletin of the American Association of Petroleum Geologists*, **66**, 750–774.

PEPPERS, R. A. 1970. Correlation and palynology of coals in the Carbondale and Spoon Formation (Pennsylvanian) of the North Eastern part of the Illinois Basin. *Illinois State Geological Survey Bulletin*, **93**, 1–173.

PHILLIPS, T. L. & PEPPERS, R. A. 1984. Changing patterns of Pennsylvanian coal swamp vegetation and implications of climatic control on coal occurrence. *International Journal of Coal Geology*, **3**, 205–255.

RAVN, R. L. 1979. An introduction to the stratigraphic palynology of the Cherokee Group (Pennsylvanian) Coals of Iowa. *Iowa Geological Survey Technical Paper*, **6**, 1–117.

—— & FITZGERALD, D. J. 1982. A Morrowan (Upper Carboniferous) miospore flora from eastern Iowa, USA. *Palaeontographica B*, **183**, 108–172.

RENTON, J. J. & CECIL, C. B. 1979. Coal composition relationships in support of a chemical coal model. *In:* DONALDSON, A. C., PRESLEY, M. M. & RENTON, J. J. (eds) *Carboniferous Coal Short Course and Guidebook*, pp. 103–128. American Association of Petroleum Geologists, Field Seminar, West Virginia Geological Economic Survey, Bulletin B-37-3.

——, ——, STANTON, R. & DULONG, F. 1979. Compositional relationships of plants and peats from modern peat swamps in support of a chemical coal model. *In:* DONALDSON, A. C., PRESLEY, M. W. & RENTON, J. J. (eds) *Carboniferous Coal, Shortcourse and Guidebook*, pp. 57–102. American Association of Petroleum Geologists, Field Seminar, West Virginia Geological Economic Survey, Bulletin B-37-3.

ROWLEY, D. B., RAYMOND, A., PARRISH, J. T., LOTTES, A. L., SCOTESE, C. R. & ZIEGLER, A. M. 1985. Carboniferous paleogeographic, phytogeographic and paleoclimatic reconstructions. *International Journal of Coal Geology*, **5**, 7–42.

SCHOPF, J. M., WILSON, L. R. & BENTALL, R. 1944. An annotated synopsis of Paleozoic fossil spores and the definition of generic groups. *Illinois Geological Survey Report of Investigation*, **91**, 1–66.

SCOTT, A. C. 1978. Sedimentological and ecological control of Westphalian B plant assemblages from West Yorkshire. *Proceedings of the Yorkshire Geological Society*, **41**, 461–508.

—— 1979. The ecology of Coal Measure floras from northern Britain. *Proceedings of the Geologists Association*, **90**, 97–116.

SHOTTON, F. W. 1928. The southern extension of the Warwickshire Coalfield. *Transactions of the Institute of Mining Engineers*, **126**, 136–147.

SMITH, A. H. V. 1957. The sequence of microspore assemblages associated with the occurrence of crassidurite in coal seams of Yorkshire. *Geological Magazine*, **94**, 345–363.

—— 1962. The palaeoecology of Carboniferous peats based on the microspores and petrography of bituminous coals. *Proceedings of the Yorkshire Geological Society*, **33**, 423–474.

—— 1964. Zur Petrologie und Palynologie der Kohlenfloze des Karbons und ihrer Begleitschichten. *Fortschritte in der Geologie von Rheinland und Westphalen*, **12**, 258–302.

—— 1968. Seam profiles and seam characters. *In:* MURCHISON, D. G. & WESTOLL, T. S. (eds) *Coal and Coal-bearing Strata*, pp. 31–40. Oliver & Boyd, Edinburgh.

—— & BUTTERWORTH, M. A. 1967. Miospores in the coal seams of the Carboniferous of Great Britain. *Special Papers in Palaeontology*, **1**, 1–324.

SMITH, A. J., BRIDEN, G. E. & DREWRY, D. 1973. Phanerozoic World Maps. *In:* HUGHES, N. F. (ed.) *Organisms and Continents Through Time*, pp. 1–42. Special Papers in Palaeontology 12.

SOMERS, Y. 1972. Revision du genre *Lycospora* Schopf, Wilson et Bentall. C.I.M.P. microfossiles organiques du Palaeozoique. *Edition du centre nationale de la Recherche scientifique*, No. 5, pp. 1–110. Paris.

SPACKMAN, W., DOLSEN, C. P. & RIEGEL, W. 1966. Phytogenic organic sediments and sedimentary environments in the Everglades–mangrove complex, Part I: evidence of a transgressing sea and its effects on environments of the Shark River area of south western Florida. *Palaeontographica B*, **117**, 135–152.

——, RIEGEL, W. & DOLSEN, C. P. 1969. Geological and biological interactions in the swamp-marsh complex of southern Florida. *In:* DAPPLES, E. C. & HOPKINS, M. E. (eds) *Environments of Coal Deposition*, pp. 1–35. Special Paper of the Geological Society of America 114.

STAUB, J. R. & COHEN, A. D. 1979. The Snuggedy Swamp of Carolina: a backbarrier coal forming environment. *Journal of Sedimentary Petrology*, **49**, 133–144.

STOPES, M. C. 1919. On the four visible ingredients in banded bituminous coals. *Proceedings of the Royal Society of London, B*, **90**, 470–487.

TAUBER, H. 1965. Differential pollen dispersion and the interpretation of pollen diagrams. *Danmarks Geologiskie Undersogelse*, **11**, 1–69.

TRICART, J. 1972. *The Landforms of the Humid Tropics, Forest and Savannas*. Longman, London.

TRUEMAN, A. E. 1947. Stratigraphical problems in the Coalfields of Great Britain. *Quarterly Journal of the Geological Society, London*, **103**, 65–104.

TURNER, P., VAUGHAN, D. J. & BESLY, B. 1985. Remanence acquisition in red beds from the Coal Measures of Central England. *Journal of the Geological Society, London*, **142**, 1015–1028.

WILLS, L. J. 1948. *The Palaeogeography of the Midlands*. University Press of Liverpool/Hodder & Stoughton, London.

WILLIAMS, E. G. & KEITH, M. L. 1963. Relationship between sulphur in coals and the occurrence of marine roof beds. *Economic Geology*, **58**, 720.

WILLIAMS, H. 1986. The sedimentology of the Upper Westphalian A in the South Derbyshire Coalfield. *Unpublished PhD thesis*. (C.N.A.A.) Oxford Polytechnic.

WORKING GROUP ON THE ORIGIN AND NATURE OF PALAEOSOLS 1971. Criteria for the recognition and classification of palaeosols. *In:* YAALON, D. H. (ed.) *Paleopedology*, pp. 153–158. International Society of Science and Israel University Press.

YOUNG, A. 1976. *Tropical Soils and Soil Survey*. Cambridge University Press.

ZIELGER, A. M., BAMBACH, R. K., PARRISH, J. T., BARRET, S. F., GIERLOWSKI, E. H., PARKER, W. C., RAYMOND, A. & SEPKOSKI, JR, J. J. 1981. Paleozoic biogeography and climatology. *In:* NIKLAS, K. Z. (ed.) *Paleobotany, Paleoecology and Evolution*, **1**, 231–266. Praeger, New York.

I. M. FULTON, British Coal, Central Area HQ, Coleorton Hall, Coleorton, Leicestershire LE6 4FA, England.

Gondwana coal basins of Australia and South Africa: tectonic setting, depositional systems and resources

D. K. Hobday

ABSTRACT: Upper Carboniferous and Permian coals of eastern Australia (Bowen, Sydney, Cooper, Galilee and satellite basins) and southern Africa (Karoo Basin) resemble one another in their general composition and associated clastic facies. The Australian examples, however, show more petrographic and geometric variability than their southern African Gondwana counterparts.

Critical controls on the distribution and quality of coal in both regions appear to have been (1) a near-polar location with accompanying periglacial conditions, (2) Permian through early Triassic warming independent of any significant change in palaeolatitude, (3) associated floral changes from tundra vegetation to swamps and deciduous *Glossopteris* forests, (4) variable subsidence and sedimentation rates in foredeep, rift, and epicratonic basin settings, (5) local palaeotopographic effects, (6) eustatically induced changes in base level and marine transgression, and (7) a range of depositional systems from proximal conglomeratic alluvial fans through fluvial, delta-plain, lake-margin, back-barrier, and blanket peat mires in sediment-starved basins.

In most cases, palaeotopography and depositional environment were the most important factors in determining coal seam geometry, and had considerable influence on coal quality, specifically its maceral content. Most of the commercial seams are contained in fluvial and delta-plain sequences, with thick, dip-elongate interchannel seams and really extensive coals typically overlying abandoned delta lobes. However, some important seams are juxtaposed with conglomerates of wet alluvial-fan origin or are associated with quartzose barrier sandstones; others originated in clastic-starved systems bordering shallow lakes or in blanket swamps that received moisture from groundwater discharge or direct rainfall.

Some seams display characteristics that were primarily controlled by other more regional factors such as rapid tectonic subsidence, changes in groundwater regime, or gradual marine transgression. While influencing the thickness and areal extent of seams, these factors were even more important in determining microlithotypes, maceral content, and inorganic composition of the coals.

Not only is coal mining crucial to the economies of both Australia and S Africa, with further expansion likely in the coming decades, but coal and dispersed coal particles have generated billions of barrels of commercial oil reserves along with large volumes of gas. Australian petroleum reserves are predominantly of land-plant origin, and source-reservoir relationships provide important analogues for exploring other non-marine basins.

Keywords: coal, Gondwana, southern hemisphere, Australia, S Africa, basin analysis, petroleum

Upper Palaeozoic coal-bearing facies of Australia and southern Africa are of typical Gondwana type. They differ from their northern hemisphere Carboniferous counterparts in their unusual inertinite composition, low sulphur content, rare seatearths, and close association with conglomeratic sediments which are partly of glaciogenic origin, particularly in the lower part of the succession. In other respects they resemble the fluvial and delta-plain facies that characterize coal basins of various ages in other parts of the world.

These Gondwana coals show marked variation with respect to tectonic setting in the foreland-epicratonic continuum of both the southern African Karoo and eastern Australia Bowen–Sydney and Cooper–Galilee Basins. Most thick seams of the Karoo Basin are present in the Transvaal, Orange Free State and Natal regions of relatively stable cratonic platform. In contrast, major coal seams are widely distributed through the Permian basins of Australia, from the tectonically active flanks to the cratonic interior.

Despite this tectonic variation, the Upper Carboniferous–Permian stratigraphic succession is generally similar in Australia, S Africa, and other Gondwana basins. Basal diamictite and glacial outwash deposits (Fig. 1) grade up into coarse fluvial and finer deltaic facies, with thick intervening lacustrine or epeiric basin shales

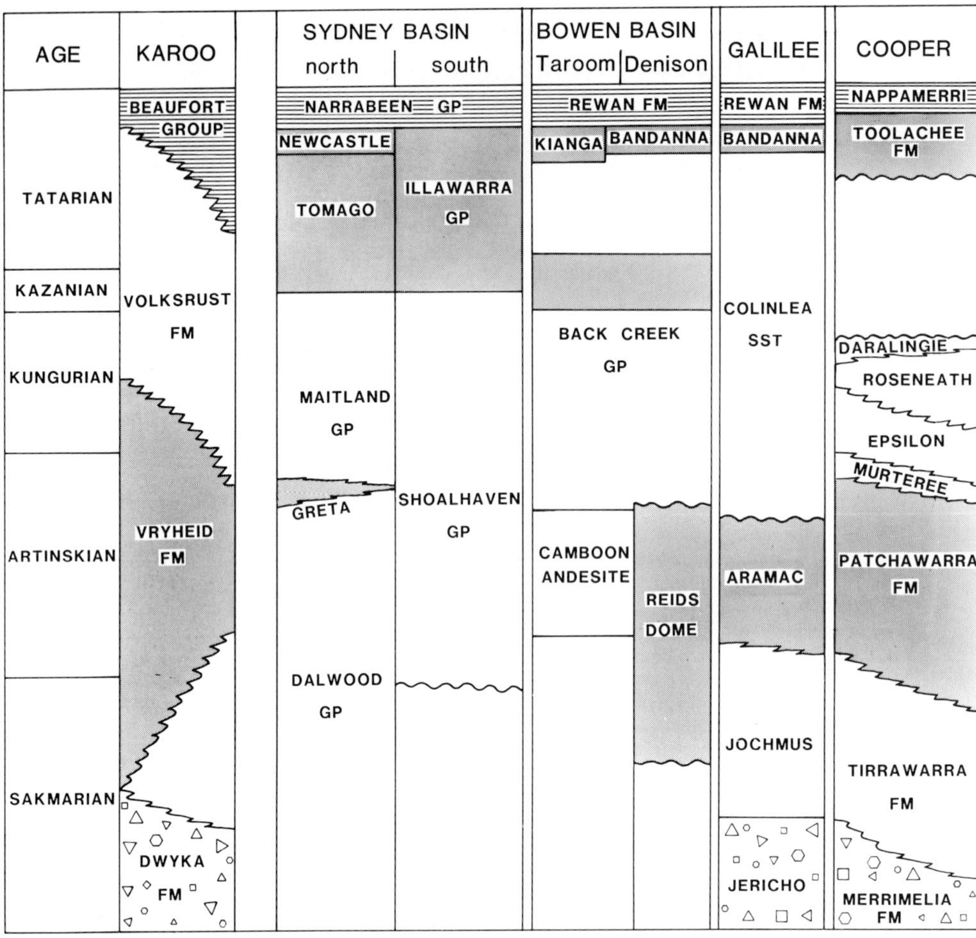

FIG. 1. Generalized Permian stratigraphy of the Karoo and Australian basins showing the vertical progression from glacial Dwyka Formation and correlatives to coal-bearing fluvial-deltaic successions (dark screen) to redbeds at the top (horizontal lines).

either succeeding the glaciogene facies, as in the Karoo Basin, or interrupting the fluvio-deltaic sequence as in the Cooper Basin.

The back-arc Sydney–Bowen Basins received coarse, gravelly sediments along their seaward margins, grading into finer-grained facies. The Karoo Basin experienced similar cratonward influx, which was accentuated in post-Permian time. Marine incursions affected post-glacial sedimentation in the western Karoo and eastern Australian basins. In the uppermost Permian or Lower Triassic the coal-bearing succession is followed, generally conformably, by highly oxidized redbeds.

Excellent outcrops, supplemented by subsurface data from coal-mining companies, have allowed detailed reconstruction of facies in the Karoo and Sydney Basins (e.g. Hobday 1973, Tavener-Smith 1980, Herbert & Helby 1980). Where petroleum exploration has occurred, for example in the Cooper and Bowen Basins, depositional environments have been reconstructed from wireline logs, cores and some seismic data (Stuart 1976, Draper & Beeston 1985). All of these studies have employed comparisons with modern depositional environments, and have demonstrated strong autocyclic controls on sedimentation patterns. Nonetheless, there are strong palaeoclimatic, tectonic, and possible eustatic overprints.

Palaeolatitude and palaeoclimate

Palaeomagnetic reconstructions of the Gondwana continents indicate a southeastward polar wandering path across the area during the Permo–

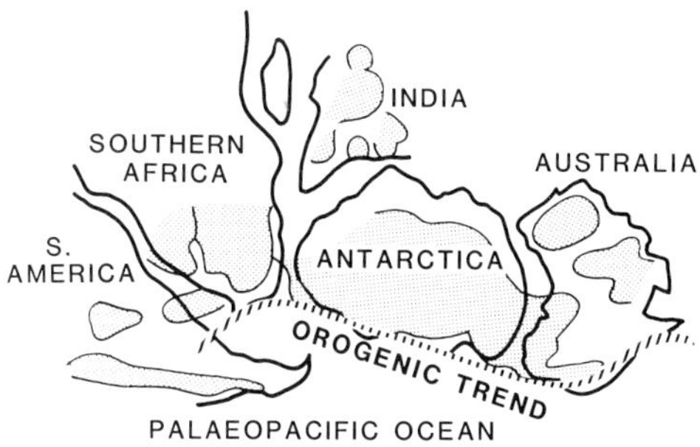

FIG. 2. Reconstruction of Gondwana continents during the late Palaeozoic showing ice-covered areas (stippled) and the orogenic trend of the Gondwana Fold Belt. Modified after Crowell & Frakes (1975).

Carboniferous (Creer 1965). Polar ice sheets (Fig. 2) converged on sedimentary basins from bordering highlands (Crowell & Frakes 1972, 1975), with widespread accumulation of subglacial till, terrestrial moraine, glaciolacustrine shales, and fluvioglacial outwash (Visser *et al.* 1978; Von Brunn 1977). Warm oceanic conditions persisted into the Carboniferous Period over the Bonaparte Gulf area of northwestern Australia, but were followed by rapid cooling as Australia drifted toward the pole. During most of the Carboniferous, Permian, and early Triassic the South Pole was in the vicinity of southeastern Australia (Veevers 1984, p. 11). The most intense glaciation may have spanned the Carboniferous–Permian boundary before terminating in mid-Permian times (Crowell & Frakes 1975). Veevers (1984, p. 239), however, favoured earlier, Australia-wide glaciation that followed regional mid-Carboniferous uplift, with deglaciation commencing in the Stephanian. Highland glaciers persisted along the eastern margin of Australia during the Early Permian. In southern Africa, ice sheets converged on the Karoo Basin from the north and east, and perhaps from the west and south. By late Permian, warm marine conditions had returned to northwestern Australia while widespread floodbasin peat accumulation characterized the basins of eastern Australia and southern Africa.

Surprisingly, the widespread Triassic Gondwana redbeds (Fig. 1) apparently do not represent a rapid migration toward lower latitudes. Indeed the South Pole was located over southeastern Australia until the Middle Triassic (Veevers 1984, p. 14). The prevailing oxidizing conditions were probably a consequence of markedly seasonal rainfall in a cool-temperate climate. Peat-swamp accumulation resumed in the late Triassic both in the southern Karoo Basin (Molteno Formation) and in southeastern Australia.

Hunt *et al.* (*in press*) ascribe some of the unique characteristics of Australian coals to progressive climatic changes. Inertinite-rich coals near the top of the Permian succession are thought to record more pervasive oxidation which heralded Triassic aridity.

Changes in vegetation

The lowermost, Upper Carboniferous coals of Gondwana are thought to have originated in a mossy tundra-type bog (Plumstead 1969), with some debris supplied directly from vegetated glaciers (Cairncross 1979). Retallack (1980) regarded the late Carboniferous *Botrychiopsis* floras as analogous to modern boreal taiga environments of Siberia; these cold-climate floras were replaced in the early Permian by swamp-dwelling gymnosperms, with the subsequent *Glossopteris* deciduous forests dominating Gondwana through most of the Permian. The Triassic *Dicroidium* flora was characterized by pteridosperms but included a great variety of sphenopsids, cycads, ginkgos, conifers, and ferns, likened by Retallack (1980) to modern coniferous forests, coastal woodlands and heath. Widespread coniferous forests of *Voltziopsis* were interpreted by Retallack (1980) as marking a lowered water table relative to the wetter Permian environments.

Palaeotopographic controls

Irregular pre-Karoo topography exercised a pronounced influence on the thickness and distribution of coals in the Witbank and Highveld Coalfields of the northern Karoo Basin (Cadle 1982, LeBlanc-Smith 1980). The lowermost seams are only present in depressions and pinch out on the flanks of basement highs (Fig. 3). Some seams attain their maximum thickness where they abut against the sloping basement surfaces, presumably because these stable areas afforded a platform for peat accumulation (Cadle 1982). Updip areas were subject to oxidation and non-preservation of peat, whereas the more depressed areas were too deeply inundated. Cadle also showed that differential compaction of palaeovalley fill caused major river systems to reoccupy routes along the palaeovalley axes, thereby truncating the upper seams.

Tectonic setting and basin development

Plate-tectonic theory recognizes a great variety of basin types, but as far as Gondwana coal

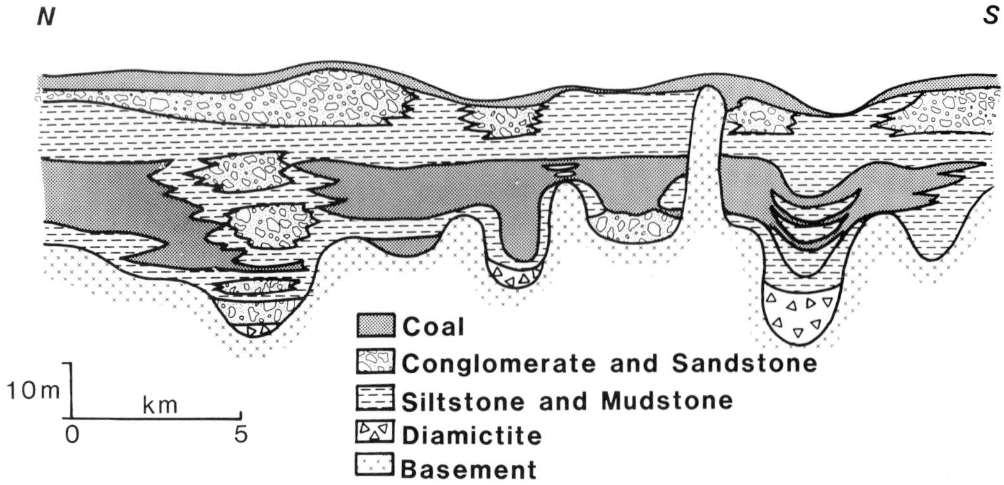

FIG. 3. Influence of palaeotopography on the No. 4 coal seams of the northern Karoo Basin, showing pinchout against basement highs and splitting and pinchout in basement valleys occupied by active fluvial channels. After Cadle (1982).

basins are concerned, three general categories stand out: foredeeps, rifts, and intra-cratonic sags. The Australian coal basins (Figs 4 & 5) include elements of all three categories, whereas the Karoo Basin (Fig. 6) coal province is essentially cratonic, bounding a major southern foredeep with only minor coals.

Tectonic activity in the Karoo foredeep, centred on the southern Cape Province, commenced in the Devonian, increased gradually through the late Carboniferous and Permian and climaxed during the Triassic, with each event accentuating the N–S asymmetry (Fig. 7).

According to Rhodes (1974) an early Palaeozoic extensional margin was subject to a reversal in plate motion, with a late Palaeozoic island arc forming south of the present coastline. Alternatively, according to De Wit (1977) the convergent margin may have been of Andean type, with the magmatic orogen located on the continental mainland. In either event the subduction zone was located some 1000 km south of the present Cape Ranges. Lock (1980) suggested that this relationship could be explained by the oceanic plate descending at a very low angle. According to this model, early Triassic deformation of the Cape Fold Belt would have occurred along zones of detachment between the superimposed plates.

During Ecca (Permian) deposition a bipolar source existed, from the northern cratonic and southern orogenic flanks, but Beaufort (Triassic) and later sedimentation was dominated by coarse clastic wedges derived from the southern highland belt.

Some coals have been mined from the Upper Triassic Molteno Formation along the southern margin of the Karoo basin, but commercial coal resources are largely restricted to the shallow northern cratonic fringe. This is in marked contrast to geologically more recent retroarc analogues, for example the Papua New Guinea–Gulf of Carpentaria region (Veevers 1984, pp. 129, 245), where the major coal-forming environments are associated with the tectonically active flank. This latter association of coal with labile, tuffaceous sediments derived from the volcanic arc is characteristic of the Bowen–Sydney Basin.

According to some reconstructions the Cape Fold Belt persisted as the Gondwana Orogenic Belt through Antarctica and eastern Australia (Lock 1980). Thus the Karoo, Sydney and Bowen Basins, together with as yet poorly known Antarctic basins, may have been linked laterally along the landward side of the orogen (Fig. 2). The gross parallelism in the stratigraphic development of these basins (Fig. 1) may be at least partly explained in terms of this unified tectonic setting, with the major differences in coal distribution reflecting stratigraphic variability in palaeoclimate and subsidence patterns. For example, the retroarc foredeep area of the southern Karoo Basin (Fig. 7) was generally a moderately deep 'flysch trough' during Permian Ecca deposition (Kingsley 1979), and during the Triassic was subaerial and oxidizing, with generally low water tables inhibiting peat preservation.

Andesitic outpourings suggest that a compres-

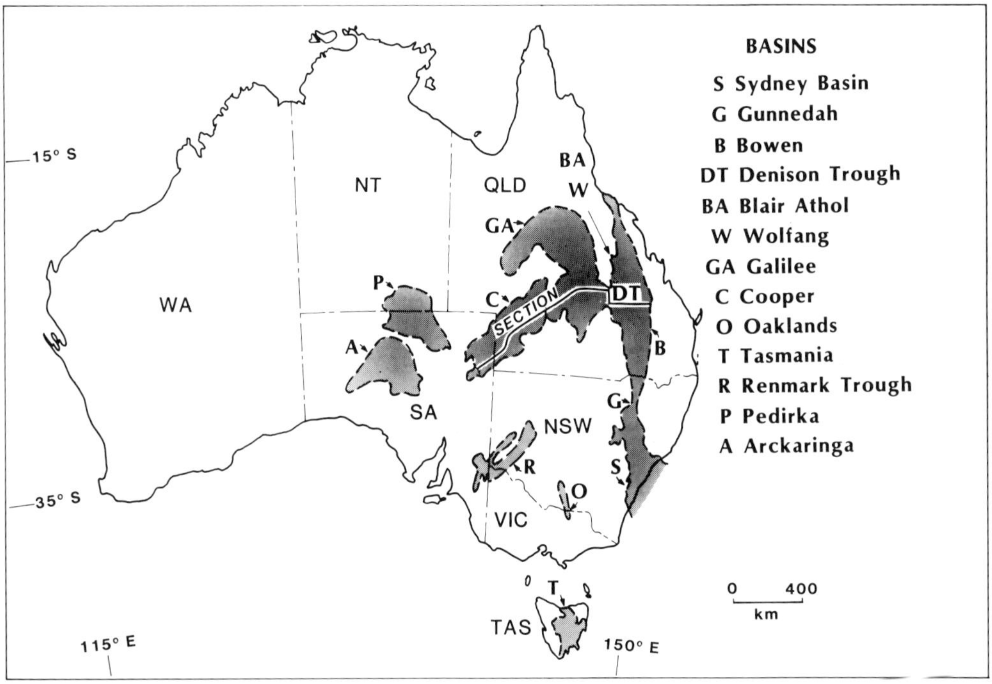

FIG. 4. Permian coal basins of central and eastern Australia with line of section Fig. 5.

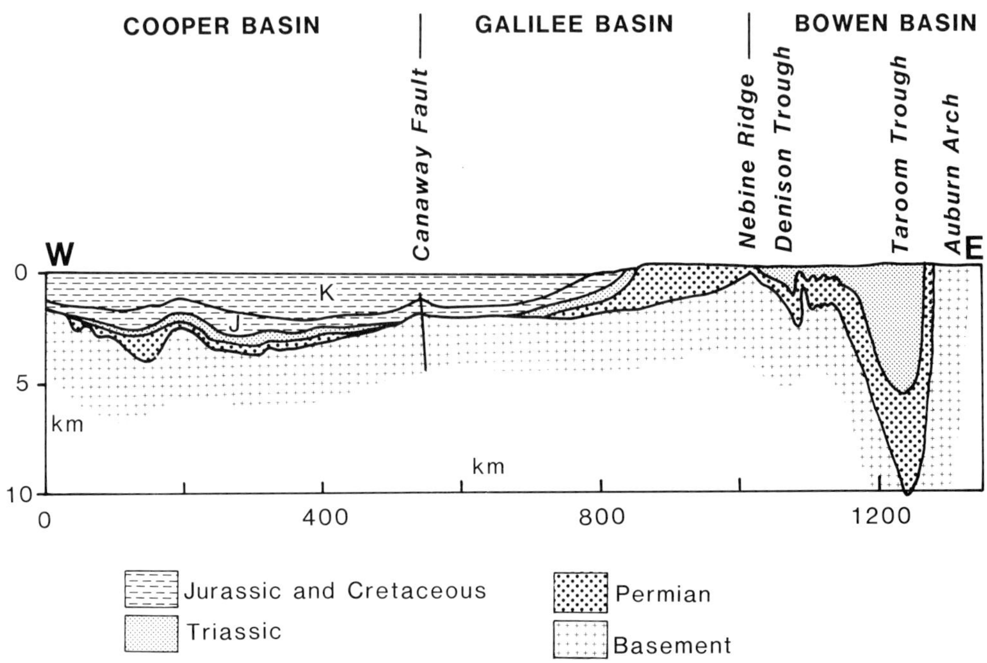

FIG. 5. Cross-section showing Permian and younger sediments of the Cooper, southern Galilee and Bowen Basins. Line of sections shown in Fig. 4. Modified after Veevers (1984).

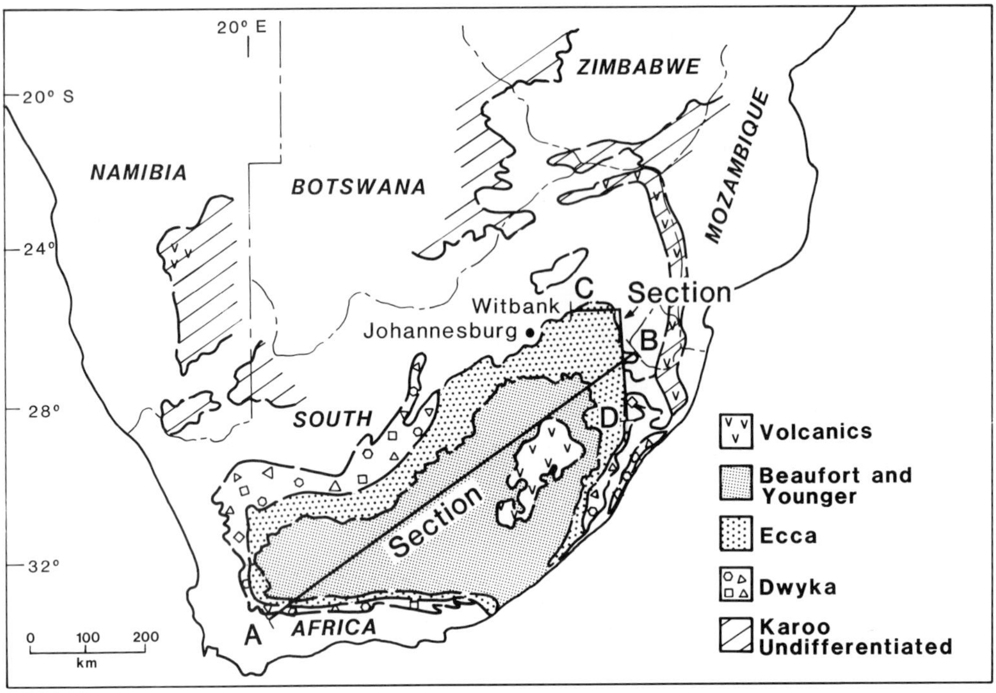

FIG. 6. The Karoo Basin of S Africa showing distribution of Ecca Group with coal-bearing facies in the north.

sive regime persisted along the eastern margin of the Australian Plate through much of Devonian to Cretaceous time. The orogenic arc complexes migrated sporadically eastward, with some reversals, and the Bowen–Sydney downwarp originated by a sudden movement of the arc at the end of the Carboniferous Period (Veevers 1984, pp. 257–266), producing an elongate back-arc basin separated from the Cooper–Galilee epicratonic basins by the Nebine Ridge (Fig. 5).

High-angle basement reverse faults are present along the eastern margins of the Bowen–Sydney Basin, including the interconnecting Gunnedah Basin (Fig. 4). A schematic interpretation of their back-arc foredeep origin is illustrated in Fig. 8. The arc complexes comprising the New England Highlands and their extension into Queensland are bounded on their seaward flanks by deformed remnants of earlier forearc basin facies (Skilbeck 1982). Despite this evidence of back-arc compression, extensional basin subsidence may have been important at times, particularly during the early Permian. Indeed, Dewey (1980) has shown that such changes in arc behaviour are to be expected as the nature and thermal properties of a subducting plate change. Old, and therefore cool,

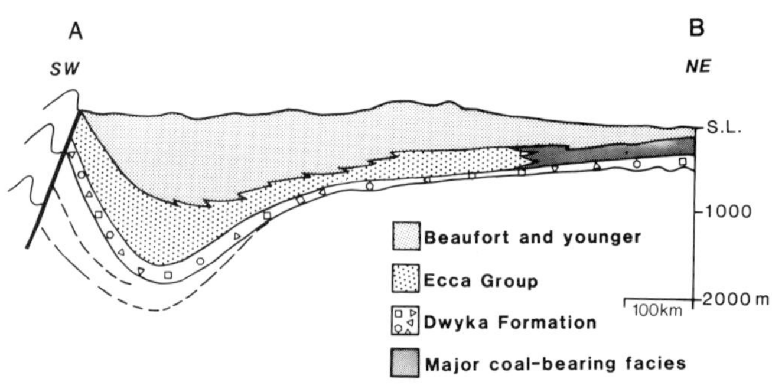

FIG. 7. Cross-section through the Karoo Basin. Line of section A–B shown in Fig. 6.

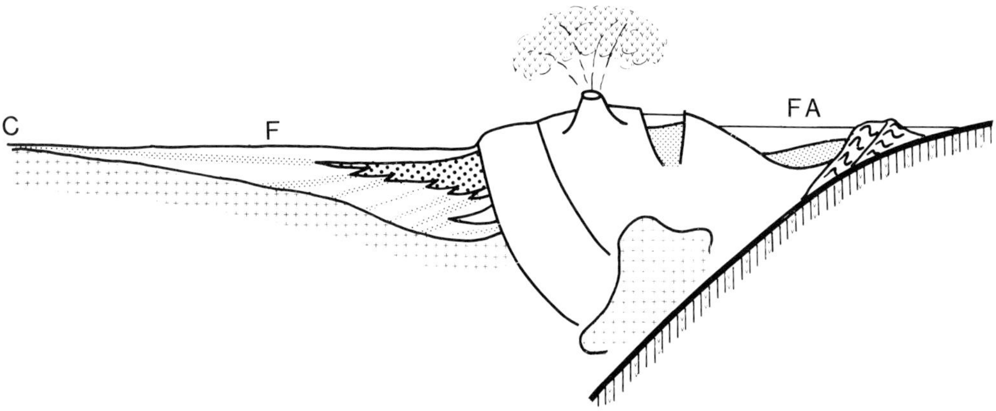

FIG. 8. Generalized model for Permian plate-tectonic origin of the Sydney Basin involving westward underriding by subducting oceanic plate, magmatic orogenic arc, and progradation of volcaniclastic sediments into foreland basin. C = craton, F = foreland, FA = forearc basin.

dense plates sink rapidly at steep angles, carrying the arc seaward and resulting in back-arc extension (Miall 1984, p. 389). This contrasts with subduction of younger, hotter crust which creates back-arc compression as occurred, for example, in the Cape Fold Belt. The strike-slip faulting that characterizes the Sydney Basin (Harrington 1982) is a feature of both compressive and extensional back-arc regimes. Wrench-related pull-apart depressions, for example in the Hunter Valley area of the northern Sydney Basin and Denison Trough, were probably accentuated by subsequent block faulting and extension.

The eastern orogenic arc extended from Cape York in the north along the eastern margins of the Bowen–Sydney Basins (Veevers 1984, p. 240), and is locally preserved on the Australian mainland, permitting a far more detailed correlation between tectonic and depositional events than is possible in the Karoo Basin. The end-Carboniferous episode of downwarping coincident with the initiation of the new arc system led to marine transgression; this was followed by volcaniclastic influx which spread across the Sydney Basin, and from the Bowen Basin westward into the Galilee Basin. Coals accumulated contemporaneously on the cratonic margins in the Cooper and Galilee Basins (Veevers 1984, p. 239). A further marine transgression during the early Permian was interrupted by uplift on the Hunter–Mooki frontal thrust along the western flank of the New England Highlands, with the Greta Coal Measures volcaniclastic wedge (Fig. 9) advancing westward into the foreland basin (Evans & Roberts 1980). This regression was shortlived, and marine conditions were re-established in the Bowen–Sydney Basin until the major tectonically-induced late Permian regression, when coal-

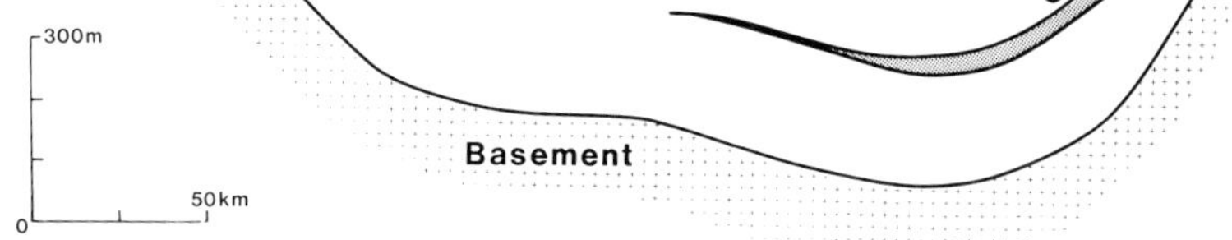

FIG. 9. NE–SW cross-section through the Sydney Basin showing the general distribution of the coal-bearing Greta, Tomago, Newcastle, and Illawarra Groups. After Shibaoka & Smyth (1975).

bearing Tomago–Newcastle–Illawarra Group volcaniclastic sediments were supplemented by minor quartz-rich sediments from the craton (Fig. 9).

Late Permian foreland subsidence was accompanied by intense volcanic activity, westward thrusting and folding (Jones & McDonnell 1981). Three distinct regressive pulses were succeeded by brief episodes of mild uplift of the foreland, followed by marine inundation (Veevers 1984, p. 259). Tectonic activity and associated volcanism ceased toward the end of the Permian Period in the Sydney Basin but persisted until Middle Triassic in the Bowen Basin. Declining clastic supply from the magmatic highlands was compensated for by eastward progradation of craton-derived quartz sands.

Coal rank in these eastern Australian basins is largely a function of maximum depth of burial (Middleton 1983), and decreases markedly from east to west. Exposed sequences in the eastern Sydney and Bowen Basins have vitrinite reflectance values of over 1.0, indicating several thousand metres of erosion. The Upper Permian of the Cooper Basin is currently within the oil-generative zone in some areas ($R_0 = 0.8$–1.3) but is overmature in the Nappamerri Trough depocentre ($R_0 = 2.0$).

The Blair Athol and Wolfang Basins west of the Bowen Basin (Fig. 4) are small rift basins that received limited volumes of clastic sediment, possibly because of elevated rims. Their Lower Permian coals are remarkably thick and deficient in clastic partings and ash. Subsidiary rifting involving essentially vertical movements was also a feature of the Bowen–Sydney Basin but is indistinguishable in detail from the extensional and wrench-related structures of the foredeep.

Depositional systems and coal distribution

The Karoo and eastern Australian basins show a comparable range of depositional systems, although their distribution varies and the individual basins each show some unique characteristics.

Alluvial fan and fan-delta systems

Wet alluvial-fan systems initiated by rapid erosion of adjacent deglaciating highlands, with high runoff of glacial meltwater, developed as extensive aprons along the tectonically active foredeep margins of eastern Australia (Diessel 1970), and on a far smaller scale on the craton. The major clastic wedges are bounded by coals which, in the Sydney Basin, converge toward the west (Fig. 10).

The Sydney Basin examples include classic Gilbertian deltas in which gravelly foresets up to 30 m high (Hunt & Hobday 1984) record progradation into lakes of corresponding depth. In some cases the foresets overlie a coal seam which may have originated by the settling of hydroponic vegetation to the floor of the lake (Conaghan 1982). Elsewhere the coals are present on the upper surface of the subaerial fan-delta plain.

On a far larger scale are lobate or digitate alluvial fans which extended considerable distances across the foreland (Fig. 11). These stream-dominated alluvial-fan systems display exhumed transverse bars containing planar-tabular foresets of coarse sandstone with imbricate pebbles deposited by braided-stream processes, but the boulder-bearing more proximal facies include some debris-flow deposits. Peatswamps devel-

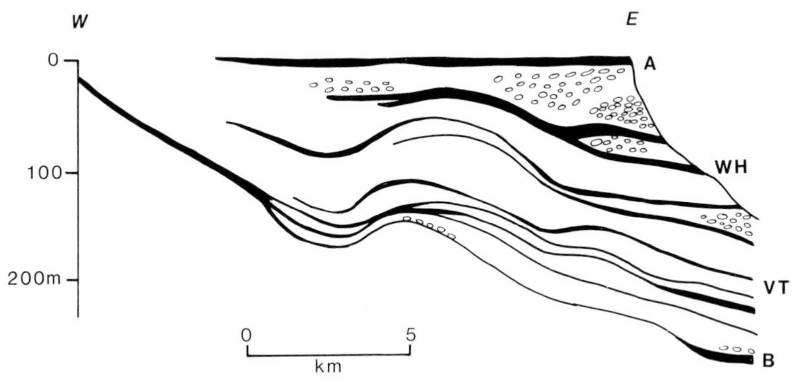

FIG. 10. Cross-section through the lower part of the Newcastle 'coal measures' showing the disposition of Borehole (B), Victoria Tunnel (VT), Wave Hill (WH) and Australasian (A) seams. Coarse-grained, volcaniclastic influx was largely from the NE and east. Modified after Shibaoka & Smyth (1975).

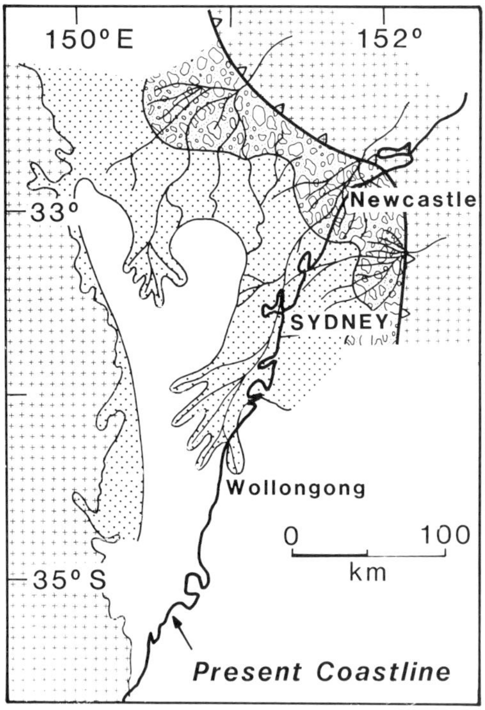

FIG. 11. Digitate patterns of conglomeratic distal alluvial fan systems grading into bedload and mixed-load fluvial systems and shaly lacustine systems of the Newcastle 'coal measures'. Modified after Hunt & Hobday (1984).

oped in depressions between adjacent fans, and along the distal fringes. Temporarily inactive fans were onlapped by the peat as the surface subsided and high groundwater levels provided ideal conditions for plant growth on the abandoned fan surfaces (Hunt & Hobday 1984). The coals thin updip and merge into oxidized palaeosols.

In broad, deglaciated palaeovalleys along the northern flank of the Karoo Basin (Witbank and Highveld Coalfields, Fig. 6) basal Ecca sediments coarsen upward from varved shale into cross-bedded sandstone and conglomerate (Le Blanc Smith 1980, Cadle 1982). This presumed fan-delta sequence is overlain by the '1' and '2' coal seams. Continuing minor discharge was confined to anastomosing fluvial channels by the aggrading peat mat (Cairncross 1980). Le Blanc Smith (1980) recognized a second upward-coarsening sequence capped by coarse-grained clastics and the '3' and '4' seams. Subsequent sequences more strongly resemble shallow-water river-dominated delta systems.

Coals associated with conglomeratic alluvial-fan facies in the Newcastle area of the Sydney Basin (Great Northern, Fassifern, and Wallarah Seams) show alternations of medium- and low-vitrinite plies (Hunt & Hobday 1984). According to Hunt (1984) these coals probably formed from raised secondary or tertiary peats in areas also characterized by groundwater seepage. The proportion of inertinite increases updip into the more proximal fan environment. On the other hand, coals associated with similar alluvial-fan systems in the Gunnedah Basin, e.g. the Maules Creek Formation, are vitrinite-rich, reflecting a wetter climate and more vigorous discharge of ground water (J. W. Hunt, personal communication, 1986).

Fluvial systems

Fluvial systems are distinguished by dip-oriented channel-fill sandstones separated by lenticular but locally very thick coal seams which are broadly elongate parallel to palaeochannels. Most of the Sydney Basin channels represent the distal continuations of alluvial-fan systems, and although finer grained have a similar lithic composition. Smaller fluvial systems originated from the western margins of the Gunnedah Basin and have a highly quartzose composition (Hamilton 1985).

Bedload and mixed-load fluvial systems are readily recognized in outcrop, but their subsurface distinction is generally difficult because of sparse data. Point-bar accretion surfaces, upward-fining, coal-capped sequences, and preserved cutbank surfaces are all characteristic of the mixed-load systems. Some of these laterally migrating channels were bounded by exceptionally thick peat which provided spectacular deformation as the channels switched position, illustrated in Moura Mine, Queensland (Mallet *et al.* 1983). The effects of differential compaction are illustrated in Galloway & Hobday (1983, p. 285).

Bedload systems, which are commonly gradational with digitate alluvial-fans (Fig. 11), do not show well-defined upward-fining patterns except near the tops of sequences which are otherwise dominated by cross-bedded sandstone, commonly pebbly. Some sequences were probably deposited by braided streams, as suggested by planar-tabular foresets, exhumed braid bars, and bar-modifying cosets of smaller cross beds. Others were coarse-grained meanderbelt systems, with bank cohesion imparted by vegetation.

Suspended-load fluvial systems commonly overlie some of the thicker coal seams in the Karoo Basin, and in some cases produce abrupt clastic splits in seams. Cairncross (1979) demonstrated that in the Van Dyks Drift Colliery of the southeastern Transvaal, minor stream courses that persisted across the peatswamps were fixed by the resiliance of the surrounding peat. These streams carried a predominance of fine-grained, suspended sediment with a small proportion of sand. In-channel aggradation proceeded contemporaneously with growth of the peat mat, resulting in narrow but remarkably thick, dip-elongate channel-fill sediments with thick, persistent coals in the interchannel lows. Comparable features are present locally in the Bowen–Sydney Basin. Thin ash bands extend laterally from the channel-fill clastics, so that the best quality coal is generally located farthest from the channels.

High ash values in coals overlying channel-fill sandstones are attributed by Cairncross (1979) to partial oxidation and degradation of the peat mat leaving a residual concentration of inherent ash.

The coals of the less humid epicratonic basins of eastern Australia have higher vitrinite content in the vicinity of channels, with their more persistently high water tables, than in the interchannel backswamp depressions where enrichment in inertinite is attributed to limited transportation of vegetal debris and variable water tables with temporary oxidation (Hunt 1984, Hunt *et al. in press*). Hunt *et al.* (*in press*) record 'band macrinite', derived from partly gelified, then oxidized, woody material as a prominent accessory. Fluvial coals of the associated foredeeps show a wide range in vitrinite content, with evidence of subaerial oxidation of woody peat. However, high-vitrinite seams such as Baralaba and Maules Creek are associated with bed-load fluvial systems that were characterized by consistent meteoric groundwater discharge and correspondingly high water tables (Hunt 1984).

Delta-plain systems

The upper delta plain is gradational with fluvial systems; any boundary is arbitrary but is generally marked by a gradual change from multistoried fluvial channel-fill, and a low proportion of shale, to discrete distributary channel-fill sequences and more widespread shale intervals (Conolly & Ferm 1971). This gradation is well exposed in outcrops around Newcastle and north of Wollongong in the Sydney Basin (Fig. 12). In the Karoo Basin the delta-plain facies are identified by their relationship to underlying thick, upward-coarsening prodelta through delta-front sequences (Hobday 1973, Van Vuuren & Cole 1979, Le Blanc Smith 1979, 1980). The associated delta-plain facies include predominantly mixed-load and suspended-load channel facies, burrowed shales, and upward-coarsening bay-fill sequences with sandy and silty crevasse splays and splay subdeltas along channel margins.

Coals typically overlie, and interfinger with, the upper parts of the bay-fill sequences, and individual seams are commonly separated by thin

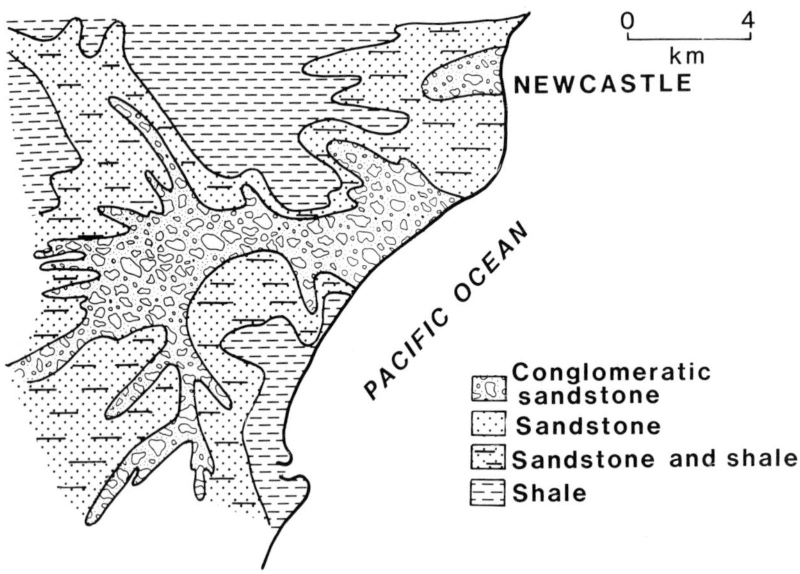

FIG. 12. Sydney Basin palaeogeography during deposition of the Newcastle 'coal measures', grading from conglomeratic alluvial-fan systems in the north through fluvial to river-dominated deltaic systems, with strandplain sands along the western margin. After Herbert (1980).

bay-fill shales and clay-rich sapropelic bands marking temporary inundation of marsh surfaces. Similar lacustrine facies are also present as 'wants' within thick seams. Lenticular sandstone splits are concentrated in the vicinity of distributary channels and provide a particular hazard in longwall mining. Thin clastic bands are present in broad belts following palaeochannels (Cairncross 1979) or alternatively form uniform 2–20 cm partings that extend over broad areas of the Sydney and Bowen Basins. These may be volcanic ash-fall layers or may represent marsh surfaces that were degraded by a change in pH occasioned by freshwater flooding.

Delta-plain coals are generally thinner but more widespread than their fluvial counterparts, and represent some of the most important commercial seams. They are broadly dip-elongate where confined by distributary channels, but may have a blanket geometry over abandoned delta lobes. Hamilton (1985) showed in the Gunnedah Basin that the distribution of lower delta plain coal corresponds closely with the outline of the underlying river-dominated deltaic systems. Hamilton has clearly distinguished between coals that originated as blanket peats over abandoned delta lobes and thinner, discontinuous coals that originated in marshes flanking active distributary channels. The blanket-like Melvilles Coal covers a variety of delta-plain clastic facies, but is thickest above areas of high net-sand ratio.

In the Karoo Basin, Cadle (1979) distinguished between thin, low-ash coals of the lower delta plain and thicker, higher ash coals of the upper delta plain. This relationship is generally obscured in eastern Australia because of the high proportion of volcanic tuffs.

Upper delta plain coals of the Sydney, Gunnedah and Bowen Basins are predominantly vitrinite-rich (Hunt 1984), with vitrinite content decreasing toward the upper delta plain and fluvial facies. There are, however, notable exceptions such as the Bayswater Seam and correlatives which span several environments, including delta plain, but have a remarkably low vitrinite content (25% mmf; Hunt 1984). Hunt ascribes this to regional epeirogenic movement or eustatic change. Galilee Basin coals show a southward increase in vitrinite content from 30 to 60% mmf, consistent with a change from fluvial to deltaic facies.

The Bulli Seam of upper delta plain origin in the southern Sydney Basin contains three petrographic cycles of upward-decreasing vitrinite content. The cycles are attributed to evolution from tree-bearing minerotrophic primary swamps to rainfall-dependent tertiary marshes followed by flooding and renewal of the cycle (Shibaoka & Smyth 1975, Hunt 1984). In contrast, upper delta plain coals of the Gunnedah Basin are vitrinite rich, consistent with a somewhat earlier, wetter climatic regime (J. W. Hunt, personal communication, 1986).

Consistent with their proximity to marine environments, the lower delta plain coals contain moderately high amounts of sulphur and phosphorus (Hunt 1984). Widened marine transgressions across the northern Karoo Basin, marked by glauconitic sediments (Cairncross 1979), have produced coals with pyritic upper surfaces. Hamilton (1985) suggested that storm inundation of peat swamps in the Gunnedah Basin caused bacterial degradation and concentration of inherent ash after the manner suggested by Renton & Cecil (1979).

Lake-margin systems

Coals of the lower and upper parts of the Patchawarra Formation, Cooper Basin, are associated with very thin shaley sequences with sporadic thin sands. These clastic sediments are thought to have been deposited in shallow, fluctuating lakes bounded by extensive but relatively shortlived marshes. The coals are thin (mainly 20–100 cm) and discontinuous, and contrast with the thick coals of the fluvial middle Patchawarra Formation. 'Reedy' lacustrine coals have also been identified in the Bowen Basin by Beeston (1982).

These lake-margin coals are low in vitrinite and semifusinite and show finely banded microlithotypes (Hunt 1982). According to Hunt (1984) periodic flooding and desiccation gave rise to hydroseral successions of peats possibly formed from reedy and herbaceous vegetation. The peats would have been subject to subaquatic oxidation in the three-dimensionally mixed waterbodies. Furthermore, large-scale fluctuation in the water table provided periodic subaerial oxidation.

Back-barrier systems

Coals that developed on a platform of infilled lagoonal sediments or in a strandplain environment are volumetrically less important but are present in most Gondwana basins. In the Karoo Basin such coals are thin, persistent along depositional strike, and locally low in ash (Cadle 1979), although many are relatively high in ash.

In the Bowen–Sydney Basin, back-barrier coal facies are generally toward the base, or near the basinward edges of deltaic successions. Their morphologic association with barrier-island facies is difficult to recognize because of transgressive reworking. The Australian back-barrier coals have a high inherent ash content and are low in vitrinite, probably because of the prevailing herbaceous vegetation. The inertodetrinite ratio is high, and relatively high sulphur and phosphorus values are typical (Hunt 1984).

A well documented example in the Gunnedah Basin (Hamilton 1985) is the Hoskissons Coal which is unusually extensive and ranges up to 18 m in thickness, implying substantial peat aggradation accompanying base-level change. The coal overlies an extensive platform of marine-reworked delta-platform sediments, probably representing a broad lagoonal embayment enclosed to seaward by barrier sands.

Starved basin peats

Coals in the Blair Athol, Wolfang, and Oaklands Basins are thick (up to 40 m) and generally independent of clastic facies. Smyth (1980) proposed that the coals represent raised bogs or mounded peats that originated as primary basal peats overlain by secondary, and finally tertiary peats fed by direct precipitation (ombrotrophic). The height to which the convex peat surface rose would have been limited primarily by the balance between evaporation and precipitation and declining nutrient supply (see Clymo, *this volume*), but may have been enhanced by seasonal freezing (Hunt 1984). Vitrinite content is exceptionally low except in the basal parts, and the coals are finely banded, with low semifusinite ratios, and remarkably free of mineral matter.

The original thickness of peat must have been enormous given the ablation of the accumulation by oxidation (Hunt 1984) plus burial compaction. There was no opportunity for the introduction of detrital mineral matter to the raised peat surface, but it is surprising that inherent ash content is negligible given the prolonged degradation of the peat surface.

Economic significance

The coal-mining industry is of crucial importance to the economies of both Australia and S Africa, and large volumes of commercial hydrocarbons are locally present in the coal-bearing succession.

Australia is the world's largest exporter of coal, some 88 million tonnes (Mt) in 1985 and expected to increase. This accounts for \$A5 billion a year in export earnings. In addition, domestic consumption amounts to 40 Mt of black coal. Some 95% of production is in New South Wales and Queensland. S Africa exports approximately 45 Mt pa and consumes vast quantities in Sasol oil-from-coal plants and in electricity generation. While coal's share of the energy market fell below 18% in the United States it remained above 75% in S Africa (Van Rensburg 1982).

Apart from coal sales, the excise, royalty and company tax on oil and gas generated and reservoired within coal basins in Australia accounted for over \$A5 billion in 1985 government revenue. The Gippsland Basin with maximum 1985 production rates of over 500 000 BOD (barrels of oil per day) and EUR (estimated ultimate reserves) of 4 billion barrels, is thought to have been internally sourced by Cretaceous–early Tertiary organic matter of land-plant origin (Thomas 1982, Thomas *et al.* 1982). The Cooper-Eromanga, Australia's most prolific onshore oil basin, is likewise sourced by coal-bearing, non-marine sequences of Permian to Jurassic age. Contrary to prevailing geochemical theory, coal and associated dispersed organic matter can be highly efficient in generating and expelling oil as well as gas (see Murchison, Khorasani, *this volume*).

Coal macerals as a petroleum source

Following Tissot *et al.* (1974), Type I kerogen of aquatic algal composition has the highest source potential for oil. Type II kerogen derived from marine phytoplankton and zooplankton has good potential for sourcing both oil and gas, whereas Type III kerogen of land-plant origin is generally gas prone. Type III kerogen comprising coal seams is also represented by dispersed organic matter (DOM) of the exinite, vitrinite, and inertinite maceral groups. The Van Krevelen diagram shows that exinite (liptinite) has a high H/C ratio capable of generating oil. The H/C ratio is lower for vitrinite and very low for inertinite, and neither maceral group is conventionally regarded as capable of generating oil. Oil generated by humic coal is distinguished by long-chain n-alkanes ($C_{22}+$), high wax content, high pour point, high ratios of pristane (C_{19}) to phytane (C_{20}) and a dominance of C_{29} steranes typical of the higher plants (Shanmugam 1985).

In a study of Cooper Basin source potential, Smyth (1983) showed that the coals and associated DOM are rich in inertinite and vitrinite with practically no exinite (Fig. 13), yet this non-marine organic matter is the undisputed source of Cooper Basin oil. These Permian rocks may have sourced some of the oil in the overlying Jurassic–Cretaceous Eromanga Basin, but Jurassic non-marine shales are regarded as at least a partial source. The Jurassic strata are coal-bearing in part, contain more exinite than the

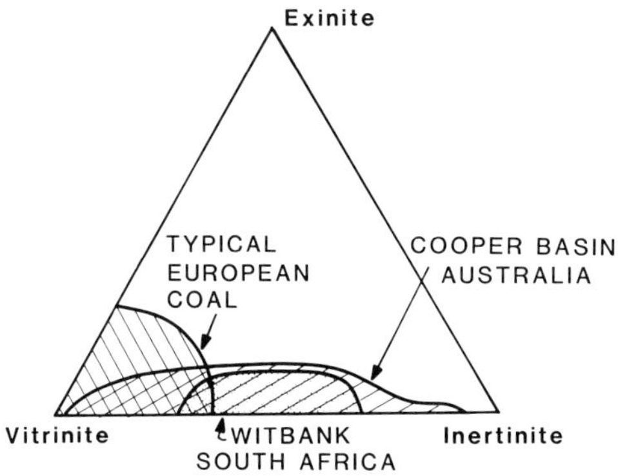

FIG. 13. Comparison of some typical maceral group compositions from Europe, Australia and S Africa. Compiled by Michelle Smyth.

Permian, and Cook (1982) proposed that oil was generated at relatively low maturation levels from vitrinite, suberinite in cell walls, bitumenite and resinite macerals of the exinite group. Oil in the Bowen and overlying Jurassic–Cretaceous Surat Basin (see Khorasani, *this volume*) was likewise regarded by Thomas (1982) as sourced from land-plant macerals of Permian age. Thomas showed that the exinite content in Jurassic and younger rocks is typically higher as a result of evolutionary expansion in the conifers and evolution of angiosperms (see also Collinson & Scott, *this volume*). In epicratonic basins such as the Cooper and Eromanga, however, these potentially prolific source rocks tend to be immature except in local depocentres or loci of higher geothermal gradient. Butcher (1984) has shown, for example, that Permian-generated oil moved up the western flank of the Bowen–Surat Basins beneath semi-regional seals to their updip edges where the oil entered successively younger stratigraphic units. Golin & Smyth (1986), however, provide evidence of some oil generation in the Jurassic sequence. Indeed, Moore *et al.* (1986) suggest that dual or multiple sourcing is typical of several Australian and Chinese non-marine petroleum basins.

The generation of Gippsland Basin oil from Upper Cretaceous–Tertiary humic coal has been long demonstrated (Brooks & Smith 1969, Cook 1982, Thomas 1982, Smith & Cook 1984), and has recently been elaborated upon by Shanmugam (1985). Vitrinite dominates, but exinite values are consistently high, with sporinite and liptodetrinite most common. Shanmugam speculates that conditions resembled the southern coniferous kauri forests of North Island, New Zealand today, with peat accumulating in extensive raised bogs.

Conclusions

Gondwana coal basins of Australia and S Africa show a comparable geological response to late Palaeozoic drift across the pole and subsequent warming, with glaciogenic sediments giving way to coal measures and then redbeds. Back-arc basins of the Gondwana Orogenic Belt include the Bowen, Gunnedah, Sydney, and southern Karoo downwarps, which merge with the epicratonic Cooper, Galilee, and northern Karoo Basins. Coals are widely distributed through all tectonic environments in Australia but are largely restricted to the northern, epicratonic portion of the Karoo Basin of S Africa.

The thickness, geometry, and quality of the coals are a product of the depositional environment, but some seams show a dominant climatic, palaeotopographic, tectonic, or eustatic overprint. All of these factors can be responsible for water-table fluctuations, ranging in scale from local (depositional environment) to regional (climate and tectonism) and global (eustatic changes).

Coals are associated with the spectrum of subaerial environments from alluvial fan to various fluvial systems, delta plain, back-barrier, lake-margin, and clastic-starved raised bog. Some alluvial-fan seams show regular alternation of dull and brighter plies, probably reflecting variations in groundwater seepage. The proportion of inertinite increases updip while vitrinite increases toward the downdip fluvial facies, which include thick, dip-elongate seams with variable ash content. Local variability in banding and maceral content is attributed to differences in channel and backswamp characteristics, floral zonation, and elevation relative to the water table. Delta-plain coals are some of the most extensive, with high vitrinite content, but may have moderately high sulphur and phosphorus content where transgression has occurred. Lake-margin and back-barrier coals are generally thin and low in vitrinite perhaps because of the dominant herbaceous vegetation and susceptibility to oxidation. Some back-barrier seams are anomalously thick, but most have above-average sulphur and ash. Thick, low-ash seams that accumulated as raised peat bogs are low in vitrinite because of pervasive oxidation.

Apart from the crucial significance of coal mining to the economies of both Australia and S Africa, large volumes of oil and gas have been sourced by coal and dispersed coal particles. Source and maturation models involving land-plant macerals have been accompanied by successful exploration for oil in several non-marine basins in Australia.

ACKNOWLEDGEMENTS: I thank the convenors of the symposium for their support which enabled me to attend. Bridge Oil Ltd staff contributed to the geological concepts, and the company kindly provided drafting and typing services. Michelle Smyth, John Hunt, Graham Lucas and Rowan Roberts commented on the manuscript.

References

BEESTON, J. W. 1982. Coal sedimentation in the Reids Dome Beds, Bowen Basin: a depositional model. *Australian Coal Geology*, **4**, 503–504.

BROOKS, J. D. & SMITH, J. W. 1969. The diagenesis of plant lipids during the formation of coal, petroleum and natural gas; II, coalification and the formation of oil and gas in the Gippsland Basin. *Geochimica et Cosmochimica Acta*, **33**, 1183–1194.

BUTCHER, P. M. 1984. The Showgrounds Formation, its setting and seal, in ATP 145P, Queensland. *Australian Petroleum Exploration Association Journal*, **24**, 336–357.

CADLE, A. B. 1979. Depositional models from the Middle Ecca and their influence on coal accumulation in the eastern Transvaal. *Geological Society of South Africa 18th Congress Abstracts*, **1**, 77–78.

—— 1982. Controls on coal distribution. *In:* CADLE, A. B. (ed.) *Coal Exploration, Economics and Assessment*, pp. I-20–I-34. University of the Witwatersrand.

CAIRNCROSS, B. C. 1979. Depositional framework and control of coal distribution and quality, Van Dijks Drift area, northern Karoo Basin. *Unpublished MSc thesis*, University of Natal, Pietermaritzburg, South Africa.

—— 1980. Anastomosing river deposits: palaeoenvironmental control on coal quality and distribution, northern Karoo Basin. *In:* CADLE, A. B. (ed.) *Karoo Basin Symposium*, pp. 301–331. Geological Society of South Africa, Special Issue.

CONAGHAN, P. J. 1982. Lacustrine (Gilbertian) deltas in the Permian coal measures of eastern Australia and India: implications for the widespread hydroponic origin and deep-water diagenesis of Gondwana coals. *16th Sydney Basin Symposium*, pp. 32–34. University of Newcastle, NSW.

CONOLLY, J. R. & FERM, J. C. 1971. Permo-Triassic sedimentation patterns, Sydney Basin, Australia. *American Association of Petroleum Geologists Bulletin*, **55**, 2018–2032.

COOK, A. C. 1982. Organic facies in the Eromanga Basin. *In:* MOORE, P. J. & MOUNT, T. J. (eds). *Eromanga Basin Symposium*, pp. 234–257. Petroleum Exploration Society of Australia and Geological Society of Australia, Adelaide.

CREER, K. M. 1965. Palaeomagnetic data from the Gondwanic continents. *Philosophical Transactions. Royal Society of London A*, **258**, 27–40.

CROWELL, J. C. & FRAKES, L. A. 1972. Late Palaeozoic glaciation: Part V, Karoo Basin, South Africa. *Geological Society of America Bulletin*, **83**, 2887–2912.

—— & —— 1975. The Late Paleozoic glaciation. *In:* CAMPBELL, K. S. W. (eds) *Gondwana Geology*, pp. 313–331. ANU Press, Canberra.

DEWEY, J. F. 1980. *In:* STRANGEWAY, D. W. (ed.) *Episodicity, Sequence and Style at Convergent Plate Boundaries*, pp. 553–573. Geological Association of Canada Special Paper 20.

DE WIT, M. J. 1977. The evolution of the Scotia Arc as a key to the reconstruction of southwestern Gondwanaland. *Tectonophysics*, **37**, 53–81.

DIESSEL, C. F. K. 1970. Paralic coal formation. *Institute of Fuel, Australia, 8th Conference*, Brisbane, 14-1–14-22.

DRAPER, J. W. & BEESTON, J. W. 1985. Depositional aspects of the Reids Dome Beds, Denison Trough. *Queensland Government Mining Journal*, May, 200–210.

EVANS, P. R. & ROBERTS, J. 1980. Evolution of central eastern Australia during the late Palaeozoic and early Mesozoic. *Geological Society of Australia Journal*, **26**, 325–340.

GALLOWAY, W. E. & HOBDAY, D. K. 1983. *Terrigenous Clastic Depositional Systems: Application to Coal, Uranium and Petroleum Exploration.* Springer-Verlag, New York.

GOLIN, V. & SMYTH, M. 1986. Depositional environments and hydrocarbon potential of the Evergreen Formation, ATP-145P, Surat Basin, Queensland. *Australian Petroleum Exploration Association Bulletin*, **26**, 156–171.

HAMILTON, D. S. 1985. Deltaic depositional systems, coal distribution and quality, and petroleum potential, Permian Gunnedah Basin, NSW, Australia. *Sedimentary Geology*, **45**, 35–75.

HARRINGTON, H. J. 1982. Tectonics of the Sydney Basin—keynote address. *In: 16th Sydney Basin Symposium*, pp. 15–19. University of Newcastle, NSW.

HERBERT, C. 1980. Depositional development of the Sydney Basin. *In:* HERBERT, C. & HELBY, R. (eds) *A Guide to the Sydney Basin*, pp. 10–53. Geological Survey of NSW Bulletin 26.

—— & HELBY, R. 1980. *A Guide to the Sydney Basin.* Geological Survey of NSW Bulletin 26.

HOBDAY, D. K. 1973. Middle Ecca deltaic deposits in the Muden-Tugela Ferry area of Natal. *Geological Society of South Africa Transactions*, **76**, 309–318.

HUNT, J. W. 1982. Relationships between microlithotype and maceral composition of coals and geological setting of coal measures in the Permian basins of eastern Australia. *Australian Coal Geology*, **4**, 484–502.

—— 1984. Ash content and composition of Permian coals. *In: Permian Coals of Eastern Australia.* National Energy Resources Development and Demonstration Council Project 78/2617.

——, BRAKEL, A. T. & SMYTH, M. in press. Origin and distribution of the Bayswater seam and correlatives in the Permian Sydney and Gunnedah Basins, Australia. *Australian Coal Geology.*

—— & HOBDAY, D. K. 1984. Petrographic composition and sulphur content of coals associated with alluvial fans in the Permian Sydney and Gunnedah Basins, eastern Australia. *In:* RAHMANI, R. A. & FLORES, R. M. (eds) *Sedimentology of Coal and Coal-bearing Sequences*, pp. 43–60. Special Publication of the International Association of Sedimentologists 7. Blackwell Scientific Publications, Oxford.

JONES, J. G. & MCDONNELL, K. L. 1981. Papua New Guinea analogue for the late Permian environment of northeastern NSW, Australia. *Palaeogeography, Palaeoclimatology, Palaeoecology*, **34**, 191–205.

KINGSLEY, C. S. 1979. A composite turbidite-delta-fluvial model for Permian sedimentation in the Ecca and lower Beaufort of the eastern Cape Province. *Geological Society of South Africa 18th Congress Abstract*, **1**, 96.

LE BLANC-SMITH, G. 1979. High-constructive deltaic sedimentation and coal accumulation, Witbank Basin. *Geological Society of South Africa, 18th Congress Abstract*, **2**, 88–92.

—— 1980. Genetic stratigraphy and palaeoenvironmental controls on coal distribution in the Witbank Basin Coalfield. *Unpublished PhD thesis*, University of the Witwatersrand, South Africa.

LOCK, B. E. 1980. Flat-plate subduction and the Cape Fold Belt of South Africa. *Geology*, **8**, 35–39.

MALLET, G. W., FLOOD, P. G. & LEDGER 1983. Sedimentation models for the Baralaba Coal Measures, Moura, Queensland. *Australian Coal Geology* 5.

MIALL, A. D. 1984. *Principles of Sedimentary Basin Analysis*. Springer-Verlag, New York.

MIDDLETON, M. F. 1983. Coal rank trends in eastern Australian Permian coal basins. *CSIRO Institute of Energy and Earth Resources Investment Report* 141.

MOORE, P. S., HOBDAY, D. K., MAI, H. & SUN, Z. C. 1986. Comparison of selected non-marine petroleum-bearing basins in Australia and China. *Australian Petroleum Exploration Association Journal*, **26**, 285–309.

PLUMSTEAD, E. P. 1969. Three thousand million years of plant life in Africa. *Geological Society of South Africa Transactions*, **72** (Annex.).

RENTON, J. J. & CECIL, C. B. 1979. The origin of mineral matter in coal. *In:* DONALDSON, A. C., PRESLEY, M. W. & RENTON, J. J. (eds) *Bulletin of the West Virginia Geological Survey*, B-37-1, 206–223.

RETALLACK, G. 1980. Late Carboniferous to Middle Triassic megafossil floras from the Sydney Basin. *Geological Survey of NSW Bulletin*, **26**, 384–430.

RHODES, R. C. 1974. Discussion of paper by A. R. Newton: A gravity folding model of the Cape Fold Belt. *Geological Society of South Africa, Transactions*, **77**, 207–209.

SHANMUGAM, G. 1985. Significance of coniferous rain forests and related organic matter in generating commercial quantities of oil, Gippsland Basin, Australia. *American Association of Petroleum Geologists Bulletin*, **69**, 1241–1254.

SHIBAOKA, M. & SMYTH, M. 1975. Coal petrology and the formation of coal seams in some Australian sedimentary basins. *Economic Geology*, **70**, 1463–1473.

SKILBECK, G. C. 1982. Carboniferous deposition systems of the Myall Lakes, northern NSW. *In:* FLOOD, P. G. & RUNNEGAR, B. (eds) *New England Geology*, pp. 121–132. University of New England.

SMYTH, M. 1980. Thick coal members; products of an inflationary environment. *Australian Coal Geology*, **2**, 53–76.

—— 1983. Nature of source material for hydrocarbons in Cooper Basin, Australia. *American Association of Petroleum Geologists Bulletin*, **67**, 1421–1428.

STUART, W. J. 1976. The genesis of Permian and Lower Triassic reservoir sandstone during phases of southern Cooper Basin development. *Australian Petroleum Exploration Association Journal*, **16**, 37–48.

TAVENER-SMITH, R. 1980. The *Skolithos*-bearing sandstone at Newlands Estate, Durban: a suspected shallow-water, offshore sand tongue in the Lower Ecca Group (Pietermaritzburg Formation). *Geological Society of South Africa Transactions*, **83**, 89–91.

THOMAS, B. M. 1982. Land-plant source rocks for oil and their significance in Australian basins. *Australian Petroleum Exploration Association Journal*, **22**, 164–178.

——, OSBORNE, D. G. & WRIGHT, A. J. 1982. Hydrocarbon habitat of the Surat/Bowen Basin. *Australian Petroleum Exploration Association Journal*, **22**, 213–226.

TISSOT, B., DURAND, B., ESPITALIE, J. & COMBAZ, A. 1974. Influence of nature and diagenesis of organic matter in formation of petroleum. *American Association of Petroleum Geologists Bulletin*, **58**, 499–506.

VAN RENSBURG, W. C. J. 1982. Coal economics. *In:* CADLE, A. B. (ed.) *Coal Exploration, Economics and Assessment*, pp. II-1–III-49. University of the Witwatersrand.

VAN VUUREN, C. J. & COLE, D. I. 1979. The stratigraphy and depositional environments of the Ecca Group in the northern part of the Karoo Basin. *Geological Society of South Africa Special Publication*, **6**, 103–111.

VEEVERS, J. J. (ed.) 1984. *Phanerozoic Earth History of Australia*. Oxford Geology Science Series 2. Clarendon Press, Oxford.

VISSER, J. N. J. *et al.* 1978. The Dwyka Formation and Ecca Group, Karoo sequence, in the northern Karoo Basin, Kimberley-Britstown area. *South African Geological Survey Annals*, **12**, 143–176.

VON BRUNN, V. 1977. A furrowed intratillite pavement in the Dwyka group of northern Natal. *Geological Society of South Africa Transactions*, **80**, 125–130.

D. K. HOBDAY, 1A Wudgong St, Mosman Bridge Oil Ltd, Sydney 2088, Australia.

Coal geology and exploration of part of the Tertiary Kutei Basin in East Kalimantan, Indonesia

D. H. Land & C. M. Jones

ABSTRACT: Geological survey of some 700 km^2 of Miocene coal-bearing strata in Borneo, near Samarinda, E Kalimantan, Indonesia, identified 1000 Mt of recoverable reserves of coal, ranging in rank from lignite A to high-volatile C bituminous, in 43 seams between 1.5 and 13 m thick. Environments of deposition are paralic and include shoreline, deltaic and alluvial flood-plain. Strata studied are divided into four newly-defined lithological formations, the Loa Duri, Loa Kulu, Prangat and Kamboja formations, in a section totalling over 3000 m thick. The survey included drilling 72 cored boreholes, between 200 and 350 m deep, of which 60 each proved at least one coal seam over 2 m thick. The coals have low ash, high moisture and generally low sulphur.

Keywords: coal, exploration, environments, lignite, bituminous coal, Tertiary, sedimentology, resources, foraminifera, Indonesia

In the early 1980s the Indonesian and British Geological Surveys undertook a joint coal exploration project in E Kalimantan, which was funded by the governments of both countries under Colombo Plan auspices. An area of Tertiary coal measures south and west of Samarinda, E Kalimantan, in the island of Borneo, was geologically surveyed and drilled, and 1000 million tonnes (Mt) of identified recoverable coal reserves were proved. Geological results of the project are fully reported in Land *et al.* (1986) and exploration techniques in Thomas *et al.* (1980). This paper summarizes the work and serves to introduce it to a wider audience.

Oil and gas supply almost all Indonesia's energy needs as well as a large proportion of the foreign exchange earnings. To safeguard these while the economy expands, alternative energy sources are sought, particularly for electricity generation. Large coal resources are known to exist (van Bemmelen 1949; Koesoemadinata *et al.* 1977), almost all in Tertiary basins which cover extensive areas of the country (Fig. 1). Katili & Johannas (1961) pointed out the need for quantitative assessment of reserves as an essential first step in planning their exploitation. Until fairly recently, with oil being relatively cheap and plentiful, there was not much economic incentive to consider coal. It is being mined at Bukit Asam and Ombilin in Sumatra, and on a very small scale elsewhere.

The E Kalimantan Coal Project was arranged to expedite the long-term study of Indonesian coal resources, by assessing a specific area, and at the same time providing the transfer of technology in coal geology and exploration. An area near Samarinda in E Kalimantan was chosen because it was known to contain coal, it was a development area, and because coal could be readily shipped, for example to Java for electricity generation. Coal had also been mined between 1880 and 1970 at Loa Kulu, 15 km west of Samarinda, annual production being up to 258 000 t.

Geography

The ground surveyed in the coal project, the 'project area', is in eastern Borneo in E Kalimantan province, south of the River Mahakam, between 50 and 110 km south of the equator, and between 20 and 50 km inland from the Makassar Strait (Figs 2 and 3). Geologically it lies within the Tertiary Kutei Basin. Few points in the area exceed 100 m in elevation above sea-level, but it is topographically complex, being deeply dissected by a dendritic network of steep sided ravines.

Climatically the area is hot and humid. Temperature varies between 23° and 37°C, while the rainfall of nearly 3000 mm y^{-1} shows little seasonal variation and falls mainly in very heavy showers. In common with the rest of eastern Kalimantan the natural vegetation is tropical rain forest. While the large timber has been felled over wide areas and some ground has been cleared for agriculture, the area is still largely forest covered. Population density is low, but development of timber and oil extraction and associated industries has led to considerable growth, and the region's two main towns of Samarinda and Balikpapan both now exceed a quarter million inhabitants. There are few roads, but logging tracks made for timber extraction provided access to many drilling sites in the

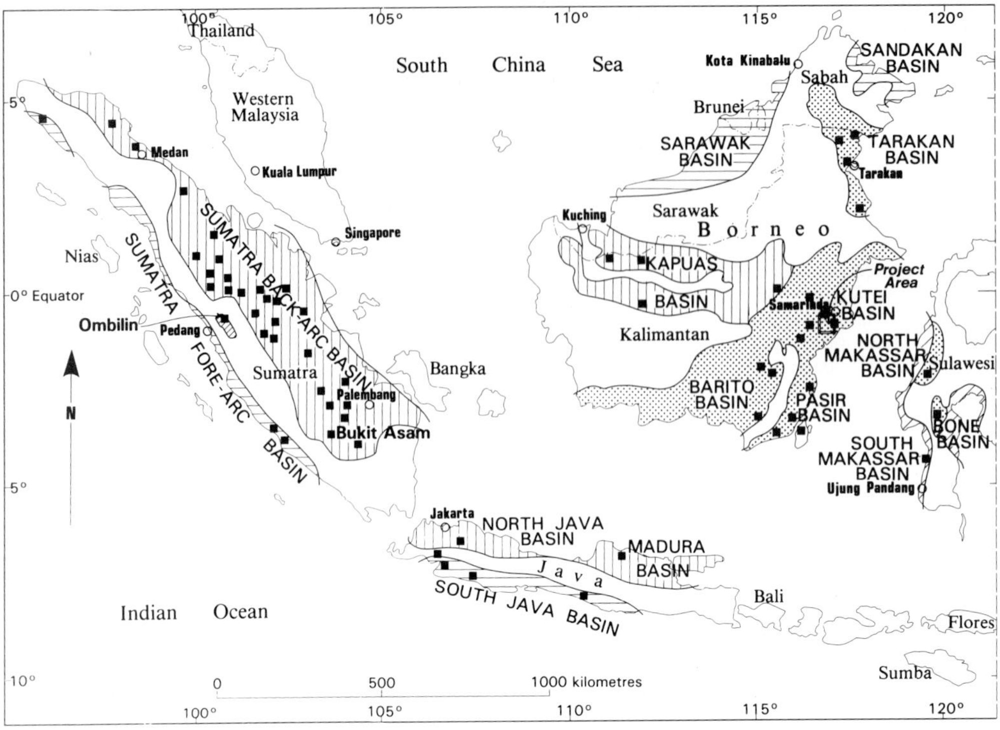

FIG. 1. Onshore Tertiary basins in Indonesia and Malaysia.

project. Most settlements are on rivers and water transport is important. Loa Kulu, for example, is on the Mahakam, here 1 km wide and about 17 m deep, 90 km from the sea and still well within tidal influence.

The geographical circumstances outlined above naturally influenced the exploration methodology, the chief constraint being difficulty of access for drilling. In this particular project it was decided to use only existing roads, and not to create additional access either by road or helicopter. With the frequent heavy rain storms, roads and tracks need constant maintenance.

Strata in the project area are mostly sandstones and mudstones, so that a scarp and dip topography is developed with sandstones making the ridges. This is dramatically marked in anticlinal areas where dips are steep: here the sandstones make long, narrow strike ridges. In synclinal areas where dips are very gentle, the strike alignment of the ridges is less marked.

The drainage pattern as a whole shows very poor correlation with the geological structure which is dominated by NNE-trending anticlines. The largest river, the Mahakam, meanders right across the NNE-trending ridges, and on a smaller scale several rivers only a few metres wide cut antecedent valleys through ridges 100 m high. The drainage is thus superimposed, and either originated on structurally simple strata now eroded away, or originated before folding with river erosion keeping pace with the rising anticlines and maintaining the stream courses. Both mechanisms appear to have operated, because there is good evidence (see below) that in the Kutei Basin, anticlines have grown during sedimentation. This means that the sediment accumulation surface is virtually flat and folds at depth have no topographic expression. As the sea regresses, a dendritic drainage pattern develops which is able to maintain itself by erosion as deeper structural levels are gradually exposed.

The project area lies within zone 50M of the Universal Transverse Mercator projection, and grid references to locations are given as UTM co-ordinates, generally to the nearest 10 m, in the form (NQ 0548 3445). These were determined by surveying from triangulation stations T9 and T28 of Roy M. Huffington Inc. who supplied co-ordinates for them:

T9	Gunung Kelapis	NQ 00632.7 37589.6
T28	Gunung Loatillo	NQ 06967.0 30223.0

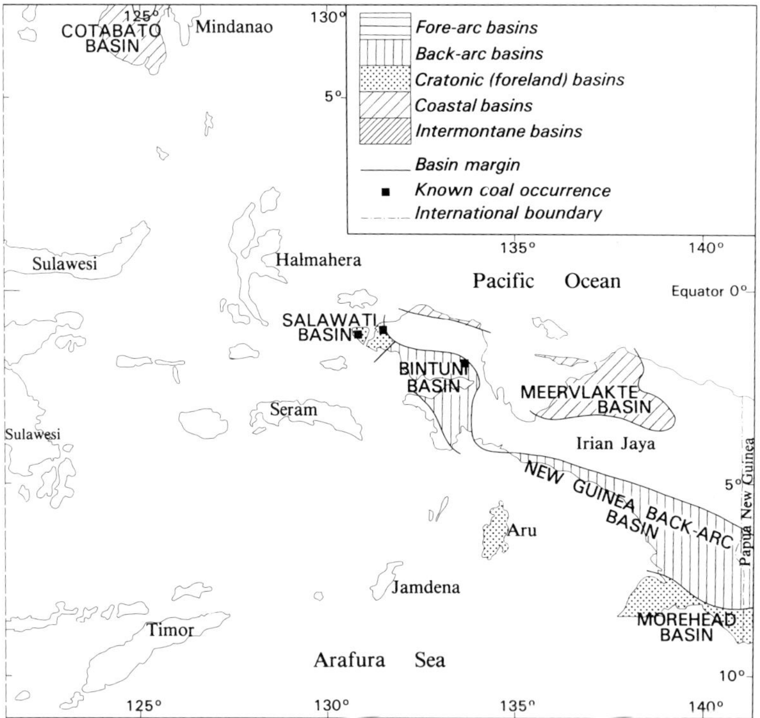

(Based on Hamilton 1979 and Koesoemadinata and others 1977)

Coal exploration

Exploration for coal is normally carried out in three stages. First, general reconnaissance proves its existence in a geological setting likely to yield appreciable resources. Second, a district survey, which includes drilling, quantifies reserves in sufficient detail for development planning. Third, detailed surveys are made for an actual mining operation. The coal project reported here falls into the second category.

Drilling was an essential element in the project and went hand in hand with geological survey. The boreholes were fully cored, which not only provided coal samples for analysis but also strata for study stratigraphically, palaeontologically and environmentally. An exploration programme concerned solely with commercial exploitation would undoubtedly have made more use of uncored boreholes coupled with geophysical logging.

In devising a drilling programme compromises have to be made between few, deep bores and many, shallow ones, and between wider and closer spacing. Deeper bores allow more certain correlation but cost more per metre than shallower bores. Wider spacing allows a larger area to be covered by a given number of holes; closer spacing allows more certain assessment of seam variability and reserve calculations. These calculations were made according to the methods of the United States Geological Survey (1976) which distinguishes between 'resources' which includes all coal known and unknown, and 'reserves' which are quantified, mineable resources. Reserves are calculated as 'measured', 'indicated' and 'inferred', in decreasing order of confidence, using cut-off parameters, which are arbitrary, as outlined in the section on reserves, below.

In this project, it was decided to aim for reserve calculations to the 'indicated' level, for which data points must be spaced no more than 2.4 km apart. Ideally these would be on a square grid to give statistically unbiassed calculations. In terrane with easy access or a dense road network an adequate approach to this statistical ideal is quite practicable. In the forested project area with only few, mostly poor roads, a less than ideal pattern of boreholes had to be accepted, and the actual pattern was quite irregular (Fig. 3). New road or helicopter access could have been made, but here again the cost of this would have reduced the budget available for actual drilling. In practice,

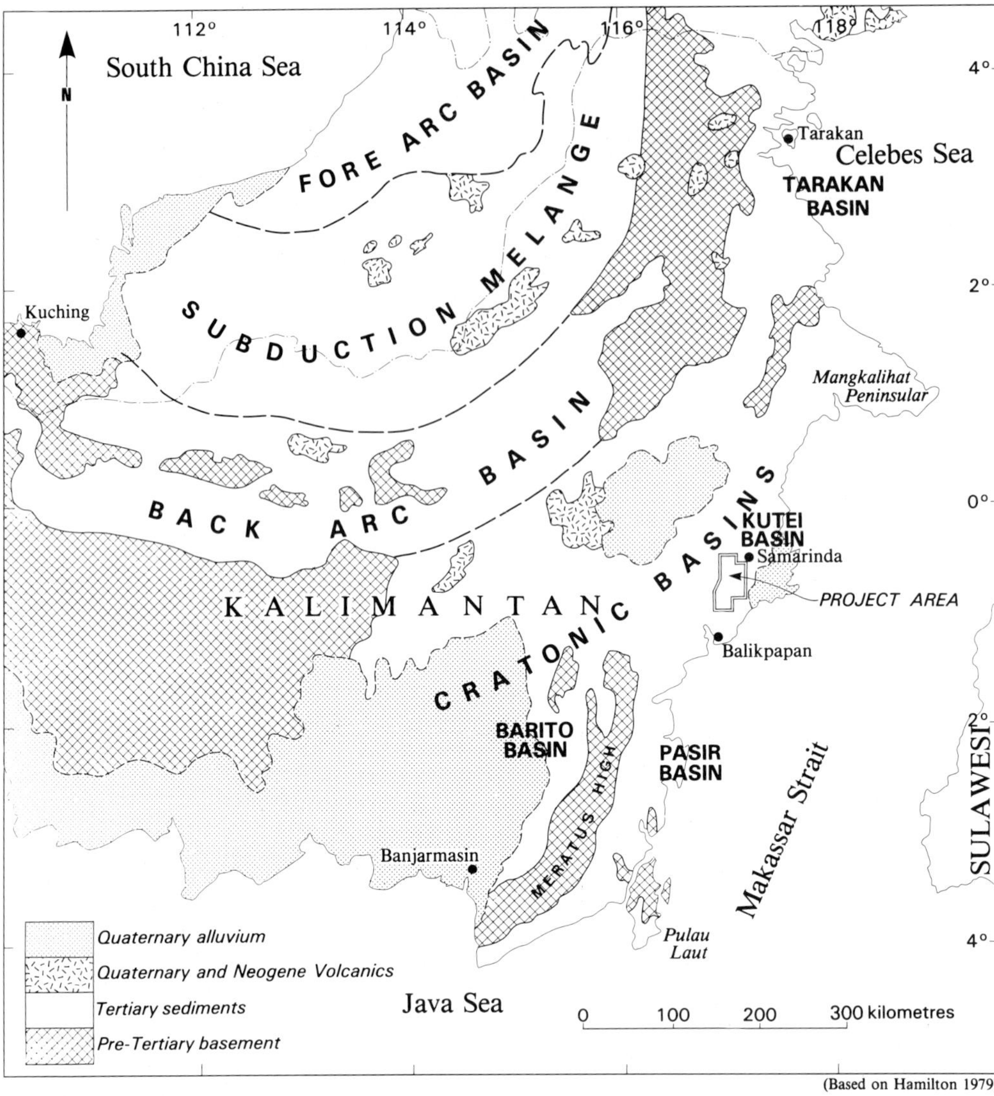

FIG. 2. Regional geological setting of the Kutei Basin.

no new access was created beyond repairing a few bridges over streams.

Project drilling was by Edeco Stratadrill 40 and Longyear 38 rigs, using wireline drilling with triple core barrels. These machines can take HQ cores (63 mm diameter) to at least 300 m depth, and NQ cores (48 mm) to 400 m. It was felt that NQ was the minimum size for adequate examination. Boreholes were generally taken to between 200 and 350 m depth. This is a compromise between few deep and many shallow holes, as mentioned above, and takes drilling to the depth (1000 feet or 305 m) to which reserve base computations are taken (US Geological Survey 1976). It also allows a reasonable stratigraphical overlap, for correlation, between adjacent bores.

Core recovery was excellent, except in the weakly cemented sandstones of the higher strata which presented most of the drilling difficulties. Good recovery helped the study and training aspects of the project, gave adequate samples for analysis and facilitated interpretation of the electric geophysical logs. The geophysical equipment used was a T450-E logger by SIE Incorp Australia, which recorded resistivity, gamma and self-potential graphically at a scale of 1 : 200.

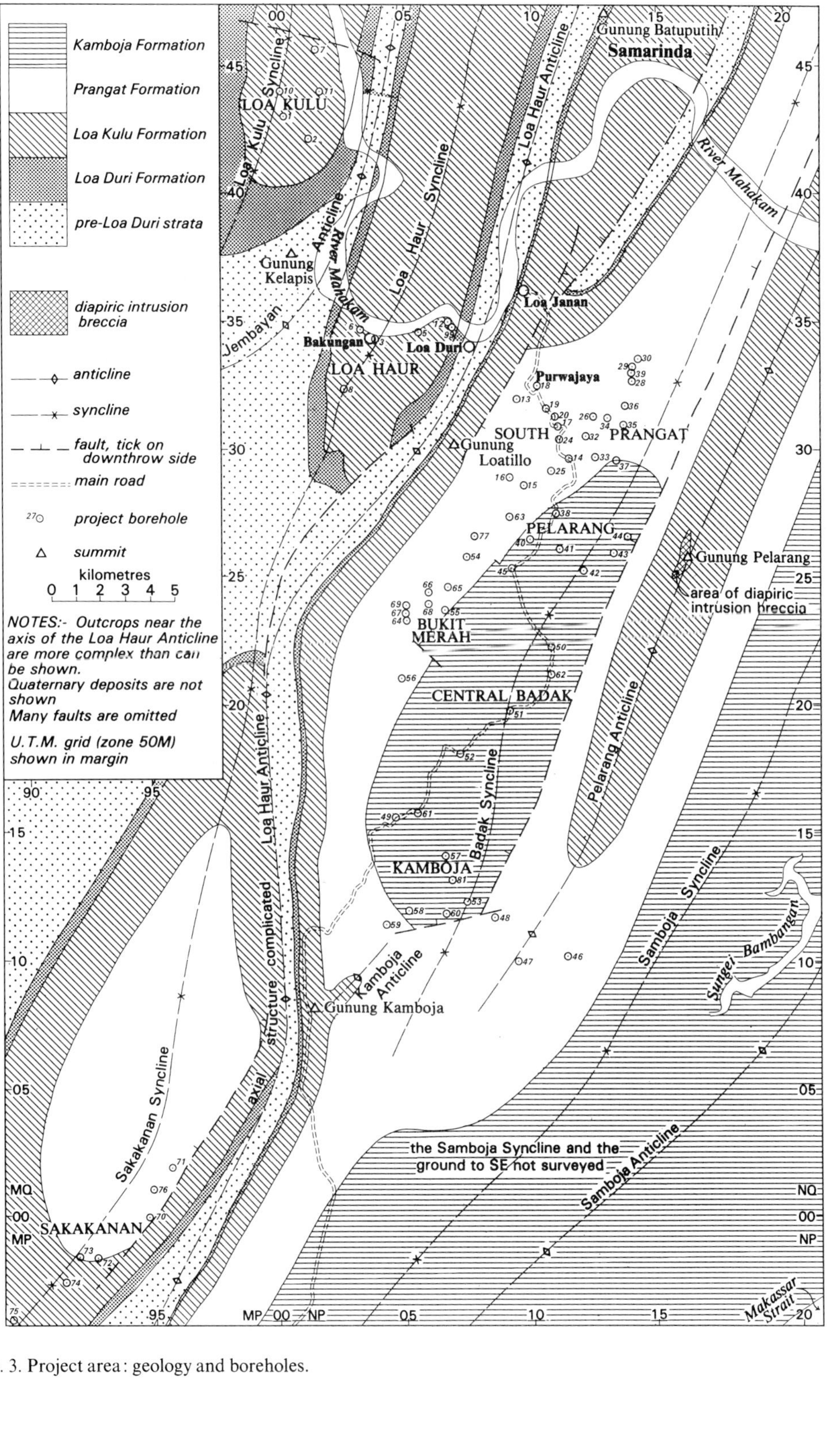

FIG. 3. Project area: geology and boreholes.

All cores were examined and described by geologists. To assist in training and computerization, and to help ensure uniformity of description of the strata, forms were devised for recording data bed by bed (Thomas *et al.* 1980, fig. 2). All the parameters necessary to cover the strata likely to be encountered are set out on the form, which is designed to be easy to use both by the geologist and by a punch-card operator. Apart from the principal lithology descriptor, no mnemonics or codes needing a handbook are used.

Drilling and geological survey progressed in parallel, results from both operations being continually assessed against each other in order to select the most appropriate site for the next borehole and to extend and confirm correlations already made. Accurate and comprehensive correlation is fundamental to any exploration programme and is essential for producing meaningful and useful results.

Because of the dearth of geographical names in the project area, coal seams were numbered in upward succession, with letter prefixes according to the area of drilling, e.g. K in Kamboja. When the base of the succession was not seen, a significant or marker horizon was chosen as datum and labelled (e.g. P50), and the seams numbered consecutively from that. The same number in different areas does not indicate correlation: for example, the lowest seam in the Kamboja Formation is P62 in the S Prangat area and K140 in the Kamboja area.

Geological maps were produced at a scale of 1:25 000 and coal reserve assessments were made on this scale. A simplification in this project for maps of areas of gentle dip was that crop lines were drawn as smooth lines without re-entrants down-dip into valleys or salients up-dip across hills. Restrictions on access and time made this necessary, but it was thought that this would not produce appreciable errors in computing areas because ground omitted by exclusion of salients would be balanced by that gained by inclusion of re-entrants.

Geological work in this project was particularly helped by using unpublished data from Pertamina, Roy M. Huffington Inc. (Huffco) and Total Indonesie. Before Independence, at the end of the Second World War, geological studies in the Kutei Basin were based on widely separated reconnaissance traverses with detailed work confined to a few near-coastal oilfields and coalfields adjacent to the River Mahakam. Van Bemmelen (1949) summarized this work. Oil exploration expanded after Independence and geological work was helped by air photography and systematic seismic traverses. New syntheses of Kutei geology were made by Samuel & Muchsin (1976) and Hamilton (1979).

Regional geology

The project area lies within the Tertiary Kutei Basin of eastern Kalimantan (Samuel & Muchsin 1976, Koesoemadinata *et al.* 1977, Hamilton 1979 and van Bemmelen 1949). This is a cratonic and back-arc basin associated with the Tertiary SE-directed subduction zone of NW Borneo (no longer active), and developed in the eastern part of the island and the Makassar Strait (Fig. 2). Tertiary rifting and crustal extension in the Strait (Katili 1978, Situmorang 1982) had an important effect on the development of the basin. Sedimentation was fairly continuous throughout the Tertiary and continues offshore today. In general terms, the sediments form a regressive sequence, with the locus of thickest sedimentation moving eastwards, although at any given locality there are several regression-transgression cycles.

Covering some 60 000 km^2 with a Neogene section up to 9000 m thick, the Kutei Basin is the largest and deepest Tertiary basin in Indonesia. Strongly deformed Cretaceous rocks form the basement on which was deposited the Tertiary sequence ranging in age from early Eocene to probably Pliocene. Strata cropping out in the project area range upward from Lower Miocene. No stratigraphically useful fossils have been found in the upper part of the succession. Quaternary alluvium occurs in the valleys, and a red-brown weathering-profile clay, generally 1–2 m thick, covers most of the surface above the alluvial flats.

Structure

Structurally, the project area is characterized by NNE–SSW trending folds, the Samarinda Foldbelt. In anticlines near the coast, which have been drilled for oil, there is evidence of growth faults, strata thinning over the anticlines, overpressured shales at depth and steeper dips with more closure at lower levels (e.g. Magnier & Samsu 1976, Schwartz *et al.* 1974) which indicates that the anticlines are growth structures which developed with sedimentation. Within the project area, the anticlines are much tighter with steep or overturned dips and steep reverse and normal faults in their complex axial regions. Mud volcanoes and diapiric intrusion breccias are further evidence that the folds grew with sedimentation. The intervening synclines are broad with wide areas of very low dips plunging variably and very gently both NNE and SSW.

The Samarinda Fold-belt structures are thought to be due to gravitational sliding with growth folds, associated with the thick sedimentation in the Kutei Basin and rifting in the Makassar Strait. A zone of decollement is provided by the over-pressured shale at depth and this also assists the rising anticlines (Chapman 1977, Hamilton 1979, van Bemmelen 1949).

There is generally a fairly abrupt transition from the gently dipping synclinal areas to the steeply dipping anticlines, commonly with large strike faults some of which are reversed.

Project drilling was confined to structurally simple areas with low dips.

Stratigraphy and sedimentology

Rutten (1914) was the first to define a stratigraphy for the Upper Tertiary of the Kutei Basin, and this was revised by Leupold & van de Vlerk (1931) whose four uppermost formations, the outcrops of which cover the project area, with their letter stages are:

Kampongbaru Beds	Th-g
Upper Balikpapan Beds	Tf_3
Lower Balikpapan Beds	Tf_2
Pulaubalang Beds	Tf_1

These formations were defined according to their assemblages of larger foraminifera, characterizing successive stages of the Indonesia Letter Classification (originated by van der Vlerk & Umbgrove 1927, and reviewed by Adams 1970). No type sections were designated and no formal definitions made. Lack of good sections, scarcity (indeed absence in higher strata) of zonally useful fossils and rapid sedimentation (giving rise to thick zones) all hinder stratigraphic studies. Correlation in Kutei 'is not easy and has led to a good deal of confusion in the nomenclature' wrote Leupold & van der Vlerk (1931): a state of affairs which persists to this day. Marks *et al.* (1982) have recently redefined formations in the east of the basin, in some cases in senses which do not coincide with earlier definitions.

The long continuous sections, through over 3000 m of strata, established by drilling during the coal project, enabled new lithostratigraphical formations (Table 1) to be proposed and formally defined for the project area by Land *et al.* (1986), and these definitions are repeated here. The new formations, in upward stratigraphical order, are the Loa Duri, Loa Kulu, Prangat and Kamboja formations.

Previous accounts of the sedimentary evolution of the Kutei Basin, based on field survey and uncored oil wells, have shown that the Tertiary sequence is broadly regressive with a dominantly offshore argillaceous sequence of Palaeocene age being replaced in the Miocene by a coal-bearing deltaic and coastal plain succession. Shoreline progradation was generally towards the east (Samuel & Muchsin 1976, Rose & Hartono 1978).

Borehole cores from the project drilling have enabled the sedimentology to be investigated in more detail than was previously possible. However, because of the small diameters of the cores (48 and 63 mm), the scarcity of natural exposures and the poor stratigraphical control, the interpretation is still tentative.

Loa Duri Formation

The Loa Duri Formation (Figs 3 and 4), which is named from Loa Duri (NQ 07 34), is essentially a turbidite sequence of mudstones and siltstones with green volcanogenic sandstones. The base of the formation is the top of the Batuputih Limestone which is exposed in quarries west and SW of Loa Duri, the type locality being the northernmost quarry (NQ 073 342). Southwestwards from here, the limestone is exposed almost continuously for 5 km; and it is also exposed east of the village (NQ 084 341) from where its outcrop runs unfaulted for over 30 km NNE through the quarries on Gunung Batuputih (NQ 129 471) NW of Samarinda. Land *et al.* (1986) defined the base of the formation at the base of the limestone, but as this belongs to a totally different facies, it is better excluded, so the formation base is here redefined to the top of the limestone.

At Loa Duri, the Batuputih Limestone contains a rich fauna of corals, molluscs and foraminifera indicating a shallow clear-water marine environment. Foraminifera include *Amphestegina sp., Flosculinella* aff. *bontangensis* (Rutten), *Lepidocyclina (Nephrolepidina) ferreroi* Provale, *L.spp., Miogypsina spp., Operculina spp.* and *Pseudotaberina malabarica* (Carter). In Dr C. G. Adam's opinion (personal communication, 1981) this indicates a Tf_1 age, that is late Lower Miocene. As the Loa Kulu Formation, which overlies the Loa Duri Formation, contains a similar foraminiferal assemblage, both formations lie within the Pulaubalang Beds of Leupold & van der Vlerk (1931).

The Loa Duri Formation is about 450 m thick at its type locality on the western side of Loa Duri, but is less than 100 m thick on the east side. Some 4 km to the SSW on the western slopes of Gunung Loatillo it seems that the formation is absent, for although the limestone is exposed

TABLE 1. *Revised stratigraphy of the project area*

Stratigraphy of Leupold and van der Vlerk (1931)	Revised stratigraphy (this paper): System		Blow zone	Formation	Thickness (m)	Environment
KAMPONG BARU BEDS Tgh	Pliocene		N19			
			N18	(top not seen)		
UPPER BALIKPAPAN BEDS Tf_3	Upper Miocene		N17	KAMBOJA FORMATION	470	Alluvial flood plain
			N16	*		
LOWER BALIKPAPAN BEDS Tf_2			N15	PRANGAT FORMATION	1375	Alluvial flood plain
	Middle Miocene		N9 to N14			
PULAU BALANG BEDS Tf_1				* unconformity		
	Lower Miocene	Late	N8	LOA KULU FORMATION	800	Deltaic to marine shelf
				LOA DURI FORMATION	450	Deep marine
				Batuputih Limestone	2-50	Shallow marine
		Early	N7			
BEBULU BEDS $Te_{4\text{-}5}$			N6			
			N5			
			N4			
PAMALUAN BEDS $Te_{1\text{-}3}$	Upper Oligocene					

* The age of the top and bottom of the Prangat Formation is uncertain

(NQ 064 303), there is no evidence of green turbiditic strata.

These thickness variations suggest the possibility of an unconformity at the base of the overlying Loa Kulu Formation. However, it is thought that large-scale faulting provides a more likely explanation, as there is no clear evidence of unconformity, the incoming of Loa Kulu shallow-water sediments resulting from a general regression.

The upper part of the Loa Duri Formation was proved in boreholes 12 and 99 near Loa Haur, and there are poor exposures on the SE bank of the Mahakam River downstream of Loa Duri. The lowest part of the sequence, overlying the Batuputih Limestone, is very poorly exposed. The limestone is thought to be about 100 m below the horizon reached at the base of borehole 99. An excellent section in a cutting (MP 9395 9610) on the Jalan Jepang, on the east flank of the Sakakanan Syncline, is correlated with the Loa Duri Formation although the relationship with the Batuputih Limestone is not clear.

All of the sandstones contain abundant volcaniclastic material giving them a characteristic green colour. Thick-bedded turbidites (1–6 m) are represented by massive sandstones with matrix-supported mudstone clasts and larger rafts up to 2 m in length, together with sub-rounded clasts of red chert, lava, siderite and coal. Minor channelling is visible in the Sakakanan outcrop. Thin-bedded turbidites (1–25 cm thick), show Bouma vertical sequences with sole marks (fine striations) at the bases of some beds. These are associated with mudstones with isolated siltstone ripples showing mean current directions towards the east. Disturbed bedding is quite common and comprises both chaotically bedded slump sediments as well as syn-sedimentary normal faulting usually showing only a few centimetres displacement. Within the sequence the only macrofossils seen are very rare thin bivalves; bioturbation is sparse. Root beds, coals and sedimentary features indicative of shallow water environments are completely absent.

In the uppermost 80 m only very rare thin-bedded turbidites occur, the succession being dominated by mudstones with starved siltstone ripples and coarser grained slumped units. This may represent a slope sequence, overlying the deeper water section containing the turbidite sandstones.

Loa Kulu Formation

The Loa Kulu Formation is named from Loa Kulu village (NQ 02 43) where several coals were worked extensively in the past (Figs 3 and 4). The base of the formation is defined at Bakungan, about 10 km SE of Loa Kulu, in project borehole no 5 (NQ 0548 3445) 100 m east of Loa Haur Bridge, with the formation base at 311.5 m depth at the base of a channel sandstone. The type section is the conjoint section proved (in upward sequence) in project boreholes 5, 3, 11, 1 and 10 which taken together give a formation thickness of over 800 m. This is not the full thickness because borehole 10 was drilled into the highest strata preserved in the Loa Kulu Syncline, and neither in this nor in the Loa Haur Syncline are any strata of the overlying Prangat formation preserved.

Loa Kulu strata consist of marine argillaceous fossiliferous limestones and mudstones, sandstones, seatearths and coals (Fig. 5). Limestones are confined to the lower part of the formation and mudstones and sandstones make up the bulk of the strata. Coals are up to 3.25 m thick and are high-volatile C bituminous and sub-bituminous A in rank. The thicker ones may be correlated and are labelled L1 to L20 in upward sequence.

In the Sakakanan area, Loa Kulu strata are much more marine and coal seams are rare.

Both the Loa Kulu and Sakakanan areas have yielded rich faunas from the Loa Kulu formation. Those from Loa Kulu include foraminifera: *Alveolopragmium* sp., *Ammobaculites* sp., *Ammodiscus* sp., *Ammonia beccarii* (Linn.), *Elphidium minimum, E.* sp., *Flosculinella bontangensis* (Rutten), *F. borneensis* (Tan), *Globigerinoides sicanus* de Stefani, *Haplopragmoides* sp., *Miliammina* sp., *Miogypsina* cf. *kotoi* Hansawa, *M.* spp., *Nonion* sp., *Operculina* sp., *Pseudorotalia* sp., *Quinqueloculina* sp., rotaliids, *Textularia* sp., *Triloculina* sp., and *Trochammina* sp.; worm: *Ditrupa* sp.; gastropods: *Acteocina* sp., Acteonidae, *Cerithidea* cf. *sucabumianus* (Martin), ?*C.*sp, *Gemmula* sp., *Lioconcha* cf. *philippinerium* (Hanley), *L.* sp., *Nassarius* s.l., ?*Natica* sp., Naticidae, *Marginella* s.l., Muricidae, Pyramidellidae, Pyrathidellidae, Turridae, Turitellidae and *Variocospira* sp.; scaphopod: *Cadula* sp.; bivalves: *Anadara* sp., Arcidae, *Bellucina* sp., *Corbula* sp., ?*Cultellus* sp., Cardiidae, *Lucina* s.l., *Nucula* s.l., *Nuculana* sp., Nuculanidae, *Macoma* sp., *Ostrea* sp., Ostreidae, ?Pectinidae, ?*Quadrans* sp., ?Solenidae and ?*Tellinella* sp.; and crabs, corals and bryozoa not determined.

Strata in Sakakanan correlated with the Loa Kulu Formation yielded the following: foraminifera: *Flosculinella bontangensis* (Rutten), *F. borneesis* (Tan), *Globigerinoides immaturus* Leroy, *G. sicanus* de Stefani, *G. subquadratus* Brönniman, *G. trilobus* (Reuss), *Globorotalia obesa* Bolli, *G. siakensis* (Leroy), *Globoquadrina dehiscens* (Chapman, Parr & Collins), *G.* sp., *Lepidocyclina (Nephrolepidina)* spp., *Miogypsina kotoi* Hanzawa, *Miogypsina* sp. and *Operculina* sp.; gastropods: *Acteocina* sp., *Architectonia* sp., Cerithiacea, ?*Cylichna* sp., *Cyllene* sp., *Gemmula* sp., ?*Hindsia* sp., *Marginella* sp., *Murex* sp., *Nassarius* sp., *Natica* sp., ?*Phos* sp., Pyramidellidae, Rissoacea, *Terebra* sp., ?*Tibia* sp., Trochidae, Turridae, *Turritella cingulifera* G. B. Sowerby, *T.* sp., ?*Vaginella* sp., *Variocospira* sp. and *Vexillum* sp.; bivalves: ?*Amusium* sp., ?*Anadara* sp., *Anomia* sp., Arcidae, *Cardita* sp., Cardiidae, *Corbula* sp., ?*Limopsis* sp., *Nucula* sp. and *Nuculana* sp.

In Dr C. G. Adam's opinion (personal communication, 1981) this funa indicates Blow Zone N8, which is also Tf_1, late Lower Miocene. Diagnostic forms include *Globigerinoides sicanus, Flosculinella bontangensis* and *Miogypsina kotoi*.

This formation is important in that it contains the highest quality coals in the sequence examined, and the only coal deposits to have previously been worked on any scale. It is also the most complex and difficult to understand from a sedimentological standpoint. There are significant differences between the sequences in the Sakakanan and Loa Haur–Loa Kulu areas. Although those sequences are believed to be roughly time equivalent, they cannot be correlated in any detail.

In both areas two distinctive sandstone lithotypes occur. Volcanogenic sandstones predominate in the north and include green litharenites with predominantly volcanic rock fragments together with red cherts, quartz and feldspar. These sandstones are closely similar to those in the underlying Loa Duri Formation. The other lithotype is a pale grey sandstone with quartz and siliceous rock fragments. This is more commonly developed in the Sakakanan sequence.

Three distinctive facies associations are recognized (Figs 5 & 6).

Association 1: This is restricted to the lowest 480 m of the Sakakanan succession in boreholes 72, 74 and 75. It contains numerous coarsening upwards cycles varying from less than 5 m to up to 40 m in thickness. The lowest part of each comprises grey mudstones with marine fossils. This grades up into a heterolithic facies of

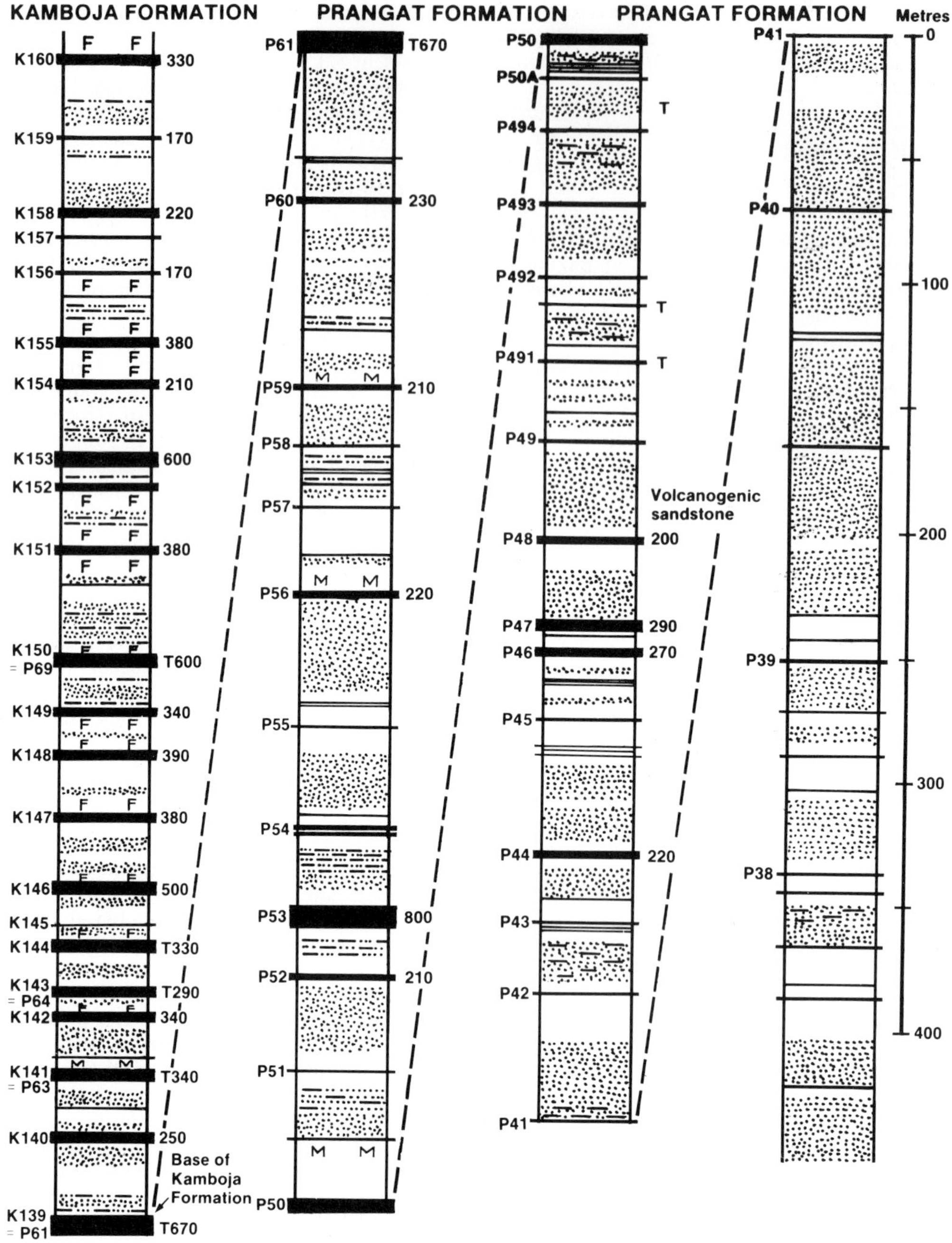

FIG. 4. Generalized vertical section.

alternating mudstones and sharp based sandstones. Wave ripple lamination (identified at outcrop) is very common and bioturbation is extensive, sometimes completely destroying most sedimentary features. The higher parts of the sequences may contain planar and hummocky cross-bedded fine to medium grained sandstones. The topmost parts of many cycles contain sharp or erosively based medium-grained sandstones, generally less than 2 m thick. These are often

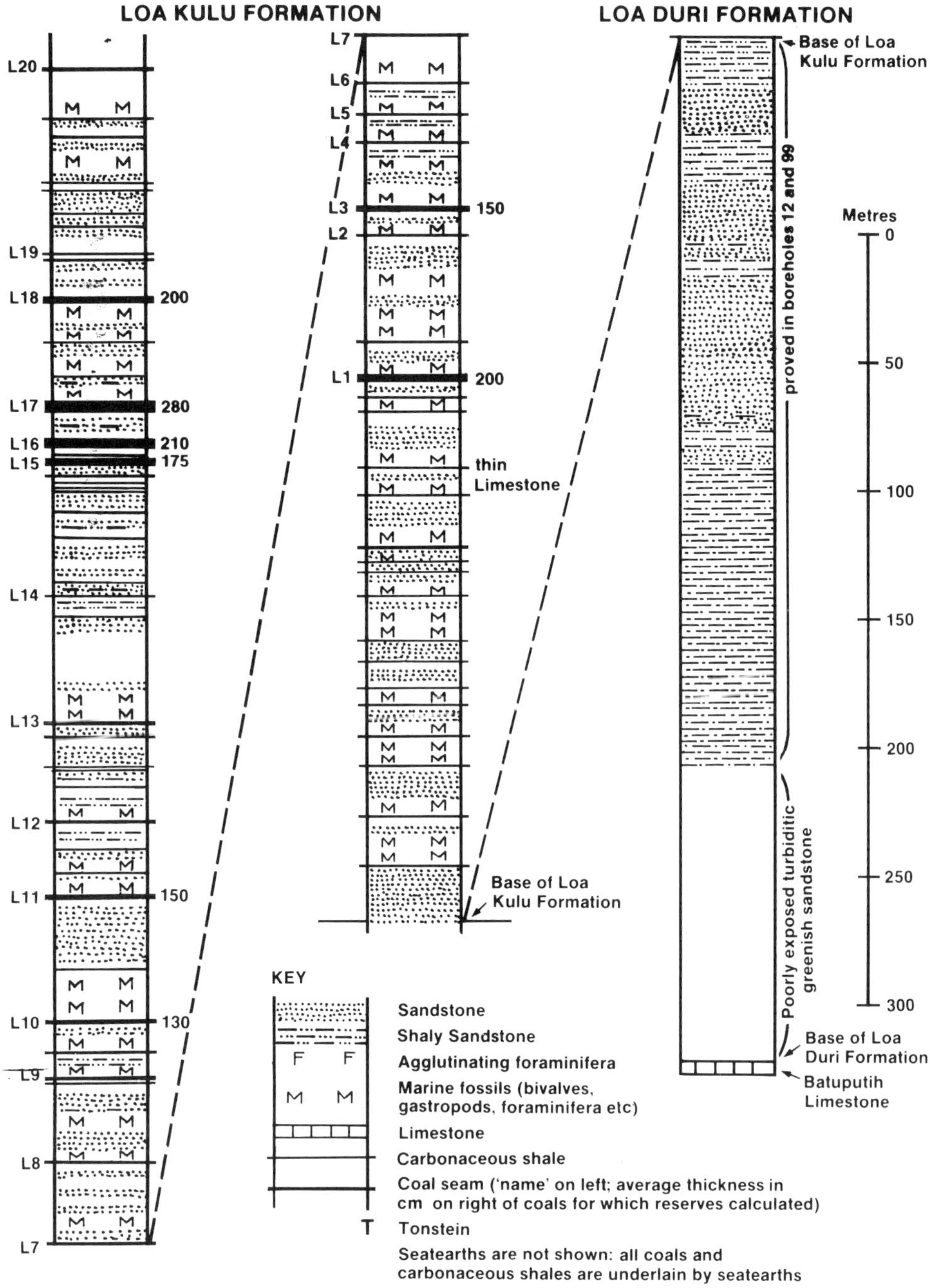

highly calcareous and also richly fossiliferous with a diverse marine fauna of bivalves, gastropods, foraminifera and rarer bryozoa, corals and algae. Bioturbation has completely destroyed all sedimentary structures. In borehole 75, the upper parts of two cycles contain thin cross-bedded fining upwards sequences with sharp or erosive bases. These are probably small channel fills. In the lowest 300 m of the Sakakanan succession rootbeds and coals are completely absent, but

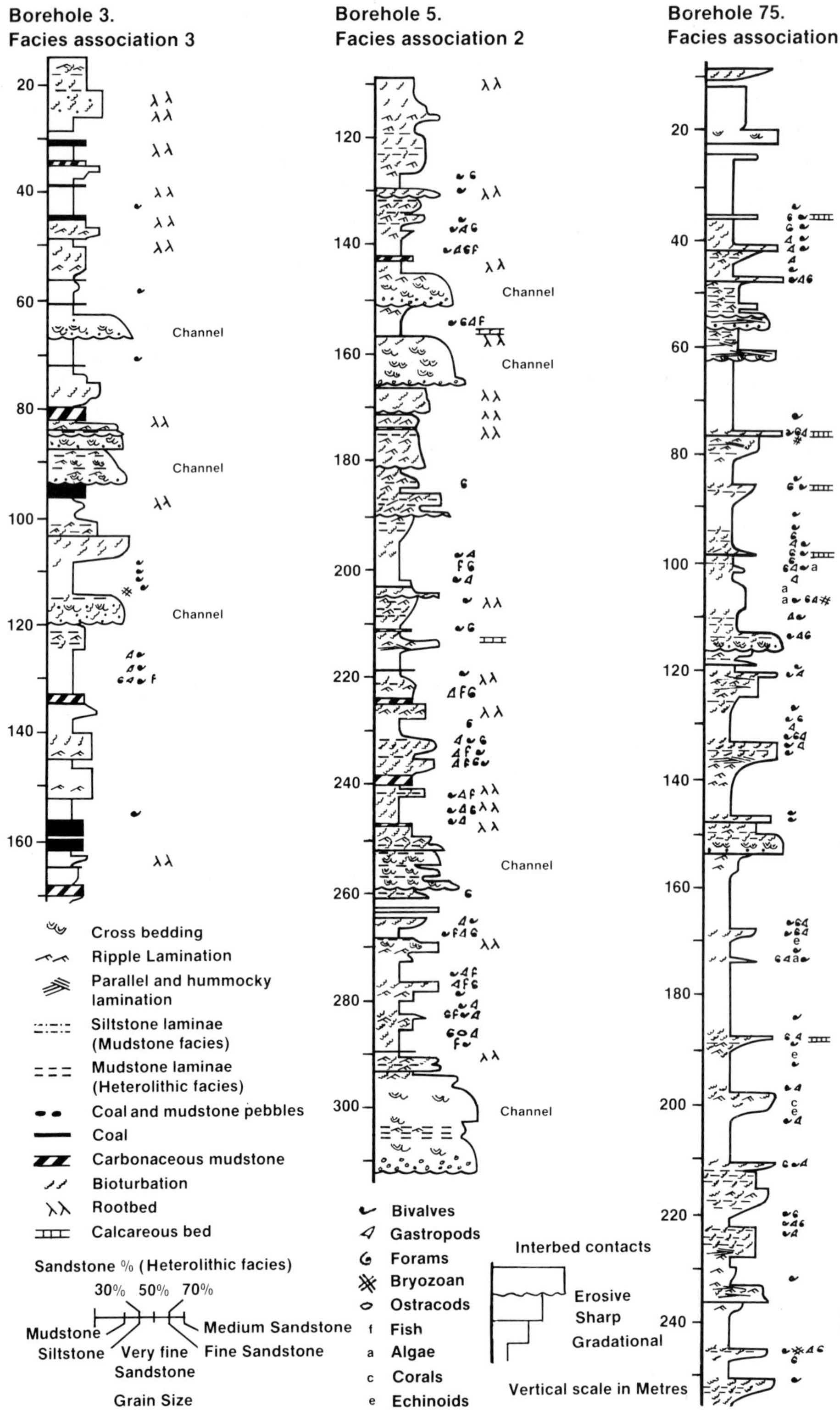

FIG. 5. Sedimentary logs in the Loa Kulu Formation.

above 188 m in borehole 74 most cycles have roots at the top and some contain very thin coals. These commonly contain numerous sand-filled burrows and always directly underlie the sharp-based sandstones.

Association 2: This association occurs in the bottom 200 m of the Loa Haur–Loa Kulu sequence in boreholes 3, 5 and 12 (Fig. 5), where it immediately overlies the turbidite-bearing Loa Duri Formation. It is also developed in the topmost 100 m of the Sakakanan succession in boreholes 70, 71 and 72. The essential feature is the presence of cross-bedded channel sandstone fills which are typically 10–15 m in thickness. These are interbedded with coarsening-upwards sequences similar to those described in Association 1, except that they tend to be thinner. Coals and carbonaceous mudstones with root beds are common, but the coals are thin, the thickest reaching 1.45 m.

Association 3: This is restricted to the upper part of the Loa Kulu Formation in boreholes 3 and 5 in Loa Haur (Fig.5) and all of the boreholes in Loa Kulu. It is not developed in Sakakanan. The essential feature of this association is the presence of economic coals. Up to 20 seams occur (Fig. 4), nine with average thicknesses greater than 1 m. However, the associated sediments are similar to those described in Association 2. Major channel sandstones are present throughout the sequence and small coarsening upwards sequences with rich marine fossil assemblages are also fairly common. Several coals are immediately overlain by marine strata.

Discussion: A reconstruction of the palaeogeography of the area during the deposition of the Loa Kulu Formation is shown in Fig. 6. A major delta frequently occupied the northern part of the area along the site of the present-day Mahakam River. To the south was a wave-dominated shoreline passing seawards into a marine shelf. Farther offshore from the shelf and delta, turbidites accumulated in deeper water.

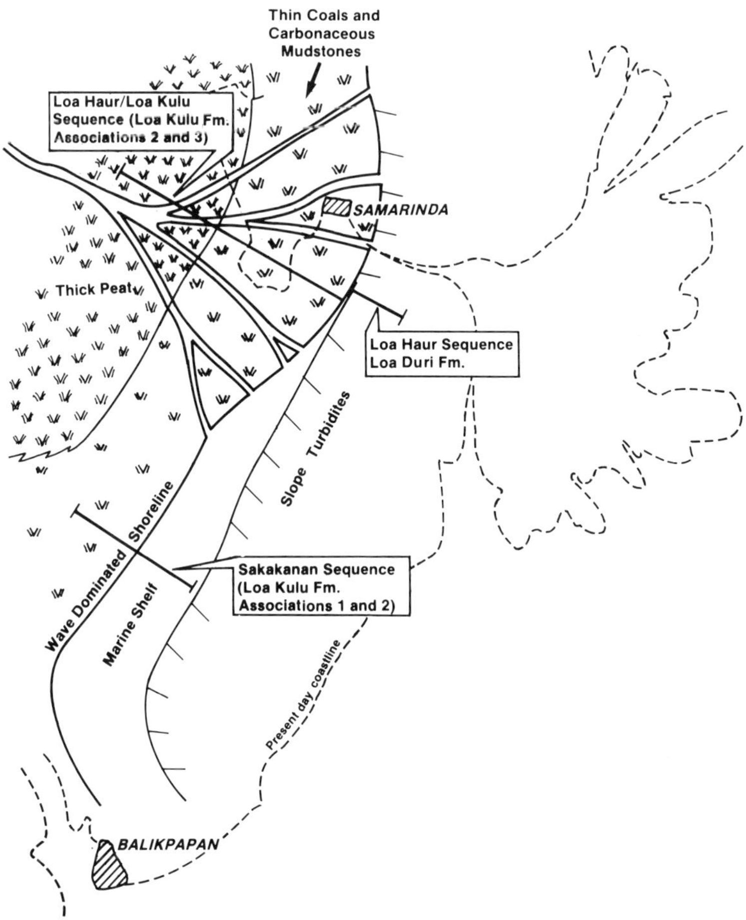

FIG. 6. Palaeogeography of the central Kutei Basin during deposition of the Loa Kulu Formation.

Association 1 was formed in the shelf and wave-dominated shoreline environment. The grey mudstones which form the greater part of many of the coarsening-upwards (cu) cycles represent marine shelf deposits. Repeated phases of shoreline progradation, probably linked to movement of the delta, formed the cu cycles. The parallel laminated and hummocky cross-bedded sandstones suggest a wave-dominated shoreline, whilst the sharp-based fossiliferous sandstones are probably storm beach deposits. Root beds developed in the higher parts of the association support the regional interpretation of a broadly regressive sequence. The thin coals are probably detrital accumulations, perhaps similar to the 'lignite beaches' of the present-day Mahakam delta (Allen *et al.* 1979).

Association 2 represents the deltaic sequence. The morphology of the delta is unknown, but the lack of fossils in the channel fill sequences argues against a tidal delta. The interbedded cu cycles with their rich marine faunas suggest that the delta was periodically abandoned and replaced by the wave-dominated shoreline environment which was prevalent farther south. Towards the top of the Sakakanan sequence the presence of channel sandstones indicate that the delta occasionally diverted southwards.

The thin coals show that, as in the present-day Mahakam delta (Allen *et al.* 1979), peat mire conditions were generally not established for long periods.

Association 3 is interpreted as an upper delta plain sequence. The well-developed fossiliferous cu sequences show that delta abandonment and marine reworking were still a common feature. At other times peat mires were established. This probably occurred during periods of maximum seawards delta progradation, as the coals are restricted to the highest (most regressive) parts of the Lower Miocene succession. This model (Figs 6 and 7) suggests that economic coal deposits will be fairly restricted in distribution within the early parts of the Miocene regressive sequence.

Prangat Formation

The Prangat Formation is named from the South Prangat area where it crops out and was first extensively drilled in the project (Figs 3 and 4). As the base is neither exposed nor intersected in any borehole in the area, it cannot be closely defined. The type section of the middle and upper part of the formation is the conjoint section proved (in upward sequence) in project bores 13 (NQ 0936 3195), 18, 19, 17, 32, 35, 14, 33 and 37 (NQ 1338, 2953) which taken together in that order show a 1300 m section of strata below the conformable base (at the top of coal P61) of the overlying Kamboja Formation.

It is likely that the Prangat Formation lies unconformably on the Loa Kula Formation. There is not enough space in the Purwajaya–Loa Duri area between the known Prangat and Loa Duri outcrops for the 800 m of Loa Kulu Formation as it is developed at Loa Kulu, 12 km to the NW. It may have thinned or may be partly cut out by faulting; but unconformity is considered most likely, because in the Sakakanan area the base of the Prangat Formation is marked by a sharp change in environmental facies with a change from marine to brackish fauna. However, because the Sakakanan succession cannot be confidently correlated in detail with that farther north, doubt remains as to the reason for the partial absence of the Loa Kulu Formation west

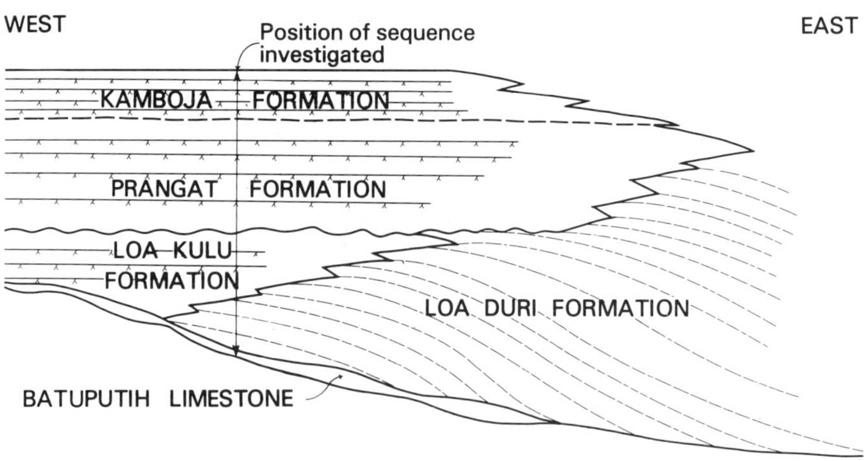

FIG. 7. Schematic cross-section showing inferred relations of the sequence investigated.

of Purwajaya. As this formation contains the highest quality coals, the reasons are of more than academic interest.

Prangat Formation strata are sandstones, siltstones and mudstones with coals and carbonaceous shales. In the lower part of the succession drilled in South Prangat about 80% of the strata are sandstones; higher up this proportion decreases to 50% and the sandstones are on the whole finer grained. Most of the sandstones are quartzitic, green volcanogenic rocks being absent except for a sandstone near the horizon of seam P48 which contains pebbles of welded dacitic tuff in a matrix of smectite. Freshness of the grains and the volcanic nature of the matrix indicate redeposition closely following eruption. Conglomerates with pebbles of coal and of red radiolarian chert are common in the lower beds, but become less frequent higher up. In the upper part of the formation there are three volcanogenic tonsteins of kaolinitic and smectitic types (Addison *et al.* 1983). Three marine bands also occur in the higher part of the formation and contain agglutinating foraminifera and bivalves which are not age diagnostic. As the underlying formation is late Lower Miocene, the Prangat may be Middle Miocene.

Coal seams range in thickness up to 13.45 m, and are sub-bituminous B and C in rank. Twenty-four seams are thick and persistent enough to be correlated across the outcrop.

In the Prangat Formation, characteristic deposits are channel sandstones and conglomerates, crevasse splay sandstones and finer overbank deposits. Typical channel fill sequences have erosive bases, fine upwards, and are 5–15 m thick. Channel lag conglomerates with coalified logs and mudstone clasts are common. The main parts of the fills are predominantly trough cross-bedded and palaeocurrents measured at outcrop showed a wide scatter with no discernable mean. In many sections, particularly in the lower parts of the sequence, multi-storey units form composite sandstones up to 60 m thick. Where the upper parts of the channel fills are preserved they show typical upper point bar sequences with ripple laminated sandstones and interbedded siltstones grading upwards into finer levee and overbank deposits.

Floodplain deposits are predominantly brown unlaminated mudstones and fine siltstones. Siderite nodules are common, as are siderite crusts, believed to be formed by groundwater movement during hiatuses in sedimentation. Bioturbation is much less common than in the underlying formation, though root horizons are widespread. Within these fairly monotonous sequences occur rare crevasse splay sandstones up to 1 m thick with sharp bases and tops. Also present are very fine mudstones with fish scales, believed to be the deposits of freshwater lakes. In Sakakanan and the southern part of the Badak syncline, brackish water faunas with molluscs and agglutinating foraminifera also occur sporadically.

Coals, all with underlying root beds, are numerous in the Prangat Formation. Seventeen are thicker than 1.5 m, the thickest being 13.5 m. This represents an original accumulation of peat of the order of 60 m (Stach *et al.* 1982, see also Collinson & Scott, *this volume*). Associated with many coals are carbonaceous mudstones containing abundant drifted plant material. These may occur above, below or within the seams. The coals have sulphur contents of less than 3%; resin lumps are common and silicified logs occur at some horizons. On maceration the coals yield a diverse fungal flora in addition to spores, pollen and cuticles (A. C. Scott, personal communication).

The Prangat Formation is interpreted as an alluvial flood plain sequence. The well-developed fining upwards channel fills and the scattered palaeocurrent directions suggest that many of the channels were meandering. The floodplain areas were probably extensively vegetated, but peat forming conditions were only periodically established. According to McCabe (1984), major coal seams are often developed well away from areas of active clastic sedimentation. This was probably the case with some of the major seams such as P61 (Fig. 4), which may be traced for over 20 km and is between 3 and 9 m thick. It also overlies a part of the sequence dominated by finer grained sediments, but generally lacking coarser-grained channel fills. On the other hand, the widespread development of carbonaceous mudstones, and the variable development from borehole to borehole of thinner seams, suggest that bogs were sometimes established fairly close to areas of active sediment input.

Kamboja Formation

The Kamboja Formation is named from the Kamboja area, NE of Gunung Kamboja, where it has an extensive outcrop (Figs 3 and 4). The base of the formation, which is conformable, is defined as the base of the mudstones overlying coal seam K139, at 266.8 m depth in project borehole 49 (NQ 0463 1551). This borehole and borehole 53 (NQ 0737 1228) constitute the type section, which is 500 m thick. Coal seam K139 correlates with seam P61 in the South Prangat area, the top of this coal being at 14.9 m in borehole 37 in the Prangat type section.

Kamboja strata are largely mudstones and siltstones with generally thin sandstones, and at least 22 thick (up to 11 m) coals, which are very variable in thickness. Four tonsteins are recorded. Agglutinating foraminifera of no use for correlation have been found in 11 inter-coal cycles, and a marine band above seam P63 = K141 has yielded bivalves and ostracods, including *Cytherelloidea rankanensis* Leroy which may indicate a Late Miocene age (Mackenzie, personal communication 1981). Coals are lignite A in rank.

The facies change from Prangat to Kamboja strata is gradational so that the base of the latter formation is somewhat arbitrary. Kamboja sandstones are thinner and more finely grained whilst the coals, as well as being more lignitic are individually more variable in thickness. Siltstones and mudstones rich in agglutinuating foraminifera are common and some contain brackish water molluscs. The relatively thin discontinuous and fine grained sandstones constitute less than 10% of the sequence drilled, compared to a minimum of 40% in the Prangat Formation.

Kamboja strata are dominated by pale grey to slightly pink coloured siltstones and silty mudstones with common well-preserved plant remains, and dark grey silty mudstones with agglutinating foraminifera. Some mudstones are distinctly reddish in colour, but contain unoxidized black plant fragments, and may have been derived from erosion of forest covered ground with a lateritic surface. Coal seams are common within the formation and in some boreholes (e.g. 49) may constitute up to 15% of the sequence drilled. Seams vary considerably in thickness, and splitting is very common so that correlation is only possible by use of tonsteins and fossil horizons. Seams associated with foraminiferal horizons (which may occur in partings within some coals) are usually relatively high in sulphur (about 2%), whilst others have a low sulphur content (<0.2%).

A small number of tonsteins occur in the sequence. The most widespread is that intercalcated with seam K150, and is a thin bed (up to 15 cm) of mainly crystalline kaolinitic clay persistent throughout the southern part of the Badak Syncline.

The Kamboja Formation differs from the underlying sequence in the much higher proportion of coal with common seam splitting, lower proportion of channel sand and more abundant occurrence of fossiliferous horizons. Like the Prangat Formation, this is also believed to be an alluvial plain succession, but here peat conditions were frequently locally interrupted by sediment input. This suggests that the alluvial plain may have been crossed by a larger number of smaller channel systems, compared to the underlying formation where fewer major rivers appeared to have been present. Standing water conditions were also frequently established. The presence of brackish water faunas proves a link with the sea, though the nature of this is uncertain. It seems quite likely that mires were developed fairly close to the sea, perhaps protected by a narrow coastal ridge, so that a slight relative sea-level rise allowed the mires to be inundated.

No petrological work (except for reflectance studies) has been carried out, so the nature of the mires is unknown. However, coals where tonsteins are preserved are unlikely to have been formed in mires with a dense tree cover as a thin continuous volcanic deposit is unlikely to have been preserved in such an environment. These coals may be lacustrine deposits. Only by a detailed integration of petrological, palaeobotanical and sedimentological features will it be possible to gain a more detailed knowledge of the Miocene coal environments.

Basin evolution

The sedimentological interpretations outlined above show that the Kutei Basin Miocene sequence was not one of uninterrupted regression. The transition from a shallow water reefal environment of the Batuputih Limestone to deeper water turbidite deposition records a significant rise in sea-level. Global sea-level curves (Vail *et al.* 1977) indicate that the late early to Middle Miocene was a period of fairly continuous sea-level rise. The upward transition into the deltaic to marginal marine Loa Kulu Formation probably marks a shoreline regression where the influx of sediment more than counterbalanced the rise in sea-level.

An unconformity probably occurs at the junction between the Loa Kulu and Prangat Formations. This is difficult to date precisely since the age of the Prangat Formation is unknown, though the overlying Kamboja Formation has tentatively been dated as Late Miocene in age. A major global sea-level fall occurs at the top of the Middle Miocene and the unconformity might therefore be correlated with that event. This would support a late Miocene age for the Prangat and Kamboja formations.

Coal quality

In the project, all coal cores were sampled for analysis. Where seams were divided by partings or into beds of different qualities, each bed was sampled separately. Any bed thicker than 15 cm was subsectioned into 15 cm lengths, unless the core was too fragmentary to determine depths, in which case the whole fragmented length was taken as one subsection. Adjacent to roof, floor or parting, a 5-cm coal subsection was taken.

Coal cores were washed clean of mud and surface moisture dried off. Each subsection was then crushed to pass a 2.5-mm mesh, weighed, and divided into two parts: one for analysis, the other for reserve. Analyses were carried out according to the American Society for Testing and Materials (ASTM) standard practices. Because of restricted facilities, samples analysed during the project were on composites, each representing successive 1-m thick sections of the coal seams, the composites being combined from subsections in proportion to the thickness each represented. Parameters measured were free moisture, air-dried moisture, volatile matter, fixed carbon, ash, and sulphur (as percentages), calorific value (cal g^{-1}) and density (g ml^{-1}), all on the air-dried basis. Some work was also started on more detailed analyses for total carbon, hydrogen, nitrogen, carbon dioxide, chlorine and forms of sulphur.

Coals in the project area have low ash, high moisture and high volatiles (see Table 2 for representative analyses). Sulphur is low (under 0.5%), except for coals directly overlain by marine strata when sulphur values may rise to around 3%. All the coals are non-caking.

In rank the coals range from lignite A to high-volatile C bituminous in the ASTM rank classification, in which lower rank coals are classed according to calorific value on the moist, mineral-matter-free basis (mmmf), as follows (converting ASTM Btu lb^{-1} values to cal g^{-1}):

High-volatile C bituminous coal	6394–7228 cal g^{-1}
Sub-bituminous A coal	5838–6394
Sub-bituminous B coal	5282–5838
Sub-bituminous C coal	4615–5282
Lignite A	3503–4615

Loa Kulu formation coals are high-volatile C and sub-bituminous A in rank. Prangat coals are mostly sub-bituminous C, with some of the lower coals in the Bukit Merah and Sakakanan areas up to sub-bituminous B in rank. Kamboja coals are of lignite A rank, with a few of the lowest seams just reaching sub-bituminous C. Loa Kulu and Prangat coals are black and lustrous with a very high proportion of huminite. Kamboja coals are duller, brownish-black and obviously lignitic in hand specimen.

In calculating values to a mineral-matter free basis it has been assumed that mineral matter equals 1.2 times the ash value. This factor of 1.2 represents a rough approximation, using the modified KMC formula (British Standards Institution 1971, p. 15), pending experimental work on Indonesian coals to derive an appropriate formula for calculating their mineral matter.

As is the case generally, coals from the project area show increasing calorific values and decreasing moisture with increasing age; but this is only a general trend, and it is not possible to use moisture content as an age indicator as was done in some early studies.

Local variations occur such as an eastward decrease in rank of seam P61 across the Badak Syncline or even more local rank increases of some seams in the Bukit Merah area. Research is needed on these as well as on factors which affect the value and usability of the coal.

Some work was done on huminite reflectance, values ranging from 0.29 to 0.44 (%R_{oil}). While reflectance generally increases with rank, the correlation between calorific value (mmmf) and reflectance was low, with a broad scatter of values.

Coal reserves and resources

Coal resources include all coal in the ground, discovered and undiscovered; and reserves are that part of the resources which are currently economically actually recoverable by mining or quarrying (US Geological Survey 1976). Reserves are calculated to several levels of confidence (measured, indicated and inferred, in decreasing order), using cut-off parameters as listed below. These are to some extent arbitrary, for example seams less than 150 cm thick or with more than 15% ash are excluded. Reserves are computed seam by seam in blocks defined by spacing of data points and geologically by outcrop, depth, dip, faulting, changes in seam thickness or quality or seam splitting. In the case of changes in a seam where there is only borehole evidence, the block boundary follows a line equidistant between the pair of boreholes showing the change. Computation is by the formula:

$$\text{Reserves} = \text{thickness} \times \text{area} \times \text{coal density} \times \text{recovery factor}.$$

Thickness in metres, area in square kilometres and density in g ml^{-1} gives millions of tonnes of coal in the ground, while the recovery factor is introduced to allow for losses in mining, to give the amount of coal actually recoverable for sale, i.e. the reserves. Computation of reserves is illustrated in Fig. 8 and the cut-off parameters used in the project are as follows. Calculations are made for seams which are over 150 cm thick, have less than15% ash (air-dried basis) and have less than 300 m cover. Recovery factors are 50%

TABLE 2. *Representative coal analyses*

Coal seam	Area	Bore-hole no.	Formation	Thickness of seam (m)	Moisture (air-dried loss, % as received)	Moisture	Volatile matter	Fixed carbon	Ash	Total sulphur	Calorific value (cal/g)		Coal rank
						per cent, air-dried basis						(mmmf.)*	
L17	Loa Kulu	10	Loa Kulu	2·47	1·50	10·88	39·31	46·69	3·12	3·39	6300	6443	High volatile C bituminous
P50	Bukit Merah	66	Prangat	3·70	1·68	16·87	37·45	42·53	3·15	0·71	5510	5625	Sub-bituminous B
P52	Bukit Merah	64	Prangat	2·15	2·50	20·94	36·26	38·75	4·05	0·12	5040	5158	Sub-bituminous C
P56	South Prangat	37	Prangat	1·79	2·22	24·31	35·69	34·77	5·23	0·10	4666	4861	Sub-bituminous C
S47	Sakakanan	71	Prangat	2·15	2·54	24·43	33·06	39·44	3·07	0·18	4982	5038	Sub-bituminous C
K150	Kamboja	53	Kamboja	6·57	5·71	26·44	30·01	27·80	5·09	0·28	4250	4253	Lignite A

* (moist, mineral-matter free)

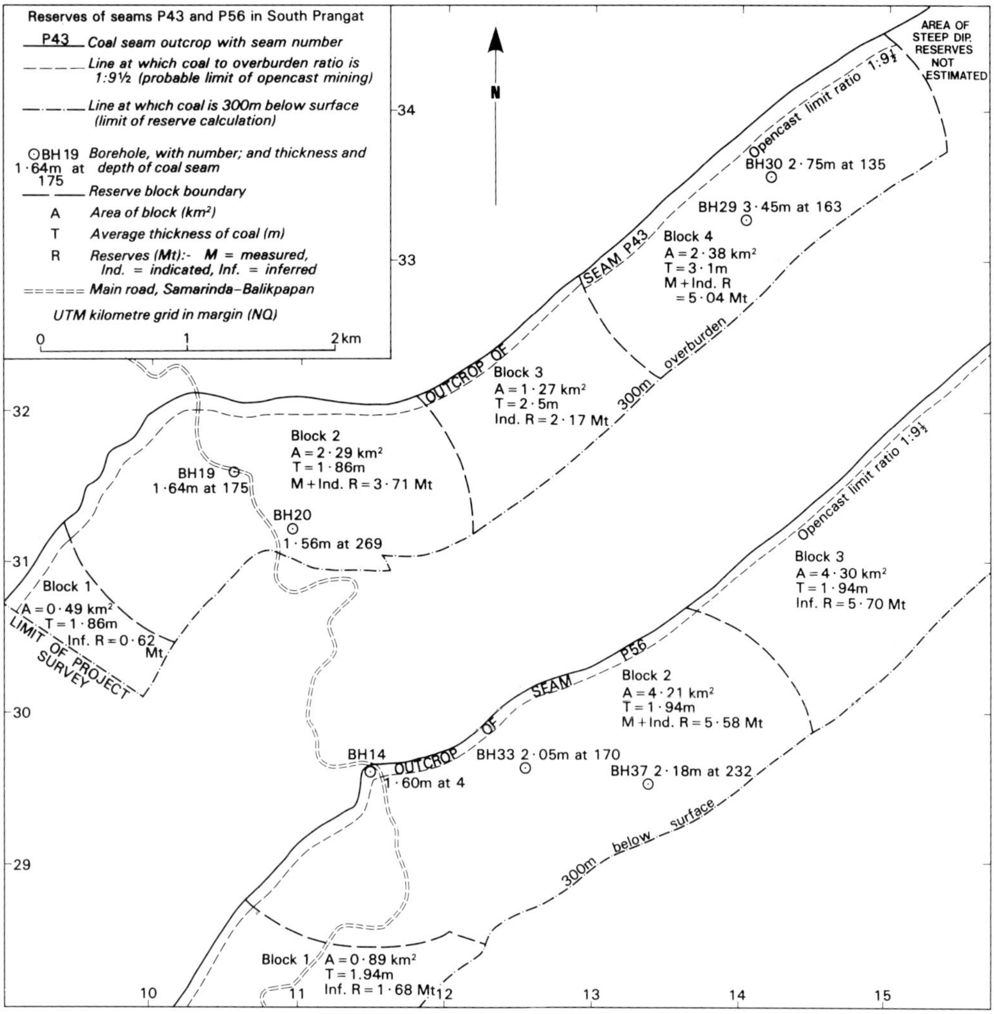

FIG. 8. Map to illustrate calculations of coal reserves.

for deep mined coal and 80% for opencast mining (US Geological Survey 1976, Averitt 1975). For the latter an overburden to coal ratio of 9.5 is taken. (This is the ratio at Tanah Hitam opencast mine at Sawahlunto, W Sumatra.) Spacing of data points for different levels of confidence is as follows: for 'measured' reserves, a spacing of less than 0.8 km with measurements extending 0.4 km from the points; for 'indicated' reserves, the spacing is less than 2.4 km with measurements extending 0.8 km beyond measured reserves; for 'inferred' reserves, the spacing is less than 9.6 km with measurements extending 3.6 km beyond indicated reserves. Measured plus indicated plus inferred reserves are described as 'identified'.

In the project area in Kalimantan, identified reserves of high-volatile C bituminous and sub-bituminous A coals from the Loa Kulu Formation at Loa Kulu and Loa Haur are 35 Mt, but these reserves are in scattered small areas. In the Prangat Formation in the Badak Syncline there are 428 Mt of reserves in 14 seams, mostly of sub-bituminous C rank coal with some sub-bituminous B; and 65 Mt were identified in the Sakakanan area. Kamboja Formation coals in the Badak Syncline include reserves of 476 Mt of lignite A rank. The project area thus shows a clear potential which would merit further study both more extended and more detailed, and which could go on to include non-geological investigations of all aspects of mining feasibility, transport, planning and land reclamation.

ACKNOWLEDGEMENTS: The authors are pleased to acknowledge the contributions and work of their colleagues in the coal project, particularly R. Addison, L. P. Thomas, Badhroom, Hardjono, Haryoko and Herudiyanto. Oil companies in the project area, namely Pertamina, Roy M. Huffington Inc. (Huffco) and Total Indonesie, gave considerable help, and Huffco generously made much unpublished data available. This paper is published with the permission of the Director of the directorate of Mineral Resources, Indonesia. D. H. Land publishes with the permission of the Director of the British Geological Survey (NERC). We thank A. C. Scott for inviting us to contribute this paper.

References

ADAMS, C. G. 1970. A reconsideration of the East Indian letter classification of the Tertiary. *Bulletin of the British Museum (Natural History), Geology*, **19**, 87–137.

ADDISON, R., HARRISON, R. K., LAND, D. H. & YOUNG, B. R. 1983. Volcanogenic tonsteins from Tertiary coal measures, East Kalimantan, Indonesia. *International Journal of Coal Geology*, **3**, 1–30.

ALLEN, G. P., LAURIER, D. & THOUVENIN, J. 1979. Etude sedimentologique du Delta de la Mahakam. Total, *Notes et Memoires*, no. 15, Paris.

AMERICAN SOCIETY FOR TESTING AND MATERIALS 1974. Standard specifications for classification of coals by rank (ASTM Designation D 338–66). *In: Gaseous Fuels, Coal and Coke; Atmospheric Analyses*, part 26, pp. 54–58. American Society for Testing and Materials.

AVERITT, P. 1975. Coal resources of the United States, January 1974. *United States Geological Survey Bulletin* 1412.

VAN BEMMELEN, R. W. 1949. *The Geology of Indonesia* (2 vols). Government Printing Office, s-Gravenhage.

BRITISH STANDARDS INSTITUTION 1971. Methods for the analysis and testing of coal and coke: Part 16, Reporting of results. *British Standard* 1016.

CHAPMAN, R. E. 1974. Subsidence and deformation of terminal regressive sequences in the Indonesian region. *Proceedings of the Indonesian Petroleum Association, Fifth annual convention (1976)*. **1**, 151–158.

HAMILTON, W. 1979. Tectonics of the Indonesian region. *United States Geological Survey Professional Paper* 1078.

KATILI, J. A. 1978. Past and present geotectonic position of Sulawesi, Indonesia. *Tectonophysics*, **45**, 289–322.

—— & JOHANNAS, A. 1961. *Review of the status of the mineral industry and the perspective of mineral exploration in Indonesia.* Institute of Technology of Bandung, Contributions of the Department of Geology no. 36.

KOESOEMADINATA, R. P., HARDJONO, USNA, I. & SUMADIRDJA, H. 1977. Tertiary coal basins of Indonesia. *United Nations Economic and Social Commission for Asia and the Pacific, Committee for Co-ordination of Joint Prospecting for Mineral Resources in Asian Off-shore Areas, Technical Bulletin*, **12**, 43–86.

LAND, D. H., ADDISON, R., BADHROOM, HARYOKO, HERUDIYANTO, JONES, C. M. & THOMAS, L. P. 1986. Geology and coal resources of coalfields west and south of Samarinda, East Kalimantan. *Report of the Directorate of Mineral Resources, Indonesia*, Bandung.

LEUPOLD, W. & VAN DER VLERK, I. M. 1931. The Tertiary. *Leidsche Geologische Mededelingen*, **5**, 611–648.

MCCABE, P. J. 1984. Depositional environments of coal and coal-bearing strata. *In*: RAHMANI, R. A. & FLORES, R. M. (eds) *Sedimentology of Coal and Coal-bearing, Sequences*, pp. 13–42. Special Publication of the International Association of Sedimentologists 7. Blackwell Scientific Publications, Oxford.

MAGNIER, PH & SAMSU, B. 1976. The Handil oilfield in East Kalimantan. *Proceedings of the Indonesian Petroleum Association, Fourth annual convention (1975)*, pp. 41–61.

MARKS, E., SUJATMIKO, SAMUEL, L., DHANUTIRTO, H., ISMOYOWATI, T. & SIDIK, B. B. 1982. Cenozoic stratigraphic nomenclature in East Kutei Basin, Kalimantan. *Proceedings of the Indonesian Petroleum Association, Eleventh annual convention (1982)*, **1**, 146–179.

ROSE, R. & HARTONO, P. 1978. Geological evolution of the Tertiary Kutei-Melawi basin, Kalimantan, Indonesia. *Proceedings of the Indonesian Petroleum Association, Seventh annual convention (1978)*.

RUTTEN, L. M. R. 1914. Neue fundestellen von tertiaren foraminiferen in Ost Borneo. *Sammlungen des geologischen Reichsmuseums in Leiden* ser. 1, **9**.

SAMUEL, L. & MUCHSIN, S. 1976. Stratigraphy and sedimentation in the Kutei Basin, Kalimantan. *Proceedings of the Indonesian Petroleum Association, Fourth annual convention (1975)*, pp. 27–39.

SCHWARTZ, C. M., LANGHBAUM, G. H., SAMSU, B. S. & ARMSTRONG, J. D. 1974. Geology of the Attaka Oilfield, East Kalimantan, Indonesia. *Proceedings of the Indonesian Petroleum Association, Second annual convention (1973)*, pp. 195–215.

SITUMORANG, B. 1982. The formation of the Makassar Basin as determined from subsidence curves. *Proceedings of the Indonesian Petroleum Association, Eleventh annual convention (1982)*, **1**, 83–107.

STACH, E., MACKOWSKY, M. TH., TEICHMÜLLER, M., TAYLOR, G. H., CHANDRA, D. & TEICHMÜLLER, R. 1982. *Stach's Textbook of Coal Petrology*, 3rd edn. Springer-Verlag, Berlin.

THOMAS, L. P., HARDJONO, LAND, D. H. & ADDISON, R. 1980. Coal exploration in East Kalimantan, Indonesia, *Revue de l'Industrie Mineral*, no. 6, 327–334.

UNITED STATES GEOLOGICAL SURVEY 1976. Coal resource classification system of the United States Bureau of Mines and United States Geological Survey. *United States Geological Survey Bulletin* 1450–B.

VAIL, P. R., MITCHUM, J. R. & THOMPSON, S. 1977. Seismic stratigraphy and global changes of sea level, Part IV: Global cycles of relative changes of sea level. *In*: PAYTON, C. (ed.) *Seismic Stratigraphy—Applications to Hydrocarbon Exploration*, pp. 85–97. American Association of Petroleum Geologists, Memoir 26.

VAN DER VLERK, I. N. & UMBGROVE, J. H. F. 1927. Tertiare Gidsforaminiferen van nederlandsch—oost Indië. *Wetenschappelijk Mededelingen Dienst Mijnbouw Nederlandsch—Oost Indië*, **6**, 1–31.

D. H. LAND, British Geological Survey, Murchison House, Edinburgh, EH9 3LA.
C. M. JONES, Britoil, Glasgow G2 5LJ, Scotland.

Recent advances in organic petrology and organic geochemistry: an overview with some reference to 'oil from coal'

D. G. Murchison

ABSTRACT: Although organic petrology and organic geochemistry differ greatly in age, it is during the last 25 years that there has been an almost exponential expansion of research and literature in both fields. Much of the support for the developments that have taken place has come from industrial sources. The steel industry underpinned applied coal petrology in the 1960s to improve carbonization practice related to the production of metallurgical cokes. The petroleum industry has generally supported both organic petrology and organic geochemistry, with a principal aim of improving methods to assess rank (level of maturity) and type (source input) of both coals and kerogens. As precise a determination as possible of both these parameters is as essential in hydrocarbon-prospect appraisal as it is in the raising of coke quality.

In organic petrology the construction of fully automated microscope fluorometric systems has been a major advance that has helped to define more precisely the low maturity levels that exist prior to and through most of the 'oil window'. The same optical systems have not been so successful in producing an acceptable single petrographic typing method for kerogens. Used in conjunction with chemical methods, however, microscope fluorometry has certainly contributed substantially towards more precise kerogen typing, allowing more satisfactory discrimination of the liptinite macerals, several of which are known 'oil formers'. Another group of potentially economic rocks, the oil shales, has also been successfully classified, principally through fluorescence methods.

Within organic geochemistry, the growth and application of analytical pyrolysis methods have been of relevance and importance. While not capable of high chemical resolution, bulk-flow pyrolysis systems (*cf.* 'Rock-Eval') are now widely used to gain a preliminary estimation of maturity level, source input and generating potential of source rocks by the screening of large numbers of samples. The most important single advance in the pyrolysis field, however, has probably been the evolution of a typing method, which, although it utilizes information from both petrography and bulk-flow pyrolysis has, as its essential core, quantitative high-resolution pyrolysis data. Kerogens can now be subdivided into eight different end-members and referred to specific depositional environments.

When these fluorometric and pyrolysis data are combined with information from biological-marker studies, which have expanded rapidly as a result of the advances in computerized gas chromatography-mass spectrometry over the past decade, it is not surprising that hydrocarbon exploration efforts are better rewarded. There is now less doubt that terrestrial sources have generated economic accumulations of oil. The principal areas where oil has been generated from land-plant material are the Australian and the Canadian Frontier sedimentary basins, and the Chinese intermonbane basins, but other similar environments have been identified and no doubt other basins with terrigenous, oil-forming source rocks will be discovered in the future.

For further rapid advances in microscopic fluorometry in organic petrology a thorough understanding of the underlying chemistry will be important. Although pyrolysis-gc systems will continue to contribute to kerogen typing, surely the principal competitor (and further stimulus) to petrographic typing methods will be pyrolysis-ms. Close collaboration between investigators in the two fields is essential. A serious reexamination of the prospects for laser-micropyrolysis systems in kerogen typing, first mooted more than 20 years ago, but then unsatisfactory because the 'state of the art' in both microscope optics and in laser and mass spectrometer technology was insufficiently refined, should be considered. Continuing sophistication and success in the detection, identification and application of biomarkers will result from the use of a variety of techniques, for example, tandem mass spectrometry, high performance liquid chromatography-mass spectrometry and gas chromatography-isotope mass spectrometry. Selected metastable ion monitoring and high-resolution gc-ms seems an inevitable core procedure for the specific identification of biomarkers in the future, probably replacing current conventional gc-ms systems.

Keywords: biomarkers, kerogen, pyrolysis, fluorescence, maturity, Australia, Canada, Indonesia, China, oil prone coal

The two fields of organic petrology and organic geochemistry differ considerably in age, organic geochemistry being much the younger. Organic petrology grew from coal petrology: both employ similar equipment and methodologies. The earliest paper in English on coal petrographical research was published in the first volume of the *Proceedings of the Geological Society of London*, when Hutton (1834) described thin sections of coal, observed by transmitted light, principally from the Northumberland and Durham coalfields. Almost 100 years were to pass, after the publication of this first paper, until the major expansion of coal petrological literature took place, although a number of important, fundamental contributions were made to the field in the early part of this century, notably by Stopes (1919) and Thiessen (1920).

Just as research in coal petrology began its expansion, what is regarded as the birth of organic geochemical literature occurred. Treibs (1934) demonstrated the presence of porphyrins in a wide range of petroleums and bitumens. Again a considerable period elapsed until there was a noticeable increase of research in organic geochemistry, which then fed the literature expansion that has been particularly pronounced over the past 10 years. There may now be an hiatus in research due to the contraction in petroleum-industry activity caused by the severe fall in the world price of oil. Bayliss (1985), however, stressed the important contribution of organic geochemistry, to the petroleum industry particularly, when he reported that 26% of all papers published in the *Bulletin of the Association of American Petroleum Geologists* over the period 1980–84 were related to this field.

A principal stimulus for the expansion of research that took place in organic petrology and organic geochemistry was industrially based and industrially financed. Coal and steel companies gave strong support to fundamental and applied coal petrology, while the broader field of organic petrology has, more recently, been well funded by the petroleum industry, which has also been an important financial mainstay for organic geochemistry. To maintain support from these sources there has had to be, understandably, a demonstration of economic benefit. For example, that applied coal petrology made a substantial technological and economic contribution to the efficient blending and beneficiation of coals used in carbonization practice, particularly in the production of metallurgical cokes, would not be questioned. Equally, petroleum companies have estimated the value of organic geochemistry to hydrocarbon-prospect appraisal: a reported increase by some companies of forecasting efficiency from 30% to greater than 60%, if organic geochemical data are included, would not be unusual. These successes in industrial fields must be seen as general advances. Many are certainly relatively recent in the petroleum field and are reviewed by Tissot (1984).

The title of this contribution, 'Recent advances in organic petrology and organic geochemistry' was originally all-embracing. Left as presented to the author, it could have resulted in an article which might at least have been regarded as idiosyncratic: and very lengthy! A most important consideration seems to be the title of the symposium, 'Coal and Coal-bearing Strata', to which contributions should bear relevance. Consequently, an appropriate constraint, 'oil from coal', has been applied to the original title. Ever since Brooks & Smith (1967, 1969) and Hedberg (1968) documented their findings, which related oil composition and oil generation to terrestrial organic matter, a large body of geological and geochemical opinion has resisted the evidence that land-plant material could generate liquid-petroleum hydrocarbons, even on a limited scale and despite oil occurrences in non-marine successions which had no close association with marine rocks. The principal arguments against a 'coal origin' for oil is first that the gross composition of coal is inappropriate to produce oil, but it will generate gas. Second, because of the ultra-fine capillary structure of coals (which ranges from pores which are fractions of a nanometer in diameter, up to channels perhaps 10 nm wide), expulsion of any generated oil would be severely impeded. Both these objections show a lack of appreciation of the well-documented information about the range of chemical composition and physical structure displayed by organic constituents (macerals) of coals laid down in differing depositional environments.

Durand & Paratte (1983) suggest, on the basis of pyrolysis data, that coals of different type can generate up to 30% by weight of hydrocarbons of oil-type components when thermally cracked. They believe that neither the microporosity nor the plastic behaviour of coals will hinder easy expulsion of generated hydrocarbons from coals. However, Jones (1979) would argue against this view on grounds of mass balance and geological constraints on oil migration from coals, while Radke *et al.* (1980) believe that any oil generated by macerals would be trapped in the micropore system and then cracked by maturation processes to produce gas. The crucial point here is surely the absorptive capacities of the macerals, notably vitrinite which has a very high microporosity, particularly at low maturity levels. What has to be accepted is that these absorptive capacities have to be overcome and this can only be achieved by sufficient hydrocarbon generation to saturate the pore spaces, leaving an excess for expulsion from the coal (Powell 1978, McAuliffe 1979, Powell 1984). Figure 1, based on data from Kantsler *et al.* (1978), Smith & Cook (1984) and Murchison *et al.* (1985), displays a relationship between different optical properties of coal macerals, their hydrocarbon generating capacities and the generalized zones of hydrocarbon generation plotted against increasing temperature.

The intention of this article is principally to discuss recent advances in techniques and methodologies that are leading to the production of unequivocal evidence that petroleum in commercially viable quantities can be generated from organic matter of terrestrial origin. Three fields that have contributed to this evidence will be considered: (i) the evolution of microscope fluorometric systems and their application in organic petrology; (ii) kerogen typing, particularly the application of pyrolysis methods and (iii) the development and use of biomarkers. Finally, based principally on the application of these techniques, some examples will be given of sedimentary basins in which the dominant source for the oils generated is most likely to be organic matter of terrestrial origin.

The fields under consideration all embrace attempts to evaluate the composition of organic matter at different levels of physical and chemical development. It will not surprise organic petrologists, who are still in pursuit of improved methods of evaluating *rank* and *type*, to know that organic geochemists have also expended much energy in following similar paths, referring only to 'maturation' instead of *rank* and 'source input' instead of *type*. Rank or maturity here is only discussed in so far as it is appropriate to the underlying qualifying theme of 'oil from coal'. Elsewhere in this volume advances in coalification studies and their geological application are considered at length by M. Teichmüller. It should become apparent from both these contributions that organic geochemists and organic petrologists have recently integrated their research efforts more closely than they have done in the past.

Finally, it is important to emphasize that the term 'coal' can be applied to deposits ranging from the autochthonous extensive and prolific forest spreads that have given rise to the coal seams of major coalfields, to the dominantly algal boghead coals which are regarded nowadays as synonymous with oil shales. The same mixes of macerals are found in minerogenic sediments, particularly in the finer grained clastics, and in this article the term 'coal' is also loosely applied to these occurrences. The principal macerals of coals are shown in Table 1.

Microscope fluorometry

Instrumentation and measurements

Probably the most widely used rank parameter over the past three decades has been oil reflectance, particularly when measured on the principal maceral of coal, vitrinite, which is also found widely distributed in sediments. The use of reflectance measurements in the field of coal petrology has been highly successful. Besides the contribution by Teichmüller (*this volume*), reviews, either in whole or in part, demonstrating the importance and applicability of reflectance measurements in scientific and technological fields have been given by Davis (1978), Neaval (1981), Rhoades *et al.* (1981), Teichmüller & Teichmüller (1981) and Murchison *et al.* (1985).

The value of vitrinite reflectance measurements is based upon two factors. First, its level of effectiveness depends upon the particular range of rank in which the measurements are made. At low rank levels, reflectance, which increases with rising rank, is principally dependent upon changes in refractive index, which only produce a slow rate of reflectance increase. This behaviour is characteristic of reflectance increase through most of the 'oil window' and at maturity levels below the 'oil window' threshold (Fig. 1). At higher levels of rank however, when absorption becomes the dominant fundamental optical property the rate of reflectance rise becomes much more rapid.

The second factor that influences the effectiveness of reflectance measurements as a rank parameter is their application to vitrinite particles disseminated in sediments. With measurements on single coals or coal blends an immediate comparison of vitrinite with other macerals is possible: identification of vitrinite is rapid and certain. Identification can be less easy when the vitrinite is distributed, often sparsely, in finely comminuted and isolated particulate form in sediment and no comparison with other macerals may be possible. Other organic matter, quite unstructured and homogeneous in appearance, can resemble vitrinite when included as isolated fragments in sediments.

These two factors were important stimuli in the search for other microscopical parameters which might allow a more effective assessment of maturity at low levels. Although reflectances can be measured on the liptinite macerals, and some of these at least have reflectance coalification

TABLE 1. *Maceral terminology (after ICCP Stopes Heerlen system, 1963, 1971, 1975; modified by Smith 1981) with maceral origins (partial list) (after Murchison* et al. *1985)*

Maceral group	Maceral	Origin
Vitrinite	Telovitrinite	Humified plant remains typically derived from woody, leaf or root tissue with well to poorly preserved cell structures
	Detrovitrinite	Humified attrital, or less commonly detrital, plant tissue with particles, typically being cell fragments
	Gelovitrinite	Humified material showing no trace of cellular structure, probably colloidal in origin
Liptinite	Sporinite	Outer casings of spores and pollens
	Cutinite	Outer waxy coating from leaves, roots and some related tissues
	Resinite	Resin filling cells and ducts in wood: resinous exudations from damaged wood
	Fluorinite	Essential oils in part; some fluorinite may be produced during physico-chemical coalification and represent non-migrated petroleum
	Suberinite	Cork cell and related tissues
	Bituminite	Uncertain but probable algal origin
	Alginite	Tests of some groups of green algae; material referred to alginite shows moderate to strong fluorescence
	Exsudatinite	Veins of bitumen-related material expelled from organic matter during coalification
Inertinite	Fusinite	Pyrofusinite—originates from wood and leaf tissues in forest fires
	Semifusinite	Wood or leaf tissues weakly altered by mouldering or by biochemical alteration
	Intertodetrinite	Similar to fusinite or semifusinite but occurring as small fragments
	Macrinite	Humic tissue probably first gelified and then oxidized by processes similar to those producing semifusinite
	Sclerotinite	Moderately reflecting tissue of fungal origin—largely restricted to Tertiary coals
	Micrinite	Largely of secondary origin and formed by disproportionation of lipid or lipid-like compounds to fluid hydrocarbons and a high reflecting granular residue of micrinite

tracks that are well correlated with similar coalification tracks of other macerals (Smith & Cook 1980, Cook 1982, Murchison *et al.* 1985), these correlations are only relatively recent achievements. Alternative maturity parameters were being sought more than a decade ago. There was also at least a suspicion that if interest was focussed on maturity levels at which oil generation occurred, then it would be more logical to concentrate on the liptinite constituents, which show greater liquid-hydrocarbon potential than vitrinite, which is regarded principally as a gas producer. Substantial effort has therefore been expended on the development of microscope fluorometric systems to allow determination of the fluorescence properties of liptinite macerals and to relate these to maturity levels. An added bonus might be that the fluorescence approach could help in the typing of coals and of organic matter in sediments through more precise differentiation of the liptinite macerals. Many macerals of the liptinite group are difficult to identify under oil immersion in reflected light because of their extremely low reflectances: some cannot be seen at all in sediments under oil immersion objectives.

The earliest papers in this field related principally to fluorescence studies on microfossils, fossil pollen and spores by van Gijzel (1966, 1967, 1971), although van Gijzel (1975), in addition to spore and pollen exines, discussed the fluorescence response of fresh plant substances, fossil wood, cuticles, extracts and a range of sedimentary organic matter. Considerable attention was given in these papers to the definition of a substantial number of parameters to categorize fluorescing samples. Less emphasis was placed on the behaviour of fluorescence parameters with increasing maturity, although the shift of the spectral maximum of the emission spectrum and the percentage of 'fading' of the entire fluorescence intensity, after 30 min exposure to irradiation, was plotted for microspores (van Gijzel 1975). Of the different parameters defined, not all are used nowadays, but those which commonly are would include, (a) the wavelength (λ_{max}) between 400 and 700 nm at which peak relative intensity of fluorescence emission occurs; (b) the spectral quotient (Q) being the ratio of the relative fluorescence intensity at 650 nm compared with that at 500 nm; (c) the spectral quotient (Q_{max}),

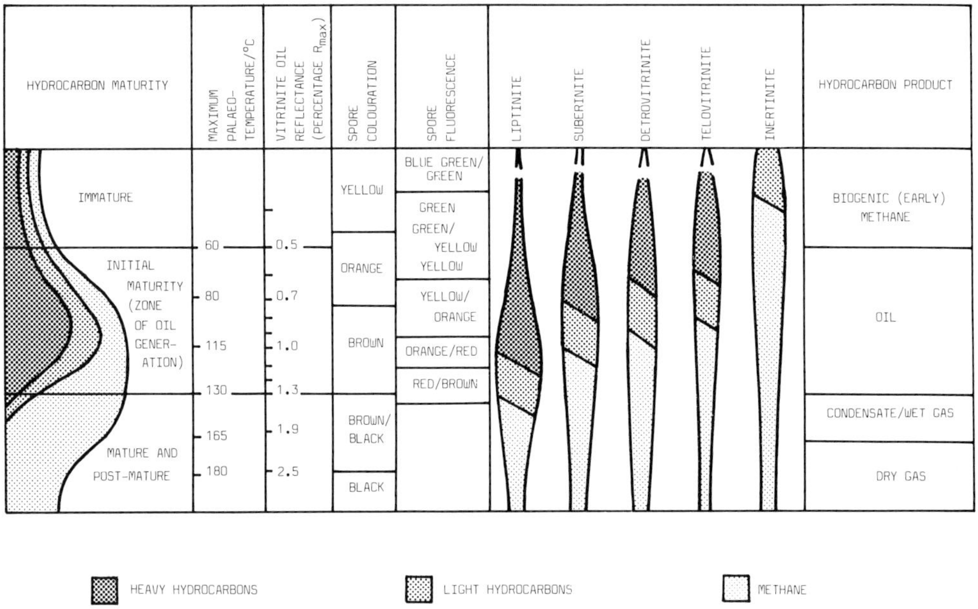

FIG. 1. Relationship between generalized zones of hydrocarbon generation, the hydrocarbon generating capacities of important maceral groups or macerals and certain optical properties of vitrinite and sporinite (based on data from Kantzler *et al.*, 1978, Smith & Cook 1984 and Murchison *et al.* 1985).

which is the ratio of the relative fluorescence intensity at λ_{max} compared with that at 500 nm and (d) A_I, the alteration of intensity ('fading' of van Gijzel 1975), at 546 nm recorded 30 min after irradiation of the field has begun. In addition, the fluorescence intensity at 546 nm, compared with that from a standard (uranyl glass) taken as 100%, can also be measured. Excitation would normally be at 365 nm for such measurements, but so-called 'blue-light excitation' at higher wavelengths is often employed for qualitative observations of fluorescence.

The first four of these parameters are employed in what van Gijzel (1975) referred to as 'polychromatic fluorescence microphotometry', which is distinguished from the 'monochromatic method' in which the fluorescence intensity is only measured at a specific wavelength. The 'polychromatic method' has proved to be much more widely applicable than the 'monochromatic method', particularly since the uranyl glass is not universally acceptable as a standard and the thickness of the irradiated macerals also governs the absolute fluorescence intensity. Full details of the 'monochromatic method' can be found in Jacob (1973, 1974). Besides the papers of van Gijzel (1966, 1967, 1971, 1975) referred to above, details of the 'polychromatic method' have been described by Ottenjann *et al.* (1974, 1975) and Teichmüller & Ottenjann (1977), the latter paper being particularly concerned with measurement of 'alteration'.

As the results of more fluorescence studies became available, it became clear that quantitative measurements of fluorescence were by no means straightforward. Ottenjann *et al.* (1975) commented that the calculation of numerical values for peak fluorescence transmission and the spectral quotient were probably easiest to achieve, but that correlation with other rank or maturity parameters was only approximate. Spectral emission curves were regarded as having greatest potential in displaying change of fluorescence characteristics with increasing maturity, but it was appreciated that a number of corrections had to be applied to such curves before they could be sensibly used, for example, corrections for background, microscope optics, dark current and other characteristics of the photomultiplier, as well as for the transmittance of dichroic mirrors in the vertical illuminator. Improvements to microfluorometric systems were made and Ottenjann (1980) modified his system to include: (a) an interference filter and a quartz collector lens to give greater excitation; (b) a photomultiplier with extended red response to allow more satisfactory evaluation of fluorescence emission at the red end of the spectrum; (c) a motorized interference filter acting as a monochromator to measure spectral fluorescence and (d) a calibrated tungsten filament lamp of known colour temperature to allow more accurate determination of correction factors throughout the spectrum. The spectral quotient could also be measured more rapidly and the method to determine 'photo-alteration' was elaborated. Further developments were reported by Ottenjann (1982). Since these improvements were made, many fully automated microfluorometric systems have been manufactured and it has become possible to measure the different fluorescence parameters rapidly and also to determine and apply corrections quickly. These successful developments exposed an entirely new area of organic petrography.

Benefits of microscope fluorometry in maturity evaluation

The first aim of the application of microscope fluorometry in organic petrology was to try to develop, if possible, a more precise and reliable determination of low levels of maturity. Van Gijzel (1975) had produced some data on sporinites, but a much more extended sequence of spectra was produced by Ottenjann *et al.* (1975) from sporinites in a range of fuels from peat to mid-bituminous rank coals. These authors also showed that satisfactory correlations existed between the carbon contents of the coals and the fluorescence emission maxima and the spectral quotients of the sporinites. It was evident that of the liptinite macerals, all of which fluoresce, many were too variable in their fluorescence properties at any particular rank level to be reliable indicators of maturity. Attention was thus principally focused on the ubiquitous spores and pollen.

The successful design of more sophisticated fluorometric systems allowed several developments. First, correlations were established between vitrinite reflectance and the fluorescence parameters of sporinite and cutinite (whose fluorescence parameters proved less variable than originally thought), and a new maceral, fluorinite (see below and Table 1). Second, relationships showing the variation of fluorescence parameters of these same macerals with depth were produced (Teichmüller 1977, 1979a, b, but see Teichmüller 1982 for many of the correlations in these earlier papers). In general, spectral maxima and spectral quotients rise with increasing reflectance and depth, while fluorescence intensity falls. So successful have been the advances in equipment development that Ottenjann (1985) has shown that fluorescence parameters indicate a rank increase with depth in bituminous coals when conventional rank parameters, including reflectance, cannot do so. Nevertheless, care is necessary. Spiro & Mukhopadhyay (1983), who maintain a satisfactory correlation between fluorescence parameters of alginite and vitrinite, do stress that the relationship breaks down if degraded alginite, which they identified in transmitted light, is used for fluorescence studies since there is a strong spectral red shift if degradation has occurred. Hutton & Cook (1980) showed that mean maximum reflectance of vitrinites was significantly lowered by the presence of alginite and they suggested that similar effects on vitrinite reflectance might be caused by other liptinite macerals or even by type differences within vitrinites. The reason for the reflectance decrease is the absorption of liquid hydrocarbons by the vitrinite at the time of their expulsion from closely associated liptinitic macerals. The vitrinites themselves may be perhydrous (hydrogen-rich).

Observations and correlations were also made in these earlier fluorescence studies (Teichmüller 1974a, Teichmüller & Wolf 1977) that were to be of direct relevance to the petroleum industry. Teichmüller (1974a, b) produced important contributions when she described highly fluorescent mobile liquids that were expelled from both liptinitic and vitrinitic macerals and then were either deposited in cleats, cracks and larger pores or absorbed into the ultra-fine capillary structure of vitrinites. She also identified two 'coalification jumps', the first occurring in both liptinites and vitrinites at a random reflectance value $R_m \approx 0.50\%$ at the time when mobile bitumens were generated and relatively highly reflecting residues of micrinite were formed. The second 'coalification jump', in the reflectance range of $R_m = 1.20$–1.60%, was characterized by complete loss of fluorescence in the macerals and an approximate coincidence of other optical properties of liptinites and vitrinites. The two coalification jumps were related to the onset of oil generation and its termination in source rocks. Later, Teichmüller (1982) identified these coalification jumps as '1' and '3', reporting a further jump, '2' at $R_m = 0.80$–0.90%, corresponding to the time of maximum oil generation, on the basis of extensive data from yields of solvent extracts from source rocks. At this stage the fluorescence parameters (intensity, peak emission and the spectral quotients) undergo sudden shifts while a secondary fluorescence in the vitrinites, which begins when bitumens are generated at the first coalification jump, increases in intensity and shows a strong red shift. Teichmüller (1982) attributes this increasing secondary fluorescence to the gradual absorption by vitrinite macerals of bitumens generated from liptinite macerals. The fluorescence properties for two important vitrinite macerals have been summarized by Cook (1986) (Table 2).

TABLE 2. *Summary of fluorescence properties of vitrinite (from Cook 1986)*

Vitrinite reflectance (% $\bar{R}_{vmax}$)	Telocollinite and precursors	Desmocollinite and precursors
<0.45	Typically none but some wood structures show fluorescence related to botanical structure	Some fluorescence at very low ranks
0.45–0.70	Typically none	Typically none
0.70–1.1	In some cases green fluorescence from globules and streams emitted from cracks, positive alteration common	Dull brown fluorescence, development of a green haze due to oil emission and positive alteration
>1.1	Typically little or no fluorescence, but oil emission occurs from vitrinite up to about 1.3% reflectance	Little or no fluorescence, oil emission may result in the development of a green haze for vitrinite up to about 1.5% reflectance

As stated earlier, Teichmüller & Durand (1983) have no doubts about the capacity of coals to generate oil. They confirmed the behaviour of parameters related to secondary fluorescence developing in vitrinites due to the formation and later fragmentation of petroleum-like hydrocarbons in coals as maturity increased. 'Rock-Eval' pyrolysis data (see also later under 'kerogen typing'), from the same samples confirmed maximum release of hydrocarbons approaching the point of the second coalification jump, with coals high in liptinites giving the highest yields of hydrocarbons. Recently, however, Radke *et al.* (1984, 1985) have examined relationships between a range of hydrocarbons and optical-rank parameters, including fluorescence properties, only to find no close correlation between chemical and fluorescence parameters, e.g. between spectral quotient and the methylphenanthrene ratio which, in contrast, was related to the behaviour of vitrinite reflectance. Earlier work by Radke *et al.* (1980), had suggested that extractable aromatics and heterocompounds could be important in producing vitrinite fluorescence, while Ottenjann *et al.* (1982), although principally concerned with relationships between fluorescence parameters and the swelling properties of a range of coals, did also note that there was a correlation between vitrinite fluorescence and yield of solvent extract.

The correlations established between fluorescence properties and increasing maturity are of interest and relevance to organic petrologists generally. It is clear that a number of fluorescence parameters are useful indicators of maturity, particularly lower levels of maturity. Not only do petroleum-like substances exude from coals, but also the correlation between fluorescence parameters and organic geochemical data suggests certain macerals are capable of generating liquid hydrocarbons. Nevertheless, probably more important to oil exploration efforts generally, is the way in which fluorescence microscopy allows a more detailed and accurate petrographic assessment to be made of both coals and other potential source rocks. In the latter particularly, the low-reflecting liptinitic macerals at last become clearly invisible. A much greater appreciation of their likely contribution to petroleum generation becomes possible, since many had previously been indistinguishable from the mineral matter of rocks: others, principally the liptinite macerals, sporinite, resinite, cutinite and alginite, are more obvious and more readily differentiated in coals. It is now also possible to make at least semi-quantitative, if not fully quantitative, assessments of the liptinite macerals present in a coal and, of more importance, in a sediment by ultra-violet or 'blue-light' excitation.

Kerogen typing

Microscopical methods

Kerogen is a complex material that varies widely in composition, depending upon its source, which can range from terrestrial to marine. It is a key intermediate product between the source material and generated petroleum and/or gas. In the strict sense, the term 'kerogen' is used to describe the insoluble organic residue from a sediment that remains after its extraction by organic solvents and with the exception of pyrite, the removal of all mineral matter, chiefly silicates and carbonates, by appropriate treatment with acids and other reagents.

The longest-standing attempts to characterize kerogens petrographically have been by using isolated kerogens spread as strew mounts on glass slides which are then viewed by transmitted light (Staplin 1969, Burgess 1974, Rogers 1979, Batten 1980, Combaz 1980, Masran & Pocock 1981).

Kerogens have generally been subdivided on this basis into a relatively small number of types (Table 3) in classifications which employ qualifying terms, e.g. 'coaly', 'woody', 'herbaceous, 'structured', 'amorphous' and 'algal' to describe the kerogens and, at the same time, to carry some implication of origin. This particular approach is limited in its application by the increasing optical absorption suffered by crustal organic matter with temperature rise, so that eventually the kerogen preparation becomes essentially opaque. Translucency is, however, maintained well through the 'oil window', so the method has been regarded as helpful in the evaluation of hydrocarbon potential of source rocks. Indeed, several maturation scales have been set up on the basis of changes in the colouration of spores in kerogens (Staplin 1969, Raynaud & Robert 1976, Barnard *et al.* 1980, Batten 1980). Smith (1983) maintains, on the basis of comparisons between transmitted-light spectral curves from standards relating to three of these scales, that there is reasonable correlation between the three sets of standards and hence the maturation scales. Grayson (1975) has measured the translucency of spores and pollens in dispersed organic matter using a fibre-optic probe connected to a photomultiplier, demonstrating a relationship between translucency and generation of hydrocarbons (cf. Fig. 1).

Kerogens can also be classified in simple chemical terms into four principal types (Tissot *et al.* 1974, Harwood 1977) that are defined by evolution paths which broadly correspond with the coalification tracks, based on plots of atomic ratios O/C : H/C, that are followed by the maceral alginite and the maceral groups, liptinite (exinite), vitrinite and inertinite (van Krevelen 1950, Dormans *et al.* 1957). Generally Type I kerogens are algal-dominated and deposited in lacustrine environments and so also do Type II kerogens consist of degraded aquatic organisms, extensively reworked, principally by bacteria, and laid down in marine environments under reducing conditions. As with Type I kerogens they can have a high oil and gas potential under appropriate geological conditions. Frequently, however, a mixture of Type I kerogen with a Type III kerogen (see below), can mimic a Type II kerogen. Terrestrial plants are the principal contributors to Type III kerogens, as they can directly be to Type IV kerogens, which also contain high-carbon residual organic matter. Type III and Type IV kerogens can be deposited under either marine or non-marine conditions, but they have been regarded principally as gas producers with much less oil potential than Type I and Type II kerogens. Type III kerogens can also arise from sources other than terrestrial plants, e.g. algal or organic matter that has been oxidized (Jackson *et al.* 1984). There are, in addition, many mixed Type II/III kerogens which occur in near-shore marine environments. In simple chemical terms, moving from Type I kerogens to Type IV kerogens, high atomic H/C ratios are gradually reduced while the low atomic O/C ratios of the Type I and II kerogens rise to high values in Type IV kerogens (Fig. 2). This chemical classification

TABLE 3. *Correlation of different kerogen classifications based on microscopy (modified from Mukhopadhyay* et al. *1985)*

Staplin (1969)	After Burgess (1974)	Combaz (1975)	Masran & Pocock (1981)
	Amorphous		Structured aqueous
Amorphous	Algal	Sapropelite	Amorphous biodegraded aqueous
Herbaceous	Herbaceous	Liptobiolites	Spores, pollen, resin, structured aqueous
Woody	Woody		Biodegraded structured terrestrial, Amorphous Spherulitic
Coaly	Coaly	Lignohumites	Charcoal, fungi

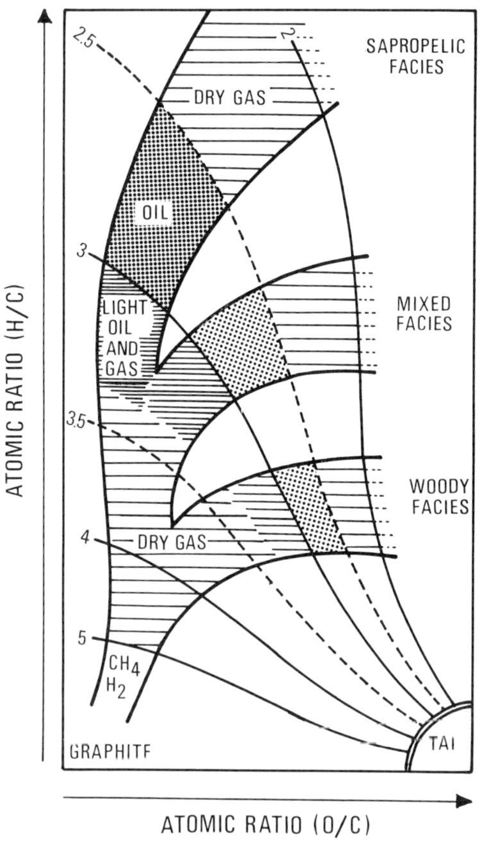

FIG. 2. Formation of hydrocarbons as a function of organic facies and thermal alteration index, based on values from Raynaud & Robert (1976). Redrawn from Combaz (1980). The three paths running from 'Sapropelic Facies', 'Mixed Facies' and 'Woody Facies' correspond broadly with the evolution paths for Type I kerogens, Type II/III kerogens and Type III/IV kerogens respectively. TAI—Thermal Alteration Index.

is necessarily somewhat crude since the analyses do not relate to specific constituents, but to mixtures of many different organic constituents, even in Type I kerogens.

Besides the application of transmitted-light microscopy to the study of kerogens, normal reflected light, ultra-violet and 'blue-light' excitation microscopy have also been used (Robert 1971, Dow 1977, Robert 1979, 1980, Dow & O'Connor 1982, Gutjahr 1983). These methods have generally been most successfully applied to Type III and Type IV kerogens. Only reflected-light, fluorescence methods have been genuinely helpful in attempting to categorize Type I and II kerogens.

Powell *et al.* (1982) have discussed the limitations of existing petrographic methods in defining oil-source rocks. Four factors were identified that influence poor correlation between petrographic data and source-rock potential: (a) unrepresentative kerogen concentrates, (b) inadequate definition of amorphous organic material, (c) poor quantitative estimation and (d) low correlation between organic-matter types and source potential. While it might be possible to correct the influence of certain of these factors that are experimentally based, the most serious problem is the precise characterization of 'amorphous kerogen', which is widely distributed and has been commonly held to be of algal origin, but in fact is a complex that can vary widely in composition. Different investigators have attempted to resolve the problem with varying degrees of success.

Although Staplin (1969) demonstrated an empirical relationship between oil occurrence and amorphous kerogen, Rogers (1979) showed that such kerogen could be an alteration product of virtually any type of organic matter, including herbaceous material. Batten (1983) discussed the identification of amorphous sedimentary matter in transmitted light and cited the problem of differentiating between amorphous substances of land-plant origin and aquatic debris, as well as the difficulty of distinguishing between non-marine and marine amorphous matter. It is not surprising, therefore, that chemical analyses of amorphous kerogens show some of them to be hydrogen-rich and typical Type II kerogens and others to be hydrogen-poor and characteristic of terrigenous plant material.

The problems of differentiating amorphous kerogens are not necessarily solved by moving away from transmitted-light examination into the reflected-light fluorescence mode. *Bituminite* was one of three macerals first described by Teichmüller (1974c), entirely on the basis of fluorescence properties, including spectral fluorescence curves. *Fluorinite*, derived from plant oils, with a close association with vitrinite, and one of the more strongly fluorescing liptinite macerals identified so far, proved to be no problem. Neither did the variably fluorescing *Exsudatinite*, expelled by different macerals into cracks, cleats and pores during the coalification process. Although not recorded as a maceral at the time, exsudatinite had certainly been observed much earlier in association with resinites in ordinary reflected light (Jones & Murchison 1963). Bituminite proved to be much less straightforward as a maceral than either fluorinite or exsudatinite. Its origin was attributed to bacterial lipids and decomposition products of fat- and protein-rich precursors by Teichmüller (1974c), although Jacob (1967, 1969) suggested an asphaltic source for such material. Bituminite displayed a reddish-brown fluorescence and, perhaps of more interest, apparently had a close relationship with the development of granular micrinite, long classed with the inertinite macerals on account of its high reflectance, but now seemingly a secondary maceral. The origin of micrinite was apparently closely related to bituminite and also, to a lesser but still pronounced degree, with resinite. It was certainly not inert in its behaviour. Teichmüller & Wolf (1977) illustrate a transition from bituminite into micrinite and refer to 'metabituminite' or 'dead kerogen' at high levels of rank.

Bituminite was soon to be shown to be a complicated maceral. Fluorescence methods in the reflected mode allowed a subdivision of 'unstructured liptinites' (Teichmüller & Ottenjann 1977, Creaney 1980) into several forms of bituminite, regarded by Teichmüller & Ottenjann (1977) as being a major source of hydrocarbons. In their study of the Posidonia Shale these authors identified three types of bituminite, all of which were dark grey with brown internal reflections in normal reflected light. Under ultra-violet excitation differentiation was possible: bituminite 1 fluoresced bright to dark brown, with no photoalteration and many liptodetrinitic inclusions; bituminite 2 displayed a homogeneous golden to reddish-brown fluorescence and weak negative photoalteration; and bituminite 3 showed no fluorescence and frequently had a finely granular structure, although this form of bituminite was relatively rare. Yet Creaney (1980), in his study of the Boundary Creek Formation, could only positively identify bituminite 1. He did identify a second, but rare (in his rocks), bituminite which did not fit into the tripartite subdivision of Teichmüller & Ottenjann (1977), and he did observe a yellow-brown fluorescence associated with the mineral matrix, which was termed 'matrix bituminite' and which was similar to material observed by Teichmüller & Ottenjann (1977) in the Posidonia Shale. Creaney (1980) also described a mesh of fluorescing bituminous material in isolated kerogens from the Boundary Creek Formation. He suggested that compositionally this substance might be similar to bituminite 1, but the presence in it of dinoflagellates on the one hand and spores and pollen on the other indicated an origin in both marine and terrestrial sources, possibly representing an anaerobic degradation product of several types of liptinite.

Powell *et al.* (1982) also examined the relationship between simple chemical parameters and the occurrence of amorphous kerogen and bituminite 1. Their results were not encouraging. Although some amorphous kerogens had high atomic H/C ratios, others did not, and the hydrogen-rich types could not consistently be distinguished microscopically from the low-hydrogen types. Samples with bituminite 1 in excess of 7.5% generally had high H/C ratios, although not exclusively so, while samples richest in matrix bituminite could display variable H/C ratios, some high others low. Matrix bituminite apparently could be finely divided bituminite 1 or finely divided material of lower hydrogen content. The same problem existed with petrographic evaluations in reflected light as with similar evaluations in transmitted light, namely, source rocks of poor oil potential could not be distinguished from those with high oil potential. There was, however, one indication which was followed by later investigators. Samples with a high-intensity fluorescence from matrix bituminite did have higher atomic H/C ratios than those displaying low-intensity fluorescence, although Powell *et al.* (1982) expressed doubts that such distinctions could be consistently made because of maturation effects.

Senftle *et al.* (1987) have discussed refinement of organic petrographic methods to characterize kerogens. They suggest grouping substances which have no discrete form or shape under the term *Amorphinite* which they can then subdivide into fluorescing *Fluoramorphinite* and non-fluorescing *Hebamorphinite*. These distinctions can be maintained up to a maturation level of vitrinite oil reflectance $R_m \approx 1.1\%$. These authors also examined relationships between simple chemical parameters and petrographic content. Poor correlations of atomic H/C ratio were obtained between total liptinite content, total amorphinite content and even with fluoramorphinite, if this constituent formed less than 30% of the sample. When, however, the %AMEX (total liptinite content plus fluoramorphinite content) (Senftle 1983) is plotted against atomic H/C ratio, a much improved correlation with atomic H/C ratio is obtained. The %AMEX parameter can therefore be used to estimate the hydrogen content of a sediment and thus its oil/gas potential. This approach would seem likely to be more consistent than that of Thompson & Dembicki (1986), who rely on differentiating four types of amorphous kerogen on the basis of textural appearance, having established oil and gas proneness through correlations with pyrolysate, infra-red and atomic-ratio data.

Whatever may be the current status of the identification of amorphous kerogens in relation to composition, the difficulties associated with their categorization have not prevented some advances in purely petrographic classification of kerogens and, particularly of a group of organic-rich rocks, the oil shales, which have considerable economic potential. Robert (1979, 1981) and Alpern (1980) have both produced classifications of kerogens in which fluorescence methods have played an important part in differentiating organic constituents. Robert's classification has a three-fold subdivision, the first containing identifiable primary biological constituents, chiefly the coal macerals; a second subdivision of primary structureless matter and alginite; and the third subdivision containing secondary products. Alpern's classification is at first sight more complicated, using a six-fold subdivision—phytoclasts, fungi, spores, lacustrine and marine organisms, bacteria, and bitumens—but the classification basically encompasses the same groups of constituents as does Robert's, merely attempting finer subdivision. Within the oil shales, Hutton *et al.* (1980), Cook *et al.* (1981) and Sherwood (1984) have probably been responsible for the most satisfactory classification of this group of rocks. They are mentioned here because several of these accumulations, chiefly the torbanites, have in the past, been considered to be allochthonous coals and not oil shales: there can be no doubt that the torbanites produce oil. What is notable about the proposed classification (Table 4) (Hutton *et al.* 1980), is that it is largely based upon the established maceral classification (ICCP 1963, 1971, 1975). Acritarchs and dinoflagellates have had to be introduced, because of their occurrence in certain oil shales. It is the extended differentiation of the liptinite macerals, again largely on the basis of fluorescence microscopy, that has allowed the maceral concept to be retained (Table 5) (Cook *et al.* 1981). The petrographic composition, world resources and evaluation of oil shales have been reported by Alpern (1981).

Pyrolysis methods

There is another approach to kerogen typing that has developed rapidly over recent years. Analytical pyrolysis methods are now widely applied in kerogen characterization studies, a sensible course, if only for the following reason. Many of the chemical data used in the evaluation of source rocks are based on (often highly sophisticated) analyses of solvent extracts from sediments and oils. Total soluble material, however, comprises only a small proportion of the total organic matter present in a sample, frequently $<2\%$. The remaining insoluble kerogen is not amenable to the methods used to analyse solvent extracts. Yet the insoluble kerogen, if appropriately analysed, certainly can contribute information about its source, depositional environment and hydrocarbon potential.

Pyrolysis methods applied to kerogen characterization have been reviewed by Larter & Douglas (1982) and more recently by Larter (1984). Most successful have been the methods falling under the headings: (a) bulk-flow pyrolysis methods which are applied to 'whole rock' samples (e.g. drill cuttings), where there is no characterization of the pyrolysate, but the rate and quantity of pyrolysate evolution are measured, and (b) high-resolution analytical pyrolysis methods in which the flash pyrolysate at *c.* 600°C is analysed either by gas chromatography (py-gc), or by mass spectrometry (py-ms), and relative amounts of specific fragments from the kerogen or 'whole rock' pyrolysate are identified. Py-gc methods are complementary to py-ms methods, but in general both should be used, although for most kerogen characterization py-gc is sufficient for routine use.

There are a number of bulk-flow pyrolysis systems, but by far the best known is the 'Rock-Eval' system (Espitalié *et al.* 1977), used widely by industry in screening procedures on drill cuttings. The instrument, rather generally, estimates maturity of organic matter in a sediment, as well as the quantity of distillable material which the organic matter can generate. The crushed rock sample is heated over a temperature range from ambient to 550°C. The products of the pyrolysis develop three chromatographic peaks (Fig. 3): those giving rise to peaks S_1 and S_2 are analysed by a flame-ionization detector, the gas-producing peak S_3 by a thermal conductivity detector. A series of parameters can be determined (Table 6).

Correlations between these parameters and other maturity and type parameters have been attempted. Teichmüller & Durand (1983) observed a close relationship between the temperature of maximum hydrocarbon release (T_{max}) and vitrinite reflectance, concluding that T_{max} could be used as a rank parameter for coals. They also showed that there was considerable variation in the amount of hydrocarbons released by pyrolysis at any given rank, but that this variation was due principally to variation in the petrographic composition of the coals. However, coals with the highest amounts of liptinites showed the highest hydrocarbon yields.

Not surprisingly, attempts have been made to correlate Rock-Eval data with other data relating to kerogens. Mukhopadhyay *et al.* (1985) developed a classification of kerogens based on maceral composition, fluorescence properties, atomic ratio H/C, range of hydrogen index from Rock-Eval assessment, reflectance level of the peak

TABLE 4. *Genetic classification of oil shales based on maceral composition and environments of deposition (slightly modified from Hutton* et al. *1980)*

	CHLOROPHYCEAE (Green algae)					CYANOPHYCEAE (Blue-green algae)	
Parent genus	*Gloecapsomorpha*	*Reinschia*	*Pila*	*Tasmanites*			
Modern day equivalent		*Botryococcus*		*Pachysphaera*			
Life form	Colonial	Colonial	Colonial	Unicellular		? Unicellular	
Maceral	ALGINITE A	ALGINITE A	ALGINITE A	ALGINITE A		ALGINITE B	
Oil shale	Kukersite	Torbanite	Torbanite	Tasmanite	Mixed oil	Lamosite	Lamosite
Non-algal organic components	?	vitrinite inertinite sporinite	vitrinite inertinite sporinite	vitrinite	sporinite vitrinite	corpohuminite sporinite	corpohuminite
Environment	marine	lacustrine associated with swamps	lacustrine	shallow marine	shallow marine	stratified lacustrine	shallow saline: lacustrine
Dominant kerogen type*	II	I	I		II	I	I–II

* Tissot *et al.* (1974), Harwood (1977).

TABLE 5. *Some characteristics of the exinite macerals in relation to their occurrence in oil shales (from Cook* et al. *1981)*

Maceral/ constituent	Origin	Form	Occurrence	Fluorescence characteristics*	
				Colours	Intensity
Alginite A	*Botryococcus Pila, Reinschia Gloeo-capsomorpha Tasmanites*	Lensoidal, fan or disc-shaped	Lamosite torbanites tasmanites kukersites	Green to yellow orange	High
Alginite B	?c.f. *Nostocopsis*	Irregular stacked sheet-like masses	Lamosites: possibly in cannels and torbanites	Yellow to orange, less commonly greenish-yellow	Moderate to high
Acritarchs and Dinoflagellates	Acritarchs and dinoflagellates	Flattened cysts	Mixed oil shales: possibly some lamosites	Greenish-yellow	High
Bituminite	Anaerobic degradation of higher plants and/or algae: may be equivalent in part to alginite B	Amorphous to lamellar: may be associated with inertinite	Characteristic groundmass of cannels	Yellow, orange, brown	Weak but with strong positive alteration
Sporinite	Spore and pollen exine and entine	Disc-shaped	Mixed oil shales, cannels, canneloid shales	Yellow to orange, rarely green	Moderate but much weaker and alginite A
Cutinite	Higher plant, typically from leaves	Flattened sheets, more rarely cylindrical	Mixed oil shales, cannels, canneloid shales	Similar to sporinite but commonly with a greater range	Variable through high moderate to weak even in one sample
Resinite	Higher plants	Lensoidal masses, cell fillings or isolated bodies	Canneloid coals, some mixed oil shales, dom	Green, yellow orange or reddish-brown	High to very weak
Fluorinite	Essentially oils of higher plants, possibly crude oil in some cases	Cell fillings or as veins	Not reported in oil shales or as dom	Blue, green yellow	Very high to high
Suberinite	Suberized cell walls of higher plants	Surrounding cells, also as massive layers	Cannels and shales (mainly of Jurassic, Cretaceous or Tertiary age)	Orange, brown, rarely yellow	Typically weak but moderate at very low ranks
Liptodetrinite	Fragments of other exinite macerals	Small fragments typically 0.01 mm	Cannels, canneloid shales, mixed oil shales, but present in most sedimentary rocks	Green, yellow, orange, brown	High to weak
Exsudatinite	Expelled hydrocarbons	Veins, pore fillings	Cannels	Orange, rarely yellow	Moderate to weak

* In oil shales which are immature to early mature—R_{0max} up to 0.75%.
dom—dispersed organic matter.

generation of hydrocarbons during Rock-Eval pyrolysis, type of generated hydrocarbons and depositional environment (Table 7). Four kerogen types are still retained, but there are two subdivisions of type II. In developing this classification an effort was made, probably not too satisfactorily, to resolve the problem of amorphous material in kerogens by replacing the terms bituminite and amorphinite with the term *Sapropelinite* (Combaz 1975), again subdivided into three types, based on its appearance in transmitted light, reflected light and under U.V.-excitation. The term cuts across the maceral concept.

A possible problem arising from the use of Rock-Eval data may affect the reliability of this classification. Mineral-matrix effects can markedly influence both the quantity and composition of pyrolysates (Larter 1984), particularly when samples have high clay, carbonate or iron-mineral

TABLE 6. *Parameters relating to 'Rock-Eval' pyrolysis (Espitalié* et al. *1977) (modified from Larter 1984) (see Fig. 3)*

Parameter	Measurement
Peak S_1	Quantity of hydrocarbons and related materials desorbed from rock sample during heating at 250°C for 5 min in helium—approximately equivalent to material extracted in solvent procedures
Peak S_2	Quantity of hydrocarbons and related material produced during pyrolysis of kerogen and in volatile bitumens over the temperature range 250–550°C at 250°C min^{-1} in helium—approximately equivalent to thermal degradation of kerogen
Peak S_3	Quantity of carbon dioxide produced during pyrolysis of the sediment over the range 250–390°C
T_{max}	Temperature of maximum kerogen decomposition rate (maximum peak S_2 evolution)—relatable to the maturity of the sample
S_1+S_2	Genetic potential
$S_1/(S_1+S_2)$	Production index
S_2/TOC	Hydrogen index
S_3/TOC	Oxygen index

TOC—total organic carbon determined on a totally decarbonated sample using a standard elemental analyser.

contents. The effect is very pronounced if the samples have total organic-carbon contents of <1 per cent. Although the influence of minerals will be most noticeable in 'whole-rock pyrolysis', there may still be matrix effects when isolated kerogen concentrates are pyrolysed. Pyrite, irremovable unless quite drastic separation procedures are employed, is likely to be present in many kerogens and is probably most abundant in kerogens whose oil potential is greatest, since these have been deposited under severely reducing conditions. Pyrite decomposes at temperatures used to pyrolyse kerogens, and it can react with the organic matter to produce substantial quantities of hydrogen sulphide, which will be added to that coming from the pyrolysis of organic sulphur compounds.

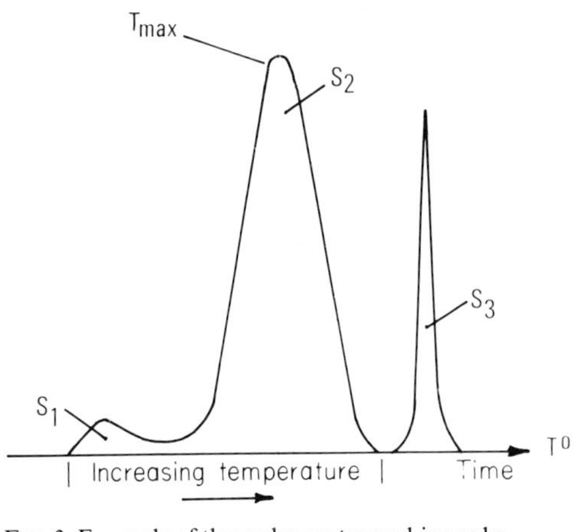

FIG. 3. Example of three chromatographic peaks generated by the products of a 'Rock-Eval' pyrolysis (see Table 6). S_1, S_2 are derived from a flame ionization detector and S_3 is derived from a thermal conductivity detector. T_{max} can be taken as a measure of maturity.

Larter & Senftle (1985) and Larter (1985) propose an innovation in high-resolution analytical pyrolysis, using py-gc and py-ms methods, which forms the core of a new kerogen typing method. It has been possible, virtually since reliable pyrolysis methods were established, to determine relative quantities of specific chemical fragments in kerogens. For instance, the more hydrogen-rich a kerogen, the less aromatic, in general terms, will be its pyrolysate. Even the use of a quite simple typing index, for example the proportion of xylene to octene (Larter 1984), measured from pyrograms to give a measure of aromaticity, shows that vitrinites and coals can be distinguished from sporinites and alginites, which themselves can also be separated. Developing the approach further involves first taking due account of not only the carbon and hydrogen present, but also the oxygen through the bulk chemical-typing approach, based on the atomic ratio H/C:O/C method of van Krevelen (1950) and Dormans *et al.* (1957), and later applied by Tissot *et al.* (1974) to kerogens. Many oxygen-containing compounds have been identified in the pyrolysates of humic materials and kerogens. Larter (1984) suggests that if abundant phenols are present in a kerogen pyrolysate, then this may indicate that the kerogen was derived from terrestrial material. So instead of a bivariate plot, involving only aliphatics, the trivariate plot of Fig. 4 which introduces phenolic compounds was

TABLE 7. *Classification of kerogens and their depositional environments (after Mukhopadhyay* et al. *1985)*

Kerogen type	Maceral composition		Fluorescence properties	Range of hydrogen index (Rock–Eval) mgHC/g^{-1} C_{org})	H/C atomic ratio	Maturity level (%R_0) at peak generation	Generated hydrocarbons	Depositional environment
	Major	Minor						
I	Alginite algodetrinite sapropelinite I	Bacterial remnants sapropelinite II	Greenish-yellow for alginite	>700	>1.5	0.6–0.9	Mainly oil	Highly anoxic lacustrine and shallow marine lagoons
II	Sapropelinite II liptodetrinite resinite (A + B)	Particulate liptinite (A + B) alginite corpocollinite	Orange to orange brown for sapropelinite II yellow for resinite (A + B)	400–700	1.1–1.5	0.6–0.9 0.3–0.7 (for resinite)	Mainly oil; major gas in higher maturation (>1.3%R_0)	Anoxic marine lagoon/ deep basin, lacustrine or supratidal
IIB	Particulate liptinite (mainly B) liptodetrinite	Alginite sapropelinite II desmocollinite humosapropelinite	Yellow for sporinite, cutinite etc.	150–400	0.8–1.3	0.7–1.1	Oil and gas	Low to medium anoxic lagoon, mangrove swamp and delta complex
III	Vitrinite (telocollinite, humosapropelinite)	Resinite inertinite sporinite	Dark brown for resinite	25–150	0.5–0.8	0.8–1.0 (oil) 1.0–1.5 (gas)	Mainly gas	Low anoxic to low oxic fresh or brackish water swamps
IV	Inertinite	Vitrinite	—	<25	0.5	No source of hydrocarbons	Minor gas (no source for liquid hydro-carbons)	Oxic swamp and oxic marine basins.

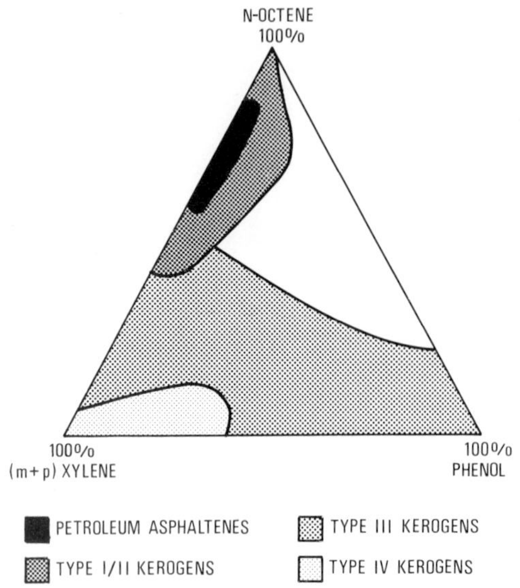

FIG. 4. Relative abundance of *n*-octene, (m + p)xylene and phenol based on pyrolysis at 600°C of a number of kerogen types (after Larter 1984).

devised. The kerogen-type areas were defined on the basis of: (a) atomic ratios H/C:O/C, (b) maturity levels and (c) for kerogen Types III and IV on the basis of petrographic observations. The petroleum asphaltenes fall within the kerogen Types I/II area. Larter (1984) stresses that not only can the four basic kerogen types be distinguished on the diagram in terms of the relative distribution of aliphatic, aromatic and phenolic species, the pyrograms themselves show these distinctions equally well. Of great importance is the fact that they are independent of rank throughout the 'oil window' (*cf.* Fig. 1). There is, however, no possibility of relating either the pyrograms or the relative distributions of pyrolysate fragments to source rock potential on a quantifiable basis in any of the illustrated relationships or from the pyrograms on which the relationships are based.

The important innovation made by Larter & Senftle (1985) and Larter (1985) is that they are now able to determine the amount of specific components in pyrolysate yields. Quantification has been achieved through the use of a polymer (polymethylstyrene) as an internal standard, of which a constant percentage weight (*c.* 0.25% by weight of the kerogen), is added to the kerogen. Quantification allows an assessment of the petroleum potential of kerogens and also finer differentiation between kerogens, some of which, on the basis of either bulk chemical data or an hydrogen/oxygen index, would previously have been grouped together. Various cross-plots are possible. For example, the ratio of aromatic to aliphatic hydrocarbons in pyrolysates always increases from kerogen Type I to Type III. This relationship is illustrated in Fig. 5 in which the total benzene, toluene and C_8 aromatics are plotted against the total C_7–C_{25} normal hydrocarbons. Figure 6 illustrates a plot of 'paraffin index' (determined from peak intensity for normal hydrocarbons and the internal standard from an 800°C/20 s pyrogram) against a calculated 'hydrogen index' based on Rock-Eval data. The subdivision into different kerogen end-members is based on the study of numerous pyrograms. Brief descriptions of the identified kerogen end-members are given in Table 8. Larter (1985) admits to some qualification of the data presented in Fig. 5 and Table 8, but feels

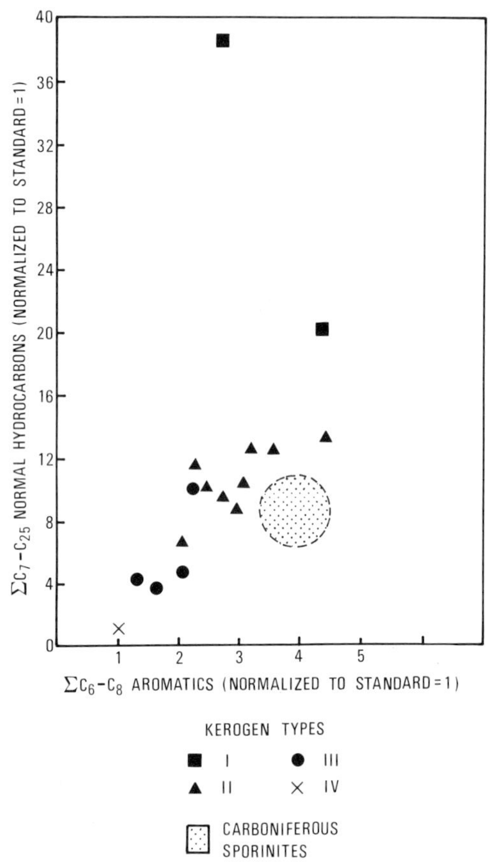

FIG. 5. Absolute yields of aromatics (benzene, toluene and C_8 aromatics) against C_7–C_{25} normal hydrocarbons (*n*-alkenes + *n*-alkanes) from pyrolysates of a range of kerogens with maturity levels < 0.8% vitrinite reflectance (after Larter 1985 and Larter & Senftle 1985).

TABLE 8. *Representative kerogen type end-members based on data from Larter (1985) and Larter & Senftle (1985) (see Fig. 7)*

Kerogen type	Petrographic content and comment	Chemical content and comment
I p (paraffinic)	Torbanite (*Botryococcus*) kerogens, rich in alginite	High hydrogen indices; highly paraffinic pyrolysates
I pn (paraffinic/naphthenic)	Tasmanite (alginite)-rich kerogens; mixed Type I kerogens	Hydrogen-rich with hydrogen distributed between paraffinic and naphthenic structures; abundant naphthenic species in pyrolysates
I n (naphthenic)	Resinite-derived materials	
II pn (paraffinic/naphthenic)	Most common Type II being amorphinite-rich kerogens from truly marine source rocks	High hydrogen indices; low paraffin contents; abundant naphthenic/naphtheno-aromatic species dominate pyrolysates
II p (paraffinic)	Substantial exinite in kerogen	Reduced hydrogen indices with relatively high paraffin content
II/III pp (paraffinic/phenolic)	Exinite-rich but with land-plant structural polymers from lignin or derived from sporopollenin/suberin	—
III pp (paraffinic/phenolic)	Vitrinite-rich kerogens deposited in marine or deltaic environments	—
III ap (aromatic/phenolic)	Vitrinite-rich kerogens from coal-swamp environment	—
IV (aromatic)	Inertinite-rich kerogens	Aromatic pyrolysates with occasional remnant phenolic species

that kerogens can be more objectively and directly compared than by using only atomic ratios or Rock-Eval data alone. The kerogens of the source rocks are defined in terms of bulk chemical composition (hydrogen index), detailed chemical composition (paraffin index) and the organic matter source input (from petrographic data). Thus, the organic facies can be characterized and, more important, the Type II/III kerogens of many near-shore marine environments can now be differentiated. The petroleum potential of the source rocks can also be evaluated if the maturity level is known. The evolution path of Fig. 6, which runs from immaturity for initially hydrogen-rich kerogens to below the 'oil window' (Fig. 1), allows: (a) an estimate of the total yield of hydrocarbons, and (b) the change in paraffin content with maturity for generated materials to be calculated.

Biomarkers

The past decade has seen a spectacular growth in the discovery and identification of biological markers or 'biomarkers', so-called 'chemical fossils' that demonstrably have a biological origin (see Mackenzie 1984 for an excellent account). A principal factor underlying this growth has been been the development of high-resolution capillary gas chromatographs which are normally coupled to relatively low-resolution mass spectrometers that incorporate sophisticated data systems (gc-ms). Such instruments are now widely used in the analysis of oils and the complex mixtures isolated in the solvent extracts from sediments. The capability of these systems is being rapidly extended.

Biomarkers are correctly hailed as being

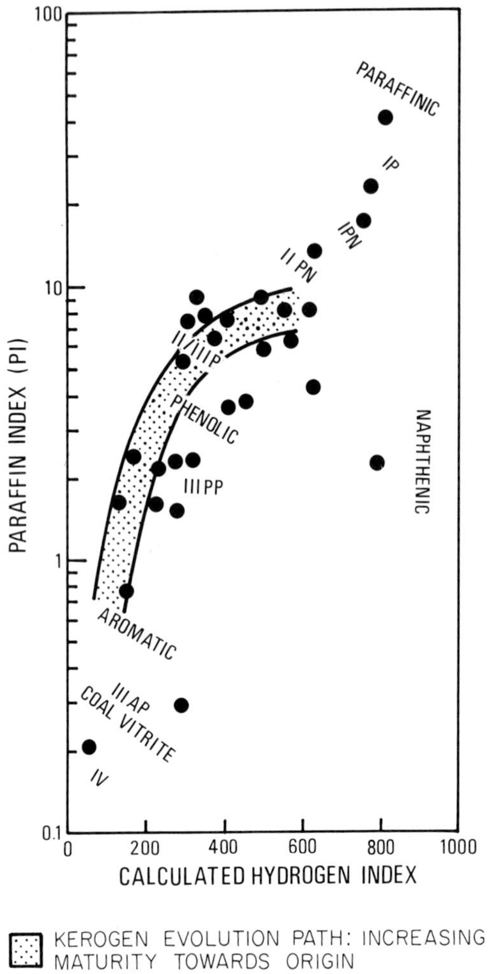

FIG. 6. 'Paraffin index' against 'hydrogen index' for a number of kerogens (after Larter 1985 and Larter & Senftle 1985). Paraffin index—from summed abundance of C_9–C_{30} *n*-alkanes + *n*-alkenes determined from pyrogram peak height data and based on the pyrolysis GC unit = 0.25% by weight. Hydrogen index—from elemental analytical data (Rock-Eval units = mg hydrocarbons per gram total organic carbon (TOC)). The kerogen evolution path is for initially hydrogen-rich kerogens extending over the range $R_o \simeq 0.5$–1.3%.

particularly important in petroleum geochemistry and to oil exploration as predictors of source inputs and maturation levels. Certain biomarkers are now employed on a routine basis in the evaluation of the oil-bearing potential of sediments. Use has even been made of biomarkers in basin-modelling studies (Mackenzie & McKenzie 1983, Mackenzie *et al.* 1985). Although many organic compounds have been used as biomarkers, the most extensively employed have been the hydrocarbons. Recent reviews of the use, application and impact of biomarkers in organic and petroleum geochemistry are given by Mackenzie *et al.* (1982), Mackenzie (1984) and Philp (1985a), while Philp (1985b) has produced a useful book on fossil-fuel biomarkers. A rather depressing but realistic review on the status of biomarkers as indicators of depositional environment is given by Moldowan *et al.* (1985). In the present context, no attempt is made to provide a general survey of the current value and the extent of application of biomarkers, but only to point to these major compound classes which may be of importance in identifying source inputs that are terrestrial and/or lacustrine in origin, with only minor comment on their use as maturity indicators.

The principal classes of biological markers that organic geochemists have investigated fall under the following headings: *n*-alkanes, branched alkanes, including isoprenoid hydrocarbons, the sesquiterpenoids and diterpenoids, the tricyclic and tetracyclic terpanoids, the pentacyclic triterpanoids and the steroids. Several of these classes have been used widely, notably the *n*-alkanes, the regular C_{13}–C_{20} isoprenoids (particularly the acyclic isoprenoids, pristane and phytane), the pentacyclic triterpanoids (especially those based on the hopane structure), and the steroids. The transition from the *n*-alkanes to the steroids shows a pronounced increase in structural complexity, but it is the very complexity of the pentacyclic triterpanoids and the steroids that gives the two compound classes a greater degree of specificity, because they carry the largest content of appropriate information. Hence, they have greater potential in petroleum exploration as source and maturity indicators, as well as being sufficiently diagnostic to aid in problems of 'oil to oil' and 'oil to source rock' correlations. Mackenzie *et al.* (1982) have demonstrated the exponential increase there has been in the number of steroid and triterpanoid biomarkers that have been identified over a ten-year period up to 1982. That number has certainly increased markedly in the meantime, probably at a similar exponential rate.

Biomarkers as indicators of terrestrial source input and depositional environment

n-alkanes

The wide distribution and the relatively high concentration of *n*-alkanes in many samples suggested that these compounds were ideal biomarkers in times when gas chromatography was essentially the principal method available for analysis of complex organic mixtures. More than 20 years ago Eglinton & Hamilton (1963) reported that *n*-alkane distributions with a high 'odd to even' carbon-number ratio in the C_{25}–C_{35} range could be attributed to plant waxes. In contrast, marine phytoplankton tended to give *n*-alkane distributions with much lower carbon numbers, *n*-C_{17} commonly being dominant, as against the *n*-alkane distributions derived from higher plants which usually maximize above *n*-C_{27}. The 'carbon preference index' which estimates the relative amounts of odd/even carbon number of *n*-alkanes is still used by some investigators to estimate maturity (Radke *et al.* 1980). Unfortunately, if the source material has been degraded by bacteria, the *n*-alkanes are susceptible compounds to attack. Even with minor degradation, more than 90% of *n*-alkanes can be removed, so altering radically the distribution of this biomarker class (Alexander *et al.* 1983). Moldowan *et al.* (1985) point to other complicating factors, for instance, weathering effects and the fact that increase of maturation can alter the high molecular weight *n*-alkanes into lower molecular weight hydrocarbons without 'odd-even' preference. There is also the problem of the transport of higher-plant material into what are essentially marine environments to give an *n*-alkane distribution biased towards the terrestrial component.

Acyclic isoprenoids and other branched hydrocarbons

Gas chromatography alone can often be used to detect the C_{13}–C_{20} regular 'head to tail' isoprenoids. Consequently these compounds have also been employed widely in petroleum geochemistry to gain information on source inputs and depositional environments, as well as for correlation studies. More sophisticated analysis, using gc-ms, allows a much wider range of isoprenoid compounds to be detected. Moldowan & Seifert (1979) demonstrated through analyses of a large number of Californian crude oils that higher concentrations of regular long-chain isoprenoids occurred in oils with a terrestrial origin than in oils from marine sources. These same authors (Moldowan & Seifert 1980) showed that the irregular branched isoprenoid, botryococcane, which arises from botryococcene in the living alga *Botryococcus braunii*, is probably a reliable source indicator for lacustrine environments.

The most extensive use of isoprenoids as source indicators has probably been through the measurement of pristane (C_{19}) to phytane (C_{20}) ratios using gas chromatograms. Brooks *et al.* (1969), then Powell & McKirdy (1973) argued that pristane, which is derived from the phytol side-chain of chlorophyll by oxidation and decarboxylation, and phytane, which also forms from phytol, but through hydrogenation and dehydration, when used as a ratio, could indicate type of depositional environment. Thus, samples with a Pr:Ph <1 were regarded as indicative of a reducing environment, while ratios of Pr:Ph >1 suggested formation under more oxidizing conditions. There can be, however, some influence from increasing maturity on the ratio. Although Rashid (1979) found that the Pr:Ph ratio tended to rise with increasing input of terrestrial material, the ratio was also higher in the more deeply buried facies. Moldowan *et al.* (1985) maintain that there is little possibility of distinguishing between oils from marine and non-marine shales on the basis of pristane: phytane ratios, but they do suggest that carbonate-sourced oils can be so characterized. Although Philp (1985a) would discount this discouraging view, there may be other origins for these compounds which would complicate the issue, for example, cell-wall ethers of archaebacteria (Brassell *et al.* 1981) contain possible percursors of phytane, while Goosens *et al.* (1984) suggest that tocopherols are also likely precursors of pristane in sediments and crude oils.

Besides pristane: phytane ratios, the pristane: *n*-C_{17} and the phytane: *n*-C_{18} ratios have also been employed to decide upon source input. Powell & McKirdy (1973, 1975), in a survey of Australian crude oils, showed that those of likely terrestrial origin displayed high pristane: *n*-C_{17} ratios. There were, however exceptions. For example, in the North Perth Basin, high-wax oils had low Pr:Ph ratios and a low proportion of isoprenoids in relation to *n*-alkanes. Powell (1984) in a subsequent study showed that these latter features were characteristic of high-wax oils generated from lacustrine sediments. McKirdy *et al.* (1986) have shown that in some cases waxy hydrocarbons may be of algal origin. Although C_{23+} *n*-alkanes are normally attributed to an origin from *n*-alkanes and wax esters of land plants, certain crude oils from the western Otway Basin, South Australia and Victoria have *n*-alkanes derived from C_{23}–C_{33} *n*-alkenes in algae, including *Botryococcus braunii*.

The Pr:*n*-C_{17} ratio can also vary considerably at a constant maturation level within a single basin. Lijmbach (1975) has suggested that in environments where there is substantial bacterial activity, phytol becomes more or less completely oxidized, leaving little available for conversion into isoprenoids, which would explain why waxy oils in lacustrine sequences are depleted in isoprenoids. It is also clear, however, that this pristane/*n*-heptadecane ratio is controlled not only by organic-matter type but by maturity rise as well. Leythauser & Schwarzkopf (1986) show that hydrocarbon-explusion effects within the maturity interval $R_m = 0.82–1.03\%$ for Type III kerogens can be followed by observing the variations in the Ph:*n*-C_{17} ratios.

Sesquiterpenoids

Relatively little attention has been given to the bicyclic sesquiterpenoids in petroleum geochemistry. Their sources are hardly known. From the restricted viewpoint of identifying 'oil from coal' sources, however, these compounds appear to have some potential. Alexander *et al.* (1983) have provided very strong evidence that one compound isolated, eudesmane, is related to eudesmanol, a sesquiterpenoid widely distributed in higher plants. Eudesmane should therefore be a good indicator of terrestrial input. In a later paper Alexander *et al.* (1984) discussed the widespread occurrence of a second sesquiterpenoid, drimane, which like eudesmane, they synthesized. Drimanes not only have a wide distribution that even competes with that of the hopanoids, they also occur in early Palaeozoic rocks and in all types of petroleum. Drimane is thus attributed to a microbial origin (Alexander *et al.* 1983). Sesquiterpenoids have only been infrequently referred to in coals, although there is increasing interest being shown in these biomarkers. Baset *et al.* (1980) reported a number of C_{15} compounds in a single low-rank coal and Grantham & Douglas (1980) also identified several sequiterpenoids in fossil resins and brown coals, suggesting they might be helpful in deciding on source input.

Diterpenoids

Oils whose formation has been attributed to a land-plant source are usually dominated by paraffinic hydrocarbons. There are, however, instances where oils of suggested terrestrial origin are low in *n*-paraffins and in which cyclic components dominate the saturate fractions. Snowdon & Powell (1982) report one such example from the Mackenzie Delta in Canada in which resinite is shown to be the dominant oil-producing constituent, confirming an earlier suggestion that resinite was the oil source (Snowdon 1980). A large aromatic component in this oil is attributed to disproportioning processes affecting unsaturated diterpenoids during diagenesis and early maturation (Snowdon & Powell 1982).

Another area in which diterpenoid-containing oils are found is Australia. Philp *et al.* (1981) found mixtures of C_{19} and C_{20} diterpenoids in oils from the Gippsland Basin and sought correlation with diterpanes in resinites of the onshore Yallourn brown-coal deposits, but without success (Philp *et al.* 1983). Indonesia is another area from which diterpenoids are reported in oils (Richardson & Miller 1982). Until now little structural information has existed for the diterpenoids, except in the case of coals (Hagemann & Hollerbach 1980). Noble *et al.* (1986) have, however, recently demonstrated the presence of nine diterpenoid hydrocarbons in the alkane fraction of an Australian crude oil.

Tricyclic and tetracyclic terpanoids

Relatively little can be said about both the tricyclic and tetracyclic terpanoids in relation to oil possibly sourced from terrestrial plant matter. Aquino Neto *et al.* (1983) studied two series of tricyclic and tetracyclic terpanes in 40 crude oils varying in age and depositional environment, with only one of the crudes unequivocally attributed to terrestrial sources. The tetracyclic terpane series was ubiquitous and was present in all samples, but with only trace amounts of the C_{24} member (in all other samples the predominant member), in the terrestrially sourced oil. A microbial origin is suggested for both series. Philp (1985b) suggests that since (a) the C_{24} tetracyclic terpane is very abundant in Australian oils and extracts of source rocks, and (b) the tricyclic terpanes are frequently absent, these compounds might be helpful as source indicators for certain Australian oils.

Pentacyclic triterpanoids

The structural complexity of triterpanoids and steroids has led to their wide application as source and maturity indicators. The simplest way in this brief account to approach what is now a very extensive literature relating to both these classes of biomarker is to refer to a recent review by Philp & Gilbert (1986). Discussion can then be confined to remarks about those triterpanoids and steroids that apparently have a specific relationship with terrestrial plant material. Triterpanoids and steroids with this connotation are shown in Table 9 which is part of a summary produced by Philp & Gilbert (1986).

The most extensively employed group of triterpanoid biomarkers in organic geochemistry are the pentacyclic terpanes, based on the hopane skeleton, and the structurally related moretanes. This large group is distinguished from a second group of pentacyclic triterpanes which are not based on the hopane structure. This second group, which includes the lupanes, oleananes, fernanes and ursanes, derived from precursors which are widely distributed in higher plants, has been identified in coals and in oils apparently derived from terrestrial sources. These compounds have not been widely used in a biomarker role, although with their relationship to higher-plants, they must have potential as source indicators.

Of the pentacyclic terpanes individually listed in Table 9, the C_{28} bisnorhopane is not of frequent occurrence. First identified in California crudes and in the Monterey Shale by Seifert *et al.* (1978), with two corresponding epimers later isolated by Moldowan *et al.* (1984), the compound is believed by some investigators to be a specific source marker although its origin has not been unambiguously decided. Seifert *et al.* (1978) felt that the compound originated in ferns, but Grantham *et al.* (1980) report the compound in North Sea cores in which no trace of fern detritus or fern spores could be detected. Katz & Elrod (1983) suggested a relationship with highly anoxic ecosystems. Whatever its origin, the C_{28} bisnorhopane certainly seems to have a close association with terrestrially derived oils, but its presence there, on the basis of current evidence, does not seem necessarily to confirm the compound as a land-plant source indicator (Philp 1985a).

Although there are other pentacyclic terpanes based on the hopane structure, referred to in Table 9, their origins and structures are not necessarily well known. In the terpane fraction monitored by the *m/z* 191 mass chromatogram, two unidentified compounds have been observed (Philp & Gilbert 1986). The first of these, referred to as 'compound X', is a probable pentacyclic C_{30}-triterpane which elutes immediately after the C_{29} 17α(H), 21β(H)-hopane (de la Grandeville 1981). Its structure is unknown but it has been seen in many Australian oils. When it does occur, there is also a peak that elutes just before the C_{27} 17α(H)-trisnorhopane which corresponds to an unidentified C_{27}-terpane (compound Y). Philp & Gilbert (1986) suggest that compound X is probably associated with a higher plant input, since it is generally absent from marine-sourced oils. It may not, however, always be present; for example, it does not occur in the oils from the Mahakam Delta (Hoffmann *et al.* 1984), which are also suggested as arising from a land-plant source. There is also the possibility that compound X may be linked with a unique bacterial input (Philp & Gilbert 1986).

The one compound that probably has the greatest specificity as a source indicator is 18α(H)-oleanane, which is a pentacyclic terpane that is not based on the hopane structure. Whitehead (1974) and Ekweozor *et al.* (1979a, b) have reported 18α(H)-oleanane in sediments and oils from the Niger Delta, while Hoffmann *et al.* (1984) also have observed high relative abundances in samples from the Mahakam Delta, Indonesia. Palmer (1984) also reports 18α(H)-oleanane in coals from the Philippines, but Philp & Gilbert (1986) note that there are few occurrences of the compound in Australian coals. All these observations relate to Tertiary oils or extracts and, as a result, it is suggested that 18α(H)-oleanane may be a specific indicator for higher land plants, particularly for the angiosperms (which did not evolve until the Cretaceous). In all the occurrences the compound is associated with high amounts of terrestrially derived organic matter. Ring-A degraded triterpanes have also been found in source rocks and crude oils from the Niger Delta (Ekweozor *et al.* 1981), often in association with 18α(H)-oleanane, but not necessarily so. Corbet *et al.* (1980) and Schmitter *et al.* (1981) found no 18α(H)-oleanane in immature coals, but possible precursors occurred, for example, olean-12(18)-ene, olean-12-ene and olean-18-ene. All these compounds are regarded as alteration products of higher-plant triterpanoids.

Of the triterpanoids unrelated to the hopanes, Rullkötter *et al.* (1982) finally confirmed the presence of lupanes in Tertiary sediments from west Greenland and from the Gulf of Suez, although Hills & Whitehead (1966) had first tentatively identified the compounds in Nigerian crude oils. The Greenland occurrence of bisnorlupanes possibly formed as a result of microbial action during early diagenesis. Compounds tentatively identified as lupanes, along with hopanes and oleananes, have been reported by Richardson & Miller (1982). The lupanes may have originated in sapotaceous trees, angiosperms which, like others in this group, did not flourish until the Tertiary, thus giving an age confirmation for the oil.

Steroids

The predominant sterane in Australian oils is the C_{29} sterane, both regular and rearranged, and this has been regarded by many investigators as a feature of oils from other areas which have received a high terrestrial contribution (Table 9). In Australian oils there are relatively low concentrations of C_{28} steranes and an absence of C_{27} steranes. Because of these relationships, oils from these basins are suggested as generating principally from terrigenous source material. The basis

TABLE 9. *A summary of some biomarker characteristics (terpanes and steranes) from Australian and other oils predominantly derived from terrestrial source material (from Philp & Gilbert 1986). ● compound present. tr = trace*

Location of reservoir	Age of reservoir	Terpanes					Predominant steranes	Terpanes/Steranes	Postulated predominant source material
		Hopanes	C_{28}-Bisnorhopane	Compound Y*	Compound X†	18α (H)-oleananes			
Gippsland Basin	Tertiary/late Cretaceous	●	●	●	●	tr	C_{29}	5	Higher plant
Sydney Basin	Permian	—	—	●	●	—	C_{29}	1	Higher plant
Surat/Bowen Basin	(1) Jurassic and Triassic	●	●	●	●	—	C_{29}	2.5	Higher plant
	(2) Permian	●	—	●	●	—	C_{27} and C_{29}	3	Marine/higher plant
Cooper/Eromanga Basin	(1) Triassic	●	●	●	●	—	C_{29}	8	Higher plant
	(2) Jurassic	●	●	●	●	—	C_{29}	2	Higher plant
	(3) Permian	●	●	●	●	—	C_{29}	3	Higher plant/marine
Mahakam Delta, Indonesia	Tertiary	●	?	—	—	●	C_{29}	10	Higher plant
Niger Delta, Nigeria	Tertiary	●	?	?	?	●	?	?	Higher plant
Taranaki, New Zealand		●	●	●	●	●	C_{29}	10	Higher plant
Beaufort/MacKenzie Canada	Tertiary	●	—	●	●	●	?	?	Resinite/higher plant

* Compound Y is an unidentified C_{27} pentacyclic terpane.
† Compound X is an unidentified C_{30} pentacyclic terpane.

of this attribution depends on work by Huang & Meinschein (1978), who proposed plotting relative amounts of C_{27}, C_{28} and C_{29} sterols on a triangular diagram and from such plots deducing the type of depositional source. The diagram is now widely used to plot sterane abundances (Fig. 7).

Probably soon after the original proposal was made by Huang & Meinschein (1978), the charge of oversimplification was laid at the use of this triangular diagram. However, in the case of the Australian oils Philp & Gilbert (1986) argue that the C_{29} sterane is so dominant, virtually to the exclusion of all other steranes, particularly the C_{27} and C_{28} steranes, it must be indicative of a terrigenous source. Similarly, Hoffmann *et al.* (1984) make a similar plea for a terrestrial source, based on high relative amounts of C_{29} sterol-derived species in a biomarker study of coals, shales and oils from the Mahakam Delta, Kalimantan, citing Huang & Meinschein (1979). From what follows there are, however, sound reasons for exercising some care when employing the Huang & Meinschein diagram.

Brassell & Eglinton (1983) point out that marine organisms contain both C_{27} and C_{29} sterols. Grantham (1986a) also has reported unusual C_{27} and C_{29} sterane predominances in two crude oils, which are contrasting in their compositions, one having a strong C_{29} sterane predominance, significant 17(αH),21(βH)-hopanes and a relatively low concentration of tricyclic terpanes. The second crude has a strong

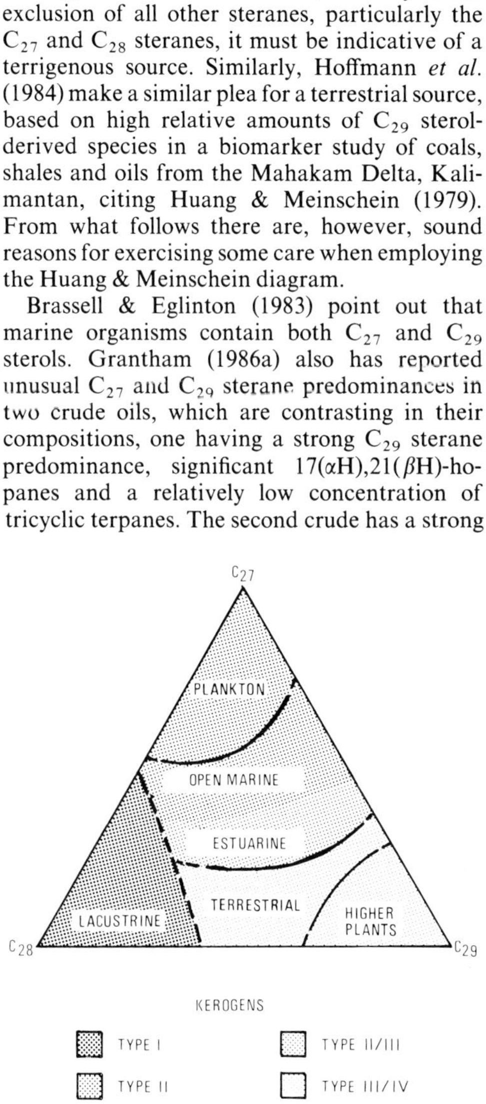

FIG. 7. Distribution of C_{27}, C_{28} and C_{29} steranes, based on a similar plot for sterols (after Huang & Meinschein 1979) (see Table 8).

C_{27} sterane predominance, lower concentrations of the 17(αH),21(βH)-hopanes and much higher concentrations of tricyclic terpanes. These compositional differences were attributed by Grantham (1986a) to an input of quite different organisms to the two oils, which was probably algal in both cases and almost certainly so for the oil with the C_{29} sterane predominance, which can be shown to have been generated from later Precambrian source rocks. Volkman (1986) also has reviewed sterol markers for marine and terrigenous organic matter, indicating that unicellular algae can be sources for a wide variety of sterols, including some of those currently relied on as indicators of a terrestrial source. Thus, a high proportion of C_{29} steranes does not necessarily confirm derivation of organic matter from vascular plants. Moldowan *et al.* (1985) show that the triangular distributions for C_{27}, C_{28} and C_{29} steranes and monaromatic steroid hydrocarbons differ because of differing origins for these sterol-derived biomarkers. Their data support the results of Huang & Meinschein (1979), suggesting that both steranes and monoaromatic steroid hydrocarbons can be useful in detecting higher-plant input, but that the latter are more effective. There is a general trend, no more than that, towards lower concentrations of monoaromatic steroids in non-marine oils.

Moldowan *et al.* (1985) also comment on sterane:hopane ratios and their applicability as source indicators. Sharp differentiation between oils from marine and non-marine sources is not achieved by these ratios, although the oils from non-marine sources tend to have low sterane:hopane ratios, as well as generally low sterane concentrations. Moldowan *et al.* (1985) consider that certain biomarkers can be used as source indicators although some are more reliable than others (Table 10).

Biomarkers as maturity parameters

If a biomarker is to have potential in the assessment of thermal maturity of sediments, a particular molecular configuration in an immature sediment should be replaced gradually by more stable molecular configurations as temperature rises. Since hydrocarbons derived from natural-product compounds can be easily recognized, Mackenzie & Maxwell (1981) argued that the relative concentrations of hydrocarbon reactants and their reaction products, as measured by gc-ms, could be employed to assess how far the thermal maturation of organic matter in a sediment has progressed. Mackenzie (1984) indicates that it is desirable, if employing biological markers as assessors of thermal maturation, that

TABLE 10. *Biomarkers in crude oils as source indicators (from Moldowan* et al. *1985)*

	Non-marine vs marine	Non-marine vs marine shale	Non-marine vs marine carbonate	Marine shale vs carbonate
Monoaromatic steroids	MI	MI	MI	WI
High mol. wt paraffins	MI	MI	MI	—
Carbon preference index	—	—	SI	WI
Pristane/phytane ratio	—	—	MI	WI
Sterane/hopane ratio	WI	WI	WI	—

SI—strong indicator.
MI—medium indicator.
WI—weak indicator.

only reactions in which the reaction product is zero, when the reaction starts, should be used. If this condition is observed, then measurements of relative concentrations of reactants and products will be independent of source input, depositional conditions and early diagenetic effects. Ideally, it would also be desirable if the reaction went to an end value of 100%. In many cases, however, particularly in isomerization reactions, for instance the isomerization at the chiral centre of C-20 in steranes, the reaction never reaches 100%, indeed may only rise from 0 to 0.5 (50%). At this point equilibrium is established and the end value attained. Then the reaction can obviously no longer be used to assess maturity.

Although porphyrins have been used in the past, more recently the hopanoids and steroids have received greatest attention as maturity indicators (Fig. 8). Both isomerization and aromatization (steroid hydrocarbons only), reactions have been employed. In the case of the hopane isomerizations, the change from 17β(H),21β(H) to 17α(H),21β(H) and 17β(H),21α(H) hopanes, and the change from 22R to 22S 17α(H),21β(H)-C_{31}-hopane, have moved to equilibrium well before the onset of oil generation in the first instance and peak oil generation in the second case (see Fig. 8). Higher activation energies are required in steroid hydrocarbon reactions, so these tend to be more related to the time of peak oil generation. Mackenzie (1984) discusses three major processes affecting steroid hydrocarbons which are thermally controlled: (a) configurational isomerization in which biologically derived 20R-steranes are converted to approximately equal proportions of 20R- and 20S-steranes, (b) aromatization when monoaromatic steroid hydrocarbons are converted with rising temperature to triaromatic steroid hydrocarbons and (c) enrichment of lower molecular weight components, the C_{20}–C_{22} steroids showing an increase in their relative abundance to C_{26}–C_{29} steroids as they pass through the 'oil window' (Fig. 8). Mackenzie *et al.* (1982) have published a detailed review of these and other processes affecting the steroids.

There seems no doubt of the potential of certain biological markers as maturity indicators, but further careful study of the different factors that may well influence rates of reaction are necessary. Apart from the possibility of catalytic effects influencing the reactions, different heating rates also affect the degree of isomerization or aromatization shown by samples (Mackenzie & McKenzie 1983). There is also independent evidence of this influence in that progression of the optical properties of vitrinites in relation to their chemical compositions varies in areas exposed to

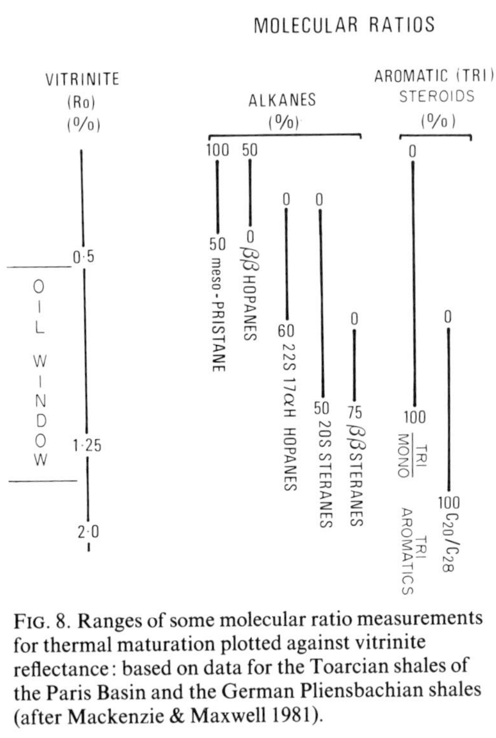

FIG. 8. Ranges of some molecular ratio measurements for thermal maturation plotted against vitrinite reflectance: based on data for the Toarcian shales of the Paris Basin and the German Pliensbachian shales (after Mackenzie & Maxwell 1981).

different heating rates (Jones *et al.* 1984). Ten Haven *et al.* (1986) have reported anomalous behaviour in the progression of both steroid and hopanoid maturity indices in hypersaline environments in which substantial amounts of relatively uncommon sterols occur. The diagenetic pathway of such sterols can lead to what is a truly immature sample having a 'mature' appearance.

The time factor may also be important in these reactions. Grantham (1986b) has recently shown, in an examination of a large number of crude oils, that sterane isomerization reactions in oils from Tertiary sources are generally incomplete and that the reactions have not progressed as far along the reaction pathway as in older oils. Mackenzie and McKenzie (1983) suggested that in Tertiary sediments, despite oil generation at relatively high temperatures, the time available for isomerization reactions to run their course in such young sediments was insufficient. Although other possible explanations were put forward by Grantham (1986b), e.g. terrestrially derived source rocks generating crude oils at relatively low temperatures, or 'immature' biomarkers taken in by the oil as it migrated through immature source rocks, Grantham still felt his data supported the view of Mackenzie & McKenzie (1983), pointing to a time–temperature constraint on sterane isomerization reactions. A similar constraint appeared to operate on moretane to hopane conversions since moretane:hopane ratios were higher in Tertiary oils with incompletely isomerized steranes than in older oils. No doubt other observations of a similar nature will be made, but they should not negate the use of biomarkers as maturity indicators if they are carefully and sensibly applied. However, maturity parameters based on reactions affecting hydrocarbons, other than hopanoids and steroids, are also useful (e.g. the Methylnaphthalene Ratio, Radke & Welte 1983). that gradual and consistent changes of the $\alpha\alpha$ to $\alpha\beta$ forms in tricyclic isoprenoid alkanes occur with depth in several sedimentary sequences. Although the isomerization occurs over a relatively narrow range, and has reached equilibrium before the onset of the main 'oil window' (see Fig. 1), the authors suggest that measurements of tricyclic triterpane isomerizations would be a useful supplement to sterane:hopane measurements.

Sedimentary basins and deltaic areas with terrestrially-sourced oils

Earlier in this paper there have been occasional references to areas in which hydrocarbon accumulations occur which are felt, on the basis of geological, petrographical and chemical evidence, to have been derived from terrestrial material. Most notable of these areas is Australia, in which substantial commercial accumulations of seemingly terrestrially sourced oils have been generated (see also Hobday, *this volume*). In terms of volume, only one other area at the present time appears to compete. There are now known extensive intermontane basins in China, in which substantial thicknesses of sediments have been deposited under fluvial and lacustrine conditions. Many of these contain substantial oil and gas on a commercial scale and there is great future potential in other similar basins. This section describes some of the basins and deltaic areas of the world where the evidence for a terrestrial origin for their hydrocarbon accumulations is strong.

Australia

The most recent appraisal for a terrestrial origin for many Australian oils, chiefly based on evidence of biomarker distributions, has been by Philp & Gilbert (1986): these authors also refer to other basins and deltaic areas in a summary (Tables 9 and 11). Few biomarkers individually seem to offer absolutely unambiguous evidence of a particular source input but, taken collectively with other evidence, biomarker characteristics are helpful in deciding if the source has been terrestrial or not. Table 11 shows biomarkers, other than the hopanes and steranes, which were discussed in a previous section, listed by Philp & Gilbert (1986), which they believe support a land-plant origin for Australian and other crude oils. This paper by Philp and Gilbert is only one of a large and expanding literature that addresses itself to this topic.

Before considering oil occurrences in several Australian basins, three general points should be made, two relating to the Australian flora, the third to the generating capacity of inertinite. First, Thomas (1982) has discussed, in general terms, evidence for the terrigenous sources of Australian oils. He stresses that there is a pronounced contrast between Australian Carboniferous–Permian coals and younger coals in Australia in their contribution to hydrocarbon accumulations. The Carboniferous–Permian coals are relatively poor in liptinitic macerals and consequently are mainly gas-prone, with little in the way of land-plant oils associated with them. The younger group of coals (and coaly sediments), from the Jurassic onwards, has a close association with waxy crude oils. This particular association, while observed world-wide in Cretaceous and

TABLE 11. *A summary of some biomarker characteristics from Australian and other oils predominantly derived from terrestrial source material (from Philp & Gilbert 1986)*

Location of reservoir	Age of reservoir	Head/tail long chain isoprenoids	Terpanes				Terpanes	Postulated predominant source material
			Sesqui-cyclic	Di-cyclic	Tri-cyclic	Sester (C_{24})	Steranes	
Gippsland Basin	Tertiary late Cretaceous	●	●	●	—	●	5	Higher plant
Sydney Basin	Permian	●	?	—	—	●	1	Higher plant
Surat/Bowen Basin	(1) Jurassic and Triassic	—	●	—	—	●	2.5	Higher plant
	(2) Permian	—	●	—	●	●	3	Marine/higher plant
Cooper/ Eromanga Basin	(1) Triassic	—	●	—	—	●	8	Higher plant
	(2) Jurassic	—	●	—	—	●	2	Higher plant
	(3) Permian	—	●	—	—	●	3	Higher plant/ marine
Mahakam Delta Indonesia	Tertiary	?	?	?	—	●	10	Higher plant
Niger Delta Nigeria	Tertiary	?	?	?	—	●	?	Higher plant
Taranaki, New Zealand		●	●	●	—	●	10	Higher plant
Beaufort/ MacKenzie, Canada	Tertiary	?	●	●	—	?	?	Resinite/higher plant

● means a particular class of biomarkers was analysed for and found to be present.
? means this compound class was not analysed for and thus it is not known whether it is present or absent.
— means compound class was analysed for and found to be absent.

Tertiary coals (based on a large number of analyses of waxy crudes from 40 regional stratigraphic sequences (Hedberg 1968)) has been a significant factor in arguing for a land-plant origin for many Australian oils, since the early discussion by Brooks *et al.* (1969) and Brooks (1970) on the origin of crudes from the Gippsland basin. A considerable number of high-wax Australian oils are documented by Powell & McKirdy (1976), many of which are characterized by high pristane/phytane ratios (>4) and which are regarded as forming under peat-swamp environments (Brooks *et al.* 1969).

Thomas (1982) makes a second point when he suggests that the high-wax contents of Australian oils are closely related to the dominance of gymnosperms in the Mesozoic and Tertiary. Such trees are particularly rich in resins and in cuticles, which are waxy, and there is a further contribution from the angiosperms, which expanded from the late Cretaceous onwards and which also are rich in wax. Indeed, Shanmugam (1985), discussing the origin of the oils of the Gippsland Basin, is prepared to be highly specific, attributing the naphthenic fraction of the oils to resins in the coals and the paraffinic fraction more generally to the remainder of the coal constituents. The principal paraffinic contributor are cuticles derived from a kauri-pine type of vegetation deposited under raised-bog conditions, with high water-table and high rainfall, and low oxygen and nutrient content. Although these attributions may be too specific, the dominance of conifers in the Australian flora in post-Triassic times (Gould 1980), would not be questioned (see discussion in Collinson & Scott, *this volume*).

The third point is rather more speculative and concerns the possibility that inertinite macerals generate liquid hydrocarbons (see Fig. 1). Such a suggestion would be regarded almost as heresy by most petroleum geochemists and even many petrographers, but in the Australian context there is sufficient evidence available to sound a note of caution to those who would be entirely dismissive.

The level of reactivity of inertinite macerals has been questioned for more than two decades. Taylor *et al.* (1967) probably felt that they had resolved the debate when they claimed from experimental evidence that semifusinite with a reflectance higher by 0.1–0.2% absolute than the mean value for vitrinite of the same coal remained entirely inert during carbonization. There was some doubt about the associated sclerotinite, which it was thought might soften at temperatures higher than 600°C. Despite, however, what seemed to be firm evidence for unreactive behaviour of all but a very small proportion of semifusinite during carbonization, many applied coal petrographers involved in the production of metallurgical cokes showed scepticism and still continued to allocate one-third of the semifusinite, recognized in a coal or coal blend, to the 'reactive' heading in predictive computations of coke strengths. More recently, Steller (1981) who hydrogenated a low-rank coal ('Gasflammkohle' in German terminology), has shown that gelified inertinite (especially semifusinite and some sclerotinite), and macrinite can be regarded as reactive, while the ungelified inertinite macerals are completely inert. Diessel (1983) identified four categories of inertinite based on the response of the maceral group to carbonization: two reactive categories, one highly so, and two non-reactive categories. These observations were supported and extended by the work of Evans *et al.* (1984), who investigated the hydrocarbon potential of sediments by hydrogenating immature prospective rocks. These authors found *inter alia* that there was essentially no difference between the hydrocarbon contents and hydrocarbon potentials of associated coal-derived vitrinites and inertinites when the vitrinite reflectance lay in the range 0.3–1.2%. The data were quite independent of the amounts of liptinite present.

Some inertinite-rich Australian sediments do display unexpected liquid-hydrocarbon potential, although the absolute level of hydrocarbon generation from inertinite has never been quantified. The available data, particularly those of Evans *et al.* (1984), do however confirm more qualitative judgements expressed by earlier workers on this point. Smith & Cook (1980), discussing the coalification tracks of the three maceral groups, vitrinite, liptinite and inertinite, suggest that 'inertinite may contribute more to hydrocarbon generation than was previously believed'. The Cooper Basin, which produces mainly gas, but also oil, is inertinite-rich and very low in liptinite macerals, leading Smyth (1983) to suggest that the only possible substantial hydrocarbon source could be inertinite. Rigby & Smith (1981) indicate pathways which provide mechanisms for the generation of wet gases and oil from inertinites dispersed in sediments. Liquefaction experiments show that the yields of liquids from inertinites and vitrinites do not differ greatly, although the yields from vitrinites are higher (Heng *et al.* 1983).

There is therefore considerable evidence for some hydrocarbon generation from inertinites and the results of recent studies are still supportive. Cook (1986) has presented data for the macerals and maceral groups that relate the initial and peak generating points of macerals and maceral groups to maturity levels denoted by vitrinite reflectance (Table 12). He comments that although the specific yield of hydrocarbons per unit mass for liptinite macerals is higher than for vitrinites, the greater abundance of vitrinite macerals will still make them important contributors. A similar comment could be made about inertinite macerals, despite their low potential, since many Australian coals are inertinite-rich. Diessel (1985) has shown that some inertinites fluoresce and those that do correspond broadly with the two reactive categories identified during carbonization (Diessel 1983). It is of course possible that these inertinites have absorbed hydrocarbons expelled from associated liptinitic macerals.

Gippsland Basin

The Gippsland Basin, a basin of Atlantic-margin type lying on- and off-shore in the SE corner of Australia (Fig. 9) is one of the outstanding world examples of commercially significant oil and gas accumulations generated from non-marine, coal-bearing sequences. Threlfall *et al.* (1976) and Smith (1982) have discussed the tectonic-sedimentary history of the basin, which is complicated, as is its burial-metamorphic history. Commercial hydrocarbon accumulations, in both structural and stratigraphic traps, are restricted to the Central Deep (Smith & Cook 1984). The onshore part of the Central Deep essentially contains large gas fields, with little oil: the off-shore part of the Central Deep has reservoirs with large undersaturated oil pools. The hydrocarbons have been generated from three distinct

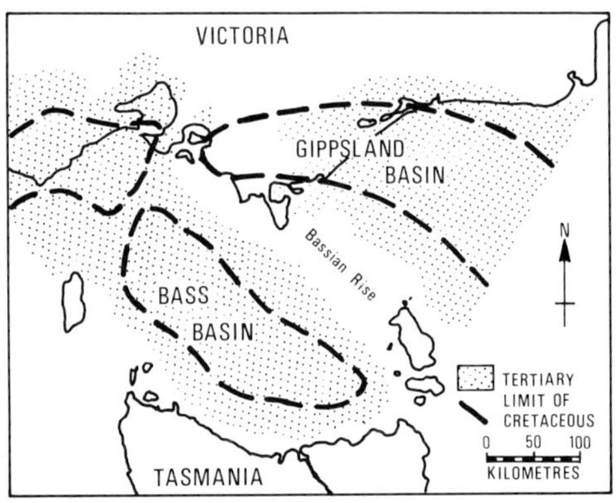

FIG. 9. Locations of Gippsland and Bass Basins in SE Australia.

TABLE 12. *Maturation ranges for oil generation by macerals (from Cook 1986) (see also Fig. 1)*

Maceral	Relative oil yield	Reflectance of co-existing vitrinite—$\%\bar{R}_{v\,max}$	
		Initial generation	Main oil generation range
Inertinite	Low—but maceral group may be important in oil migration	0.4	—
Resinite	High	0.4–0.45	0.5–0.8
Suberinite	Moderate to high	0.45	0.5–0.8
Bituminite	Moderate to high	0.4	0.5–0.8
Vitrinite	Moderate	0.45	0.5–1.0
Sporinite	High	0.6	0.7–0.9
Cutinite	High	0.6	0.7–0.9
Alginite	Very high	0.7	1.1

organic facies in the Latrobe Group which extends in age from the Middle Cretaceous virtually to the present day. The generative capacity of the organic sediments increases upwards through the Group. Although there are algae present at some horizons, geological and petrographic, particularly fluorometric, evidence suggests that many of the hydrocarbons have originated in perhydrous vitrinites and liptinites of coals which are relatively free from clastics. Biomarker evidence is generally supportive of a land-plant origin for the oils (Philp & Gilbert 1986) (Tables 9 and 11). Stainforth (1984) cites high carbon preference indices, very high pristane/phytane ratios and high pristane/nC_{17} ratios as evidence of a terrestrial origin for the oils.

Bass Basin

The Bass Basin lies immediately to the SW of the Gippsland Basin across the Bassian Rise (Fig. 9). The two basins have been compared by Davidson *et al.* (1984) who, while noting some geological similarities, show that there are structural differences. The Bass Basin has normal fault blocks, which are variably modified by wrench and reverse faults, the Gippsland having compressional anticlines which give large traps, the comparable structures in the Bass being smaller. Nicholas *et al.* (1981) have reported on the maceral compositions of cores from wells in the Bass, recording that the sediments carry half to one-third less liptinite than the equivalent section in the Gippsland Basin. If, however, the source rocks are mature and they contain more than 1.5% liptinite, then generation of wet gas will have occurred in substantial quantities. Sometimes the gas is so wet that it could be described as 'light oil'. The geochemical and petrographic data for the Bass sediments are less comprehensive than for the Gippsland sections. Despite hopes that careful structural and seismic studies may lead to oil and gas discoveries, the prognosis for the basin is much less attractive than for the Gippsland Basin.

Cooper/Eromanga Basins

Wholly onshore in eastern Australia there are several basins containing oil and gas. By far the largest of these is the Eromanga Basin which underlies parts of the states of Queensland, New South Wales, South Australia and the New Territories (Fig. 10). The Eromanga is a part of the Great Artesian Basin, one of the greatest artesian systems of the world, with an area occupying 20% of the Australian landmass. Other related constituents of the Eromanga are the Surat Basin immediately to the east (Fig. 10) and the associated Bowen Basin.

The Eromanga Basin consists of Jurassic and Cretaceous sediments and it overlies the smaller, more complicated Permo–Triassic intracratonic Cooper Basin (Fig. 10) which is a major generator of gas while containing only small amounts of oil. The source of the hydrocarbons has been long known (Brooks *et al.* 1971, Battersby 1976), and lies in a thick, coal-rich, lacustrine and fluviatile Permian succession in which the sediments are liptinite-poor and rich in vitrinite and inertinite (Shibaoka *et al.* 1978, Smyth 1979). There are areas with local concentrations of liptinite macerals and degraded vitrinite that have generated oil (Kantsler *et al.* 1983). The coal deposits (*c.* 1×10^{12} tonnes) of the southern Cooper basin, when considered with the total amount of disseminated organic matter in shales and siltstones (Schwebel *et al.* 1980), constitute a vast hydrocarbon source. There is much less total organic and pyrolysable carbon in the thinner sequence in the northern

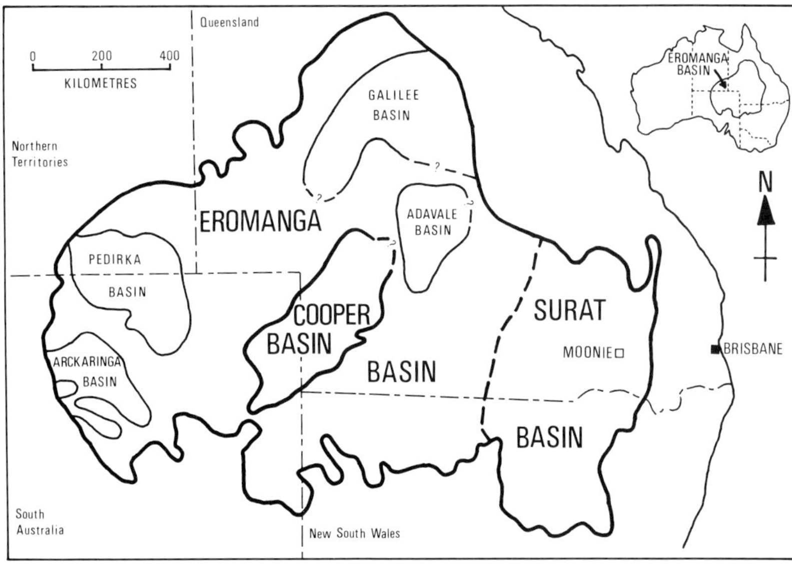

FIG. 10. Locations of Eromanga, Cooper and Surat Basins in central eastern Australia.

Cooper Basin (Thomas 1982). The nature and distribution of the hydrocarbons in the Cooper Basin are closely related to geothermal and rank gradients (Kantsler *et al.* 1978, 1983). Major gas fields are associated with areas that have suffered a high degree of coalification due to temperature rise through some telemagmatic event.

Knowledge of the Eromanga Basin is probably more extensive than for the Cooper Basin, not least because of the recent publication of a major text on the geology, organic petrology, organic geochemistry and other aspects of the basin (Gravestock *et al.* 1986). Numerous earlier publications on the basin exist, but the more recent major contributions have been by Kantsler *et al.* (1983), Smyth *et al.* (1984), Vincent *et al.* (1985) and Passmore & Boreham (1986). In contrast to the Cooper Basin, hydrocarbon discoveries have principally been of oil at eight different stratigraphic horizons. The discoveries lie in non-marine Jurassic to early Cretaceous rocks, generally fluvial sandstones, with the oil held mainly in anticlinal traps, although some are stratigraphic (Moore 1986). Productive wells are sited principally in an area of the Eromanga which either overlies or immediately surrounds the extent of the underlying Cooper Basin, which led Kantsler *et al.* (1983) to suggest a contribution from a Permian source. Powell (1986), however, indicates that current evidence points to the oils being sourced from the Jurassic. Oils of the Eromanga Basin display all the characteristics commonly attributed to a terrigenous origin—light, waxy, high pristane to phytane ratios, relatively high proportions of these isoprenoids relative to *n*-alkanes and steranes dominated by C_{29} components (Powell 1986). Kantsler *et al.* (1986), in a further contribution, reiterate that on the basis of modelling maturity data, host rocks for the majority of oil and gas discoveries in the Eromanga lie either in mature coal-bearing Permian sediments or in downdip mature Jurassic rocks. Biomarker distribution (Philp & Gilbert 1986) (Tables 9 and 11) supports the view of Kantsler *et al.* (1983, 1986) that at least the oils in Triassic reservoirs were probably sourced from the underlying Permian coal measures.

Cook (1986) has identified a total of five organic facies in the Eromanga Basin, four dominated by coals and coal-related organic matter and one marine facies. With one exception, the Triassic–Jurassic organic matter of the non-marine facies contains abundant liptinite, with varying proportions of inertinite. The fourth coal-dominated facies, of Upper Cretaceous age, has only small amounts of liptinite, but abundant band vitrinite (texto-ulminite). The marine facies is referred to below.

Three horizons in the Eromanga Basin have been investigated for source potential in some detail. First, the Birkhead Formation, of Middle to late Jurassic age, is an organic-rich shale unit of fluvial, lacustrine origin which contains sandstone reservoirs of poor to fair quality, mainly in the form of stratigraphic traps (Paton 1986). Smyth *et al.* (1984) have described the organic-matter composition in terms of coals and dispersed material. Pyrolysis products from py-gc studies are dominated by C_{20}–C_{30} hydrocarbons, indicating that waxy oils would probably be generated on maturation (Smyth *et al.* 1984).

The Bulldog Shale is a proximal marine sediment of early Cretaceous age, with an anomalous kerogen. McKirdy *et al.* (1986) have recognized three distinct organic geochemical facies on the basis of kerogen composition, Rock-Eval parameters and alkane patterns. Vitrinite is a minor component, while liptinite forms between 25 and 75% of the dispersed organic matter, the balance being inertinite. Despite this promising petrographic composition for hydrocarbon generation, chemical assessment shows the kerogen to be Type III, trending to Type IV at the top of the section. This anomalous petrographic-chemical relationship is attributed to the kerogen being of mixed aquatic-terrestrial origin, and principally composed of oxidized instead of fresh liptinite, so making it gas-prone.

The Toolebuc Formation has an areal extent of greater than 600,000 km^2 with possible *in situ* resources more than 2000×10^9 barrels of shale oil (Sherwood & Cook 1986). The Formation consists of 'mixed oil shales' (see Table 5) and as such is not strictly 'coal', although within the facies boghead coals and algal cannels almost certainly occur. The mixed maceral assemblage of the shales can be placed in a two-component system: bituminite/micrinite on the one hand and lamalginite, with associated liptodetrinite, on the other. The major precursor of the bituminite is unknown, although Glikson & Taylor (1986) suggest that irregularly spaced calcareous and organic-rich laminae in the shales are suggestive of a contribution from cyanobacterial mats. Most of the lamalginite apparently arises from cysts of dinoflagellates and acritarchs.

Surat and Bowen Basin

Rather less information is available for the Surat and Bowen Basin (Fig. 10), although oil and gas have been known in the area for more than 25 years. Approximately 30 oil fields have already been found, but accumulations are generally small. The Moonie oilfield, the earliest discovered, is the largest. The Surat Basin contains mainly Jurassic and Cretaceous sediments (up to a thickness of 2500 m), deposited in an intracratonic downwarp, which overlies an elongate, northward plunging, half-graben known as the Taroom Trough which is filled with Permian and Triassic sediments. This trough is a southerly extension of the Bowen Basin which extends northwards into Queensland beyond the Surat Basin (see Hobday, *this volume*).

The most recent summary of the stratigraphy, structure, hydrocarbon occurrences and migration is given by Thomas *et al.* (1982) (see also Khorasani, *this volume*). There is a wide distribution of terrestrial organic matter throughout the Jurassic of the Surat Basin and also in the Permian and Triassic sediments of the Taroom Trough (Thomas 1982). Thomas *et al.* (1982) indicate that the Walloon Coal Measures (Middle Jurassic), in organic petrographic terms similar to the Latrobe Group of the Gippsland Basin, contain coal-rich sequences which are good potential sources. There are systematic differences shown up by petrographic analysis, for example, the Jurassic coals and carbonaceous shales are rich in resinite and cutinite, while the Triassic rocks have much less resinite, but are notable for thick cutinite. Vitrinite and inertinite predominate in the Permian rocks which also carry much structureless organic matter. Thus, the Permian looks to be more gas-prone and have less oil potential than the Triassic and Jurassic sediments. Unfortunately, only the Permian and the lowermost Triassic in the Taroom Trough have been buried sufficiently to reach generative temperatures, while the more oil-prone Jurassic coal sequences still lie in the immature zone.

Most of the oils from these basins show biomarker characteristics that would be associated with oils derived from a terrigenous source, although oil from the first discovery in 1960 also shows biomarker compounds from a marine source (Philp & Gilbert 1986). Thomas *et al.* (1982) comment that there are variations in wax content of the oils, reflecting variations in source input, or possibly level of maturity, at the time the oil was expelled. Philp & Gilbert (1986) drew attention to an absence of both 18α(H)-oleanane and diterpenoids in these Triassic and Jurassic oils which probably can be explained by the age and type of plant material entering the depositional area.

Perth and Carnarvon Basins and Exmouth Plateau

Mesozoic rifting and continental break-up have been responsible for the production of the 'Atlantic-type' continental margin of Western Australia. The Perth Basin is a N–S elongate trough, approaching 900 km in length, containing clastic sedimentary rocks of Silurian to Recent age. To the north the Carnarvon Basin has suffered a similar depositional and structural history to the Perth Basin. It also parallels the west coast of Australia and comprises an elongate series of depressions containing sediments of

Silurian to late Tertiary age. It has an area approximately twice that of the Perth Basin, and like the Perth Basin, approximately half is onshore.

Oils and condensates throughout the Perth Basin are extremely waxy and mainly paraffinic (Powell & McKirdy 1976), strongly indicating a source from low-rank terrestrial matter. Kantsler & Cook (1979) have demonstrated that potential source rocks are widespread, particularly in the Permian and Mesozoic. With there being so much terrestrial material, the Perth Basin has been thought of as essentially gas-prone, but much of the organic matter is perhydrous (i.e. relatively high hydrogen), rather than subhydrous, so there is oil potential in coals and sediments of this type. Some coaly horizons with up to 30% liptinite and 50% suberinite must have oil potential.

Light waxy crudes are also found in the late Jurassic–early Cretaceous of the Carnarvon Basin. Cook & Kantsler (1979) give brief descriptions of organic-matter occurrences through the Permian, Mesozoic and Tertiary. Most of the organic-rich sediments lie in the marginally mature to mature zones and again there is a significant amount of land-plant material throughout the succession, including liptinite and bitumens. Despite lying in a marine-deltaic sequence, the crude oils show evidence of a strong terrestrial contribution.

Immediately to the north of the Carnarvon Basin lies the Exmouth Plateau containing rocks ranging in age from Upper Permian to Miocene–Pliocene age. Cook *et al.* (1985) have discussed the source potential of the fluviodeltaic systems of the Upper Triassic Mungaroo Formation. Mainly a lower-delta plain depositional environment, the coals and dispersed organic matter in sediments have very high and high vitrinite contents respectively. When the environment changes to alluvial plain, there is an incoming of inertinite. Cook *et al.* (1985) suggest that the Mungaroo Formation is a potential source rock for oil. In this study it was possible to distinguish distributary-mouth bar coals, which are richer in liptinite, from crevasse-splay coals on the basis of maceral content.

Other possible Australian basins with terrestrially-sourced hydrocarbons

There may well be other Australian basins in which commercial accumulations of crude oil originating from terrestrial organic matter will be found. Philp & Gilbert (1986) comment that extensive Permian coal measures occur in the Sydney Basin, yet there is little oil production, although analysed oil seeps suggesting a terrigenous origin are known. The Clarence–Moreton Basin lying due east of the Surat Basin, is also viewed as a possible prospect (Ties *et al.* 1985). From preliminary analytical data the Walloon Coal Measures apparently contain Type I and Type II kerogens, whereas the underlying Marbarg Formation (early Jurassic) has Type III kerogens which are gas-prone.

Canada

A considerable amount of information has become available over the past decade about Canadian onshore and offshore hydrocarbon-bearing basins, although Powell & Snowdon (1980) noted that there was still little, if any, geochemical data released for a number of important Canadian sedimentary basins. Much of the work of organic geochemists and petrologists in Canada has been aimed at defining source inputs for Canadian oils and, as these have been identified, attempting to develop generation models for the different depositional environments, after comparison with the standard generation model that relates to Type I and Type II kerogens (Vassoevich *et al.* 1974) (see Fig. 1).

The principal basins considered by Powell & Snowdon (1980) are the Alberta, the Beaufort–Mackenzie and, in the Arctic Islands, the Sverdrup Basin and the Franklinian Geosyncline, and finally the Scotian Shelf and the Labrador Shelf. Optical and chemical methods have been applied, including petrographic kerogen typing, determination of reflectance, atomic ratios, paraffin indices, hydrocarbon distributions and stable hydrogen and carbon isotopic contents. Of the basins considered, the Beaufort–Mackenzie, the Scotian Shelf and the Labrador Shelf have low-maturity sequences in which terrestrially-derived organic matter has been identified as the dominant organic component. The generalized 'normal' hydrocarbon-generation model (Fig. 1) is probably most closely related to a marine organic source where the generating sequence with depth is oil → gas condensate → dry gas, with the gas condensate generally generating towards over-maturity. In these three basins, however, gas condensate has been trapped at much lower levels of maturity than in the 'normal' case.

Beaufort–Mackenzie Basin

Minor but significant amounts of light oil occur in the Beaufort–Mackenzie Basin, as well as gas condensate. The oils and condensates at the Upper Cretaceous–Tertiary boundary in this basin are the least mature, based on pristane:n-C_{17} ratios, while those of the Lower Cretaceous are most mature, although they have apparently been generated in the zone of vitrinite reflectance $R_o \leqslant 0.6\%$. Powell *et al.* (1978) and Snowdon (1980) have suggested that, since the rocks here contain dominantly coal and herbaceous debris, the hydrocarbons have been derived from resinite, which, in the Beaufort–Mackenzie delta rises to between 10 and 15% of the total organic matter. Besides the presence of resinite in the kerogens, identified by petrography, analyses of extracts provide other lines of evidence for the presence of resinite which include (a) a diterpane 'hump' of unresolved branched/cyclic compounds in gas chromatograms of saturate fractions and (b) the occurrence of diterpenoid derivatives, structurally similar to resin acids, which are in high concentration in some extracts. Snowdon & Powell (1982) have also identified many tricyclic diterpanes in crude oils from the area. Very recently Brooks (1986) has reported that the sterane and triterpane distributions of Tertiary oils and source-rock extracts from this basin suggest a land-plant origin for the oils. The major triterpane in the oils is 18(α)H-oleanane, attributed in other studies (see earlier section of this paper on pentacyclic triterpanoids), as being a specific indicator for land plants, in particular, for the angiosperms.

Snowdon (1980) commented that the oils recovered from the Mackenzie delta are naphthenic in character and low in wax. They do not fit the more generally accepted high-wax image for terrestrially sourced oils (*cf.* the Gippsland Basin). The gas in the Beaufort–Mackenzie Basin is not biogenic gas but thermogenic gas, but it generated very early, in the vitrinite reflectance zone $R_o = 0.45$–0.60%. Monnier *et al.* (1983) show that there is at least a five-fold increase in gas yield, per unit weight of organic carbon, from Type III kerogen in the reflectance range 0.55–0.65%. This increase in gas yield immediately precedes the onset of liquid-hydrocarbon generation, the reverse situation of marine-sourcing.

Scotian Shelf and Labrador Shelf

In the other basins regarded as containing mainly terrestrially derived organic matter, accumulations of gas condensates and light oils have been found on the Scotian Shelf. Three 'families' of oils were identified in this area by Powell & Snowdon (1979), two occurring in the Cretaceous, and the third spanning the Cretaceous–Jurassic boundary, lying in or just above the zone of maturity. The oils of the two lower zones are thought to be terrestrially sourced. Powell (1982), in a study of the Verrill Canyon Formation, showed that it contains terrestrially-derived organic matter, which is largely gas-prone, although minor condensate occurs and there is oil potential in the more mature Sable Sub-basin. In contrast to the Beaufort–Mackenzie Basin, in which the oil formed at relatively low levels of maturation, the oil in the Sable Sub-basin has formed in the main zone of hydrocarbon generation. Condensates found in thermally immature reservoirs on the Labrador Shelf have also formed from gas-prone Type III kerogens (Rashid *et al.* 1980), but the condensates have undergone extensive vertical migration after generation in the mature to over-mature zones.

Alberta Basin and Franklinian Geosyncline

Of the two remaining basins considered by Powell & Snowdon (1980), in the Alberta, normal Type II kerogen appears to be the source for petroleum deposits which range from heavy oil to dry gas in the Devonian (Deroo *et al.* 1977) up to the Cretaceous. Gas has, however, been generated from terrestrial organic matter in the Cretaceous Deep Basin of Alberta (Welte *et al.* 1982). The generation model for the Sverdrup Basin agrees substantially with that of Vassoevich *et al.* (1974), supporting derivation of the oils from a Type II kerogen, but there are anomalous gas deposits occurring in part of the Basin where there are low-maturity zones (Powell & Snowdon 1980). Oil in the Franklinian Geosyncline in the Devonian of the southern flank of the Sverdrup Basin fits the normal marine-source generation model.

Generation models

Running through the work of many authors cited in this section is the theme of a hydrocarbon generation model for terrestrial organic matter. Snowdon & Powell (1982) and Powell & Snowdon (1983) have developed such a model (Fig. 11), which they compare with a generation model for marine organic matter (Fig. 12). To a large extent Fig. 11 summarizes much of what is said in many of the references cited in this section. Figure 1 shows that different macerals or maceral groups begin to generate oil and gas at different maturation levels. This varying behaviour must have an influence on the different kerogen types derived from terrigenous sources: thus Fig. 11 is in many respects analagous to Fig. 1. Whether the kerogen is resinite-, liptinite- or vitrinite-rich will govern (a) the particular threshold maturation level at which liquid-hydrocarbon generation begins and

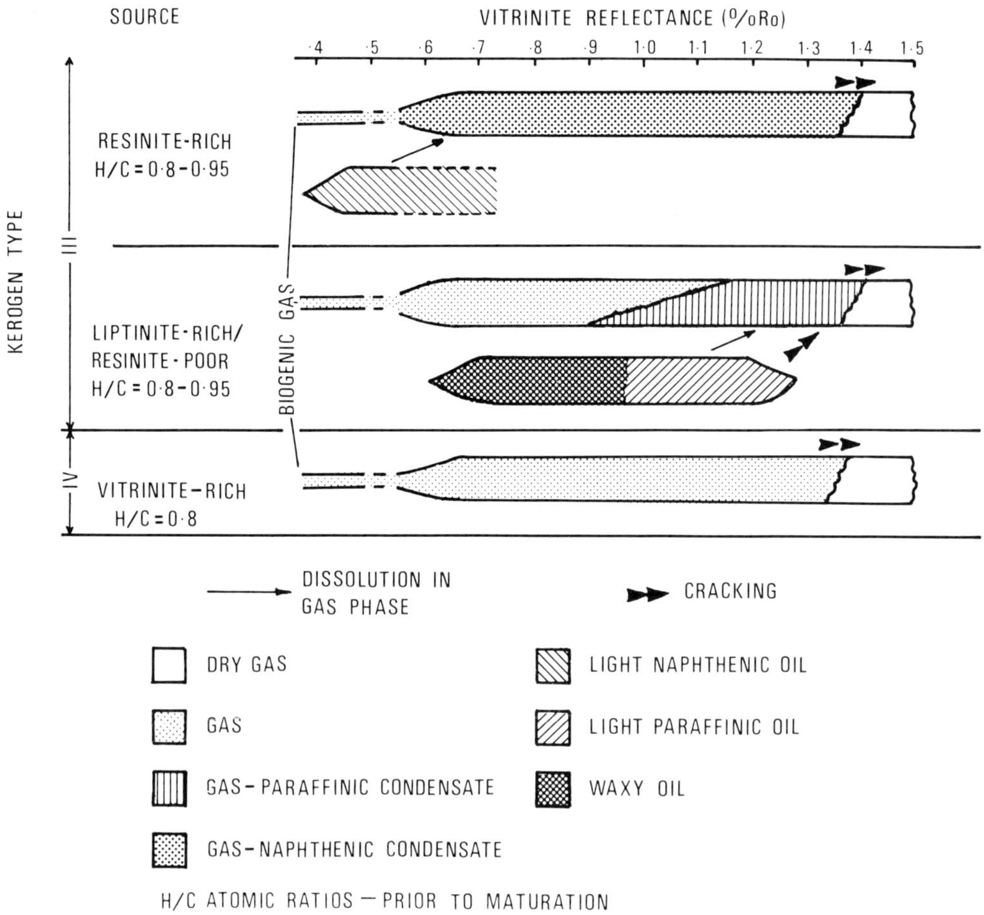

FIG. 11. Oil and gas generation model for different types of terrestrially-derived organic matter (from Powell & Snowdon 1983).

(b) the type of hydrocarbons formed. Gas in significant amounts is generated from certain terrestrial organic matter at lower levels of maturation than it is from marine organic matter. An early naphthenic oil and/or condensate phase develops if plant resins are predominant in the kerogen. There is thus a marked contrast between marine and terrestrial sources, not only in the timing of generation, but also in the type of oil and condensate generated (Figs 11 and 12).

The Far East

Tables 9 and 11 also show areas, other than those in Australia and Canada, in which terrestrially-sourced oil and gas have been identified (see also the section of this paper dealing with biomarkers).

Many Indonesian sedimentary basins contain coals and shales with terrestrially derived organic matter which are apparently potential oil-source rocks of good quality. Oil, condensate and gas fields are found in Tertiary deltaic sediments, which range from Miocene to Pliocene in age, including those of the Mahakam Delta which overlies the Kutai Basin in Kalimantan (see Land & Jones, *this volume*). The background geography and geology and some of the earlier organic geochemical data for this Delta have been reported by Magnier *et al.* (1975), Combaz & De Matherel (1978), Durand & Oudin (1979) and Oudin & Picard (1982). The more recent studies have been essentially chemical. Virtually all these studies have confirmed a terrestrial source for the oils based on many of the parameters discussed earlier. Schoell *et al.* (1983) found only Type III and humic kerogens in their study and related biomarker and stable isotope variations to oil-source rock correlations. They compared the

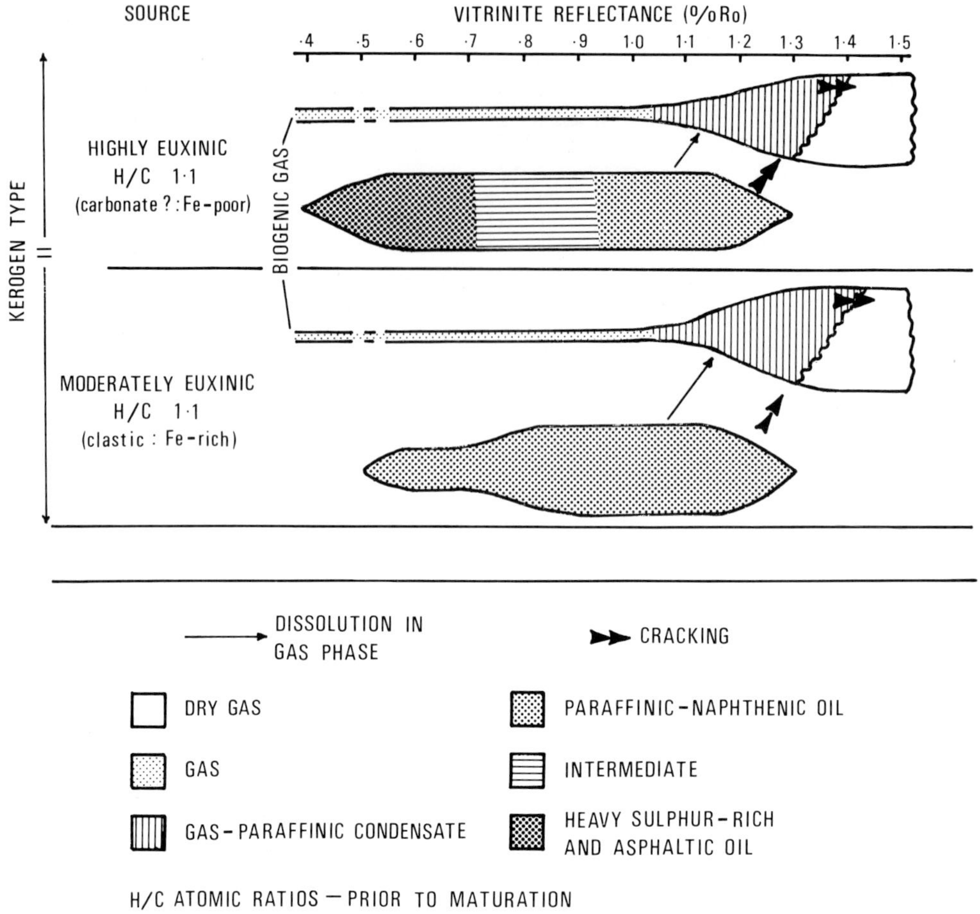

FIG. 12. Oil and gas generation model for different sources of marine-derived organic matter (from Powell & Snowdon 1983).

sterane ($m/z = 217$) and triterpane ($m/z = 191$) patterns of mature extracts and oils from wells in the area. The patterns showed great similarity and supported the view that mature Type III kerogens in the area are potential sources for oil and not just gas (Durand & Oudin 1979). Schoell *et al.* (1983) also used a hopane/moretane maturation index (M_H) to show that the oils originated from a source rock at a maturity level between $T_{max} = 440°C$ and 450°C. The observations of Schoell *et al.* (1983) were supported by Hoffman *et al.* (1984) who came to similar conclusions, based on analysis of oils, coals and shales (principally by gc-ms), demonstrating that all the fractions contained high relative abundances of compounds from land plants.

In a specific study of triterpanes in crude oils from the Mahakam Delta, Grantham *et al.* (1983) recognized the triterpanes 18α(H)-oleanane and 'compound J' (Hills & Whitehead 1966), both of which point strongly to the presence of terrestrial organic matter in the source rocks. Three unusual C_{30} pentacyclic compounds (but no further identification) also occurred in some of the crudes. The C_{30} hydrocarbons were also found in the extract of a heated resin from Lumapas in Brunei and in a South Sumatran sediment known from maceral analysis to contain resins (Grantham *et al.* 1983), and so the oils were suggested as originating in plant resins which are, in any case, widespread in Tertiary sediments. All five compounds, 18(αH)-oleanane, 'compound J' and the three C_{30} hydrocarbons from resins, vary considerably in the proportions in which they occur at different localities, a variation which must reflect differences in the land-plant source input. Thus, in a crude oil from the South China Sea, two of the three unidentified C_{30} pentacyclic compounds are major components (i.e. Lumapas type), whilst a West Irian (Java) crude contained much more of the precursors of oleanane and 'compound J' derived from other resins (Grantham *et al.* 1983).

What is surprising is that such a large amount of published information is chemical, with a lack of supporting petrographic data. Thompson *et al.* (1985) have, however, produced petrographic analyses in terms of inertinite, vitrinite, algal sapropel and 'waxy (Types I+II) kerogen' for coals from the Kutai Basin, NW Java and Sudan. Although these authors state that in reflected light the coals appear to be vitrinite-rich, in 10 of the 14 petrographic analyses quoted, inertinite exceeds vitrinite, often by substantial amounts. In correlations between Rock-Eval parameters and petrographic data, only the most generalized relationship exists between, on the one hand, the production index and potential yield and, on the other, algal sapropel and waxy kerogen, taken either individually or in total from the maceral analyses.

The Niger Delta

The Niger Delta is another important hydrocarbon habitat (Evamy *et al.* 1978) in which oils have almost certainly been sourced from land-plant material. There is a high relative abundance of 18α(H)-oleanane in both crude oils and in sediments from the Delta (Pym *et al.* 1975, Ekweozor *et al.* 1979a, b). Grantham *et al.* (1983) cite a reaction mechanism which would account for the occurrence of 18α(H)-oleanane in Nigerian crude oil, which is supported by the discovery of an intermediate compound, olean-13(18)-ene, in an immature shale from the Agbada Formation in the Delta (Ekweozor *et al.* 1979a, b). In the Nigerian crudes only 18α(H)-oleanane and 'compound J' are present, with no trace of the C_{30} pentacyclic hydrocarbons found by Grantham *et al.* (1983) in Far East crudes, testifying to yet another variation of the resin source. Other triterpanes have been identified (Hills *et al.* 1968); for example, relatively large abundances of the very common 17α(H),21β(H)-hopanes are present in Nigerian crudes compared with Far Eastern crudes.

China

Relatively recent information from China refers to continental lake basins which provide most of the 100 million tonnes of oil produced annually in that country. Zhai *et al.* (1984) and Huang *et al.* (1984) have developed a theory for petroleum generation and occurrence which is based on extensive and long-lived lakes lying in intermontane basins in which rich oil and gas source rocks have developed. There are 12 such productive basins in China, the Songliao Basin in NE China being the most significant and containing the Daqing field which is the largest oil producer. The lake that occupied the Songliao Basin in mid-Cretaceous times is estimated to have covered an area of up to 100,000 km^2, with 4500 m of fluvio-lacustrine sediments deposited in it (Yang *et al.* 1985) (*cf.* Lake Victoria with an area of 70,000 km^2 and a depth of 80 m). At the other end of the spectrum the Biyang Rift Lake Basin of Henau Province only covered an area of 1000 km^2, but contains within it 8300 m of non-marine strata and 100 million tonnes of oil reserves (Li *et al.* 1984).

Powell (1986) has discussed the petroleum geochemistry and depositional setting of lacustrine source rocks, much of the data based on Chinese locations. He acknowledges the wide variety of rock types that can develop as a result of differing depositional environments, which can range from: (a) freshwater lakes in a humid climate through (b) fresh-to brackish water lakes in semi-humid climates and (c) saline lakes in semi-arid climates to (d) hypersaline lakes in arid climates (Tian *et al.* 1983, Huang *et al.* 1984). Not surprisingly, kerogens in such environments range from Types I to III, very often with mixtures of Types II and III. The input of land-plant material can be substantial. Understandably, Powell (1986) points out that there can be no uniform maturation model applying to such a wide variety of source rocks, which certainly can contain autochthonous coals, cannels and bogheads. Depending on the depositional setting, there will also be sediments in which the organic matter has a similar varied mix to these coal types. Most lacustrine source rocks appear to have formed, however, in lakes varying in size from small to very large, which are freshwater or slightly saline, and so therefore tend to contain a high proportion of terrestrially derived organic matter, supplemented by bacterial biomass to varying degrees. Thus, excluding source rocks forming under hypersaline conditions, and presumably kerogens with substantial resin contents (Powell & Snowdon 1983 and Fig. 9), maturation levels above a vitrinite reflectance of 0.7% will be needed to generate substantial quantities of hydrocarbons (Powell 1986).

The vast majority of the crude oils generated from lacustrine source rocks are highly paraffinic in character, they have low sulphur contents and high wax contents which may even leave the oil solid at ambient temperature. In these respects the oils are similar to terrestrial crudes. But there are differences because they may have low pristane : phytane ratios (< 1) and an even to odd predominance in the *n*-alkanes which makes them appear similar to oils from hypersaline marine environments. Powell (1986) regards extremely high wax oils of the type found throughout most of the Chinese fields (and elsewhere: *cf.* certain Uinta Basin oils), as a special instance of non-marine oils. These appear to generate from Type I kerogens, and they are dominated by degraded terrestrial organic matter with a varying contribution from bacterial and algal remains. Such kerogens are regarded as the basis for the best source rocks in the Songliao Basin and they are deposited in a deep-water lacustrine facies. Powell (1986) also points to the paradox that these lacustrine oils are often higher in wax content than oils generated from land-plant material in coal-measure sequences. There is still no certainty that the wax content is due entirely to land-plant lipids and lipids from the microbial biomass or whether there has been a substantial freshwater algal contribution (McKirdy *et al.* 1986).

Other areas

There are other known petroleum occurrences which have almost certainly been sourced by terrigenous material, but where petrographic and/or geochemical evidence is less substantial than in the examples discussed earlier. In general these occurrences are deltaic sequences in which coal seams and shales with terrestrial organic matter are interbedded, for example, the Douala Basin of Upper Cretaceous age in the Cameroon, or the Westphalian Delta in the eastern part of the Paris Basin. There are other possible areas. For example, in the North Sea graben system it is widely accepted that reservoirs, ranging from Devonian to Eocene in age, have been largely sourced from anoxic sapropelic shales of the late Jurassic–early Cretaceous Kimmeridge Clay or its lateral equivalent in the Norwegian sector. Similar high-quality source rocks may occur at other stratigraphic levels, for instance, the Liassic Dunlin Group, and there may also be subsidiary sourcing of different type at other horizons. Thompson *et al.* (1985) comment on the coaly, carbonaceous deltaic sequences of early and middle Jurassic age in the Norwegian sector of the North Sea, suggesting that, while the hydrocarbon potential of these rocks is not high in the Haltenbanken and Tromso areas, the late Triassic–early Jurassic coaly sequence is buried to a sufficient depth to generate hydrocarbons, while the Kimmeridge is not and, furthermore, it is here a thinner section than elsewhere. Oil generation could come from allochthonous coals around the delta margins. Pittion & Gouadain (1985) have undertaken reflectance, fluorescence and pyrolysis studies and believe that the Lower Jurassic 'Coal Unit' of the Haltenbanken area probably has 'fair' source-rock potential, particularly when lying at a depth that places the 'Unit' within the 'oil window'. Liptinite levels are quite high, vitrinite is dominant in the coals, and hydrogen indices from Rock-Eval pyrolysis, lying between 100 and 350 mg-g, are comparable with those from coals and sediments with oil and gas potential in the Mahakam Delta.

The future

Within the three fields of microscope fluorometry, high-resolution analytical pyrolysis techniques and biomarkers, in which there have been substantial advances, further developments of ways to assess type and rank of sedimentary organic matter are clearly possible, so improving further the chances of differentiating between source inputs for oils. Larter (1984) makes a critical assessment of how pyrolysis methods might develop, particularly in relation to other petrographical and chemical methods of appraisal. Bulk-flow pyrolysis methods, for example Rock-Eval, while most useful for preliminary screening purposes, do not have a resolution that compares with current microscopical methods used to assess maturity. Thus, reflectance and different fluorescence parameters will continue to be employed for ranking purposes. In the area of kerogen typing, however, Larter (1984) believes that high-resolution analytical pyrolysis methods hold a greater potential for accurate type assessment of kerogens than does the microscopical approach.

The recent work of Larter & Senftle (1985) and Larter (1985) is of great value. Their approach allows the definition of a larger number of kerogen end-members than could previously be categorized, through the identification of specific chemical species in pyrolysates, as well as quantitation of pyrolysis yields using pyrolysis gas chromatography. These data must form a useful basis for the future advance of micro-scale pyrolysis methods. Larter (1984) feels that the ultimate kerogen typing method will be found through pyrolysis mass spectrometry, simply because of the speed of analysis, the amount of data produced and the removal of operator subjectivity. This view may be too radical. There is much evidence which points to greatest success in kerogen typing stemming from a multi-parameter approach.

Although high-resolution analytical pyrolysis methods may have a dominant rôle to play in kerogen typing in the future, there must be further scope for microscope fluorometry to contribute to assessment of low maturity levels and kerogen typing. Many of the correlations between fluorometric parameters and other optical and chemical data in maturity studies, although useful, are basically of a 'finger-printing' nature. Indeed, some authors (Thompson-Rizer & Woods 1987) are sceptical that the current spectral parameters, used to define the fluorescence behaviour of macerals, suffer from a lack of standardization between laboratories and so therefore are difficult to compare. They proposed reporting fluorescence data in terms of 'equivalent vitrinite reflectance values' ('R_F values'), based on fluorescence and other data, maintaining this approach would have more immediate meaning in designating a rank or maturation level. Adopting this suggestion, however, will not explain observed fluorescence changes, although it may render fluorescence data more immediately intelligible. What is required is that greater attention should be given to the relationships that exist between fluorescence phenomena and the underlying chemistry, which will have to be better understood if greater insight is to be gained into the factors controlling variations in fluorescence intensity and spectral shifts from macerals and extracts excited by UV radiation.

Some progress in this direction has already been made. For example, Bertrand *et al.* (1986) have reported on factors that control fluorescence. They list these as being: (a) the structure of the inherent chromophores which are generally associated with the aromatic systems, (b) the concentration of the chromophores and (c) matrix-solvent effects which allow energy exchanges over inter- or intra-molecular interactions, with the absorbed energy diffusing away outside of the visible spectrum or in some other form of energy. Lin & Davis (1986) have considered primary fluorescence—remnant fluorescence from the original properties of tissues, e.g. xylem cell walls and contents (Stout & Bensley 1987). They attribute decrease of fluorescence intensity and spectral red-shift in vitrinites to delocalization of the π electron system that is brought about by biochemical and geochemical processes causing cross-linking and condensation of the aromatic lamellae. Secondary fluorescence of vitrinites is now also an acknowledged phenomenon. Lin *et al.* (1986) suggest, from investigation of the fluorescence spectra of fractions from solvent extracts of vitrinites, that secondary fluorescence develops through a mobile phase containing fluorophors. This phase builds up due to (a) smaller molecules being trapped within the vitrinite structure as the aromatic structure condenses and (b) migration into the vitrinite of petroleum-like substances after their expulsion from liptinitic macerals (see Teichmüller, 1974a, b). Khorasani (1987) has also investigated the fluorescence properties of solvent extracts from fossil fuels and organic-rich sediments, as well as the fluorescence behaviour of model compounds. From an examination of spectral ranges (and their assignment to specific chemical structures), red and blue spectral shifts, and variations of the spectral quotient (Q), qualitative determination of the level of concentration of saturates, highly alkylated aromatics and aromatic systems in extracts and geopolymers is possible.

Biomarkers have still much to offer. At present a rigorous view of the reliability and specificity of biomarkers in maturation and source-input studies might conjure up an unpromising future. Marine-derived organic matter and crude oils can be categorically identified by C_{30} steranes (Moldowan 1984, Moldowan *et al.* 1985), with the mono-aromatic steroids giving support in the differentiation of oils derived from non-marine and marine sources. Botryococcane (Moldowan & Seifert 1980) seems to be highly characteristic of input from the alga *Botryococcus* and there is strong evidence that 18α(H)-oleanane is specific for land plants, probably for the angiosperms, as is eudesmane (Alexander *et al.* 1983). The diterpenoids which have a wide distribution in higher-plant compounds, particularly the resins, are also reliable indicators of a terrestrial input (Snowdon 1980). For less specific biomarker parameters all is not lost since, by examination of a group of such parameters, surely a satisfactory general indication of source can be obtained, particularly when the biomarker data are used in conjunction with information from pyrolysis and optical studies? Much the same can be said of inferences about level of maturity based upon biomarkers: several biomarker-parameter trends with depth should be used.

In the immediate future, the way forward for biomarker studies is probably still through low-resolution mass spectrometers coupled to gas chromatographs fitted with high-resolution capillary columns. These instruments will continue to be the principal analytical tool of the organic geochemist, but they will be supported by tandem mass spectrometry and a variety of ionization techniques, including fast-atom bombardment, which may well have a substantial rôle to play in the search for high molecular weight precursors of biomarkers. Although high-resolution, selected-ion monitoring, mass spectrometry has been used for analysis of pentacyclic terpanes and steranes, the technique which seems to offer greatest potential for fine characterization of these structurally complex biomarker classes is selective metastable ion monitoring. Warburton & Zumberge (1983) successfully introduced this technique to increase resolution in the analysis of C_{27}–C_{29} steranes because the use of conventional gc-ms to monitor the *m/z* 217 fragment ion often produces complicated and poorly resolved mixtures on the mass fragmentograms. Moldowan (1984) applied the method to the C_{30} steranes and Brooks *et al.* (1984a, b) extended the method to include, not only the steranes, but also the C_{27}–C_{35} pentacyclic triterpanes and other biomarkers. With its capacity to provide (euphemistically) generous quantities of information and high specificity, selective metastable ion monitoring gc-ms may well replace the now conventional high-resolution capillary gc-low resolution ms methodology in organic geochemistry and petroleum geochemistry.

Further advances will come in organic petrology and organic geochemistry, outside of the fields discussed here, but not only through the more established approaches. For example, relatively little use is made of electron microscopy in either organic petrology or organic geochemistry. Glikson (1984) and Glikson & Taylor (1986), however, have illustrated the application of transmission electron microscopy and scanning electron microscopy in the investigation of source input to the Toolebuc Formation, particularly to the way the organic matter has been affected by microbial degradation. These authors suggest that vitrinite-like material in the oil shales may have formed from compressed cyanobacteria and possibly also from sulphate-reducing bacteria. They have identified filamentous bundles in the shales that resemble *Oscillatoria*-type cyanobacteria. Since oil generation and oil biodegradation are so much bound up with bacterial activity, here is yet another area of study that might prove immediately fruitful in the more precise categorization of organic matter, which has oil-bearing potential, as well as its source. There should also be a further serious attempt to assess the prospects for laser-micropyrolysis systems in typing and maturity evaluations. First pioneered in organic petrology by Vastola *et al.* (1970), the systems were unsatisfactory because of the relatively low resolution of the reflected-light microscope optics then available. A 'time-of-flight' mass spectrometer was linked to the laser optics system and this also proved to be unsatisfactory. The advances that have been made in the technology of mass spectrometers and in the quality of microscope optics should offer much greater opportunities to develop a satisfactory analytical tool of this kind.

ACKNOWLEDGEMENTS: The author would like to thank Yvonne Hall for her cheerful resilience in typing repeated drafts of the manuscript and Alexia Summerbell for preparing the line diagrams. He has also benefited from much advice and critical comment from several of his colleagues, notably Anne Raymond, Timothy Eglinton and Martin Jones in this Unit to whom he offers warm appreciation. Particular gratitude must go to Andrew Scott, the editor of this volume, for his continued and long-standing patience and forebearance with the author over many months.

References

ALEXANDER, R., KAGI, R. & NOBLE, R. 1983. Identification of the bicyclic sesquiterpenes drimane and eudesmane in petroleum. *Journal of the Chemical Society Chemical Communications*, pp. 226–228.

——, ——, NOBLE, R. & VOLKMAN, J. K. 1984. Identification of some bicyclic alkanes in petroleum. *Organic Geochemistry*, **6**, 63–70.

——, ——, WOODHOUSE, G. W. & VOLKMAN, J. K. 1983. The geochemistry of some biodegraded Australian oils. *Journal of the Australian Petroleum Exploration Association*, **23**, 53–63.

ALPERN, B. 1980. Pétrographie du kérogène. *In*: DURAND, B. (ed.) *Kerogen*, pp. 339–84. Éditions Technip, Paris.

—— 1981. Les schistes bitumineux: constitution, réserves, valorisation. *Bulletin des Centres de Recherches Exploration-Production Elf-Aquitaine*, **5**, 319–52.

AQUINO NETO, F. R., CARDOSO, J. N., RODRIQUES, R. & TRINDADE, L. A. F. 1986. Evolution of tricyclic alkanes in the Espirito Santo Basin, Brazil. *Geochimica et Cosmochimica Acta*, **50**, 2069–72.

——, TRENDEL, J. M., RESTLE, A., CONNAN, J. & ALBRECHT, P. A. 1983. Occurrence and formation of tricyclic and tetracyclic terpanes in sediments and petroleums. *In*: BJORØY, M., *et al.* (eds) *Advances in Organic Geochemistry 1981*, pp. 659–67. Wiley, Chichester.

BARNARD, P. C., COOPER, B. S. & FISHER, M. J. 1980. Organic maturation and hydrocarbon generation in the Mesozoic sediments of the Sverdrup basin, Arctic Canada. *Proceedings of the IV International Palynological Conference, Lucknow (1976–1977)*, **2**, 581–8.

BASET, Z. H., PANCIROV, R. J. & ASHE, T. R. 1980. Organic compounds in coal: structure and origins.

In: DOUGLAS, A. G. & MAXWELL, J. R. (eds) *Advances in Organic Geochemistry 1979*, pp. 619–30. Pergamon Press, Oxford.

BATTEN, D. J. 1980. Use of transmitted light microscopy of sedimentary organic matter for evaluation of hydrocarbon source potential. *Proceedings of the IV International Palynological Conference, Lucknow (1976–1977)*, **2**, 589–94.

—— 1983. Identification of amorphous sedimentary organic matter by transmitted light microscopy. *In*: BROOKS, J. (ed.) *Petroleum Geochemistry and Exploration of Europe*, pp. 275–87. Geological Society Special Publication No. 12. Blackwell Scientific Publications, Oxford.

BATTERSBY, D. G. 1976. Cooper Basin gas and oil fields. *In*: LESLIE, R. B., EVANS, H. J. & KNIGHT, C. L. (eds) *Economic Geology of Australia and Papua New Guinea—3. Petroleum*, pp. 321–68. Australasian Institute of Mining and Metallurgy, Monograph 7.

BAYLISS, G. 1985. Geochemistry in oil and gas exploration: a review. *World Oil*, August, 41–45.

BERTRAND, P. PITTION, J.-L. & BERNAUD, C. 1986. Fluorescence of sedimentary organic matter in relation to its chemical composition. *Organic Geochemistry*, **10**, 641–7.

BRASSELL, S. C. & EGLINTON, G. 1983. Steroids and triterpenoids in deep sea sediments as environmental and diagenetic indicators. *In*: BJORØY, M. *et al.* (eds) *Advances in Organic Geochemistry 1981*, pp. 684–97. Wiley, Chichester.

——, WARDROPER, A. M. K., THOMPSON, I. D., MAXWELL, J. R. & EGLINTON, G. 1981. Specific acyclic isoprenoids as biological markers of methanogenic bacteria in marine sediments. *Nature*, **290**, 693–6.

BROOKS, J. D. 1970. The use of coals as indicators of the occurrence of oil and gas. *Journal of the Australian Petroleum Exploration Association*, **10**, 35–40.

——, GOULD, K. & SMITH, J. W. 1969. Isoprenoid hydrocarbons in coal and petroleum. *Nature*, **222**, 257–9.

——, HESP, W. R. & RIGBY, D. 1971. The natural conversion of oil to gas in sediments in the Cooper Basin. *Journal of the Australian Petroleum Exploration Association*, **11**, 121–5.

—— & SMITH, J. W. 1967. The diagenesis of plant lipids during the formation of coal, petroleum and natural gas—I. Changes in the *n*-paraffin hydrocarbons. *Geochimica et Cosmochimica Acta*, **31**, 2389–97.

—— & —— 1969. The diagenesis of plant lipids during the formation of coal, petroleum and natural gas—II. Coalification and the formation of oil and gas in the Gippsland Basin. *Geochimica et Cosmochimica Acta*, **33**, 1183–94.

BROOKS, P. W. 1986. Unusual biological marker geochemistry of oils and possible source rocks, offshore Beaufort–Mackenzie Delta, Canada. *Organic Geochemistry*, **10**, 401–6.

——, MEYER, T. & CHRISTIE, O. H. J. 1984a. Aspects of biomarker analysis by selected metastable ion monitoring GC/MS. I. Experimental techniques. *Analytica Chimica Acta*, **161**, 65–82.

——, —— & —— 1984b. A comparison of high resolution GC–MS and selected metastable ion monitoring of polycyclic steranes and triterpanes. *Organic Geochemistry*, **6**, 813–16.

BURGESS, J. D. 1974. Microscopic examination of kerogen (dispersed organic matter) in petroleum exploration. *In*: DUTCHER, R. R., HACQUEBARD, P. A., SCHOPF, J. M. & SIMON, J. A. (eds) *Carbonaceous Materials as Indicators of Metamorphism*, pp. 19–30. Special Paper 153. The Geological Society of America, Boulder.

COMBAZ, A. 1975. Essai de classification des roches carbonées et des constituents organiques des roches sedimentaires. *In*: ALPERN, B. (ed.) *Pétrographie de la matière organique des sédiments, relations avec la paléotempérature et le potentiel petrolier*, pp. 93–102. Centre National de la Recherche Scientifique, Paris.

—— 1980. Les kérogènes vus au microscope. *In*: DURAND, B. (ed.) *Kerogen*, pp. 55–112. Éditions Technip, Paris.

—— & DE MATHAREL, M. 1978. Organic sedimentation and genesis of petroleum in Mahakam Delta, Borneo. *Bulletin of the American Association of Petroleum Geologists*, **62**, 1684–95.

COOK, A. C. 1982. *The origin and petrology of organic matter in coals, oil shales and petroleum source-rocks*. The University of Wollongong, New South Wales.

—— 1986. The nature and significance of the organic facies in the Eromanga Basin. *In*: GRAVESTOCK, D. I., MOORE, P. S. & PITT, G. M. (eds) *Contributions to the Geology and Hydrocarbon Potential of the Eromanga Basin*, pp. 203–19. Geological Society of Australia, Special Publication No. 12.

—— HUTTON, A. C. & SHERWOOD, N. R. 1981. Classification of oil shales. *Bulletin des Centres de Recherches Exploration–Production Elf-Aquitane*, **5**, 353–81.

—— & KANTSLER, A. J. 1979. The maturation history of the epicontinental basins of Western Australia. *United Nations ESCAP/Committee for Coordination of Joint Prospecting for Mineral Resources in Asian Offshore Areas/SOPAC. Technical Bulletin*, **3**, 171–95.

——, SMYTH, M. & VOS, R. G. 1985. Source potential of Upper Triassic fluvio–deltaic systems of the Exmouth Plateau. *Journal of the Australian Petroleum Exploration Association*, **25**, 204–15.

CORBERT, B., ALBRECHT, P. & OURISSON, G. 1980. Photochemical or photomimetic fossil triterpenoids in sediments and petroleums. *Journal of the American Chemical Society*, **102**, 1171–3.

CREANEY, S. 1980. The organic petrology of the Upper Cretaceous Boundary Creek Formation, Beaufort–Mackenzie Basin. *Bulletin of Canadian Petroleum Geology*, **28**, 112–29.

DAVIDSON, J. K., BLACKBURN, G. J. & MORRISON, K. C. 1984. Bass and Gippsland Basins: a comparison. *Journal of the Australian Petroleum Exploration Association*, **24**, 101–9.

DAVIS, A 1978. The reflectance of coal. *In*: KARR, C. (ed.) *Analytical Methods for Coal and Coal Products, Vol 1*, pp. 27–81. Academic Press, New York.

DEROO, G. POWELL, T. G., TISSOT, B. & MCCROSSAN, R. G. 1977. The origin and migration of petroleum in the Western Canadian Sedimentary Basin, Alberta. *Bulletin of the Geological Survey of Canada*, **262**.

DIESSEL, C. F. K. 1983. Carbonization reactions of inertinite macerals in Australian coals. *Fuel*, **62**, 883–92.

—— 1985. Fluorometric analysis of inertinite. *Fuel*, **64**, 1542–6.

DORMANS, H. N. M., HUNTJENS, F. J. & KREVELEN, D. W. VAN 1957. Chemical structure and properties of coal xx—composition of the individual macerals (vitrinites, fusinites, micrinites and exinites). *Fuel*, **36**, 321–39.

DOW, W. G. 1977. Kerogen studies and geological interpretations. *Journal of Geochemical Exploration*, **7**, 79–99.

—— & O'CONNOR, D. I. 1982. Kerogen maturity and type by reflected light microscopy applied to petroleum exploration. *In*: *How to Assess Maturation and Paleotemperatures*, pp. 133–57. Society of Economic Paleontologists and Mineralogists Short Course 7.

DURAND, B. & OUDIN, J. L. 1979. Exemple de migration des hydrocarbures dans une série deltaique: le Delta de la Mahakam, Kalimantan, Indonesia. *Proceedings of the 10th World Petroleum Congress*, **2**, 3–11.

—— & PARATTE, M. 1983. Oil potential of coals: a geochemical approach. *In*: BROOKS, J. (ed.) *Petroleum Geochemistry and Exploration of Europe*, pp. 255–65. Geological Society Special Publication No. 12. Blackwell Scientific Publications, Oxford.

EGLINTON, G. & HAMILTON, R. J. 1963. The distribution of alkanes. *In*: SWAIN, T. (ed.) *Chemical Plant Taxonomy*, pp. 187–208. Academic Press, London.

EKWEOZOR, C. M., OKOGUN, J. I., EKONG, D. E. U. & MAXWELL, J. R. 1979a. Preliminary organic geochemical studies of samples from the Niger Delta (Nigeria), I. Analyses of crude oils for triterpanes. *Chemical Geology*, **27**, 11–28.

——, ——, —— & —— 1979b. Preliminary organic geochemical studies of samples from the Niger Delta (Nigeria), II. Analyses of shale for triterpenoid derivatives. *Chemical Geology*, **27**, 29–37.

——, ——, —— & —— 1981. C_{24}-C_{27} degraded triterpanes in Nigerian petroleums—novel molecular markers of source input or organic maturation. *Journal of Geochemical Exploration*, **15**, 653–62.

ESPITALIÉ, J. LAPORTE, J. L., MADEC, M., MARQUIS, E., LEPLAT, P., PAULET, J. & BOUTEFEU, A. 1977. Méthode rapid de caractérisation des roches mères, de leur potential pétrolier et de leur degré d'évolution. *Revue de l'Institut Français du Pétrole*, **32**, 23–42.

EVAMY, B. D., HAREMBOURE, J., KAMERLING, P., KNAPP, W. A., MOLLOY, F. A. & ROWLANDS, P. H. 1978. Hydrocarbon habitat of Tertiary Niger Delta. *Bulletin of the American Association of Petroleum Geologists*, **62**, 1–37.

EVANS, E. J., BATTS, B. D. & SMITH, J. W. 1984. Determination of the hydrocarbon prospectivity of sediments by hydrogenation. *Journal of the Australian Petroleum Exploration Association*, **24**, 222–9.

GIJZEL, P. VAN 1966. Die Fluoreszenz-Mikrophotometrie von Mikrofossilien mit dem Zweistrahl-Mikroskopphotometer von Berek. *Leitz Mitteilungen für Wissenschaft und Technik*, **3**, 206–14.

—— 1967. Autofluorescence of fossil pollen and spores with special reference to age determination and coalification. *Leidse Geologische Mededelingen*, **40**, 263–317.

—— 1971. Review of the UV-fluorescence microphotometry of fresh and fossil exines and exosporia. *In*: BROOKS, J., GRANT, P. R., MUIR, M. D., GIJZEL, P. VAN & SHAW, G. (eds) *Sporopollenin*, pp. 659–85. Academic Press, London.

—— 1975. Polychromatic UV-fluorescence microphotometry of fresh and fossil plant substances with special reference to the location and identification of dispersed organic material in rocks. *In*: ALPERN, B. (ed.) *Pétrographie de la matière organique des sédiments, relations avec la paléotempérature et le potentiel pétrolier*, pp. 67–91. Centre National de la Recherche Scientifique, Paris.

GLIKSON, M. 1984. Further studies on torbanites and coorongite using transmission electron microscopy and C-isotope analysis. *Organic Geochemistry*, **7**, 161–8.

—— & TAYLOR, G. H. 1986. Cyanobacterial mats: major contributors to the organic matter in Toolebuc Formation oil shales. *In*: GRAVESTOCK, D. I., MOORE, P. S. & PITT, G. M. (eds) *Contributions to the Geology and Hydrocarbon Potential of the Eromanga Basin*, pp. 273–86. Geological Society of Australia Special Publication No. 12.

GOOSENS, H., LEEUW, J. W. DE, SCHENCK, P. A. & BRASSELL, S. C. 1984. Tocopherols as likely precursors of pristane in ancient sediments and crude oils. *Nature*, **312**, 440–2.

GOULD, R. E. 1980. The coal-forming flora of the Walloon coal measures. *Coal Geology*, **1**, 83–105.

GRANDEVILLE, B. F. de la 1981. Appraisal and development of a structural and stratigraphic trap oil field with reservoirs in glacial to periglacial clastics. *In*: HALBOUTY, M. J. (ed.) *The Deliberate Search for the Subtle Trap*, pp. 267–286. American Association of Petroleum Geologists, Memoir 32. Tulsa, Oklahoma.

GRANTHAM, P. J. 1986a. The occurrence of unusual C_{27} and C_{29} sterane predominances in two types of Oman crude oil. *Organic Geochemistry*, **9**, 1–10.

—— 1986b. Sterane isomerisation and moretane/hopane ratios in crude oils derived from Tertiary source rocks. *Organic Geochemistry*, **9**, 293–304.

—— & DOUGLAS, A. G. 1980. The nature and origin of sesquiterpenoids in some Tertiary fossil resins. *Geochimica et Cosmochimica Acta*, **44**, 1801–10.

——, POSTHUMA, J. & GROOT, K. DE 1980. Variation and significance of the C_{27} and C_{28} triterpane content of a North Sea core and various North Sea crude oils. *In*: DOUGLAS, A. G. & MAXWELL, J. R. (eds) *Advances in Organic Geochemistry 1979*, pp. 29–38. Pergamon Press, Oxford.

——, —— & BAAK, A. 1983. Triterpanes in a number

of Far-Eastern crude oils. *In*: BJORØY, M. *et al.* (eds) *Advances in Organic Geochemistry 1981*, pp. 675–83. Wiley, Chichester.

GRAVESTOCK, D. I., MOORE, P. S. & PITT, G. M. (eds) 1986. *Contributions to the Geology and Hydrocarbon Potential of the Eromanga Basin*. Geological Society of Australia, Special Publication No. 12.

GRAYSON, J. F. 1975. Relationship of palynomorph translucency to carbon and hydrocarbons in clastic sediments. *In*: ALPERN, B. (ed.) *Pétrographie de la matière organique des sédiments, relations avec la paléotemperature et le potentiel petrolier*, pp. 261–73. Centre National de la Recherche Scientifique, Paris.

GUTJAHR, C. C. M. 1983. Introduction to incident light microscopy of oil and gas source rocks. *Geologie en Mijnbouw*, **62**, 417–25.

HAGEMANN, H. W. & HOLLERBACH, A. 1980. Relationship between the macropetrographic and organic geochemical composition of lignites. *In*: DOUGLAS, A. G. & MAXWELL, J. R. (eds) *Advances in Organic Geochemistry 1979*, pp. 631–8. Pergamon Press, Oxford.

HARWOOD, R. J. 1977. Oil and gas generation by laboratory pyrolysis of kerogen. *Bulletin of the American Association of Petroleum Geologists*, **61**, 2082–102.

HAVEN, H. L. TEN, LEEUW, J. W. DE, PEAKMAN, T. M. & MAXWELL, J. R. 1986. Anomalies in steroid and hopanoid maturity indices. *Geochimica et Cosmochimica Acta*, **50**, 853–5.

HEDBERG, H. D. 1968. Significance of high-wax oils with respect to genesis of petroleum. *Bulletin of the American Association of Petroleum Geologists*, **52**, 736–50.

HENG, S., COLLIN, P. J. & WILSON, M. A. 1983. Hydrogenation of maceral concentrates from Bayswater coal: effect of temperature on the yield and mean chemical composition of the product. *Fuel*, **62**, 1359–68.

HILLS, I. R., SMITH, G. W. & WHITEHEAD, E. V. 1968. Optically active spirotriterpane in petroleum distillates. *Nature*, **219**, 243–6.

—— & WHITEHEAD, E. V. 1966. Triterpanes in optically active petroleum distillates. *Nature*, **209**, 977–9.

HOFFMANN, C. F., MACKENZIE, A. S., LEWIS, C. A., MAXWELL, J. R., OUDIN, J. L., DURAND, B. & VANDENBROUCKE, M. 1984. A biological marker study of coals, shales and oils from the Mahakam Delta, Kalimantan, Indonesia. *Chemical Geology*, **42**, 1–23.

HUANG, D., SHANG, H. & LI, J. 1984. Advances in theoretical research on continental oil generation in China. *Proceedings of the Beijing Petroleum Symposium*, Beijing, China.

HUANG, W.-Y. & MEINSCHIEN, W. G. 1978. Sterols in sediments from Baffin Bay, Texas. *Geochimica et Cosmochimica Acta*, **42**, 1391–6.

—— & —— 1979. Sterols as ecological indicators. *Geochimica et Cosmochimica Acta*, **43**, 739–45.

HUTTON, A. C. & COOK, A. C. 1980. Influence of alginite on the reflectance of vitrinite from Joadja, NSW, and some other coals and oil shales containing alginite. *Fuel*, **59**, 711–14.

——, KANTSLER, A. J. COOK, A. C. & MCKIRDY, D. M. 1980. Organic matter in oil shales. *Journal of the Australian Petroleum Exploration Association*, **20**, 44–67.

HUTTON, W. 1834. Observations on coal. *Proceedings of the Geological Society of London*, **1**, 415–17.

INTERNATIONAL COMMITTEE FOR COAL PETROLOGY 1963, 1971, 1975. *International Handbook of Coal Petrography*, 2nd edn (1963), 1st Supplement (1971), 2nd Supplement (1975), Centre National de la Recherche Scientifique, Paris.

JACOB, H. 1967. Petrologie von Asphaltiten und asphaltischen Pyrobitumina. *Erdöl und Kohle*, **20**, 393–400.

—— 1969. Lumineszenz-Mikroskopie der organopetrographischen Bestandteile von Sedimentgesteinen. *Leitz Mitteilungen für Wissenschaft und Technik*, **4**, 250–4.

—— 1973. Kombination von Fluoreszenz- und Reflexions-Mikroskopphotometrie der organischen Stoffe von Sedimenten und Böden. *Leitz Mitteilungen für Wissenschaft und Technik*, **6**, 21–7.

—— 1974. Fluoreszenz-Mikroskopie und-Photometrie der organischen Substanz von Sedimenten und Böden. *In*: FREUND, H. (ed.) *Handbuch der Mikroskopie in der Technik*, **4**, part 2, pp. 369–91. Umschau Verlag, Frankfurt.

JACKSON, K. S., MCKIRDY, D. M. & DECKELMAN, J. A. 1984. Hydrocarbon generation in the Amadeus Basin, Central Australia. *Journal of the Australian Petroleum Exploration Association*, **24**, 42–65.

JONES, J. M., DAVIS, A., COOK, A. C., MURCHISON, D. G. & SCOTT, E. 1984. Provincialism and correlations between some properties of vitrinites. *International Journal of Coal Geology*, **3**, 315–31.

—— & MURCHISON, D. G. 1963. The occurrence of resinite in bituminous coals. *Economic Geology*, **58**, 263–73.

JONES, R. W. 1979. Some mass balance and geological constraints on migration mechanisms. *Proceedings of the Indonesian Petroleum Association. Eighth Annual Convention*, pp. 143–57.

KANTSLER, A. J. & COOK, A. C. 1979. Maturation patterns in the Perth Basin. *Journal of the Australian Petroleum Exploration Association*, **19**, 94–107.

——, —— & ZWIGULIS, M. 1986. Organic maturation in the Eromanga Basin. *In*: GRAVESTOCK, D. I., MOORE, P. S. & PITT, G. M. (eds) *Contributions to the Geology and Hydrocarbon Potential of the Eromanga Basin*, pp. 305–22. Geological Society of Australia, Special Publication No. 12.

——, PRUDENCE, T. J. C., COOK, A. C. & ZWIGULIS, M. 1983. Hydrocarbon habitat of the Cooper/Eromanga Basin, Australia. *Journal of the Australian Petroleum Exploration Association*, **23**, 75–92.

——, SMITH, G. C. & COOK, A. C. 1978. Lateral and vertical rank variation: implications for hydrocarbon exploration. *Journal of the Australian Petroleum Exploration Association*, **18**, 143–56.

KATZ, B. J. & ELROD, L. W. 1983. Organic geochemistry of DSDP site 467, offshore California, Middle Miocene to Lower Pliocene strata. *Geochimica et Cosmochimica Acta*, **47**, 389–96.

KHORASANI, G. K. 1987. Novel development in fluorescence microscopy of complex organic mixtures: application in petroleum geochemistry. *Organic Geochemistry* (in press).

KREVELEN, D. W. VAN 1950. Graphical-statistical method for the study of structure and reaction processes of coal. *Fuel*, **29**, 296–83.

LARTER, S. R. 1984. Application of analytical pyrolysis techniques to kerogen characterisation and fossil fuel exploration/exploitation. *In*: VOORHEES, K. (ed.) *Analytical Pyrolysis—Methods and Applications*, pp. 212–75. Butterworths, London.

—— 1985. Integrated kerogen typing in the recognition and quantitative assessment of petroleum source rocks. *In*: THOMAS, B. M. *et al.* (eds) *Proceedings of a Norwegian Petroleum Society Conference*, 'Organic Geochemistry in Exploration of the Norwegian Shelf', pp. 269–86. Graham & Trotman, London.

—— & DOUGLAS, A. G. 1982. Pyrolysis methods in organic geochemistry: an overview. *Journal of Applied and Analytical Pyrolysis*, **4**, 1–19.

—— & SENFTLE, J. T. 1985. Improved kerogen typing for petroleum source rock analysis. *Nature*, **318**, 277–80.

LEYTHAUSER, D. & SCHWARZKOPF, TH. 1986. The pristane/*n*-heptadecane ratio as an indicator for recognition of hydrocarbon migration effects. *Organic Geochemistry*, **10**, 191–7.

LI, C., ZHU, S. & ZHU, S. 1984. Sedimentary system and oil-bearing features of the small intermontane rift basin of Biyang. *Proceedings of the Beijing Petroleum Symposium*, Beijing, China.

LIJMBACH, G. W. M. 1975. On the origin of petroleum. *Proceedings of the Ninth World Petroleum Congress*, **2**, 357–69.

LIN, R. & DAVIS, A. 1986. Fluorogeochemistry of coal macerals. *Society of Organic Petrologists*, 1987 Meeting, Lexington, Kentucky (extended abstract).

——, ——, BENSLEY, D. F. & DERBYSHIRE, F. J. 1986. Vitrinite secondary fluorescence: its chemistry and relation to the development of a mobile phase and thermoplasticity in coal. *International Journal of Coal Geology*, **6**, 215–28.

MACKENZIE, A. S. 1984. Applications of biological markers in petroleum geochemistry. *In*: BROOKS, J. & WELTE, D. (eds) *Advances in Petroleum Geochemistry, Volume 1*, pp. 115–214. Academic Press, London.

——, BEAUMONT, C., BOUTILIER, R. & RULLKÖTTER, J. 1985. The aromatization and isomerization of hydrocarbons and the thermal and subsidence history of the Nova Scotia margin. *In*: EGLINTON, G., CURTIS, C. D., MCKENZIE, D. P. & MURCHISON, D. G. (eds) *Geochemistry of Buried Sediments*, pp. 203–32. Royal Society of London.

——, BRASSELL, S. C., EGLINTON, G. & MAXWELL, J. R. 1982. Chemical fossils: the geological fate of steroids. *Science*, **217**, 491–504.

—— & MAXWELL, J. R. 1981. Assessment of thermal maturation in sedimentary rocks by molecular measurements. *In*: BROOKS, J. (ed.) *Organic Maturation Studies and Fossil Fuel Exploration*, pp. 239–54. Academic Press, London.

—— & MCKENZIE, D. P. 1983. Isomerization and aromatization of hydrocarbons in sedimentary basins formed by extension. *Geological Magazine*, **120**, 417–528.

MCAULIFFE, C. D. 1979. Oil and gas migration: chemical and physical constraints. *Bulletin of the American Association of Petroleum Geologists*, **63**, 761–81.

MCKIRDY, D. M., COX, R. E., VOLKMAN, J. K. & HOWELL, V. J. 1986. Botryococcane in a new class of Australian non-marine crude oils. *Nature*, **320**, 57–9.

——, EMMETT, J. K., MOONEY, B. A., COX, R. E. & WATSON, B. L. 1986. Organic geochemical facies of the Cretaceous Bulldog Shales, Western Eromanga Basin, South Australia. *In*: GRAVESTOCK, D. I., MOORE, P. S. & PITT, G. M. (eds) *Contributions to the Geology and Hydrocarbon Potential of the Eromanga Basin*, pp. 287–304. Geological Society of Australia, Special Publication No. 12.

MAGNIER, P. L., OKI, T. & KARTAAIPOETRA, L. 1975. The Mahakam Delta, Kalimantan, Indonesia. *Proceedings of the 9th World Petroleum Congress*, **2**, 239–50.

MASRAN, TH. C. & POCOCK, S. A. 1981. The classification of plant-derived particulate organic matter. *In*: BROOKS, J. (ed.) *Proceedings of the 5th International Palynological Conference*, pp. 145–75. Academic Press, London.

MOLDOWAN, J. M. 1984. C_{30}-steranes, novel markers for marine petroleums and sedimentary rocks. *Geochimica et Cosmochimica Acta*, **48**, 2767–8.

—— & SEIFERT, W. K. 1979. Head-to-head linked isoprenoid hydrocarbons in petroleum. *Science*, **204**, 169–71.

—— & —— 1980. First discovery of botryococcane in petroleum. *Journal of the Chemical Society Chemical Communications*, 912–14.

——, ——, ARNOLD, E. & CLARDY, J. 1984. Structure, proof and significance of stereoisomeric 28,30-bisnorphanes in petroleum and petroleum source rocks. *Geochimica et Cosmochimica Acta*, **48**, 1651–61.

——, —— & GALLEGOS, E. J. 1985. Relationship between petroleum composition and depositional environment of petroleum source rocks. *Bulletin of the American Association of Petroleum Geologists*, **69**, 1255–68.

MONNIER, F., POWELL, T. G. & SNOWDON, L. R. 1983. Qualitiative and quantitative aspects of gas generation during maturation of sedimentary organic matter. Examples from Canadian Frontier Basins. *In*: BJORØY, M. *et al.* (eds) *Advances in Organic Geochemistry 1981*, pp. 487–95. Wiley, Chichester.

MOORE, P. S. 1986. An exploration overview of the Eromanga Basin. *In*: GRAVESTOCK, D. I., MOORE, P. S. & PITT, G. M. (eds) *Contributions to the Geology and Hydrocarbon Potential of the Eromanga Basin*, pp. 1–8. Geological Society of Australia, Special Publication No. 12.

MUKHOPADHYAY, P. K., HAGEMANN, H. W. & GORMLY, J. R. 1985. Characterisation of kerogens as seen under the aspect of maturation and hydrocarbon generation. *Erdöl Kohle*, **38**, 7–18.

MURCHISON, D. G., COOK, A. C. & RAYMOND, A. C. 1985. Optical properties of organic matter in relation to thermal gradients and structural deformation. *In:* EGLINTON, G., CURTIS, C. D., MCKENZIE, D. P. & MURCHISON, D. G. (eds) *Geochemistry of Buried Sediments*, pp. 157–86. Royal Society of London.

NEAVAL, R. C. 1981. Origin, petrography and classification of coal. *In:* ELLIOTT, M. A. (ed.) *Chemistry of Coal Utilization.* 2nd supplementary volume, pp. 91–158. Wiley, New York.

NICHOLAS, E., LOCKWOOD, K. L., MARTIN, A. R. & JACKSON, K. S. 1981. Petroleum potential of the Bass Basin. *Bureau of Mineral Resources Journal of Australian Geology and Geophysics*, **6**, 199–212.

NOBLE, R. A., ALEXANDER, R., KAGI, R. I. & KNOX, J. 1986. Identification of some diterpenoid hydrocarbons in petroleum. *Organic Geochemistry*, **10**, 825–9.

OTTENJANN, K. 1980. Spektrale Fluoreszenz-Mikrophotometrie von Kohlen und Ölschiefern. *Leitz Mitteilungen für Wissenschaft und Technik*, **7**, 262–72.

—— 1982. Verbesserungen bei der mikroskopphotometrischen Fluoreszenzmessung an Kohlenmaceralan. *Zeiss Information*, **26**, 40–6.

—— 1985. Fluoreszenzeigenschaften von Flözproben aus stratigraphisch jungen Schichten des Oberkarbons. *Fortschritte in der Geologie von Rheinland und Westfalen*, **33**, 169–95.

——, TEICHMÜLLER, M. & WOLF, M. 1974. Spektrale Fluoreszenzmessungen an Sporiniten mit Auflicht-Anregung, eine mikroskopische Methode zur Bestimmung des Inkohlungsgrades gering inkohlter Kohlen. *In:* WIEGEL, E. (ed.) *Inkohlung und Erdöl*, pp. 1–35. Fortshcritte in der Geologie von Rheinland und Westfalen, **24**. Geologisches Landesamt Nordrhein Westfalen, Krefeld.

——, —— & —— 1975. Spectral fluorescence measurements of sporinites in reflected light and their applicability for coalification studies. *In:* ALPERN, B. (ed.) *Pétrographie de la matière organique des sédiments, relations avec la paléotempérature et le potentiel pétrolier*, pp. 67–91. Centre National de la Recherche Scientifique, Paris.

——, WOLF, M. & WOLFF-FISCHER, E. 1982. Das Fluoreszenzverhalten der Vitrinite zur Kennzeichnung der Kokungseigenschaften von Steinkohlen. *Glückauf-Forschungshefte*, **43**, 173–9.

OUDIN, J. L. & PICARD, P. F. 1982. Genesis of hydrocarbons in the Mahakam Delta and the relationship between their distribution and the overpressured zones. *Proceedings of the Indonesia Petroleum Association XIth Annual Convention*, 1982, pp. 181–202.

PALMER, S. E. 1984. Effect of water washing on C_{15+} hydrocarbon fraction of crude oils from Northwest Palawan, Philippines. *Bulletin of the American Association of Petroleum Geologists*, **68**, 137–49.

PASSMORE, V. L. & BOREHAM, C. J. 1986. Source rock evaluation and maturation history of the central Eromanga Basin. *In:* GRAVESTOCK, D. I., MOORE, P. S. & PITT, G. M. (eds) *Contributions to the Geology and Hydrocarbon Potential of the Eromanga Basin*, pp. 221–40. Geological Society of Australia, Special Publication No. 12.

PATON, I. M. 1986. The Birkhead Formation: a Jurassic petroleum reservoir. *In:* GRAVESTOCK, D. I., MOORE, P. S. & PITT, G. M. (eds) *Contributions to the Geology and Hydrocarbon Potential of the Eromanga Basin*, pp. 195–201. Geological Society of Australia, Special Publication No. 12.

PHILP, R. P. 1985a. Biological markers in fossil fuel production. *Mass Spectrometry Reviews*, **4**, 1–54.

—— 1985b. *Fossil Fuel Biomarkers.* Elsevier, Amsterdam.

—— & GILBERT, T. D. 1986. Biomarker distributions in Australian oils predominantly derived from terrigenous source material. *Organic Geochemistry*, **10**, 73–84.

——, —— & FRIEDRICH, J. 1981. Bicyclic sesquiterpenoids and diterpenoids in Australian crude oils. *Geochimica et Cosmochimica Acta*, **45**, 1173–80.

——, SIMONEIT, B. R. T. & GILBERT, T. D. 1983. Diterpenoids in crude oils and coals of south-eastern Australia. *In:* BJORØY, M. *et al.* (eds) *Advances in Organic Geochemistry 1981*, pp. 698–704. Wiley, Chichester.

PITTION, J. L. & GOUADAIN, J. 1985. Maturity studies of the Jurassic 'Coal Unit' in three wells from the Haltenbanken area. *In:* THOMAS, B. M. *et al.* (eds) *Petroleum Geochemistry of the Norwegian Shelf*, pp. 205–11. Graham & Trotman, London.

POWELL, T. G. 1978. An assessment of the hydrocarbon source rock potential of the Canadian Arctic Islands. *Geological Survey of Canada, Paper 78–12.*

—— 1982. Petroleum geochemistry of the Verrill Canyon Formation: a source for Scotian Shelf hydrocarbons. *Bulletin of Canadian Petroleum Geology*, **30**, 167–9.

—— 1984. Developments in concepts of hydrocarbon generation from terrestrial organic matter. *Proceedings of the Beijing Petroleum Symposium*, pp. 1–36. Beijing, China.

—— 1986. Petroleum geochemistry and depositional setting of lacustrine source rocks. *Marine and Petroleum Geology*, **3**, 200–19.

——, CREANEY, S. & SNOWDON, L. R. 1982. Limitations of use of organic petrographic techniques for identification of petroleum source rocks. *Bulletin of the American Association of Petroleum Geologists*, **66**, 430–5.

——, FOSCOLOS, A. E., GUNTHER, P. R. & SNOWDON, L. R. 1978. Diagenesis of organic matter and fine clay minerals: a comparative study. *Geochimica et Cosmochimica Acta*, **42**, 1181–97.

—— & MCKIRDY, D. M. 1973. Relationship between ratio of pristane to phytane, crude oil composition and geological environments in Australia. *Nature*, **243**, 37–9.

—— & —— 1975. Geological factors controlling crude oil composition in Australia and Papua New Guinea. *Bulletin of the American Association of Petroleum Geologists*, **59**, 1176–97.

—— & —— 1976. Geochemical character of crude oils from Australia and New Guinea. *In:* LESLIE, R. B., EVANS, H. J. & KNIGHT, C. L. (eds) *Economic*

Geology of Australia and Papua New Guinea, 3. Petroleum, pp. 18–29. Australasian Institute of Mining and Metallurgy, Monograph 7.

—— & SNOWDON, L. R. 1979. Geochemistry of crude oils and condensates: offshore eastern Canada. *Bulletin of Canadian Petroleum Geology*, **27**, 453–66.

—— & —— 1980. Geochemical controls on hydrocarbon generation in Canadian sedimentary basins. *In:* MIALL, A. D. (ed.) *Facts and Principles of World Oil Occurrence*, pp. 421–46. Canadian Society of Petroleum Geologists Memoir 6, Calgary.

—— & —— 1983. A composite hydrocarbon generation model: implications for evaluation of basins for oil and gas. *Erdöl Kohle*, **36**, 163–70.

PYM, J. G., RAY, J. E., SMITH, G. W. & WHITEHEAD, E. V. 1975. Petroleum triterpane fingerprinting of crude oils. *Analytical Chemistry*, **47**, 1617–22.

RADKE, M., LEYTHAEUSER, D. & TEICHMÜLLER, M. 1984. Relationship between rank and composition of aromatic hydrocarbons for coals of different origins. *Organic Geochemistry*, **6**, 423–30.

——, SCHAEFER, R. G., LEYTHAEUSER, D. & TEICHMÜLLER, M. 1980. Composition of soluble organic matter in coals: relation to rank and liptinite fluorescence. *Geochimica et Cosmochimica Acta*, **44**, 1787–800.

——, ——, —— & —— 1985. Bildung und Umwandlung schwerflüchtiger Kohlenwasserstoffe in Kohlen des Ruhrkarbons in Abhängigkeit vom Inkohlungsgrad. *Fortschritte in der Geologie von Rheinland und Westfalen*, **33**, 197–217.

—— & WELTE, D. H. 1983. The Methylphenanthrene Index (MPI): a maturity parameter based on aromatic hydrocarbons. *In:* BJORØY, M. *et al.* (eds) *Advances in Organic Geochemistry 1981*, pp. 504–12. Wiley, Chichester.

RASHID, M. A. 1979. Pristane-phytane ratios in relation to source and diagenesis of ancient sediments from the Labrador Shelf. *Chemical Geology*, **25**, 109–22.

——, PURCELL, L. P. & HARDY, I. A. 1980. Source rock potential for oil and gas of the East Newfoundland and Labrador Shelf areas. *In:* MIALL, A. D. (ed.) *Facts and Principles of World Oil Occurrence*, pp. 589–608. Canadian Society of Petroleum Geologists Memoir 6, Calgary.

RAYNAUD, J.-F. & ROBERT, P. 1976. Les méthodes d'étude optique de la matière organique. *Bulletin du Centre de Recherches Pau-SNPA*, **10**, 109–27.

RHOADES, A. H., GRAY, R. J. & HUNTINGTON, H. D. 1981. Application of microscopic techniques in the evaluation of coal-coke and related products. *In:* HANSEN, D. M. & PARK, W. C. (eds) *Process Mineralogy*, pp. 453–78. Proceedings of a Symposium at the 110th American Institute of Mechanical Engineers Annual Meeting, Chicago, Illinois.

RICHARDSON, J. S. & MILLER, D. E. 1982. Identification of dicyclic and tricyclic hydrocarbons in the saturate fraction of a crude oil by gas chromatography-mass spectrometry. *Analytical Chemistry*, **54**, 765–8.

RIGBY, O. & SMITH, J. W. 1981. An isotopic study of gases and hydrocarbons in the Cooper Basin. *Journal of the Australian Petroleum Exploration Association*, **21**, 222–9.

ROBERT, P. 1971. Etude pétrographique des matières organiques insolubles par la mesure de leur pouvoir reflecteur. Contribution à l'exploration petrolière et à la connaissare des basins sédimentaires. *Revue Institut Français Pétrole*, **26**, 105–36.

—— 1979. Classification des matières organiques en fluoresecence: application aux roches-mères pétrolières. *Bulletin des Centres de Recherches Exploration-Production Elf-Aquitaine*, **3**, 223–63.

—— 1980. The optical evolution of kerogen and geothermal histories applied to oil and gas exploration. *In:* DURAND, B. (ed.) *Kerogen*, pp. 368–414. Éditions Technip, Paris.

—— 1981. Classification of organic matter by means of fluorescence: application to hydrocarbon source rocks. *International Journal of Coal Geology*, **1**, 101–37.

ROGERS, M. A. 1979. Application of organic facies concept to hydrocarbon source evaluation. *In: Exploration, Supply and Demand*, **2**, pp. 23–30. Proceedings of the 10th World Petroleum Congress, Bucharest.

RULLKÖTTER, J., LEYTHAEUSER, D. & WENDISCH, D. 1982. Novel 23,28 bisnorlupanes in Tertiary sediments. Widespread occurrence of nuclear demethylated triterpanes. *Geochimica et Cosmochimica Acta*, **46**, 2501–9.

SCHMITTER, J. M., ARPINO, P. J. & GUIOCHON, G. 1981. Isolation of degraded pentacyclic triterpenoid acids in a Nigerian crude oil and their identification as tetracyclic carboxylic acids resulting from ring A_1 cleavage. *Geochimica et Cosmochimica Acta*, **45**, 1951–6.

SCHOELL, M., TESCHNER, M., WEHNER, H., DURAND, B. & OUDIN, J. L. 1983. Maturity related biomarker and stable isotope variations and their application to oil/source rock correlation in the Mahakam Delta, Kalimantan. *In:* BJORØY, M. *et al.* (eds) *Advances in Organic Geochemistry 1981*, pp. 156–63. Wiley, Chichester.

SCHWEBEL, D. A., DEVINE, S. B. & RILEY, M. 1980. Source, maturity and gas composition relationships in the southern Cooper Basin. *Journal of the Australian Petroleum Exploration Association*, **20**, 191–200.

SEIFERT, W. K., MOLDOWAN, J. M., SMITH, G. W. & WHITEHEAD, E. V. 1978. First proof of structure of a C-28 pentacyclic triterpane in petroleum. *Nature*, **271**, 436–7.

SENFTLE, J. T. 1983. Optical analysis of kerogen—its rôle in an integrated approach for kerogen typing. Paper and Abstract presented at the *North American Coal Petrographical Meeting*, Merrilville, Indiana.

——, BROWN, J. H. & LARTER, S. R. 1987. Refinement of organic petrographic methods for kerogen characterisation. *International Journal of Coal Geology*, **7**, 105–17.

SHANMUGAM, G. 1985. Significance of coniferous rain forests and related organic matter in generating commercial quantities of oil, Gippsland Basin,

Australia. *Bulletin of the American Association of Petroleum Geologists*, **69**, 1241–54.

SHERWOOD, N. R. 1984. Low rank oil shales: Part I—organic petrology. *Final Project Report to the Department of National Development and Energy on NERD & D Program Project 79/9014*. The University of Wollongong Printery, Keiraville, NSW 2500.

—— & COOK, A. C. 1986. Organic matter in the Toolebuc Formation. *In:* GRAVESTOCK, D. I., MOORE, P. S. & PITT, G. M. (eds) *Contribution to the Geology and Hydrocarbon Potential of the Eromanga Basin*, pp. 255–65. Geological Society of Australia, Special Publication No. 12.

SHIBAOKA, M. SAXBY, J. D. & TAYLOR, G. H. 1978. Hydrocarbon generation in Gippsland Basin, Australia—comparison with Cooper Basin, Australia. *Bulletin of the American Association of Petroleum Geologists*, **62**, 1151–8.

SMITH, G. C. 1982. A review of the Tertiary–Cretaceous tectonic history of the Gippsland Basin and its control on coal measure sedimentation. *Australian Coal Geology*, **4**, 1–38.

—— & COOK, A. C. 1980. Coalification paths of exinite, vitrinite and inertinite. *Fuel*, **59**, 641–6.

—— & —— 1984. Petroleum occurrence in the Gippsland Basin and its relationship to rank and organic matter type. *Journal of the Australian Petroleum Exploration Association*, **24**, 196–216.

SMITH, P. M. R. 1983. Spectral correlation of spore coloration standards. *In*: BROOKS, J. (ed.) *Petroleum Geochemistry and Exploration of Europe*, pp. 289–94. Geological Society Special Publication No. 12. Blackwell Scientific Publications, Oxford.

SMYTH, M. 1979. Hydrocarbon generation in the Fly Lake–Brolga area of the Cooper Basin. *Journal of the Australian Petroleum Exploration Association*, **19**, 108–14.

—— 1983. Nature of source material for hydrocarbons in Cooper Basin, Australia. *Bulletin of the American Association of Petroleum Geologists*, **67**, 1422–8.

——, COOK, A. C. & PHILP, R. P. 1984. Birkhead revisited: petrological and geochemical studies of the Birkhead Formation, Eromanga Basin. *Journal of the Australian Petroleum Exploration Association*, **24**, 230–42.

SNOWDON, L. R. 1980. Resinite—a potential petroleum source in the Upper Cretaceous/Tertiary of the Beaufort Mackenzie basin. *In*: MIALL, A. D. (ed.) *Facts and Principles of World Oil Occurrence*, pp. 509–21. Canadian Society of Petroleum Geologists Memoir 6, Calgary.

—— & POWELL, T. G. 1982. Immature oil and condensate—modification of hydrocarbon generation model for terrestrial organic matter. *Bulletin of the American Association of Petroleum Geologists*, **66**, 775–88.

SPIRO, B. & MUKHOPADHYAY, P. K. 1983. Effects of degradation on spectral fluorescence of alginite. *Erdöl Kohle*, **36**, 297–9.

STAINFORTH, J. G. 1984. Gippsland hydrocarbons—a perspective from the basin edge. *Journal of the Australian Petroleum Exploration Association*, **24**, 91–100.

STAPLIN, F. L. 1969. Sedimentary organic matter, organic metamorphism, and oil and gas occurrence. *Bulletin of the Canadian Petroleum Geologists*, **17**, 47–66.

STELLER, M. 1981. Untersuchungen an Hydrierrückständen einer Gasflammkohle. *Dissertation Technische Hochschule*, Aachen.

STOPES, M. C. 1919. On the four visible ingredients in banded bituminous coal. *Proceedings of the Royal Society of London B*, **90**, 470–87.

STOUT, S. A. & BENSLEY, D. F. 1987. Fluorescing macerals from wood precursors. *International Journal of Coal Geology*, **7**, 119–33.

TAYLOR, G. H., MACKOWSKY, M.-TH. & ALPERN, B. 1967. The behaviour of inertinite during carbonization. *Fuel*, **46**, 431–40.

TEICHMÜLLER, M. 1974a. Enstehung und Veränderung bituminöser Substanzen in Kohlen in Beziehung zur Enstehung und Umwandlung des Erdöls. *In*: WIEGEL, E (ed.) *Inkohlung und Erdöl*, pp. 65–112. Fortschritte in der Geologie von Rheinland und Westfalen, **24**, Geologisches Landesamt Nordrhein Westfalen, Krefeld.

—— 1974b. Generation of petroleum-like substances in coal seams as seen under the microscope. *In*: TISSOT, B. & BIENNER, F. (eds) *Advances in Organic Geochemistry 1973*, pp. 379-408. Éditions Technip, Paris.

—— 1974c. Über neue Macerale der Liptinit-Gruppe und die Enstehung von Micrinit. *In*: WIEGEL, E. (ed.) *Inkohlung und Erdöl*, pp. 37–64. Fortschritte in der Geologie von Rheinland und Westfalen, **24**, Geologisches Landesamt Nordrhein Westfalen, Krefeld.

—— 1977. Inkohlungsuntersuchungen in der Bohrung Uedem 1. Archive, *Geologisches Landesamt Nordrhein-Westfalen*, Krefeld (unpublished).

—— 1979a. Die Liptinit-Diagenese in der Bohrung Kohlhaus 1 und ihre Beziehungen zur Bitumen-Neubildung und Verkokbarkeit, *Geologisches Landesamt Nordrhein-Westfalen*, Krefeld (unpublished).

—— 1979b. Chemische und mikroskopische Untersuchungen zur Liptinit-Diagenese an Mattkohlen-Proben aus der Bohrung Uedem 1. *Geologisches Landesamt Nordrhein-Westfalen*, Krefeld (unpublished).

—— 1982. Fluoreszenz von Liptiniten und Vitriniten in Beziehung zu Inkohlungsgrad und Verkokungsverhalten. *Geologisches Landesamt Nordrhein-Westfalen*, Krefeld.

—— & DURAND, B. 1983. Fluorescence microscopical rank studies on liptinites and vitrinites in peat and coals and comparison with results of the Rock-Eval pyrolysis. *International Journal of Coal Geology*, **2**, 197–230.

—— & OTTENJANN, K. 1977. Liptinite und lipoide Stoffe in einem Erdölmuttergestein. *Erdöl Kohle*, **30**, 387–98.

—— & TEICHMÜLLER, R. 1981. The significance of coalification studies to geology. *Bulletin des Centres de Recherches Exploration—Production Elf-Aquitaine*, **5**, 491–534.

—— & WOLF, M. 1977. Application of fluorescence microscopy in coal petrology and oil exploration. *Journal of Microscopy*, **109**, 49–74.

THIESSEN, R. 1920. Structure in Paleozoic bituminous coals. *United States Bureau of Mines Bulletin*, **117**.

THOMAS, B. M. 1982. Land-plant source rocks for oil and their significance in Australian basins. *Journal of the Australian Petroleum Exploration Association*, **22**, 164–78.

——, OSBORNE, D. G. & WRIGHT, A. J. 1982. Hydrocarbon habitat of the Surat/Bowen Basin. *Journal of the Australian Petroleum Exploration Association*, **22**, 213–26.

THOMPSON, C. L. & DEMBICKI, H. JR 1986. Optical characteristics of amorphous kerogens and the hydrocarbon-generating potential of source rocks. *International Journal of Coal Geology*, **6**, 229–49.

THOMPSON, S., COOPER, B. S., MORLEY, R. J. & BARNARD, P. C. 1985. Oil-generating coals. *In*: THOMAS, B. M. *et al.* (eds) *Petroleum Geochemistry of the Norwegian Shelf*, pp. 59–73. Graham & Trotman, London.

THOMPSON-RIZER, C. L. & WOODS, R. A. 1987. Microspectrofluorescence measurements of coals and petroleum source rocks. *International Journal of Coal Geology*, **7**, 85–104.

THRELFALL, W. F., BROWN, B. R. & GRIFFITH, B. R. 1976. Gippsland Basin, offshore. *In*: LESLIE, R. B., EVANS, H. G. & KNIGHT, C. L. (eds) *Economic Geology of Australia and Papua New Guinea, 3. Petroleum*, pp. 41–67. Australian Institute of Mining and Metallurgy, Monograph 7, Melbourne.

TIAN, Z., CHANG, C., HUANG, D. & WU, C. 1983. Sedimentary facies, oil generation in Meso–Cenozoic continental basins in China. *Oil and Gas Journal*, **81**, 120–6.

TIES, P., SHAW, R. D. & GEARY, G. C. 1985. The petroleum prospectivity of the Clarence-Moreton Basin in New South Wales. *Journal of the Australian Petroleum Exploration Association*, **25**, 15–33.

TISSOT, B. P. 1984. Recent advances in petroleum geochemistry applied to hydrocarbon exploration. *Bulletin of the American Association of Petroleum Geologists*, **68**, 545–63.

——, DURAND, B., ESPITALIÉ, J. & COMBAZ, A. 1974. Influence of the nature and diagenesis of organic matter in the formation of petroleum. *Bulletin of the American Association of Petroleum Geologists*, **58**, 499–506.

TREIBS, A. 1934. Chlorophyll-und Häminderivate in bituminösen Gesteinen, Erdölen, Erdwachsen und Asphalten. *Annalen der Chemie*, **510**, 42–62.

VASSOEVICH, N. B., AKRAMKHODZHAEV, A. M. & GEODEKYAN, A. A. 1974. Principal zone of oil formation. *In*: TISSOT, B. & BIENNER, F. (eds) *Advances in Organic Geochemistry, 1973*, pp. 309–14. Éditions Technip, Paris.

VASTOLA, F. J., GIVEN, P. H., DUTCHER, R. R. & PIRONE, A. J. 1970. A laser-micropyrolysis system for the study of organic sediments and inclusions. *In*: HOBSON, G. D. & SPEERS, G. C. (eds) *Advances in Organic Geochemistry, 1966*, p. 427 (abstract). Pergamon Press, Oxford.

VINCENT, P. W., MORTIMORE, I. R. & MCKIRDY, D. M. 1985. Hydrocarbon generation, migration and entrapment in the Jackson-Naccowlah area ATP259P, south-western Queensland. *Journal of the Australian Petroleum Exploration Association*, **25**, 62–84.

VOLKMAN, J. K. 1986. A review of sterol markers for marine and terrigenous organic matter. *Organic Geochemistry*, **9**, 83–99.

WARBURTON, G. A. & ZUMBERGE, J. E. 1983. Determination of petroleum sterane distributions by mass spectrometry with selective metastable ion monitoring. *Analytical Chemistry*, **55**, 123–6.

WELTE, D. H., SCHAEFER, R. G., RADKE, M. & WEISS, H. M. 1982. *Association of American Petroleum Geologists: abstracts*, p. 123.

WHITEHEAD, E. V. 1974. The structure of petroleum pentacyclanes. *In*: TISSOT, B. & BIENNER, F. (eds) *Advances in Organic Geochemistry 1973*, pp. 225–43. Éditions Technip, Paris.

YANG, W., LI, Y. & GAO, R. 1985. Formation and evolution of non-marine petroleum in Songliao Basin, China. *Bulletin of the American Association of Petroleum Geologists*, **69**, 1112–22.

ZHAI, G., WANG, S. & LI, G. 1984. Characteristics of sedimentary basins in China and its oil and gas distribution. *Proceedings of the Beijing Petroleum Symposium*, Beijing, China.

D. G. MURCHISON, Organic Geochemistry Unit, Department of Geology, University of Newcastle, Newcastle upon Tyne NE1 7RV, England.

Oil-prone coals of the Walloon Coal Measures, Surat Basin, Australia

Ganjavar Khavari Khorasani

ABSTRACT: Studies of autochthonous coals from the Walloon Coal Measures, Surat Basin, Australia, suggest that the process of liptinite enrichment and the formation of perhydrous vitrinites have given rise to oil-prone coals. These processes were essentially controlled by the type of flora contributing to the peat accumulation and the biochemical transformation of the flora under dysaerobic conditions. These processes lead to the coals being hydrogen-rich and oil prone, and contrast with the process described by other authors for Indonesian source rocks. The mature phase of oil generation from hydrogen-rich coals such as those of the Walloon Coal Measures may be preceded by several earlier phases, during which hydrocarbons are likely to be generated from subereous components and terpene resinites in low rank lignites and sub-bituminous coals.

Keywords: liptinite, vitrinite, oil generation, hydrocarbon

The possibility that coals or non-marine coaly sediments may have substantial potential to generate oil has been discussed and/or evaluated in recent years (Roe & Polito 1970, Thomas 1982, Durand & Paratte 1983, Thompson *et al.* 1985). A process of separation of water-soluble and insoluble organic matter in tidal dominated, coastal plain environments, resulting in the accumulation of hydrogen-rich, drifted coals and carbonaceous shales in SE Asia has recently been proposed and discussed by Thompson *et al.* (1985). These authors have also suggested that unusually high rates of water flow may flush out humic acid gels from *in situ* peats of the upper coastal plain deposits, leading to enrichment of the peats in liptinitic kerogen. However, this mechanism does not explain the formation of many other oil-prone autochthonous coals.

The Middle Jurrasic Walloon Coal Measures of Surat Basin, Australia (Fig. 1), present an example of an autochthonous deposit, in which the nature of the coal-forming flora and the accumulation of peat under dysaerobic conditions have resulted in the development of unusually hydrogen-rich, oil-prone coals. In this paper the coal facies and its decisive influence on the oil-generating potential of the Walloon deposit is discussed. The effect of petrographic composition on the yield of hydrocarbons already generated from the coals is demonstrated.

Results and discussion

Coal facies

The coal seams of the Walloon Coal Measures are autochthonous in origin. The predominant microlithotype comprises liptinite-rich clarite. Vitrite is much less dominant and inertite is virtually absent. The inertinite content does not exceed 4–5% (by vol.), when it essentially comprises semifusinite and micrinite. Liptinite amounts to 14–30% of the coal composition. The primary macerals lipid and terpene resinites, cutinite, suberinite and liptodetrinite and the secondary, neo-formed exudatinite are the normal liptinite macerals of the Walloon Coal Measures (Plate 1).

Vitrinite, the major coal maceral, is perhydrous and for the most part closely associated with subereous components. The perhydrous nature of the Walloon vitrinites is substantiated by their atomic H/C ratios which are consistently higher than the ratios expected for normal vitrinites of similar rank (Tables 1 and 2). This is also confirmed by X-ray analysis of these vitrinites. The X-ray diffraction pattern of the Walloon vitrinites display a predominant diffuse band of alicyclic nature and an additional 002 carbon reflection (Fig. 2A). The latter, which is not resolved from the diffuse band that appears at lower angles, arises from the packing of aromatic layers. In normal vitrinites of similar rank, the band of alicyclic nature is much less predominant and the 002 carbon reflection displays a higher degree of resolution (Fig. 2B) compared to those of the Walloon-type vitrinites. These differences imply that the latter contains larger amounts of saturated structures than the former, i.e. the Walloon vitrinites are perhydrous.

Compared with classical coal-facies types, described by Teichmüller & Teichmüller (1968), these coals are comparable to a mixed facies type formed from forest and herbaceous swamp environments. The unusually low inertinite con-

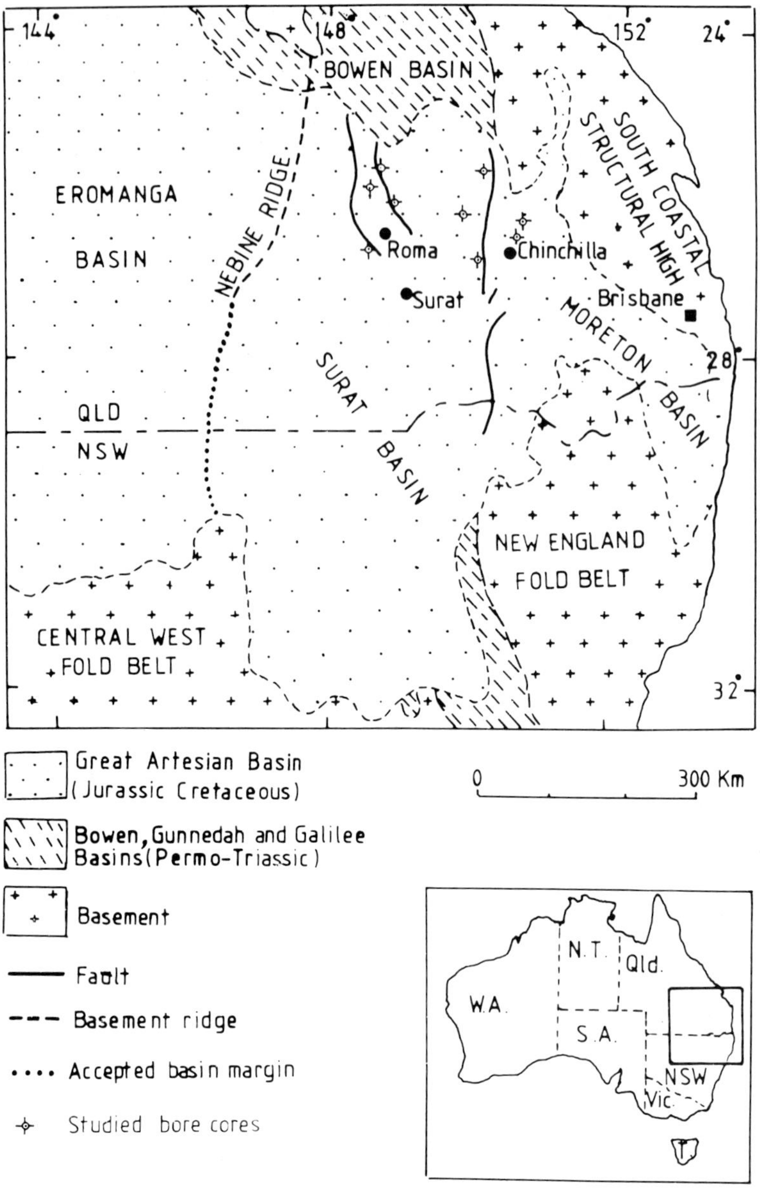

FIG. 1. Locality map of Surat Basin, southeastern Queensland and northeastern New South Wales, Australia.

tent of the coals and the fact that vitrinites are perhydrous strongly suggest that the original peats were formed under dysaerobic conditions, where mouldering and/or dehydration-oxidation of humic matter was not favoured. The pristane/phytane ratios of these coals (Pr/Ph: 3.2 to 4.6) are much lower than the values expected for a typical coal swamp environment (Pr/Ph: 7 to 10), which also suggests a dysaerobic condition during the peat accumulation.

The bulk of the coal appears to be derived from conifers with a significant contribution from ferns and cycadeoids, while the nature and structure of floral elements are suggestive of a moist-temperate climate (Gould 1980).

Oil-generating potential

The hydrogen indices of the Walloon coals are much higher than those expected for the classical

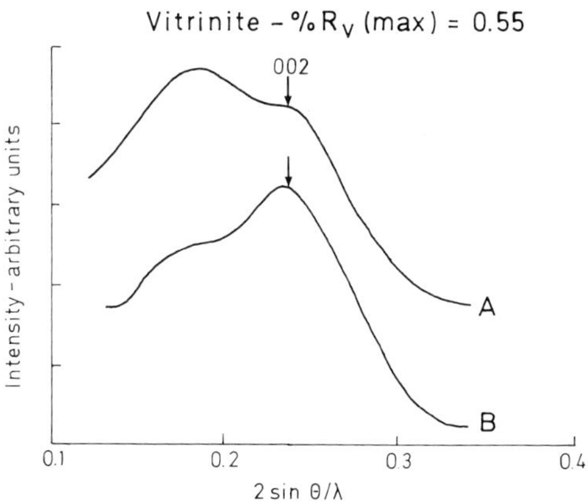

FIG. 2. Typical low-angle scattering intensities of vitrinites from Walloon Coal Measures. 2Θ = angle of scattering and λ = the X-ray wavelength (Cu-Kα radiation). The intensities are corrected for PLG factors.

humic coals or woody-herbaceous type III kerogen (Fig. 3). Some of these indices are even higher than those reported for the oil-productive coaly source rocks of Indonesia in the same range of maturity (Thompson *et al.* 1985). Coals with very high hydrogen indices have also been recorded in the Eocene of Kyushui (Japan), the Oligocene of Jiului Paroseni (Rumania) and the Carboniferous of Illinois (USA) (Teichmüller & Durand 1983).

Expulsion of liquid hydrocarbons from the coals, when samples are observed in the fluorescence mode (UV excitation), is inevitable (Plate 1e). Results of Rock-Eval analysis, S_1/Org. C index, which is indicative of the amount of hydrocarbons already generated in the coals, is shown in Fig. 4. In view of the low maturity of the Walloon coals, discussed below, these S_1/Org. C values, which are verified to be due to indigenous hydrocarbons, are unusually high.

TABLE 1. *Variations in atomic H/C ratios of sub-bituminous coals for ortho- and perhydrous vitrinites*

Origin of vitrinite	Atomic H/C	Type of vitrinite
Vitrinites from classical humic coals*	0.80–1.00	Ortho-hydrous
Vitrinites from Walloon Coal Measures†	1.10–1.28	Per-hydrous

* From Van Krevelen (1961). † From present study.

The vitrinite reflectance range of the unit varies from anything between R_v(max) = 0.50% at the top of the unit to a maximum of R_v(max) = 0.60% at the bottom of the unit, which is consistent with sub-bituminous coal rank. The saturated fractions of the coal extracts display a very strong odd predominance in the C_{23}–C_{33} range (Fig. 5 and Table 2). This odd predominance, which is inherited from long chain waxes synthesized by the higher plant precursors, suggests that the coals are immature. This immaturity is further confirmed by the extent of isomerization at C-14, C-17 and C-20 in steranes (Fig. 6 and Table 2). The geologically stable $\alpha\alpha\alpha$-sterane with 20S stereochemistry (the first eluting epimer) is much less abundant than the $\alpha\alpha\alpha$-sterane with 20R stereochemistry (the second eluting epimer) inherited from the biological precursors. Further, the content of stable $\alpha\beta\beta$-steranes relative to the $\alpha\alpha\alpha$-sterane with 20R stereochemistry is low, again an indication that the coals are thermally immature (Mackenzie *et al.* 1980).

Having established that the Walloon coals are immature by geochemical means, the significant amounts of generated hydrocarbons from the coals and the extensive variations observed in the

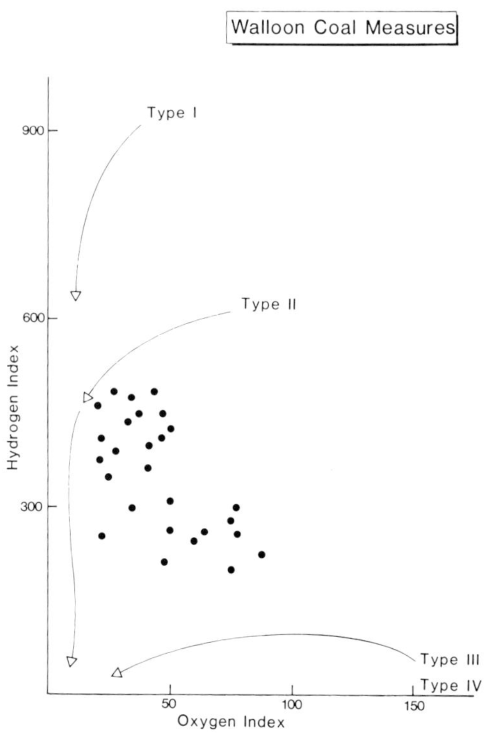

FIG. 3. Kerogen typing of the Walloon Coals, based on Rock-eval analysis.

TABLE 2. *Presentation of part of data base for samples varying in reflectance and/or petrographic composition*

Sample code	R_v (max)	CPI_{23-33}	C_{29} steranes $\alpha\beta\beta/(\alpha\alpha\alpha+\alpha\beta\beta)$	C_{29} steranes 20S/(20S+20R)	TOC%	S_1 (mg/gr)	HI	OI	H/C (atomic)	total liptinite (vol)%	S + T res. / tot. lipt. x 100 ***
S-01	0.50	1.91	0.28	0.15	57	3.74	200	75	1.22	14	5
S-02	0.50	1.88	0.26	0.12	65	52.91	350	25	1.26	27	46
S-03	0.52	1.92	n.d	n.d	72	2.90	245	61	1.13	16	10
S-04	0.53	1.86	0.29	0.14	66	4.73	387	58	n.d	18	0
S-05	0.53	1.93	n.d	n.d	62	56.10	425	52	n.d	26	62
S-06	0.53	n.d	n.d	n.d	58	35.16	375	22	n.d	16	37
S-07	0.53	1.89	0.27	0.15	67	12.14	422	54	n.d	28	10
S-08	0.55	n.d	n.d	n.d	59	13.22	300	35	1.24	22	9
S-09	0.55	1.89	n.d	n.d	62	13.14	365	41	1.19	25	19
S-10	0.55	n.d	n.d	n.d	69	38.05	315	50	1.15	22	27
S-11	0.55	1.90	0.29	0.14	63	26.72	405	47	1.22	28	21
S-12	0.55	1.88	0.26	0.15	55	33.50	400	41	1.18	27	19
S-13	0.55	n.d	n.d	n.d	72	10.30	490	45	n.d	30	14
S-14	0.55	n.d	n.d	n.d	65	42.92	390	29	1.20	29	40
S-15	0.55	1.92	0.26	0.12	58	7.26	435	52	n.d	28	0
S-16	0.55	n.d	n.d	n.d	70	54.19	300	78	1.20	18	62
S-17	0.55	n.d	n.d	n.d	66	75.05	450	38	1.28	30	90
S-18	0.56	n.d	n.d	n.d	62	5.73	225	88	1.12	17	11
S-19	0.56	n.d	0.27	0.14	59	75.10	410	23	n.d	28	84
S-20	0.56	1.87	0.28	0.14	67	3.10	260	67	1.14	14	3
S-21	0.56	1.91	0.29	0.15	53	4.79	225	78	n.d	14	0
S-22	0.58	1.87	n.d	n.d	n.d	n.d	n.d	n.d	1.16	21	13
S-23	0.58	1.91	n.d	n.d	n.d	n.d	n.d	n.d	1.21	17	25
S-24	0.58	1.85	0.26	0.14	62	7.38	215	48	1.13	15	9
S-25	0.58	1.87	n.d	n.d	75	6.28	247	61	1.16	Vt.con.	0
S-26	0.58	1.90	0.27	0.15	58	67.76	485	28	1.28	29	73
S-27	0.54	1.87	0.29	0.13	62	37.36	205	68	1.14	15	54
S-28	0.54	1.91	n.d	n.d	53	4.19	450	48	1.19	18	0
S-29	0.54	1.89	0.26	0.13	66	3.32	251	23	n.d	14	3
S-30**	0.56	1.93	n.d	n.d	71	2.73	312	51	n.d	22	7
S-31**	0.56	1.88	0.30	0.14	64	21.46	245	61	1.14	15	29
S-32**	0.56	1.90	n.d	n.d	55	12.86	475	34	1.23	26	13

* Inertinite content of all samples < 5%
** Composite samples
*** (Suberinite + terpene resinite / total liptinite) x 100
Vt.con. = Vitrinite concentrate
n.d = Not determined

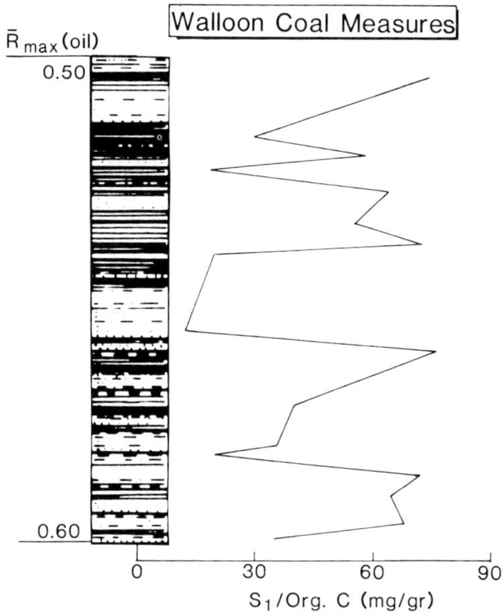

FIG. 4. A common example of the results of Rock-Eval analysis, S_1/Org. C, for sequences of the Walloon coal Measures.

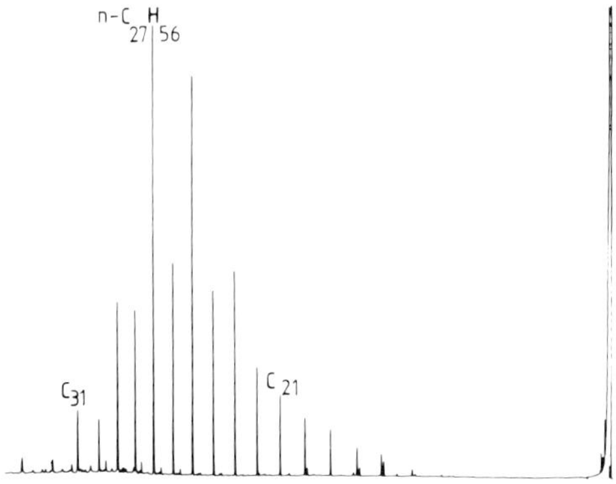

FIG. 5. The distribution of normal paraffins in coal extracts of the Walloon Coal Measures.

quantities of free hydrocarbons present in coals of similar rank (Fig. 4) require an explanation. In earlier studies, unusually high extract yields found in law rank lignites (Durand *et al.* 1975) and large values of S_1/Org. C found in some of the immature Haltenbanken coals (Pittion & Gouadain 1985), have been regarded as abnormal occurrences, for which no satisfactory explanation could be given. In the present investigation it was found that high extract levels and large amounts of indigenous hydrocarbons generated from immature coals, or coaly source rocks, have a positive relationship with alteration of certain liptinite macerals at fairly low maturation levels (Figs 7 and 8 and Table 2).

Subereous components, which are common waxy constituents of both forest and herbaceous plant communities and which are usual in many Jurassic, Cretaceous and Tertiary coals, are

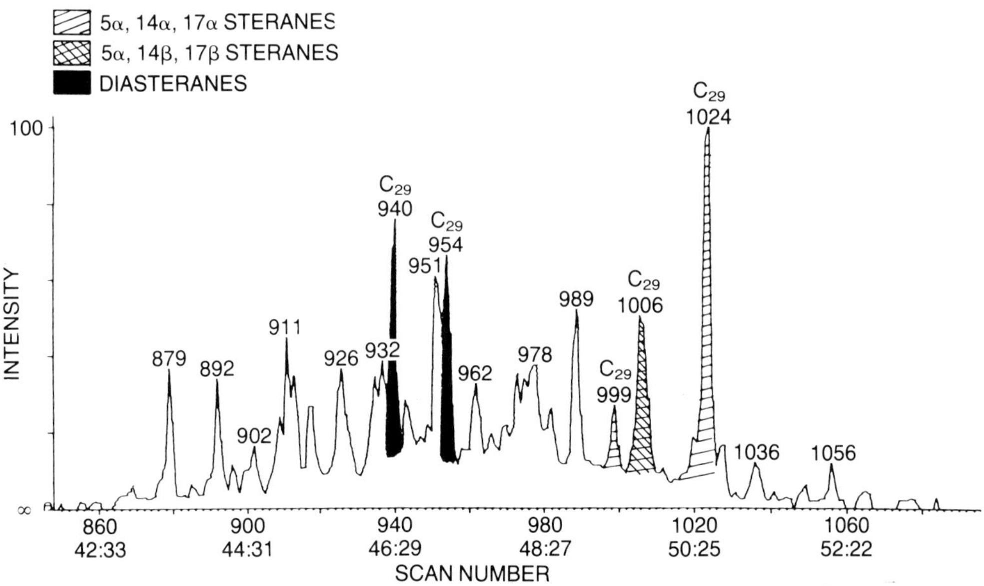

FIG. 6. Mass fragmentogram for *m*/*z* 217, characteristic for steranes, from GC–MS analyses of the Walloon coals.

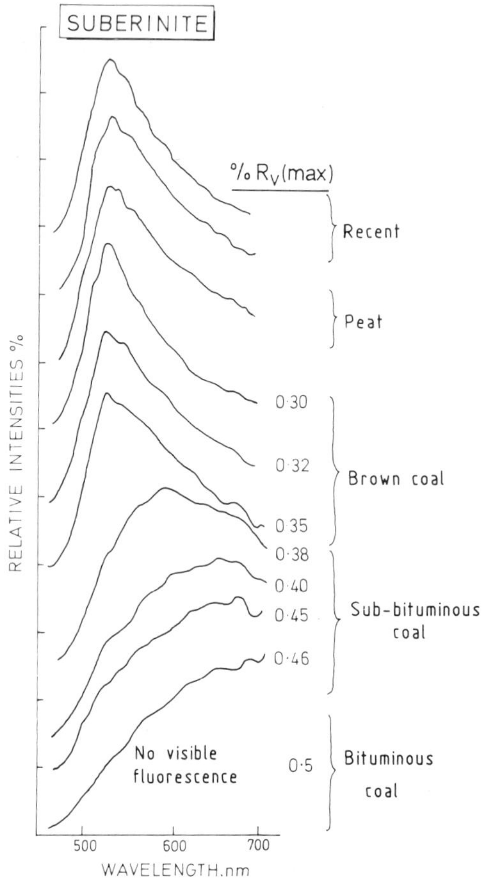

FIG. 7. Fluorescence spectral response of subereous components with increasing rank in brown coal and in sub-bituminous stages of coalification.

common in coals of the Walloon Coal Measures. Suberin has been considered to be a polymer containing polyesters and aromatics (Kolattukudy 1980). Recent suberins display spectral fluorescence maxima corresponding to blue-green or short-wavelength green regions of the spectrum (Fig. 7). This is a distinguishing feature of organic mixtures in which a) the concentration of saturated structures is high and b) the aromatic molecules are essentially non-condensed (Khavari Khorasani 1987).

Fluorescence spectral variations of numerous samples, some examples of which are shown in Fig. 7, show that up to the brown coal stage of coalification suberinite remains virtually unchanged or alters very little (Fig. 7). But in the reflectance range of R_v(max)=0.35–0.45% it appears to undergo severe alterations to generate substantial amounts of highly waxy hydrocarbons in low rank lignites and sub-bituminous coals (Fig. 7). The extreme red shift of the fluorescence spectra of suberinites, which occurs in early stages of coalification (Fig. 7), indicates an increasing aromatic character and a marked reduction in concentration of saturated structures in the mass composition of suberinite (Khavari Khorasani 1987). As early as 1948/49 Szádeczky-Kardoss detected that suberinite passes through a 'coalification jump' in low-rank, sub-bituminous/high volatile bituminous coals, which is consistent with the early changes in chemical character of this maceral.

Early thermal decomposition of subereous components is followed by considerable change in certain resinites (R_v(max)=0.45–0.60%). According to Powell *et al.* (1978) resinous components yield especially high amounts of hydrocarbons at relatively low reflectance levels: R_v(max)=0.50–0.60%. The results of the present study indicate that only terpene resinites are capable of generating hydrocarbons in thermally immature coals. Generation from lipid resinite-fluorinites will not occur until they are subjected to much higher activation energies. In the immature coals of the Walloon Coal Measures severe thermal transformation of other liptinite macerals and vitrinites is not evident, suggesting that large S_1/Org. C indices are essentially the result of thermal decomposition of the suberinites and terpene resinites.

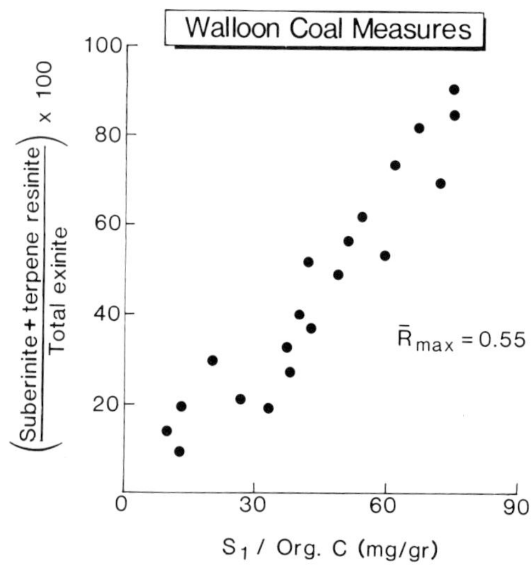

FIG. 8. Variation of the S_1/Org. C index with an increase in the total suberinite and terpene resinite fraction, for liptinite-rich, immature, iso-rank coals of the Walloon Coal Measures (liptinite content = 22–30%; inertinite content = < 5%).

The positive relationship between the S_1/Org. C index and the total suberinite and terpene resinite fraction (as a percentage of the total liptinite), found for the iso-rank coals of the Walloon Coal Measures, is shown in Fig. 8. Such a relationship not only confirms that suberinites and terpene resinites are the likely candidates for generating hydrocarbons in low rank, immature coals, but also explains the variations in S_1/Org. C indices observed in coals which are of similar rank and which contain similar proportion of vitrinite, liptinite and inertinite macerals (Fig. 8). With respect to generation of oil from similar coals or coaly source rocks, the mature phase of oil generation may then be preceded by several earlier phases, during which large amounts of hydrocarbons are likely to be generated through low-temperature decomposition of the thermally unstable liptinite macerals.

Although high S_1/Org. C ratios, observed for the Walloon coals, display a positive relationship with increasing concentration of suberinites and terpene resinites (Fig. 8), the overall oil-proneness of the coals, reflected by high hydrogen indices (Fig. 3), is governed not only by enrichment of the coals by different liptinite macerals but also by the perhydrous nature of the vitrinites and the lack of inertinite macerals (Tables 1 and 2). In the autochthonous deposit under consideration, the concentration of the subereous and resinous components is determined by the character of the peat-forming plant communities. According to Gould (1980) the Walloon flora can be roughly divided into two groups: (a) herbaceous plants, including lycopods, equisetaleans and ferns, (b) woody plants, including seed ferns, cycadeoids and conifers.

While terpene and lipid resinites associated with the Walloon coals are mainly derived from highly resinous conifers similar to the *Araucarian*, with additional contributions from cycadeoids, the subereous components are derived from both woody and herbaceous plants. Due to proneness of the subereous constituents to mechanical degradation, a large proportion of the non-fluorescing liptodetrinites in the Walloon coals are considered to be derived from the suberin-containing elements. Like suberinite, these liptodetrinites may have already undergone thermal transformation to become largely non-fluorescing.

Numerous thick conifer cuticles are another feature of the Walloon coals, and their contribution has probably added to the oil proneness of the coals. Furthermore, preservation of suberins and some of the other liptinite precursors, which can be prone to oxidative polymerization in dry, oxygen-rich environments, is favoured under dysaerobic conditions such as those prevailing in the accumulation of the Walloon coals.

Conclusions

Enrichment of autochthonous coals in oil-prone organic matter can be governed by the type of flora contributing to the peat accumulation and by biochemical transformation of the flora under dysaerobic conditions.

The coal seams of the Walloon Coal Measures are characteristically rich in liptinites and perhydrous vitrinites and poor in inertinites. The formation of perhydrous vitrinites and the lack of inertinites is related to the very wet, dysaerobic conditions, under which the peats accumulated.

Suberinites and terpene resinites are the likely candidates for the generation of large quantities of hydrocarbons in low rank lignites and subbituminous coals.

The mature phase of oil generation from coals or coaly source rocks may be preceded by several earlier phases, during which large quantities of hydrocarbons may be generated by transformation of thermally unstable liptinites at fairly low activation energies.

ACKNOWLEDGEMENTS: I wish to express my thanks to Dr S. Thompson for his constructive comments and suggestions on the manuscript. I also wish to thank E. Irgens and J. Ellingsen of the Geological Institute, University of Bergen, for preparing the illustrations. The financial support of the Norwegian Academy of Science and Statoil for the publication of this work is gratefully acknowledged.

References

DURAND, B., NICAISE, G., ROUCACHE, J., VANDENBROUCKE, M. & HAGEMANN, H. W. 1975. Etude géochimique d'une séries de charbons. *In:* CAMPOS, R. & GONI, J. (eds) *Advances in Organic Geochemistry*, pp. 601–632. Enadimsa, Madrid.

—— & PARATTE, M. 1983. Oil potential of coals. *In:* BROOKS, J. (ed.) *Petroleum Geochemistry and Exploration of Europe*, pp. 285–292. Blackwell Scientific Publications, Oxford.

GOULD, R. 1980. The coal-forming flora of the Walloon Coal Measures. *Coal Geology (Australia)*, **1** (3), 83–105.

KHAVARI KHORASANI, G. 1987. Novel development in fluorescence microscopy of complex organic mixtures: application to petroleum geochemistry. *Organic Geochemistry*, in press.

KOLATTUKUDY, P. E. 1980. Biopolyester membrance of plants: cutin and suberin. *Science*, **208**, 990–1000.

MACKENZIE, A. S., PATIENCE, R. L., MAXWELL, J. R., VAN DENBROUCKE, M. & DURAND, B. 1980. Molecular parameters of maturation in the Toarcian shales, Paris Basin, France—I. Changes in the configurations of acyclic isoprenoid alkanes, steranes and triterpanes. *Geochimica et Cosmochimica Acta*, **44**, 1709–1721.

PITTION, J. L. & GOUADAIN, J. 1985. Maturity studies of the Jurassic 'Coal Unit' in three wells from the Haltenbanken area. *In:* THOMAS, B. *et al.* (eds) *Petroleum Geochemistry in Exploration of the Norwegian Shelf*, pp. 205–211. Graham & Trotman, London.

POWELL, T. G., FOSCOLUS, A. E., GUNTHER, P. R. & SNOWDON, L. R. 1978. Diagenesis of organic matter and fine clay minerals. *Geochimica et Cosmochimica Acta*, **42**, 1181–1197.

ROE, G. D. & POLITO, L. J. 1970. Source rocks for oils in Ardjuna sub-basin of the Northwest Java Basin, Indonesia. *Proceeding Seminar on Generation and Maturation of Hydrocarbons in Sedimentary Basins.* Manila, 1977, pp. 180–194. UN/ESCAP, CCOP Technical Publication 6, Bangkok.

SZÁDECZKY-KARDOSS, E. 1948/49. Über systematik und Umwandlung der Kohlengemengteile. *Universität Technisches Wissenschaft, Soron, Mitteilung Berg und Hüttenmänn. Abteilung*, **17**, 176–193.

TEICHMÜLLER, M. & DURAND, B. 1983. Fluorescence microscopical rank studies on liptinites and vitrinites in peat and coals, and comparison with results of the Rock-eval prolysis. *International Journal of Coal Geology*, **2**, 197–230.

—— & TEICHMÜLLER, R. 1968. Cainozoic and Mesozoic coal deposits of Germany. *In:* MURCHISON, D. G. & WESTOLL, T. S. (eds) *Coal and Coal-bearing Strata*, pp. 347–379. Oliver & Boyd, Edinburgh.

THOMAS, B. M. 1982. Land-plant source rocks for oil and their significance in Australian basins. *Australian Petroleum Exploration Association Journal*, **22**, 164–178.

THOMPSON S., COOPER, B. S., MORLEY, R. J. & BARNARD, P. C. 1985. Oil-generating coals. *In:* THOMAS, B. *et al.* (eds) *Petroleum Geochemistry in Exploration of the Norwegian Shelf*, pp. 59–73. Graham & Trotman, London.

VAN KREVELEN, D. W. 1961. *Coal.* Elsevier, Amsterdam.

GANJAVAR KHAVARI KHORASANI, Geological Institute, Department A, University of Bergen, Allégt. 41, 5014 Bergen, Norway.

Index

Pages on which figures appear are printed in *italic*, and those with tables in **bold**